Springer Textbooks in Earth Sciences, Geography and Environment

The Springer Textbooks series publishes a broad portfolio of textbooks on Earth Sciences, Geography and Environmental Science. Springer textbooks provide comprehensive introductions as well as in-depth knowledge for advanced studies. A clear, reader-friendly layout and features such as end-of-chapter summaries, work examples, exercises, and glossaries help the reader to access the subject. Springer textbooks are essential for students, researchers and applied scientists.

Amanda Reichelt-Brushett

(Editor)

Marine Pollution— Monitoring, Management and Mitigation

Springer

Editor
Amanda Reichelt-Brushett
Southern Cross University
Lismore, NSW, Australia

Southern Cross University, Australia
Centre for Environment, Fisheries and Aquaculture Science, United Kingdom
The Editor, Amanda Reichelt-Brushett, Australia

ISSN 2510-1307   ISSN 2510-1315   (electronic)
Springer Textbooks in Earth Sciences, Geography and Environment
ISBN 978-3-031-10129-8   ISBN 978-3-031-10127-4   (eBook)
https://doi.org/10.1007/978-3-031-10127-4

This Springer imprint is published by the registered company Springer Nature Switzerland AG
The registered company address is: Gewerbestrasse 11, 6330 Cham, Switzerland

Prologue

When I embarked on the journey to write this text book it didn't take long to realise the magnitude of the task ahead, and that I needed help. I reached out to colleagues and contacts of colleges and as a result have broadened my network considerably. Interacting with authors and co-authors was refreshing and inspiring, particularly through 2020 and 2021. What a wonderful collaboration it has turned out to be. My sincere thanks to all the contributors for your engagement and patience.

This book aims to provide multidisciplinary contexts for understanding marine pollution. There are three Parts to the book. Part I introduces you to practical approaches and methodologies for studying marine pollution. Part II explores the main types of marine pollution and Part III considers multiple stressors, mitigation and restoration and describes the international regulatory frameworks relevant to marine pollution. It is not expected that you read the book from front to back, but I do suggest that you read ▶ Chap. 1 first to provide you with some general background understanding.

Each chapter draws upon the specific expertise of the authors and they bring to you a digestible and current interpretation of the literature. You are in good hands in your learning. Indeed, I have learnt a lot putting this book together. We all stand on the shoulders of our predecessors and trailblazers in our quest to understand the problem of marine pollution and learn how to measure, manage and mitigate. As a student of marine pollution, you are the future. The world's oceans and coastal ecosystems need smart, considered minds and unwavering enthusiasm.

There are many hundreds of exciting career pathways that, in some way, help to address the problem of marine pollution and there is literally something to inspire everyone's interests. The application of new technologies, mitigation solutions, research, science informed regulation and management and education are just a few fields you might find yourself in. My advice to you: never stop being a learner. I wish you the best of success in your journey.

My thanks to Dr. Alexis Vizcaino, Senior Editor at Springer, who inspired me to maintain momentum with regular meetings and enjoyable discussions.

I welcome feedback and suggestions to improve future editions of this book.

Amanda Reichelt-Brushett
Lismore, Australia

Contents

1   **Marine Pollution in Context** 1
Amanda Reichelt-Brushett

2   **Collecting, Measuring, and Understanding Contaminant Concentrations in the Marine Environment** 23
Amanda Reichelt-Brushett

3   **Assessing Organism and Community Responses** 53
Amanda Reichelt-Brushett, Pelli L. Howe, Anthony A. Chariton and Michael St. J. Warne

4   **Nutrients and Eutrophication** 75
Michelle Devlin and Jon Brodie

5   **Metals and Metalloids** 101
Amanda Reichelt-Brushett and Graeme Batley

6   **Oil and Gas** 129
Angela Carpenter and Amanda Reichelt-Brushett

7   **Pesticides and Biocides** 155
Michael St. J. Warne and Amanda Reichelt-Brushett

8   **Persistent Organic Pollutants (POPs)** 185
Munro Mortimer and Amanda Reichelt-Brushett

9   **Plastics** 207
Kathryn L.E. Berry, Nora Hall, Kay Critchell, Kayi Chan, Beaudin Bennett, Munro Mortimer and Phoebe J. Lewis

10   **Radioactivity** 229
Amanda Reichelt-Brushett and Joanne M. Oakes

11   **Atmospheric Carbon Dioxide and Changing Ocean Chemistry** 247
Kai G. Schulz and Damien T. Maher

12   **Other Important Marine Pollutants** 261
Amanda Reichelt-Brushett and Sofia B. Shah

13   **Marine Contaminants of Emerging Concern** 285
Munro Mortimer and Graeme Batley

14   **Multiple Stressors** 305
Allyson L. O'Brien, Katherine Dafforn, Anthony Chariton, Laura Airoldi, Ralf B. Schäfer and Mariana Mayer-Pinto

15   **Pollution Mitigation and Ecological Restoration** 317
Amanda Reichelt-Brushett

16   **Regulation, Legislation and Policy—An International Perspective** 339
Edward Kleverlaan and Amanda Reichelt-Brushett

Supplementary Information 359
Appendix I 360
Appendix II 362
Index 363

Editor and Contributors

About the Editor

Prof. Amanda Reichelt-Brushett
Mandy is dedicated to enhancing our understanding of marine pollution and protecting marine biodiversity through research and education. She teaches coursework and research students and conducts research in the fields of environmental chemistry, ecotoxicology, catchment management and marine and aquatic pollution.
► Chapters 1–3, 5–8, 10, 12, 15 and 16
Faculty of Science and Engineering, Southern Cross University, New South Wales, Australia. amanda.reichelt-brushett@scu.edu.au

Contributors

Dr. Laura Airoldi
► Chapter 14
Chioggia Hydrobiological Station "Umberto D'Ancona", Departmento of Biology, University of Padova, Italy; Department for the Cultural Heritage, University of Bologna, Italy.
Laura's research focuses on designing conservation and restoration strategies that maximise environmental and societal benefits.

Dr. Graeme Batley
► Chapters 5 and 13
Centre for Environmental Contaminants Research, CSIRO Land and Water, Lucas Heights, New South Wales, Australia.
Graeme has been undertaking leading research on contaminants in the aquatic environment for over 40 years in the fields of environmental chemistry and ecotoxicology. Graeme.Batley@csiro.au

Beaudin Bennett
► Chapter 9
Beaudin is a writer, traveller, and committed civil servant. He lives on the West Coast of Canada, within walking distance of the Pacific, and aims to pick up at least one piece of trash every day.

Dr. Kathryn L. E. Berry
▶ Chapter 9

Applied Technology, Fisheries and Oceans Canada, British Columbia, Canada

Kathryn is a marine scientist with a special interest in understanding and quantifying the impacts of known and emerging contaminants on marine organisms and ecosystems. Kathryn.Berry@dfo-mpo.gc.ca

Dr. Jon Brodie
▶ Chapter 4

The late Jon Brodie was a Professorial Fellow at the ARC Centre of Excellence for Coral Reef Studies (Coral CoE) at James Cook University, Australia. He made significant contributions to protecting the Great Barrier Reef, leading many research programs that broadened public knowledge and scientific understanding of the Australian icon. He was fearless, often speaking out forcefully and truthfully about environmental issues. He was a world authority on water quality, investigating the sources of pollutants in catchments, their transport to marine environments and effects on ecosystems.

Dr. Angela Carpenter
▶ Chapter 6

Visiting Researcher, School of Earth and Environment, University of Leeds, Leeds, United Kingdom.

Angela is a social scientist who specialises in research into governance and policy measures relating to marine pollution, marine environmental protection, and also maritime ports, installations, and shipping. Angela is widely published in academic journals and, among other activities, has co-edited volumes on Oil Pollution in the North Sea and Oil Pollution in the Mediterranean Sea for the Springer-Handbook of Environmental Chemistry series. A.Carpenter@leeds.ac.uk

Kayi Chan
▶ Chapter 9

Department of Earth Sciences, The University of Hong Kong, Hong Kong.

I am a marine ecologist investigating the state of plastic pollution in Hong Kong. kca2@hku.hk

Assoc. Prof. Anthony Chariton
► Chapters 3 and 14
School of Natural Sciences, Macquarie University, New South Wales, Australia.
Anthony is a molecular ecologist whose research focuses on understanding how natural and anthropogenic stressors alter the structure, function and connectivity of aquatic systems. anthony.chariton@mq.edu.au

Dr. Kay Critchell
► Chapter 9
School of Life and Environmental Sciences, and the Centre for Integrative Ecology, Deakin University, Victoria, Australia.
Kay is a biological oceanographer and spatial scientist who is experienced in modelling the transport and accumulation of marine debris

Assoc. Prof. Katherine Dafforn
► Chapter 14
School of Natural Sciences, Macquarie University, New South Wales, Australia.
I am an environmental scientist and science communicator focused on understanding and managing urban impacts in aquatic systems
katherine.dafforn@mq.edu.au

Dr. Michelle Devlin
► Chapter 4
Centre for Environment, Fisheries and Aquaculture Science (CEFAS), Lowestoft, United Kingdom; Centre for Sustainable Use of our Seas, University of East Anglia, Norwich, United Kingdom; TropWater, James Cook University, Queensland, Australia.
Michelle has worked in the field of eutrophication and impacts across many coastal ecosystems, with a strong background on monitoring and assessment of those coastal systems. michelle.devlin@cefas.co.uk

Nora Hall
► Chapter 9
College of Science and Engineering, James Cook University, Queensland, Australia.
Nora currently works in communications and previously co-founded the nonprofit The Hydrous.

Dr. Pelli Howe
► Chapter 3
Pelli is an environmental scientist and aquatic ecotoxicologist currently working as a freelance scientific editor. pellihowe@gmail.com

Edward Kleverlaan
► Chapter 16
Enviro-Seas Pty Ltd, Sydney, New South Wales, Australia
Edward Kleverlaan (B.Sc. (Physics/Geophysics) and post graduate degrees in Physical Oceanography and Meteorology), has worked on marine environment protection matters and climate change policy since the mid 1980's. Most recently, Head, Office for the London Convention/Protocol and Ocean Affairs at the International Maritime Organization. Currently an Independent Marine Environmental Advisor to governments across the globe. edward.kleverlaan@gmail.com

Dr. Phoebe J, Lewis
► Chapter 9
School of Science, RMIT University, Victoria, Australia.
Phoebe is an environmental scientist with a special interest in the chemicals (mainly affiliated with plastics and persistent organic pollutants, POPs) found in polar regions and their effects on seabirds. phoebe.lewis@rmit.edu.au

Prof. Damien Maher
► Chapter 11
Faculty of Science and Engineering, Southern Cross University, New South Wales, Australia.
Professor Maher undertakes research across various aspects of the global carbon, nutrient and hydrological cycles. damien.maher@scu.edu.au

Dr. Mariana Mayer Pinto
► Chapter 14
Centre for Marine Science and Innovation, Evolution and Ecology Research Centre, School of Biological, Earth and Environmental Sciences, The University of New South Wales, Australia.
I am interested in how anthropogenic stressors, such as contamination and urbanisation, affect the marine environment with the ultimate goal of developing evidence-based solutions for not only mitigating their impacts, but also restoring and rehabilitating marine ecosystems. m.mayerpinto@unsw.edu.au

Dr. Munro Mortimer
► Chapters 8, 9 and 13
Queensland Alliance for Environmental Health Sciences at The University of Queensland, Australia.
Munro R Mortimer is a former investigative scientist with the EPA (Queensland, Australia). He holds an honorary appointment as Senior Fellow at the University of Queensland. His research interests are in the fields of environmental sampling and analysis, and contaminant fate and transport, particularly in respect to persistent organic pollutants. munro@ozemail.com.au

Assoc. Prof. Joanne Oakes
► Chapter 10
Centre for Coastal Biogeochemistry, Faculty of Science and Engineering, Southern Cross University, New South Wales, Australia.
Joanne is an isotope enthusiast with research interests in biogeochemical cycling and over 20 years of experience teaching introductory and advanced science to undergraduate students. joanne.oakes@scu.edu.au

Dr. Allyson O'Brien
► Chapter 14
School of BioSciences, University of Melbourne, Australia.
Allyson is a marine ecologist and teaching specialist. She teaches marine biology, environmental science and statistics and conducts research in the fields of marine pollution, ecotoxicology and ecology. allyson.obrien@unimelb.edu.au

Sofi a B. Shah
► Chapter 12
Chemistry Department, Fiji National University, Fiji.
Sofia is an academic engaged in teaching Foundation Chemistry, Environmental Chemistry, Marine Chemistry and Advanced Inorganic Chemistry. My research areas include environmental analytical and toxicology chemistry. Sofia.Shah@fnu.ac.fj

Prof. Ralf B. Schäfer
► Chapter 14
iES Landau, Institute for Environmental Sciences, University Koblenz-Landau, Germany.
Quantitative Landscape Ecologist and working on the response of freshwater and riparian ecosystems to anthropogenic stressors. schaefer-ralf@uni-landau.de

Assoc. Prof. Kai Schulz
► Chapter 11
Centre for Coastal Biogeochemistry, Faculty of Science and Engineering, Southern Cross University, New South Wales, Australia.
Kai is a biological oceanographer interested on the effects of climate change on the base of marine foodwebs and potential climate mitigation approaches. kai.schulz@scu.edu.au

Assoc. Prof. Michael St. J. Warne
► Chapters 3 and 7
School of Earth and Environmental Sciences, University of Queensland, Australia; Queensland Department of Environment and Science, Australia; Centre for Agroecology, Water and Resilience, Coventry University, United Kingdom. Michael's research focuses on the fate and effects of pollutants in aquatic and terrestrial ecosystems. He places great importance on converting science into policy and regulation in order for environmental management to have a sound scientific basis.
michael.warne@uq.edu.au

Many thanks to others who made contributions to the book by providing images, text boxes and assistance: Tom Alletson, Kirsten Benkendorff, Andrea Brushett, Don Brushett, Daniele Cagnazzi, Maxine Dawes, Dexter dela Cruz, Joanne Green, Peter Harrison, Kirsten Michalek-Wagner, Andrew Negri, Kym Petersen, Suhaylah Shah, Dean Senti, Steve Smith, Kate Summer and Alexis Vizcaino. It is my responsibility if I have missed someone and I sincerely apologise. Please know that I have valued every contribution. Amanda Reichelt-Brushett.

Marine Pollution in Context

Amanda Reichelt-Brushett

Contents

1.1   Introduction – 2
1.1.1   Intentional, Accidental, and Uncontrollable Pollution – 3

1.2   Properties of Seawater – 6

1.3   Water in the Mixing Zone Between Rivers and the Ocean – 11

1.4   A Brief Social History of Pollution – 11
1.4.1   Contamination and Pollution – 12

1.5   Organism Exposure to Contamination – 12

1.6   Contaminant Behaviour – 15

1.7   A Multidisciplinary Approach to Understanding Pollution and Polluting Activities – 15

1.8   Polluting Substances—Local and Global Considerations – 15

1.9   Summary – 20

1.10   Study Questions and Activities – 21

        References – 21

© The Author(s) 2023
A. Reichelt-Brushett (ed.), *Marine Pollution—Monitoring, Management and Mitigation*,
Springer Textbooks in Earth Sciences, Geography and Environment,
https://doi.org/10.1007/978-3-031-10127-4_1

1

Acronyms and Abbreviations

DDT	Dichloro-diphenyl-trichloroethane
GESAMP	Group of Experts on the Scientific Aspects of Marine Environmental Protection
MARPOL	International Convention for the Prevention of Pollution from Ships
PCBs	Polychlorinated biphenyls
PFOS	Perfluorooctanesulphonic acid
PFOA	Perfluorooctanoic acid
TBT	Tributyltin

1.1 Introduction

You have opened this book because you have an interest in the ocean and the impact of humans upon it. This is a serious issue that gains plenty of media attention, but prior to the early 1950s it was generally considered that oceans were so expansive that they could absorb waste inputs indefinitely. Early concerns were raised specifically in response to the dumping of radioactive wastesin the ocean. Other globally recognisable events, such as mercury poisoning in Minamata Bay—Japan, oil spill disasters from vessels such as the Torrey Canyon in Great Britain in 1967, and the Oceanic Grandeur in Torres Strait in 1970, further highlighted the vulnerability of oceans to pollution. The highly visual impacts of large oil spills provided the initial direction for marine pollution research, and publications in the decade between 1970 and 1980 were dominated by studies on oil pollution. The risk of oil spills still exists today and incidences such as the Exxon Valdez Spill in 1989 and *Deepwater Horizon* (British Petroleum) in 2010 have challenged even the best available oil spill response programs and strategies. Periods after both events saw a further proliferation of research publications on oil pollution, expanding our knowledge and challenging our management capabilities.

Pollution of the marine environment is caused by a wide range of activities, and it is commonly reported that as much as 80% of marine pollution is a result of land-based activities (□ Figure 1.1). Fertiliser runoff from agricultural land has been highlighted as the cause of the **dead zone** in the Gulf of Mexico. The Northern Pacific Gyre Garbage Patch is an example of the consequences of poor solid waste management on a global scale. The Fukushima Daiichi nuclear accident in 2011, ocean acidification, ports and shipping, and multiple other land-based point and non-point source (also known as diffuse source) inputs highlight the challenges for minimising the threat of marine pollution. Contemporary research publications related to marine pollution not only cover the traditional focus on oil spills and radioactive waste dumping, but include a wide range of existing and emerging chemicals and substances of concern such as pesticides, pharmaceuticals, phthalates, metals, fire retardants, nano- and micro-particles, and mixtures of these. The growing body of knowledge has aided our understanding of how these substances behave in the marine environment and how organisms interact with them, helping to define the study of marine pollution.

Marine pollution is a challenging field of study requiring a **multidisciplinary approach** to assessment and management that incorporates social, environmental, economic, and political considerations (e.g. Ducrotoy and Elliott 2008). Importantly, we must also consider the impacts of pollution in combination with other stressors that affect the health of marine ecosystems, such as over-exploitation and harvesting of marine species, natural disasters, diseases, and exotic species.

Each chapter of this book has been touched upon in the above paragraphs, and the more pages you explore the more informed you will become about marine pollution. With 70% of the Earth's surface covered by oceans, marine pollution is unfortunately a large local and global issue and will be for many years to come. *Homo sapiens* have inhabited the Earth for around 150,000 years, and over this time our species has vastly influenced chemical, physical, and biological processes. However, the greatest anthropogenic impacts of pollution have occurred in the last 100 years. The fact that our population has more than quadrupled in this time, increasing from 1.9 billion in 1918 to over 8.0 billion today (2023) (Worldometer, 2023), highlights the scale of human influence. This human population expansion has no doubt contributed to the proposition of a new geological epoch, the **Anthropocene**, which represents the period in Earth's history dominated by humans, commencing around the start of the Industrial Revolution (Steffen et al. 2007).

Our consumption as individuals and communities has inevitably contributed to global-scale demands for raw materials, industrial chemicals, pharmaceuticals, and the associated waste production and pollution caused by the way we live in modern society. Indeed, our human footprint varies between different social and cultural circumstances across the globe. On a per capita basis, higher income countries are generally the greatest consumers of resources. At the same time, there is a large increase in consumerism in middle-income countries with the expansion of a middle-class population who have more money to spend on prod-

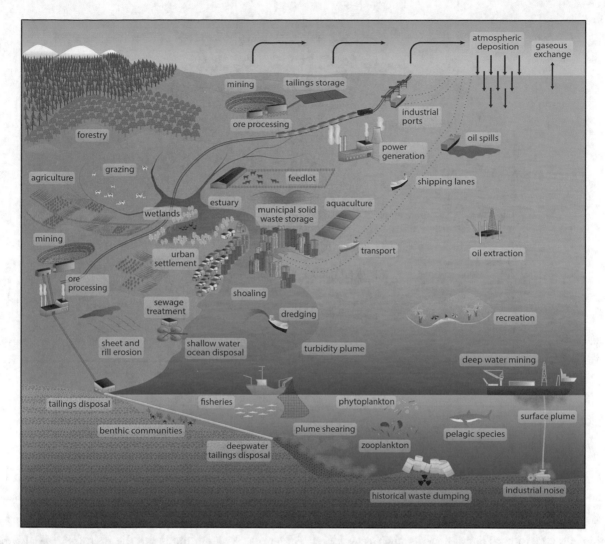

Figure 1.1 Consider the different causes of pollution and how they might be managed differently. *Image*: designed by A. Reichelt-Brushett created by K. Petersen

ucts and services (Balatsky et al. 2015). Low-income countries tend to have a per capita lower contribution to consumption, but also have fewer resources to manage the waste that is produced. Low-income countries also accept waste from high-income countries for payment and recycling, sometimes in working conditions that are harmful to human and environmental health (e.g. e-waste and plastics) (e.g. Makam 2018).

1.1.1 Intentional, Accidental, and Uncontrollable Pollution

Pollution is not always deliberate; a distinction can be made between intentional, accidental, and uncontrollable pollution (**Figure 1.2**). Furthermore, marine pollution may be slow and chronic or sudden and more acute (**Figure 1.3**). Sudden pollution events tend to be unintentional, and include accidents and natural disasters. Chronic pollution is often intentional and controlled, and may have a direct point source or non-point sources.

There may be many reasons for intentional pollution such as shipping practices (▶ Box 1.1), a lack of alternative options (such as the availability of waste collection services for litter in low- and middle-income countries), and simply a disregard for regulations. Intentional pollution can be addressed by creating an enabling environment for change and facilitating pollution reduction measures (e.g. Robinson 2012). Incentive and disincentive schemes (colloquially known as carrot and stick approaches) such as encouraging the development of pollution reduction technologies, creating local- to global-scale law, policy and penalties have been shown to reduce polluting behavior (Hawkins 1984).

Accidents are usually caused by factors or events that were unforeseen in risk assessment and/or are a result of inadequate risk minimisation strategies (Garrick 2008). This generally reflects a lack of knowledge and/or poor

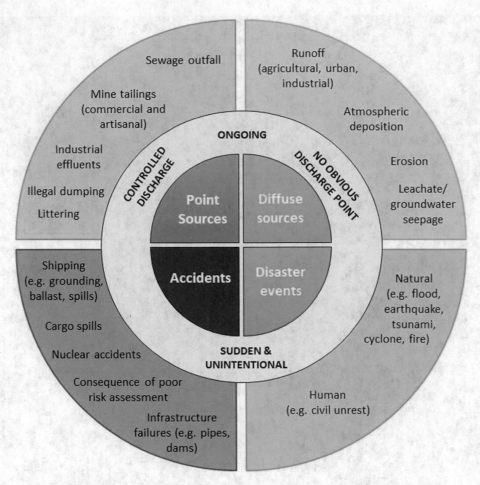

Figure 1.2   Summary of some of the primary sources of marine pollution. *Image*: designed by A. Reichelt-Brushett, created by K. Summer and A. Reichelt-Brushett

contingency provisions. Disaster events, such as cyclones/hurricanes/typhoons (e.g. Hurricane Irma, Cyclone Debbie, and Typhoon Hato all in 2017), floods (e.g. monsoon floods in India and Bangladesh in 2017, flooding and mudflows in California in 2018), and tsunamis (e.g. affecting Thailand and Indonesia in 2004, Japan in 2010, and Haiti in 2010) are largely uncontrollable, but can be major generators of pollution (Figure 1.3d). Extreme events such as these create large volumes of marine debris and cause the breakdown of urban infrastructure such as sewage systems, and waste disposal and storage facilities. Extensive flooding during these events transports polluting substances from activities on land into marine environments.

The environmental consequences of pollution do not distinguish between intentional and unintentional causes, but understanding the nature of the causes is important for minimising future risks and repeated in-

cidences. While humans are the polluters, we also hold the key to solutions and this is where we can focus positive energy and create beneficial outcomes for marine ecosystems and the environment in general. For example, rather than just feeling disappointed about the number of plastic containers washed up on a beach, and rather than just picking up that plastic or doing surveys to measure the amount of debris washed up on beaches, we can develop and implement solutions to reduce the production of litter at the source.

Throughout the world, experts and non-experts alike have invested themselves in managing and understanding pollution. Some people's careers are dedicated to reducing the impacts of marine pollution, and the rise of citizen science and volunteer programs highlights the community interest in pollution reduction. The imagery of pollution such as Figure 1.4 (see also ▶ Box 1.2) evokes emotion and enhances public con-

□ **Figure 1.3** **a** Intentional plastic discarded in Eastern Indonesia, an area of poorly developed waste management infrastructure and limited land resources (*Photo*: A. Reichelt-Brushett), **b** Clean-up following an oil spill in 2007, the Cosco Busan, a container ship, dumped 58,000 gallons of oil after striking the San Francisco Bay bridge in Calif (*Photo*: "Clean Up After a Big Oil Spill" by NOAA's National Ocean Service is licensed under CC BY 2.0), **c** Uncontrollable algae bloom from septic waste seepage into the ocean in Indonesia (*Photo*: A. Reichelt-Brushett), and **d** debris and damage from the 2004 Tsunami, Aceh, Indonesia (*Photo*: "Tsunami 2004: Aceh, Indonesia" by RNW.org is licensed under CC BY-ND 2.0)

cern, which in turn drives the demand for clean-up operations, prosecutions (where applicable), and legislative change. Popular science books such as *Toxic* Fish *and Sewer Surfing* (1989) by Sharon Beder and *Moby Duck* (2011) by Donovan Hohn have also contributed to raising awareness of marine pollution issues.

We are all part of the problem, but you are also an essential part of the solution. I hope this book provides you with guidance and enhances your passion to make a difference. An important place to start is to develop our understanding of the natural systems we are living and working in.

Box 1.1: Example of Intentional Contaminant Release

In 1973, 1.5 million tonnes of crude oil were intentionally released from the tanker *Zoe Colocotronis* when the ship ran aground just off the southwest coast of Puerto Rico. Along with the jettison of cargo, this oil release was ordered to help the ship get off the reef. Three years later, a cargo ship ran aground on the Nantucket Shoals, but this time jettison of cargo was suggested but rejected. This ship broke apart and all the cargo was lost to sea. The United States National Academy of Sciences developed a lengthy report, "Purposeful Jettison of Petroleum Cargo", in 1996, to provide clarification on when cargo jettison is appropriate and may prevent a larger incident. In the past, many vessels were required to slowly release oil. The lifeboats on board the Titanic were required to carry oil for "use in stormy weather", under the British Merchant Shipping Act 1894, and United States Coast Guard regulations also required "storm oil" to be carried on lifeboats. This is because the thin slick that oil forms on the water surface absorbs energy and dampens waves. The regulations requiring the carrying of "storm oil" were removed in 1983. For further details: ▶ https://response.restoration.noaa.gov/about/media/some-situations-ships-dump-oil-purpose.html.

◘ **Figure 1.4** Every year thousands of young albatrosses die a slow and painful death on the Midway Atoll, a small coral- and sand bank in the North Pacific. They are fed by their parents with plastic waste floating on the sea—3000 km from the nearest continent. In the end, they starve from too much plastic in their stomachs. Midway Atoll National Wildlife Refuge in Papahānaumokuākea Marine National Monument. *Photo*: "Raise your Voice (2010): Midway—Message from the Gyre (2009)/Chris Jordan" by Ars Electronica CC BY-NC-ND 2.0 ► https://creativecommons.org/licenses/by-nc-nd/2.0/legalcode

Box 1.2: Plastics, Microplastics, and Nanoplastics

For some of the most problematic contaminants, it is not possible to simplify the cause of the pollution. All drains lead to the sea, and littering and dumping of plastic waste have resulted in the increasing accumulation of plastic in the ocean, to the extent that rubbish islands have formed due to gyres in the ocean.

Plastics can cause toxicity to organisms upon exposure or ingestion, and contain additives which themselves can be toxic. Because plastic takes a long time to break down, and many types of plastic are less dense than water and hence are easily transported in currents, we have created a new vector for transporting not only contaminants but also pathogens and invasive species over long distances. A 2018 report of a supermarket plastic bag located in the Mariana Trench at a depth of 10,898 m highlights an emerging threat of plastic pollution in the ocean (Chiba et al. 2018). As you can imagine, organisms attached to such debris would not normally find themselves in such ecosystems, and the ecological consequences of such introductions are completely unknown.

We have been hearing about microplastics in the ocean for several years in mainstream media. There has been some effort to remove microplastics from some products. Nanoplastics are small microplastics, generally defined as between 1 and 100 nm. Ultimately, all plastic will eventually be broken down into nanoparticles, so the nanoplastic concentration in the ocean is only going to increase as the large amounts of plastic debris in the ocean disintegrate. Remediation of microplastic pollution is not currently possible; while nets can be adapted to remove large plastic debris, once plastic has broken into small pieces there is no practical way to remove them from the environment.

► Chapter 9 is specifically focused on plastics in the marine environment.

1.2 Properties of Seawater

Seawater makes up 97% of water on Earth. It supports around 20% of the currently known species but two-thirds of the predicted total of ~8.7 million species (Mora et al. 2011). Although estimates of unknown species vary between studies, the point is that we clearly lack in our understanding of the immense biodiversity of marine ecosystems. Nonetheless, we do have a good understanding of the general chemistry of the system that supports the abundance of marine life. There is much more to learn about biogeochemical variability throughout the world's oceans, how organisms adapt to local conditions, and how these conditions influence the behaviour, bioavailability, and toxicity of contaminants.

Among all molecules, water stands out for its diversity (i.e. it is found naturally in solid, liquid, and gaseous states), occurrence throughout the environment, vast variety of uses, and its role as a medium for life (Manahan 2009). Water is an important chemical transport medium and an excellent solvent; it has a high latent heat capacity, is transparent and penetrable by light, and is prone to pollution but totally recyclable. Central to the behaviour of water is **hydrogen bonding**. Hydrogen bonding is a weak electrostatic force that influences the orientation of individual water molecules as the hydrogen atoms of one water molecule are attracted to the oxygen atom of other water molecules close by. These bonds are about 10 times weaker than a covalent O–H bond but strong enough to be maintained during temperature change. Therefore, water can resist changes in temperature by absorbing energy that would otherwise increase the motion of H_2O molecules (► Box 1.3). All of the water molecules in solid ice have formed the maximum four hydrogen bonds with a heat capacity of 0.5 cal/g/°C compared to liquid water which has by definition a heat capacity of exactly 1 cal/g/°C (i.e. 1 g of water is increased by 1 °C for every calorie of added heat energy). This is extremely high compared to other liquids and solids (second only to liquid ammonia) and effectively causes fresh and seawater bodies to withstand great changes in temperature compared to atmospheric temperatures, enabling the large oceans to act as climate moderators where summer heat is stored and radiated back to the atmosphere in winter. Libes (2009) elaborates in several excellent chapters that explain the detailed physical chemistry of seawater and the biogeochemistry of marine systems.

In general terms, the chemistry of seawater is quite stable and has some very similar properties to fresh water. You can consider it fresh water with increased quantities of specific dissolved ions which influence its properties (◘ Table 1.1 compares the composition of seawater to fresh water). For example, fresh water freezes at 0 °C whilst seawater freezes at around -2 °C, due to differences in surface density (seawater 1.02 g/cm^3 compared to fresh water 1.00 g/cm^3 at 25 °C [Libes, 2009]), which also slightly influences the solubility of gases and dissolved ions. **Salinity** is generally referred to as being 35 g/kg (or 35 parts per thousand) but ranges from 31 to 38 g/kg, being influenced by precipitation and evaporation.

Box 1.3: Water, Solvation, and Energy

Dr. Don Brushett, Chemist, Southern Cross University.

You may have heard the statement that "*water is life*". While that statement may be hyperbole, it is true that without water, there is no life.

The water molecule has a number of characteristics that ensures its important role on our planet. Firstly, let us investigate the Polarity **Polarity** of the water molecule. Polarity is a term used to describe the unequal sharing of electrons within a molecule. Atoms are imbued with a unique characteristic known as electronegativity; this can be described as the strength with which each atom attracts electrons. Oxygen is much more electronegative than hydrogen and therefore the electrons spend more time around the oxygen atom. ◘ Figure 1.5 shows that electron density, the regions where the electrons spend 90% of the time. We can think of this polarised molecule as somewhat like a bar magnet which can attract or repel other polar molecules or ions.

The innate Polarity of water is what makes it such a powerful solvent. The average salinity of seawater is approximately 35 (‰). That is to say that approximately 3.5% of seawater by mass of seawater is dissolved salts. This ability to

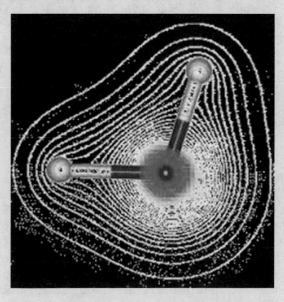

◘ **Figure 1.5**   ► Box 1.3: The water molecule and electron cloud. Licenced under CC BY-NC-ND 2.0

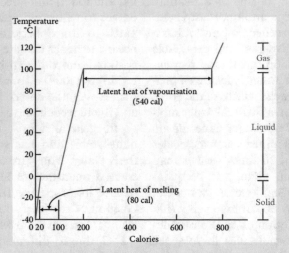

◻ Figure 1.6 ▶ Box 1.3: The phase transitions of water that are caused by changing heat content. The slopes of the lines indicate heat capacity. Adapted from Libes (2009)

dissolved charged particles has important implications when we consider ionic pollutants. Water has the ability to dissolve large quantities of ions and to transport them over large distances. Conversely, water does not solve non-polar molecules. As they say, "*oil and water don't mix*". Although many organic pollutants are not soluble in water, the polar nature of water forces them to associate with other non-polar substances such as soils, clays, and organic material. In effect, this can concentrate the organic pollutants to the detriment of benthic and filter feeding organisms.

Directly associated with the polar nature of the water molecules is the phenomena called **hydrogen bonding**. Hydrogen bonds are transient and form between the oxygen of one molecule and the hydrogen of another molecule. These polar-polar interactions are about one-tenth as strong as a covalent. This is a significant force when you are vast numbers of molecules. Consider water (18amu) and carbon dioxide (44amu). Carbon dioxide is more than twice the mass of water, yet it only exists as a gas on Earth. Hydrogen bonding makes water **sticky** and allows the water molecule to exist as a solid, liquid and gas in the temperature present on our planet.

This stickiness, due to hydrogen bonding, has important implications for the physical properties of water. Water has a relatively large latent heat of fusion and latent heat of vaporation. The former is the amount of heat required to transform 1 g of ice into liquid water or the amount of heat that must be removed to transform 1 g of liquid water into ice (Libes 2009) (◻ Figure 1.6). The latent heat of evaporation is comparable to the latent heat of fusion, but refers to the liquid–gas phase transition (Libes, 2009). These relatively high latent heat are another consequence of hydrogen bonding.

It takes nearly 4 times as much energy to raise the oceans 1 °C compared to the surrounding land. This moderates the temperature of coastal regions in both hot and cold climates. Water absorbs energy in hot regions which is transferred to cooler polar regions by ocean currents and warms them. Consider the Gulf Stream which makes the climate of Europe pleasant, compared to equivalent latitudes in Siberia. The **stickiness** associated with hydrogen bonding also endows water with large heat of phase change. This also influences global temperatures, for example, water in the tropics absorbs energy when it changes phase from a liquid to a gas. The gaseous water moves through the atmosphere and energy is released when water condenses and falls as rain or snow in cooler parts of the globe. These phenomena moderate the climate around the planet.

Compared to freshwater systems, the pH of seawater is generally fairly constant. However, there is evidence that the changing **carbon dioxide** concentration in the atmosphere is affecting the natural bicarbonate/carbonate buffer system of seawater. Carbon dioxide dissolution in the ocean acts to reduce available carbonate ions, impacting calcification rates of organisms, and releasing hydrogen ions that influence pH and calcium carbonate solubility (Doney et al. 2009). Even small coral reef islands have been shown to influence the local pH of seawater through the exchange of tidal waters seeping into and reacting with calcareous sands (Santos et al. 2011). Local temperatures may increase to extreme levels in rock pools cut off from the ocean during low tides and become hypersaline through evaporation, reaching salinity levels over

Table. 1.1 Comparison of the properties and major components of natural seawater and freshwater systems

Component	Examples/dominant forms	Seawater	Fresh water
Physicochemistry			
Density		1.02 (g/cm³) Solutes contribute to increased mass; sits below fresh water at saltwater wedge in estuaries, etc.	1.00 (g/cm³) (pure water)
Conductivity		53.0 (mS/cm) or 35 ppt Proportionate to ionic charge (i.e. salinity)	0.05–1.5 (mS/cm) (= 5–1500 µS/cm)
pH	[H⁺] [OH⁻]	pH 7.5–8.4 Bicarbonate buffer (alkalinity) resists overall pH changes	pH 5.5–7.5 Variable buffering capacity; more easily influenced (e.g. by organic acids and carbonic acid in rainfall)
Alkalinity	HCO_3^-, CO_3^{2-}, H_2CO_3	100–150 (mg/L, total)	5–500 (mg/L, total) Related to softness/hardness (i.e. Ca^{2+}, Mg^{2+}, which typically contribute carbonate [CO_3^{2-}])
Freezing point		-1.91 °C Salts between H_2O molecules reduce rate of ice crystal formation	0.00 °C (pure water)
Specific heat		3.898 (j/g/°C) Intermolecular salts reduce number of H bonds and thus potential energy to be overcome	4.182 (j/g/°C) All H bonds must be disrupted before heat can be raised
Overall stability		Expansive, deep, well-mixed, chemically and physically stable systems; less spatial and temporal variability	Smaller (often closed), shallower, less stable, and more variable catchment scale biogeochemical/anthropogenic influences
Elements/ions			
Major elements/ions	Na^+, Cl^-	~35,000 ppm total salts > 500 mM (~86% total salinity)	Highly variable depending on local geology; dominant elements/ions: HCO_3^-, (48% of total), Ca^{2+}, Mg^{2+}, K^+, SiO_2, Fe^{3+}
	Mg^{2+}, SO_4^{2-}, Ca^{2+}, K^+	10–50 mM (~13.8% total salinity)	
Minor elements/ions	Br^-, Sr^{2+}, F^-, B^{2+}, Ba^{2+}	0.1–100 mM (<0.2% total salinity)	
Trace elements	Fe, Mn, Cu^{2+}, Li^+, Ni, Zn, Cr, Al, Co, etc.	<1–100 mM (<0.2% total salinity)	
Gases			
	N_2, O_2, Ar, CO_2, N_2O	nM to mM Surface waters in equilibrium with atmosphere; saturation changes with depth (pressure), temperature, sediment, oxygenation, and redox reactions	nM to mM Greater variation between locations; varies with physicochemistry as well as plant/microbial communities, local atmospheric deposition, etc

(continued)

◻ **Table. 1.1** (continued)

Component	Examples/dominant forms	Seawater	Fresh water
Nutrients			
	NO_3^-, NO_2^-, NH_4^+, PO_4^{3-}, K^+, SO_4^{2-}	µg/L to mg/L Oligotrophic—low (e.g. coral reefs), mesotrophic—moderate (e.g. coastal waters, upwelling), eutrophic—excessive	µg/L to mg/L Variable depending on land use, rainfall, hydro-dynamics, organisms and nutrient cycling, closed/open system
Dissolved organic compounds			
	Humic materials (humic and fulvic acids), tannins, amino acids, lipids, organometallic compounds	ng/L to µg/L Generally very low; different water solubilities among compounds—potentially affected by salts	ng/L to mg/L Variable between locations; can be naturally very low or high depending on vegetation, land use, seasonal differences
Particulate material			
	Inorganic (sand, clay, metal hydroxides); organic (hydrogen atoms biomass, faeces, moults, and dead tissue)	Concentration range µg/L–mg/L, increases with proximity to coastlines (related to river inputs, turbulence, depth), generally low in open water column; smaller particle sizes = slow settling velocity	Concentration range µg/L–g/L, variable between locations; influenced by hydrology, geology, topography, land use, trophic structure, seasonal rainfall

Sources: Libes (1992); Pilson (1998); Riley and Chester (1971) and prepared by K. Summer

50 ppt. Marine organisms have adapted on an evolutionary timeline to cope with these locally dynamic conditions. ► Chapter 11 is dedicated to further understanding atmospheric carbon dioxide and changing ocean chemistry.

The natural composition of **coastal seawater** is more variable than the open ocean due to influences from activities on adjacent land and river systems draining into the oceans (◘ Figure 1.1). Water quality is affected by the array of associated catchment activities in river systems that drain into coastal waters. Globally, there are numerous examples of inputs of contaminants to the marine environment from catchment activities such as agriculture, deforestation, aquaculture, mining, manufacturing industries, shipping, urban settlements, landscape modification, and the like (e.g. Edinger et al. 1998; Brodie et al. 2012; Vikas and Dwarasish 2015). Point sources and non-point sources of pollution come from both land- and sea-based activities (◘ Figure 1.1). Point sources are far easier to manage and legislate compared to non-point sources.

1.3 Water in the Mixing Zone Between Rivers and the Ocean

The transition zones between freshwater **catchment** areas and saline oceans are known as **estuaries**. Here, the physicochemical conditions naturally vary both temporally and spatially. During flood events, rivers may flow with fresh water to their mouths, drastically reducing local ocean salinity. Drought conditions may see the influences of ocean salinity extend far upstream in low-lying river systems. Historically, estuaries were some of the earliest settled areas on many continents and are now among the most heavily exploited natural systems in the world; with that comes a legacy of the impacts of human activities (Barbier et al. 2011). Importantly, estuaries are highly productive systems and breeding grounds for many marine pelagic species (Meynecke et al. 2008; Pasquaud et al. 2015). Estuaries provide extensive ecosystem services and are valued for their raw materials, coastal protection, fisheriesFisheries, nutrient cycling, along with tourism, recreation, education, and research. However, the health of estuaries has been in decline for many years and this is recognised on a global scale. Water quality decline is one of the major threats to the health of estuaries throughout the world (e.g. Kennish 2002; Karydis and Kitsiou 2013).

The mixing between fresh and seawater is a complex zone of chemical interactions that have important influences on the behaviour of contaminants, particulates, and their potential toxicity. Competing ions in seawater influence adsorption and deposition of contaminants onto and off fine sediments. At the **saltwater wedge** (where seawater meets less dense fresh water in an estuary), flocculation occurs whereby suspended particles settle out of the water column along with associated bound contaminants, only to be later redistributed through the system in high rainfall events and periods of fast-flowing water. A detailed perspective of these interactions in estuaries can be found in Reichelt-Brushett et al. (2017).

1.4 A Brief Social History of Pollution

Defining **pollution** is not easy and the word has shifted its dominant meaning considerably over time. Nagle (2009) provides an interesting legal perspective on the "*Idea of Pollution*", and some background context from this helps set the scene for understanding marine pollution. The word pollution was used as early as 1611 in The King James translation of the Bible, and mostly referred to disgust related to a judgement with broad reference to effects or harm upon humans or human environments. In legal cases decided before 1800, English courts used the word pollution in the context of harm to family, church, government, or other human institutions. Pollution occurred in the context of sexual or spiritual harm, newspapers have been referred to as "*polluted vehicles that lacked truth*", and corrupt legal or political processes were considered polluted processes. In 1820, the act of slavery was described as the "*pollution of slavery*". In 1878, the Louisiana Supreme Court described money earned from the sale of slaves as "*polluted gold*". This human focus on the meaning of pollution still exists and is used in moral, ethical, and cultural contexts. Reference to environmental pollution is not really mentioned in political debates until the end of the nineteenth century. Nagle (2009) suggests that river pollution was a key to transforming the meaning and context of the word. Importantly, the **judgement** connotation was removed, and instead pollution was more descriptive and perhaps technical.

Sometimes, the meanings of words are defined for a very specific purpose. Indeed, the definition of pollution means different things under different legalisations, even within a single country, so the meaning of the word becomes relative to the context in which it is used. People have tried to create broad definitions of pollution, only to come to a realisation that activities such as children blowing bubbles would be deemed as pollution. Even though the concept of pollution eludes a precise definition, there is a strong argument in the environmental science literature that differentiates between contamination and pollution (c.g. Chapman 2007; Walker et al. 2012). As an ecotoxicologist, I value this differentiation and have found it useful when reporting and publishing research findings be-

1

cause it helps to focus attention on research needs, and sites and situations of high concern and risk. However, the distinction is limited by the current scientific understanding, exposure concentration, and defining what an adverse effect is (Walker et al. 2012). The following text provides some further insights into defining contamination and pollution.

1.4.1 Contamination and Pollution

When considering marine **contamination**, we make an immediate link to substances present in the marine environment that should not be there at all, or are present in excessive concentrations that are not natural or normal. Importantly, the natural background level of any given substance will vary between and within locations around the world. You should also recognise that there are no normal background levels for synthetic substances. With this in mind, we may work with the following definition:

"Marine contamination occurs when the input of a substance from human and human-related activities results in the concentration of that substance in the marine environment becoming elevated above the naturally occurring concentration of that substance in that location".

Missing from the definition of contamination is the fact that there is no clarity about how a contaminant affects organisms and what concentrations are harmful, and this is what differentiates contamination from pollution (Chapman 2007). We can measure a substance and find that it is elevated compared to background concetrations, but what does that mean for the health of different species, ecosystem function, and services that are exposed to it? At what concentrations and forms should different contaminants concern us? How do we assess situations where more than one type of contaminant is present? We also have to consider the impacts of these contaminants on receptors that are not distinctly marine but interact with the marine environment (i.e. those organisms that feed on marine biota including humans, birds, polar bears, and other wildlife). A weight of evidence approach (i.e. using a combination of information and independent sources to provide sufficient evidence to support decision-making) can be applied to gain a fuller understanding of when and how contamination causes pollution (Chapman 2007). Once we gain this understanding, it is possible to identify if a contaminant is actually detrimental and polluting. Chapman (2007) highlights that all pollutants are contaminants, but not all contaminants are pollutants. The distinction also infers that pollution is more serious and through this, it has become a more emotive word than contamination.

According to the joint Group of Experts on the Scientific Aspects of Marine Environmental Protection (GESAMP):

» *"Pollution is the addition by* human activity, *directly or indirectly, of substances or energy to the marine environment which results in detrimental effects, for example hazards to human health, hindrance to amenity use,* recreational *activities and* fishing.*"*

Put simply, a contaminant is a substance present in the environment where it should not normally occur, or at concentrations above background levels, and a pollutant is a contaminant that causes adverse effects in the natural environment. Many subtle variations in these definitions exist in the literature. A key point to consider is that the word **pollution** should be used when a detrimental effect has been determined. Indeed, it may just be a matter of further research to prove a contaminant is a pollutant. Remember that pollution is socially constructed in all contexts, so there will always be grey areas, particularly when considering natural causes of pollution (e.g. do extreme weather events cause marine pollution even in natural landscapes? Should sedimentation from a landslide be deemed as pollution?).

1.5 Organism Exposure to Contamination

The definition of pollution by GESAMP uses broad examples of **detrimental impacts** that are largely human-focused. Importantly, these detrimental impacts are linked to changes in organism and ecosystem health after exposure to contaminants through biotic and abiotic factors. The degree to which marine species may be exposed depends on the chemical behaviour of contaminants (e.g. speciation, complexation) in different environmental conditions (e.g. physicochemistry), along with the physiology of a given species and how it interacts with the surrounding environment. Environmental interactions are dictated by the ecological niche of the species including the resources it uses and where it sits in the trophic structure. The species behaviour, mobility, metabolic processes, strategies of feeding and reproduction, and lifespan influence environmental interactions and the potential pathways of chemical uptake, storage, and elimination. Even organisms that are taxonomically similar may have vastly different exposure pathways. ◻ Table 1.2 highlights how differently some species of mollusc interact with their environment.

Sessile organisms do not move at all, whilst **sedentary** organisms tend to have very limited movement, and are thus both are favourable for biomonitoring and in situ studies. Albeit, we must keep in mind that the early life stages of many marine species may be free moving and transported extensive distances by winds, currents, tides, and wave action. By comparison, free-moving species have the potential to actively avoid unfavourable condi-

Table. 1.2 Various species of marine mollusc (Phylum: Mollusca) interact differently with the environment, despite being taxonomically similar

Genus/species name	Habitat	Morphology	Feeding habit and trophic status	Mobility and behaviour	Life cycle/other
Austrocochlea porcata (zebra top snail)	Wide range of habitats, exposed rocky shores, mid-high intertidal, sand, seagrass, mangroves	Rounded black and white-yellow striped shell, ~2.5 cm; diet and environmental conditions affect colouration and width of stripes	Herbivore; diatoms, and algae scraped from rock/substrate	Occurs in large aggregations	Dioecious; external fertilisation
Cabestana spengleri (triton; predatory whelk)	Sheltered moderately exposed reef to 20 m depth; among rocks and cunjevoi (*Pyura stolifera*) on exposed intertidal rocky shores	Medium-large (~15 cm), thick, brown trumpet-shaped shell with series of ribs	Preys almost exclusively on ascidians (*Pyura stolonifera*); locate by chemoreception of prey substances in water	Intermittent, relatively rapid foraging movements	Reproduction—egg fertilisation by mating, association in pairs, females remain on egg mass for one month
Conus anemone (cone shell)	Sheltered, moderately exposed reef, lagoons, 0–50 m depth	Conical-shaped shell 6–16 cm, harpoon-like radula connected to specialised venom sacs	Carnivore; uses venom to paralyse fish and invertebrates which are eaten live	Usually hidden under rocks/in sand during day, emerge at night to feed	Lays egg clusters on hard substrate; few offspring survive to adulthood
Donax deltoides (pipi/Goolwa cockle)	Exposed, high-energy sandy beaches, below intertidal sand surface	Triangular bivalve, shell ~5 cm with white-light pink exterior, purple interior	Filter feeder; water passed across gills where small particles are extracted; retract fully to avoid predation	Burrowing	Simultaneous hermaphrodites—serial broadcast spawning; 4–5 yr lifespan
Haliotis rubra (black lip abalone)	Crevices, caves, vertical rock surfaces; intertidal zone to 40 m depth	Large (up to 20 cm), oval foot and shell; rapid growth; respire by virtue of holes in shell through which water flows	Herbivore; graze on drift algae, kelp, and algae on rock surface	Move to more exposed locations as they grow to avoid predation; little movement	15 yr lifespan, broadcast spawning
Ischnochiton (chitons)	Attached to wave-swept intertidal rock surfaces and in crevices	Muscular foot; oval shell 3–9 cm with overlapping plates; lack eyes though have light sensitive organs in shell; respire by beating water under body with cilia	Omnivore; use radula teeth to rasp encrusting algae/animal material from rock	Move slowly after dusk to avoid predation	Dioecious; females deposit eggs on rock surfaces after males release sperm into water

(continued)

1

■ **Table. 1.2**   (continued)

Genus/species name	Habitat	Morphology	Feeding habit and trophic status	Mobility and behaviour	Life cycle/other
Morula marginalba (mulberry whelk)	Rock crevices in mid-intertidal zone	Off-white shell ~2 cm with raised black bumps; radula used to drill hole in shell of prey, assisted by production of sulphuric acid which helps dissolve calcium carbonate	Carnivore; important predator of barnacles and other invertebrates; feeds on oysters in estuaries where it is considered a pest	Often occur in clusters, little movement	Internal fertilisation; attaches eggs individually to substrate
Mytilus edulis (blue mussel)	Sheltered-moderately exposed reef, engineered structures; wide thermal tolerance	Bivalve, up to 12 cm; fibrous byssal threads attach to substrate; complex gill system	Filter feeder; bioaccumulate significant metal loads (commonly used as a biomonitor for this reason)	Permanently attached (sessile)	Broadcast spawning; range limited by drift of larval/juvenile stages; 18–24 yr lifespan—varies significantly with attachment location
Nautilus pompilius (chambered/pearly nautilus)	Depths up to 600 m—buoyancy controlled by changing shell chamber fluid volume/density	Well-known logarithmic spiral shell cross section; diameter up to 26 cm; primitive eye, ~90 cirri (tentacles)	Carnivore; feeds on suspended carrion/detritus, living shellfish; find prey by olfaction and chemotaxis; eat infrequently due to low energy expenditure	Freely swimming; propelled by drawing water in and pushing out of chamber; can also crawl or land on seafloor	Dioecious, annual mating; ~20 yr lifespan

Sources: Beesley et al. (1998), Edgar (2000). Table created by K. Summer

tions. For this reason, they are poor **biomonitors** because we usually do not understand their history of exposure. Additionally, accumulation of contaminants in some species and magnification in higher organisms may be specifically of interest for human health reasons given that we are top-order consumers and rely on oceans for food (in some areas more so than others). There is more discussion in ► Chapter 3 along with other chapters about organism interactions with contaminants, measuring toxic effects, food chain transfer, etc.

1.6 Contaminant Behaviour

All contaminants, whether inorganic or organic (◻ Table 1.3), will ultimately be distributed through ecosystems and stored in various compartments (e.g. sediments and body tissues). The environmental fate of contaminants results from their chemical properties, and it is the fugacity (sometimes described as the potential to move between media) of a substance that determines its likely distribution after its release into the environment (e.g. water or lipid solubility, vapour pressure) as well as the hydrodynamic and physiological processes occurring in those ecosystems (e.g. winds, currents, flow rate, upwelling, and sedimentation). We tend to focus our sampling on the various compartments of water, biota, and sediment. Once compartmentalised, the duration of storage will depend on the stability of the conditions in that compartment. For example, when sediments are disturbed, stored contaminants can be remobilised back into the water column, or if an organism dies the contaminants that were taken up and stored in its body will become available to detritivores and through trophic levels thereafter.

Most organic compounds break down over time; metals, however, are elements and as with other elements they cannot be broken down further. For this reason, they tend to sequester in different environmental compartments. Plants and animals vary widely in their ability to regulate their metal content, and how organisms respond will depend on the type of metal, type of organism, and physicochemical conditions that define the metal species (complex). Ecotoxicological studies help us to understand how an organism interacts with a contaminant and identify measurable stress responses.

1.7 A Multidisciplinary Approach to Understanding Pollution and Polluting Activities

Consideration must be given to the various exposure pathways, distribution processes, contaminant behaviour, and organism interactions to effectively manage marine pollution. A combination of applied sciences including chemistry, biology, ecology, hydrodynamics, toxicology, statistics, and oceanography should be used in the monitoring, management, and mitigation of marine pollution. ◻ Figure 1.7 provides a conceptual approach that highlights the interacting factors associated with understanding marine pollution for research and management. Undesirable outcomes of past polluting activities highlight the need for social research to also be included in the multidisciplinary approach for decision-making.

Community expectations have changed over the years in many parts of the world, particularly for mining and other potentially polluting industries. Resource extraction projects and infrastructure developments both require community-engaged decision-making during planning, construction, and operation. In some countries, these industries are working with the concept of gaining a **social licence to operate** (e.g. Prno 2013; Kelly et al. 2017) to gain community endorsement. Interestingly, this reintroduces the judgement in the historical use of the word pollution (Nagle 2009). Such research helps to identify the social acceptability of biodiversity offsets and trade-offs as tools to protect marine environments (e.g. Richert et al. 2015), and helps define how a development or project may be accepted by a community.

1.8 Polluting Substances—Local and Global Considerations

When we consider polluting substances, there are distinctly different threats to coastal marine ecosystems compared to open ocean ecosystems. We have already noted that around 80% of marine pollution is from land-based sources. The extent to which these reach the open ocean generally decreases with distance from land (e.g. Vikas and Dwaraskish 2015), although floating pollutants such as plastics can travel 1000s of kilometres across the ocean. In the context of marine-based sources of polluting substances, Tornero and Hanke (2016) provide a detailed review of sources in European seas. They highlight shipping, mariculture, offshore gas exploration and production, seabed mining, dredging and dumping, and legacy sites as major sea-based activities that release contaminants (Tornero and Hanke, 2016). These activities are globally relevant as marine-based sources of polluting substances.

Ocean dumping of wastes have in the most part been addressed by international conventions and protocols (► Chapter 16), but legacy problems remain. The Convention on the Prevention of Marine Pollution by Dumping of Wastes and Other Matter 1972, commonly known as the **London Convention**, is one of

1

□ Table. 1.3 Contaminants in the marine environment: types, common sources, and general factors influencing impacts and potential as pollutants and useful references/examples

Contaminant class	Examples	Examples of sources	General impacts and influencing factors	Reference
Trace elements ▶ Chapter 5	Arsenic (As), cadmium (Cd), copper (Cu), iron (Fe), nickel (Ni), silver (Ag), zinc (Zn); rare earth metals of increasing concern, e.g. lithium (Li), caesium (Cs), and yttrium (Y)	Antifouling paints, batteries, electronics, fuels, building materials, mining wastes (small and large scale, terrestrial, deep sea; marine tailings disposal), agricultural and urban runoff, industrial effluents, landfill leachate, erosion, atmospheric deposition, shipping accidents, and operational ship discharges	Many essential at low concentrations, highly toxic at elevated levels; natural variation due to local geology; solubility and toxicity strongly influenced by water physicochemistry (i.e. pH, salinity, complexing capacity, and dissolved oxygen), competition with essential elements affects structure and function of biomolecules and physiological processes. Biological detoxification and depuration processes exist	Walker et al. (2012), Reichelt-Brushett (2012)
Organometallic compounds Chapters 4 and 7	Mono, di and tri-butyltin (TBT), methylmercury (CH_3Hg^+), and some pesticides	Agricultural runoff and waste, industrial effluents, mining wastes, and antifouling paints	Organic forms of some metals are more toxic than their inorganic counterparts; may be direct inputs or naturally formed after metal introduction—influenced by type and abundance of organic ligands, microbial communities (i.e. methylating bacteria), and oxygenation	Renzoni et al. (1998)
Agricultural chemicals Chapters 7 and 8	Insecticides, herbicides, fungicides, and other specific pesticides and disinfectants	Agricultural runoff and waste, aquaculture, and mariculture	Synthesised for biocidal properties (many non-specific) similar activity in receiving environment; continuing legacy of older formulations (e.g. DDT) that are extremely persistent compared to most modern formulations	Walker et al. (2012)
Nutrients ▶ Chapter 4	Nitrogen (N) and phosphorous (P)	Fertilisers, sewage, urban and agricultural runoff, aquaculture, and mariculture	Enrichment in plant nutrients increases primary productivity; eutrophication and associated changes to ecosystem structure and function (e.g. increased crown of thorns starfish outbreaks, algal blooms produce toxins and cause deoxygenation); influenced by water exchange, limiting nutrients, nutrient cycling, and type and resilience of ecosystem (e.g. reefs requiring oligotrophic conditions will be more significantly affected than a mesotrophic system)	Brodie et al. (2012)
Persistent organic and halogenated chemicals Chapters 7 and 8	Dioxin, fluorocarbons, organochlorines, organophosphates, organobromines, etc	By-products of fuel and waste combustion and industrial processes (e.g. paper and plastic manufacturing), flame retardants, fire-fighting foam, non-stick coatings, landfill leachates, agricultural runoff, and atmospheric deposition	Extremely Persistent chemical structures with many now banned in many countries. They are mostly lipophilic and bioaccumulate and biomagnify. There are limited biological detoxification and degradation pathways	Arpin-Pont et al. (2016)

(continued)

Table. 1.3 (continued)

Contaminant class	Examples	Examples of sources	General impacts and influencing factors	Reference
Hydrocarbons and dispersants ▶ Chapter 6	Oil, gas, and some petroleum products	Shipping operations and accidents and off-shore extraction and processing	Comprised of C and H with high molecular stability and lipophilicity and are physical and chemical toxicity. The extent of pollution depends on amount, type of hydrocarbon (i.e. volatility, viscosity) and the hydrodynamics of the spill area	Walker et al. (2012); Kujawinski et al. (2011)
	Dispersants	Oil spill remediation	Used to break oil into droplets which can be easily dispersed in water, and enhance biodegradation; toxic to pelagic/benthic organisms; should not be used in shallow or ecologically sensitive waters (e.g. reefs, fishing areas, and marine parks)	Kujawinski et al. (2011)
Dissolved gases ▶ Chapter 11	Carbon dioxide (CO_2)	Fossil fuel emissions, agriculture, and atmospheric equilibration	Ocean acidification depending on carbonate biogeochemistry (pH buffering capacity). May cause decreased carbonate saturation and thus reduce biogenic calcification	Anthony et al. (2008)
Pharmaceuticals Chapters 12 and 13	Antibiotics, antidepressants, blood pressure and diabetes medication, birth control, anti-bacterial agents, illicit drugs, and veterinary products	Sewage outfall (treated and untreated) and biosolids, agricultural runoff, wastewater discharges, aquaculture, and mariculture	Sewage treatment methods often ineffective. Concentrations in the environment reflect variable spatial and temporal usage trends. May cause endocrine disruption (e.g. feminisation of male organisms). The pollution potential depends on the mode of action, concentration, and stability of the compound as well as potential mixtures	Walker et al. (2012); Arpin-Pont et al. (2016)
Personal care and household products Chapters 12 and 13	Soaps; sunscreens; disinfectants; surfactants; nanoparticles (e.g. zinc and titanium dioxide)	Sewage; recreation; wastewater discharges; industrial cleaning; operational ship discharge	High inputs into coastal zones which are sensitive breeding/nursery grounds for many species (not significant in open ocean). Generally slow to degrade with potential photochemical reactions and toxic degradation products. Specific persistence and toxicity vary between compounds	Walker et al. (2012)
Pathogens ▶ Chapter 12	Parasites, viruses, protozoans, and bacteria	Untreated sewage, agricultural runoff, aquaculture, and mariculture	Introduction of new diseases, shifts in symbiotic microbial populations, and reduced resilience to existing diseases which can be exacerbated by deteriorating water quality	Lafferty et al. (2004)

(continued)

◘ Table. 1.3 (continued)

Contaminant class	Examples	Examples of sources	General impacts and influencing factors	Reference
Plastics and debris ▶ Chapter 9	Plastic bottles; packaging and general rubbish; microplastics; fishing nets and lines; timber; metal; other building materials	Dumping, litter, natural disasters, commercial fishing, aquaculture, and shipping accidents	Pollution potential depends on the size, shape, and buoyancy of the material. They are typically persistent impacting organisms by ingestion, entanglement, and release of toxicants (e.g. PCBs). May cause reduced air–water and sediment–water oxygen exchange and act as hosts for translocation and introduction of species attached to drifting plastics	Derraik (2002)
Sediment Chapters 5 and 7	Soil; clay; sand; fine sediment	Dredging, port construction, direct dumping of dredged sediment, catchment clearing, poor quality riparian vegetation, high rainfall, and erosion	Various particle sizes and settling rates. Causes high turbidity, smothering, and reduced light penetration. The impacts are related to the frequency, intensity, and duration of exposure and associated contaminant toxicity	Brodie et al. (2012); Erftemeijer et al. (2012)
Other Chapters 10 and 12	Radioactivity and explosives,	Conflict, accidents, weapons testing and use, and emissions from historical ocean dumping	May be non-ionising (low-energy photons) (e.g. light, radio waves) or ionising (high energy alpha, beta, gamma particles). Results in free radical production (e.g. OH−) that is highly reactive. The impacts are influenced by radiation type and decay rate (half-life)	Livingston and Povinec (2000)
	Noise	Shipping, military operations, seismic surveys; deep sea mining	Impact on communication and navigation signals of marine species may also cause chronic stress. The density of water means that sound propagates much further and faster in water than in air	Williams et al. (2015)
	Thermal pollution	Heated industrial effluents; increasing sea surface temperature (global warming)	Heat causes oxygen depletion resulting in changes in species distribution and metabolism, possible breakdown of symbiosis (e.g. coral bleaching). Heat may also increase the toxicity of chemical contaminants noting that tropical species tend to live closer to their upper temperature thresholds than temperate species but some species adapt	Baker et al. (2008)

Table preparation assisted by K. Summer

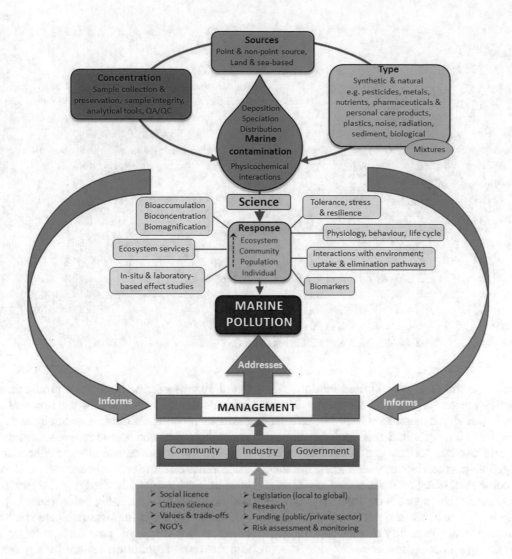

Sources
Point & non-point source,
Land & sea-based

Concentration
Sample collection &
preservation, sample integrity,
analytical tools, QA/QC

Type
Synthetic & natural
e.g. pesticides, metals,
nutrients, pharmaceuticals &
personal care products,
plastics, noise, radiation,
sediment, biological

Mixtures

Deposition
Speciation
Distribution
**Marine
contamination**

Physicochemical
interactions

Science

Bioaccumulation
Bioconcentration
Biomagnification

Tolerance, stress
& resilience

Response
Ecosystem
Community
Population
Individual

Physiology, behaviour, life cycle

Ecosystem services

Interactions with environment;
uptake & elimination pathways

In-situ & laboratory-
based effect studies

Biomarkers

**MARINE
POLLUTION**

Addresses

Informs   **MANAGEMENT**   Informs

Community   Industry   Government

➢ Social licence   ➢ Legislation (local to global)
➢ Citizen science   ➢ Research
➢ Values & trade-offs   ➢ Funding (public/private sector)
➢ NGO's   ➢ Risk assessment & monitoring

◻ **Figure 1.7** The multidisciplinary nature of marine pollution studies. *Image*: designed by A. Reichelt-Brushett and created by K. Summer

the first global conventions to protect the marine environment from human activities and has been in force since 1975. Its main purpose or objective is to promote the effective control of all sources of marine pollution and to take all practicable steps to prevent pollution of the sea by dumping of wastes and other matter. Currently, 87 States are signatories to this convention. In 1996, the **London Protocol** was agreed upon to modernise the Convention and, eventually, replace it. The International Convention for the Prevention of Pollution from Ships (MARPOL) is the main international convention covering the prevention of pollution of the marine environment by ships from opera-

tional or accidental causes and has had various modifications over the years (▶ Chapter 16). Importantly, these conventions and protocols do not cover discharges from land-based point and non-point sources. Consequently, management of land-based marine pollution and practices such as submarine tailing disposal and sewage discharge rely on local and national legislation for approval, operations, and control. There are other relevant conventions such as Minamata Convention on Mercury that have reduced the serious consequences of marine pollution (▶ Box 1.4) (see also ▶ Chapter 16).

1

Box 1.4: The Minamata Disaster

Between 1932 and 1968, a chemical production plant run by Chisso Co. Ltd. intentionally and knowingly discharged untreated methylmercury (MeHg)-laden wastewater into fresh water and marine environments surrounding what is now Minamata City in Japan. MeHg is a deadly neurotoxin, and in the 1950s and 1960s, people who consumed local seafood developed mysterious neurological symptoms including sensory disturbances, visual field constriction, and ataxia. More than 200 infants were born with Minamata disease between 1955 and 1959.

Also, in the 1950s, people in the vicinity of the chemical plant noticed fish floating on the water surface, barnacles appearing unable to stick to boat hulls, huge numbers of shellfish being washed on shore with open shells, and the seaweed appeared to have stopped growing.

By March 2001, thousands of people had died, and more than 10 000 had received financial compensation from Chisso Co. Ltd and the Japanese government. In 2010, there were 2271 official Minamata disease patients in the Minamata area, and more than 40,000 people exhibited partial symptoms.

A criminal trial in 1988 found the chief of the acetaldehyde plant and the president of Chisso Co. Ltd. guilty of intentionally diverting the wastewater drainage channel towards the river without treating the effluent, despite being (perhaps only somewhat) aware of its extremely high toxicity. Both were imprisoned for 2 years in 1979.

This tragic event led to the Minamata Convention on Mercury, which was signed on 16 August 2017. In 2019, there were 102 member countries of this convention.

(See Hachiya, 2012; Yorifuji, 2013; UNEP, 2018, 2019; Yokohama, 2018.)

On a final note the seventeen United Nations sustainability goals are an urgent call for action by all countries in a global partnership. Goal 14, Life Below the Water, has 10 targets that this book has import relevance to and can support people in realising these goals. This global partnership includes low-, middle-, and high- income and, being an open access resource, it is freely accessible to the anyone with Internet access. It will hopefully provide benefit to those wanting to learn about improving the sustainability of their marine environment and acting on their knowledge.

Targets for Goal 14 -Life below the water:

- 14.1 Reduce marine pollution
- 14.2 Protect and restore ecosystems
- 14.3 Reduce ocean acidification
- 14.4 Sustainable fishing
- 14.5 Conserve coastal and marine areas
- 14.6 End subsidies contributing to overfishing
- 14.7 Increase the economic benefits from sustainable use of marine resources
- 14.a Increase scientific knowledge, research and technology for ocean health
- 14.b Support small scale fishers
- 14.c Implement and Enforce international sea law
  For more information: ▶ https://www.globalgoals.org/ and ▶ https://sdgs.un.org/goals/goal14

1.9 Summary

Marine pollution has been created by human activities both on land and in/on the ocean. As consumers, we are all contributors to the increasing global resource demand, manufacturing, and waste production. By definition, a contaminant is a substance present in the environment where it should not normally occur, or at concentrations above background levels, while a pollutant is a contaminant that causes adverse effects in the natural environment. In order to understand how contaminants become pollutants, knowledge of seawater chemistry and how this influences the behaviour of contaminants and their toxicity is important. There is a wide and ever-expanding range of potential polluting substances, and the risk of pollution caused by any one or combination of these will depend on their sources, transport, bioavailability, and fate. Furthermore, organism interactions with their surrounding biotic and abiotic environment influences their exposure to contaminants and the subsequent potential impacts on their health.

Scientific research linked to a weight of evidence approach can be used to inform decisions that reduce the risk of environmental impacts associated with human activities. Importantly, the coastal environment has far different challenges associated with pollution reduction compared to the open ocean. Major pollution incidents have raised the public, political, and scientific profiles of marine pollution, and legislative frameworks now address ocean dumping. We are still faced with the challenge of how to reduce the incidence of pollution and manage the impacts and improve degraded systems. This is a challenging, multidisciplinary field of study requiring collaboration between scientists, governments, industries, and communities to enhance our understanding and knowledge, and develop solutions to reduce waste production, improve management capability, and therefore reduce the threat of marine pollution now and into the future.

1.10 Study Questions and Activities

1. Research an accidental pollution incident that greatly impacted marine environments. Write a paragraph that includes information as to how, where, and when the accident occurred, what happened, what the immediate consequences were, and what the reported long-term consequences have been (if any). Investigate whether any recent follow-up studies have been done to assess effects.
2. Explain (in your own words) the difference between contamination and pollution. Describe the advantages and disadvantages of the two definitions.
3. Identify an important land-based source of marine pollution, state the contaminant(s) that are associated with it, and briefly describe what is known about the effects on marine ecosystems.

References

Anthony KRN, Kline DI, Diaz-Pulido G, Dove S, Hoegh-Guldberg O (2008) Ocean acidification causes bleaching and productivity loss in coral reef builders. Proc Natl Acad Sci USA 105(45):17442–17446

Arpin-Pont L, Martínez-Bueno MJ, Gomez E, Fenet H (2016) Occurrence of PPCPs in the marine environment: a review. Environ Sci Pollut Res 23(6):4978–4991

Baker AC, Glynn PW, Reigl B (2008) Climate change and coral reef bleaching: an ecological assessment of long-term impacts, recovery trends and future outlook. Estuar Coast Shelf Sci 80(4):435–471

Balatsky AV, Balatsky GI, Borysov SS (2015) Resource demand growth and sustainability due to increased world consumption. Sustainability (switzerland) 7(3):3430–3440

Barbier EB, Hacker SD, Kennedy C, Koch EW, Stier AC, Silliman BR (2011) The value of estuarine and coastal ecosystem services. Ecol Monogr 81(2):169–193

Beesley PL, Ross GJB, Wells A (eds) (1998) Mollusca: The southern synthesis. Fauna of Australia, vol 5, Part A. CSIRO Publishing, Melbourne, p 563

Brodie JE, Kroon FJ, Schaffelke B, Wolanski EC, Lewis SE, Devlin MJ, Bohnet IC, Bainbridge ZT, Waterhouse J, Davis AM (2012) Terrestrial pollutant runoff to the Great Barrier Reef: an update of issues, priorities and management responses. Mar Pollut Bull 65(4–9):81–100

Chapman PM (2007) Determining when contamination is pollution—weight of evidence determinations for sediments and effluents. Environ Int 33:492–501

Chiba S, Saito H, Fletcher R, Yogi T, Kayo M, Miyagi S, Ogido M, Fujikura K (2018) Human footprint in the abyss: 30 year records of deep-sea plastic debris. Mar Policy 96:204–212

Derraik JGB (2002) The pollution of the marine environment by plastic debris: a review. Mar Pollut Bull 44(9):842–852

Doney SC, Fabry VJ, Feely RA, Kleypas JA (2009) Ocean acidification: the other CO_2 problem. Ann Rev Mar Sci 1:169–192

Ducrotoy JP, Elliott M (2008) The science and management of the North Sea and the Baltic Sea: natural history, present threats and future challenges. Mar Pollut Bull 57(1–5):8–21

Edgar GJ (2000) Australian marine life: The plants and animals of temperate waters, 2nd edn. New Holland, Reed, p 544

Edinger EN, Jompa J, Limmon GV, Widjatmoko W, Risk MJ (1998) Reef degradation and coral biodiversity in Indonesia: effects of land-based pollution, destructive fishing practices and changes over time. Mar Pollut Bull 36(8):617–630

Erftemeijer PLA, Riegl B, Hoeksema BW, Todd PA (2012) Environmental impacts of dredging and other sediment disturbances on corals: a review. Mar Pollut Bull 64(9):1737–1765

Garrick BJ (2008) Quantifying and controlling catastrophic risks. Academic Press, California, p 376

Hawkins K (1984) Environment and enforcement: regulation and the social definition of pollution. Clarendon Press, New York, p 253

Hachiya N (2012) Epidemiological update of methylmercury and Minamata disease. In: Ceccatelli S, Aschner M (eds) Methylmercury and neurotoxicity. Springer, Boston, pp 1–11

Karydis M, Kitsiou D (2013) Marine water quality monitoring: a review. Mar Pollut Bull 77:23–36

Kelly R, Pecl GT, Fleming A (2017) Social licence in the marine sector: a review of understanding and application. Mar Policy 81:21–28

Kennish MJ (2002) Environmental threats and environmental future of estuaries. Environ Conserv 29(1):78–107

Kujawinski EB, Kido Soule MC, Valentine DL, Boysen AK, Longnecker K, Redmond MC (2011) Fate of dispersants associated with the Deepwater Horizon oil spill. Environ Sci Technol 45(4):1298–1306

Libes SM (1992) An introduction to marine biogeochemistry. John Wiley and Sons, Inc., Singapore, p 734

Libes SM (2009) Introduction to marine biogeochemistry, 2nd edn. Academic Press, New York, p 928

Lafferty KD, Porter JW, Ford SE (2004) Are diseases increasing in the ocean? Annu Rev Ecol Evol Syst 35:31–54

Livingston HD, Povinec PP (2000) Anthropogenic marine radioactivity. Ocean Coast Manage 43(8–9):689–712

Makam AN, Puneeth MK, Varalakshmi, Jayarekha P (2018) E-waste management methods in Bangalore. In: Proceedings of 2nd International conference on green computing and internet of things, ICGCIoT, pp. 6–10, 8572976

Manahan S (2009) Environmental chemistry, 9th edn. CRC, Boca Raton, p 783

Meynecke JO, Lee SY, Duke NC (2008) Linking spatial metrics and fish catch reveals the importance of coastal wetland connectivity to inshore fisheries in Queensland, Australia. Biol Conserv 141(4):981–996

Mora C, Tittensor DP, Adl S, Simpson AGB, Worm B (2011) How many species are there on earth and in the ocean? PLoS Biol 9(8):e1001127

Nagle CJ (2009) The idea of pollution. Univ Calif Davis School Law Rev 43(1):1–78

Pasquaud S, Vasconcelos RP, França S, Henriques S, Costa MJ, Cabral H (2015) Worldwide patterns of fish biodiversity in estuaries: effect of global vs. local factors. Estuar Coast Shelf Sci 154:122–128

Pilson MEQ (1998) An introduction to the chemistry of the sea. Prentice-Hall, Upper Saddle River, p 529

Prno J (2013) An analysis of factors leading to the establishment of a social licence to operate in the mining industry. Resour Policy 38(4):577–590

Reichelt-Brushett A (2012) Risk assessment and ecotoxicology limitations and recommendations for ocean disposal of mine waste in the Coral Triangle. Oceanography 25(4):40–51

Reichelt-Brushett A, Clark M, Birch GF (2017) Physical and chemical factors to consider when studying historical contamination and pollution in estuaries. In: Weckström K, Saunders K, Gell P, Skilbeck C (eds) Applications of paleoenvironmental techniques in estuarine studies. Springer, Netherlands, pp 239–276

Renzoni A, Zino F, Franchi E (1998) Mercury levels along the food chain and risk for exposed populations. Environ Res 77(2):68–72

1

Richert C, Rogers A, Burton M (2015) Measuring the extent of a social license to operate: the influence of marine biodiversity offsets in the oil and gas sector in Western Australia. Resour Policy 43:121–129

Riley JP, Chester R (1971) Introduction to marine chemistry. Academic Press, San Diego, p 465

Robinson L (2012) Changeology—how to enable groups, communities, and societies to do things they've never done before. Scribe, Brunswick, p 272

Santos IR, Glud RN, Maher D, Erler D, Eyre BD (2011) Diel coral reef acidification driven by porewater advection in permeable carbonate sands, Heron Island, Great Barrier Reef. Geophys Res Lett 38(3):1–5

Steffen W, Crutzen PJ, McNeill JR (2007) The anthropocene: are humans now overwhelming the great forces of nature? Ambio 36(8):614–621

Tornero V, Hanke G (2016) Chemical contaminants entering the marine environment from sea-based sources: a review with a focus on European seas. Mar Pollut Bull 112(1–2):17–38

UNEP (United Nations Environment Programme) (2018) The Minamata Convention on Mercury celebrates its first anniversary. UN Environment. ▶ https://www.unenvironment.org/news-and-stories/story/minamata-convention-mercury-celebrates-its-first-anniversary. Accessed 21 Jan 2019

UNEP (United Nations Environment Program) (2019) The Minamata Convention on Mercury. ▶ http://www.mercuryconvention.org/News/FromtheExecutiveSecretary/tabid/6352/language/en-US/Default.aspx. Accessed 12 Feb 2019

Vikas M, Dwarakish GS (2015) Coastal pollution: a review. Aquat Procedia 4:381–388

Walker CH, Sibly RM, Hopkin SP, Peakall DB (2012) Principles of ecotoxicology, 4th edn. CRC, Florida, p 386

Williams R, Wright AJ, Ashe E, Blight LK, Bruintjes R, Canessa R, Clark CW, Cullis-Suzuki S, Dakin DT, Erbe C, Hammond PS, Merchant ND, O'hara PD, Purser J, Radford AN, Simpson SD, Thomas L, Wale MA (2015) Impacts of anthropogenic noise on marine life: publication patterns, new discoveries, and future directions in research and management. Ocean Coastal Manage 11517–11524

Worldometer (2023) Current world population. Avaialbe at: ▶ https://www.worldometers.info/world-population/. Accessed 28 Feb 2023

Yokohama H (2018) Mercury pollution in Minamata. Springer, Cham, p 67

Yorifuji T, Tsuda T, Harada M (2013) Minamata disease: a challenge for democracy and justice. In: Late lessons from early warnings: science, precaution, innovation. European Environment Agency, Copenhagen, Denmark, pp 92–130. Available at: ▶ https://www.eea.europa.eu/publications/late-lessons-2. Accessed 14 Jan 2022

Collecting, Measuring, and Understanding Contaminant Concentrations in the Marine Environment

Amanda Reichelt-Brushett

Contents

2.1   Introduction – 25

2.2   Defining the Purpose of the Research – 25

2.3   Transport and Storage of Contaminants – 26

2.4   Developing a Sampling Program – 27
2.4.1   Define Locations, Sites, and Replicates – 27
2.4.2   Sampling Plan – 27

2.5   Units of Measurement – 28

2.6   Water Sampling and Analysis – 29
2.6.1   Surface Water – 29
2.6.2   Water from Depth – 32
2.6.3   Pore Water and Groundwater – 32

2.7   Sediment Sampling and Analysis – 33
2.7.1   Surface Sediments – 33
2.7.2   Sediment Cores – 33
2.7.3   Suspended Particulate Matter (SPM) – 38

2.8   Biota Sampling – 39
2.8.1   Tissue Sampling – 39
2.8.2   Biomonitors – 42
2.8.3   Collecting Pelagic Species – 42
2.8.4   Collecting Benthic Species – 42

2.9   Quality Assurance and Quality Control – 43
2.9.1   NATA Registration and Other Global Systems – 43

© The Author(s) 2023
A. Reichelt-Brushett (ed.), *Marine Pollution—Monitoring, Management and Mitigation*,
Springer Textbooks in Earth Sciences, Geography and Environment,
https://doi.org/10.1007/978-3-031-10127-4_2

2.9.2   Chain of Custody – 44

2.9.3   Sample Storage and Integrity – 44

2.9.4   Step to Ensure Analytical Certainty – 44

2.9.5   Detection Limits – 45

2.9.6   Dealing with Difficult Samples – 45

2.9.7   Dealing with Novel Contaminants – 45

2.10   Identifying Contamination – 46

2.10.1   Determining Background Concentrations – 46

2.10.2   Normalising Techniques – 47

2.10.3   Understanding Degradation – 47

2.10.4   Using Guideline Values – 47

2.10.5   Development of Guidelines for New and Emerging
         Contaminants – 48

2.11   Summary – 48

2.12   Study Questions and Activities – 48

         References – 48

Acronyms and Abbreviation

AFRAC	African Accreditation Cooperation
APAC	Asia-Pacific Accreditation Cooperation
ARAC	Arab Accreditation Cooperation
BAF	Bioaccumulation factors
BCF	Bioconcentration factors
CCCs	Criteria continuous concentrations
CMCs	Criteria maximum concentrations
DO	Dissolved oxygen
GLP	Good laboratory practise
EA	European Accreditation
ERLs	Environmental risk limits
EQS	Environmental quality standards
IAAC	Inter-America Accreditation Cooperation
ICP-MS	Inductively coupled plasma mass spectrometry
ILAC	International Laboratory Accreditation Cooperation
LOD	Limit of detection
LOR	Limit of reporting
MRA	Mutual recognition agreement
NATA	National Association of Testing Authorities Australia
NOAA	National Oceanic and Atmospheric Administration
NTAC	National Technical Advisory Committee
OECD	Organisation of Economic and Cooperation and Development
PFOS	Perfluorooctanesulfonic acid
PFOA	Perfluorooctanoic acid
QA	Quality assurance
QC	Quality control
QCs	Quality control samples
RHTs	Recommended holding times
SADCA	South African Development Community Cooperation in Accreditation
SPM	Suspended particulate matter
US EPA	United States Environmental Protection Authority
USA	United States of America
WQC	Water quality criteria

2.1 Introduction

A large part of marine pollution studies is about collecting, analysing, and interpreting the concentrations of contaminants in the environment. This involves field and laboratory work to collect and analyse the samples. Some analyses are completed directly in the field (i.e. in situ). From the process of collection through to the final analyses, there are many **quality assurance and quality control (QA/QC)** steps that are required, which, when used properly, ensure sample integrity and the reliability of results, therefore, resulting in meaningful interpretations and conclusions. Ultimately, following correct **sampling** and **analytical procedures** facilitates an accurate understanding of risk to marine organisms, ecosystems, and human health. This chapter provides a general introduction to procedures for identifying contamination in marine environments. This may not be a chapter that you read from beginning to end, but rather one that you will use as a valuable resource when putting together a

sampling program. It will also help you understand what analyses may be complementary to the sampling effort in order to interpret the behaviour of contaminants, and how concentrations relate to guideline values. ■ Table 2.1 highlights some useful references that provide expanded detail for you to investigate further if you find yourself needing detailed knowledge of sampling and analytical protocols. Information on determining organism, population, and ecosystem responses to contaminants (ecotoxicology) is also provided in ▶ Chapter 3.

2.2 Defining the Purpose of the Research

The approach used for site selection during field sampling and assessment should be **informed by the research question**. For this reason, sampling programs may differ between studies for logical and justifiable reasons. However, mistakes can be made, particularly in interpreting results if the sampling program is not well designed. Im-

2

▣ Table 2.1   Selected text resources regarding environmental sampling and analyses

Reference details
Baird, R. and Bridgewater. L. (2017). *Standard Methods for the Examination of Water and Wastewater*. 23rd Edition. Washington, American Public Health Association. P 1 545
Blasco, J. and Tovar-Sanchez. A. eds., (2023). *Marine Analytical Chemistry*. Cham, Springer Nature. P 455
US Department of the Interior and US Geologic Survey. (2015). *National Field Manual for the Collection of Water-Quality Data*. Available at: ▶ http://water.usgs.gov/owq/FieldManual/. [Accessed June 14 2021]
Mudroch, A. and Macknight, S.D. (1994). *Handbook of Techniques for Aquatic Sediments Sampling*. 2nd Edition. Boca Raton: CRC Press. P 256
Simpson S.L. and Batley G.E. eds., (2016). *Sediment Quality Assessment – A Practical Guide*. 2nd Edition. Clayton South: CSIRO Publishing. P 360

agine designing a program that aims to determine the extent of contamination from a mine tailings disposal pipeline. You would likely consider designing a spatial sampling program that investigates sites at various and increasing distances from the source (pipeline outfall point) in a grid or radial pattern and include replicate samples of the waters, sediments, and/or biota. You could thereby gain an understanding of the overall footprint of the contamination impact. In choosing the sites, it would also be useful to consider potential environmental factors and temporal changes (e.g. where and how currents and tides move water and factors affecting the transport, settlement, and resuspension of sediments). It is useful to complete literature searches during the initial phase of the research project to gain insights into the locations of interest and site-specific detail which will inform the design of a good study program. Useful information may include historical data on contamination events and recorded concentrations, previous studies at the location and/or similar locations, as well as physicochemical and biological information including bathymetry, current and tide patterns, sediment characteristics, habitat types, and species composition.

When biota is included in a sampling program, consideration must be given to selecting a suitable range of target species. Species with sessile or sedentary life cycles (e.g. most adult bivalves) will prove useful when relating body concentrations of contaminants to the chronic exposure regime at particular sites. By contrast, a species with a large range (e.g. tuna) might be just passing through a contaminated site and we would have limited understanding of past exposure, thus making it difficult to identify the source of the body burden (amount of contaminant measured in the tissue of the organism). Such data would, therefore, not be representative of the actual sampling site.

Setting up a sampling program requires questions or hypotheses to be established that can be answered with planned data collection. I have seen sampling programs with serious effort committed to sampling contaminants in pelagic fish from sites at selected distances from a con-

tamination source. These types of data are costly to obtain and the results are difficult to interpret because they are not clearly linked to site conditions since most pelagic fish have a relatively large area in which they roam.

In some studies, part of the interest in sampling fish is related to contaminant consumption and human health impacts and there are very good reasons for being concerned with contaminant concentrations in edible fish. A good place to obtain samples of fish intended for human consumption in communities close to impact sites is from local markets or from local fishermen who collect fish from their fishing grounds. These samples may be quite different from sites selected in a sampling program focussed on distance from a contaminant source.

2.3 Transport and Storage of Contaminants

All contaminants originate from a source and are distributed around the environment by physical forces such as winds, tides, upwelling, downwelling, currents, ocean circulation, rainfall, groundwater movement, and surface runoff. Throughout these distribution processes, biota come into contact with contaminants and may accumulate and transport them. The size and density of **solid particles** influence their distribution and the contaminants that bind to them. In general, heavy or larger particles settle out first and deposit in higher energy environments compared to fine particles like clay. Clays (diameter $< 2\,\mu m$) may remain in suspension for months or years and have a high surface area to volume ratio, and, therfore greater adsorptive capacity for contaminants (Reichelt-Brushett et al. 2017). Fine particles can potentially transport bound contaminants many kilometres from the original source (▣ Figure 2.1).

The **solubility** of a contaminant also influences its distribution. Water-soluble substances will move readily in the marine environment as they will be dissolved

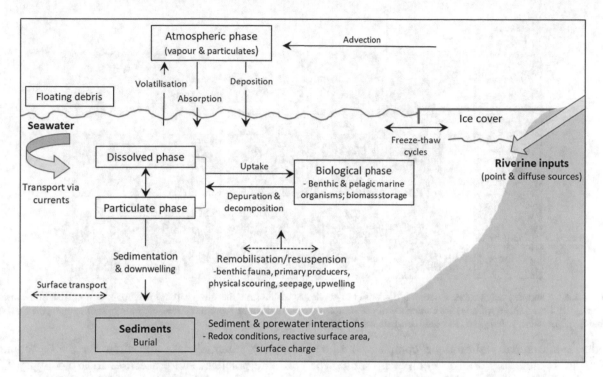

Figure 2.1   Marine contaminant distribution and storage in the environment. *Image*: A. Reichelt-Brushett and K. Summer

throughout the water column and transported with currents and tides. Some water-soluble substances, such as metal ions, are charged and will adsorb to particulates (► Chapter 5). Similarly, insoluble substances will more commonly be associated with sediments, or float on the water surface.

Organisms interact with contaminants from different **physical compartments** (e.g. water, sediment, and biota) how they do so will influence the rate and pathway of contaminant uptake. For example, a filter-feeding organism such as an oyster will filter large volumes of water and suspended particles through its system, whilst a polychaete worm will have a close affiliation with the sediments in which it burrows. The **trophic transfer** of some contaminants is also an important consideration. Different organisms will take up contaminants via different uptake pathways; some contaminants may be stored in body tissue, whilst others may be metabolised and excreted.

2.4   Developing a Sampling Program

2.4.1   Define Locations, Sites, and Replicates

A sampling program should define terms related to the sampling effort, and usually, a design is set up to ensure the suitability of the collected data for statistical anal-

yses. Consider the approach in ▣ Figure 2.2; the study program has five locations, within each location, there are three sites, and within each site, there are three replicate samples (e.g. replicate sediment samples). In other words, the term **replicate** is explained as a function of **site**, the site is explained as a function of **location**, and the locations are within a study program. The definition of these words (or similar) should be locked in as part of the program. Ideally, there should be a minimum of three replicates and these must be true replicates (i.e. sediments collected from three different grab samples taken at each site, not three sediment samples from a single grab sample (this is called **pseudo replication**). Likewise, three tissue samples from a single fish is not true field replication (repeat samples like this may be useful in another context such as to ensure QA/QC in the laboratory). Upon collection, although it may seem obvious and be simple, it is critical that samples are clearly and correctly labelled to ensure that this detail is not lost and can be interpreted later.

2.4.2   Sampling Plan

Care needs to be taken to ensure that **sample integrity** is not compromised by the process of collection, transport, and storage. Suitable and appropriate equipment needs to be available to collect different types of samples. Equipment used in the field should be appropriately cleaned and calibrated; calibration, maintenance

2

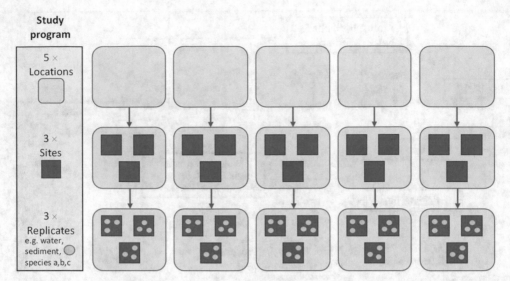

◘ Figure 2.2 A simple study program design is comprised of replicates within sampling sites, within locations-use of correct and consistent terminology is important. Replicates are comprised of one sample type and should be reproduced for different types of samples such as water, sediment, and organisms. *Image*: A. Reichelt-Brushett and K. Summer

schedules, and repairs should be completed and recorded in related log books. Equipment should be stored clean and according to manufacturer recommendations. During sampling trips, planning for enough **storage space** is important. Much of the time samples may be stored on ice and frozen immediately on return to the laboratory and prior to analysis. Some samples may require more rigorous **storage protocols**, such as; storing at -80 °C, fixing in acetone, immediate acidification, or storage in the dark. Text and tables later in this chapter provide details related to sample storage requirements for different types of analysis.

It is essential to determine the amount of material required for all intended analyses, and that samples are stored appropriately for the needs of specific analyses or pre-treated in the field if required. There are similar but different sampling protocols for different environmental compartments (i.e. waters, sediment, and biota). These protocols are put in place to avoid contaminating a sample with field equipment, compromising the physicochemistry, and to ensure the stability of a sample matrix. These procedures prevent contaminant loss, sample degradation, and transformation of chemical matrixes.

Sampling is costly, in terms of both time and resources, and needs to be carefully considered within a **budget** that is most often limited and defined. For this reason, it is important to determine the number of specialised people and support staff required to complete the sampling in a safe manner (i.e. estimate the amount of time required for each task at each site, and consider how much sampling is achievable each day).

Most organisations require risk assessments to be completed prior to field work. These assessments help identify potential risks and allow for the development of **safe practices and procedures** to avoid injury. Consider staff qualifications such as a boat licence, coxswain licence, level of SCUBA diving qualifications, first aid, and resuscitation certifications. These types of qualifications are a valuable part of a marine scientist's capabilities and employability. On large research vessels, there may be specialist teams attached to the operation of a given vessel that support the logistics of expert sampling.

2.5 Units of Measurement

As technology advances, so do instrument capabilities and the speed and accuracy of analyses. It is easier now than ever before to obtain fast and accurate measurements of very low contaminant concentrations, with limits of detection for some elements now in the part per trillion (ng/L) range. It is essential to know your **units** of concentration and what they mean (◘ Table 2.2) and remember:

One part per million (ppm) is equal to 1/1000000 of the whole:

1 ppm = $1/1000000 = 0.000001 = 1 \times 10^{-6}$

1 ppm is equal to 0.0001%

1 ppm is equal to 1000 ppb (part per billion)

At times, concentrations may be reported in **molarity**. To convert from molarity to ppm (mg/L), take molarity (with units mol/L), and multiply it by the molar mass (with units g/mol) you get g/L. Just multiply g/L by 1000 to convert g to mg, and you have ppm (in mg/L of water).

◻ **Table 2.2** Common units used for reporting concentrations of contaminants (liquids use per L and solids use per kg) (see also Appendix I). Created by K. Summer

Extended unit	Abbreviation option 1	Abbreviation option 2	Scientific notation	Context
Percent	%	–	1×10^{-2}	Parts per hundred
1 g per Litre 1 g per kilogram	%	–	1×10^{-3}	Parts per thousand Per-mille
1 mg per Litre 1 mg per kilogram	1 mg/L 1 mg/kg	1 mg L^{-1} 1 mg kg^{-1}	1×10^{-6}	Parts per million (ppm)
1 µg per Litre 1 µg per kilogram	1 µg/L 1 µg/kg	1 µg L^{-1} 1 µg kg^{-1}	1×10^{-9}	Parts per billion (ppb)
1 nanogram per Litre 1 nanogram per kilogram	1 ng/L 1 ng/kg	1 ng L^{-1} 1 ng kg^{-1}	1×10^{-12}	Parts per trillion (ppt)
1 picogram per Litre 1 picogram per kilogram	1 pg/L 1 pg/kg	1 pg L^{-1} 1 pg L^{-1}	1×10^{-15}	Parts per quadrillion (ppq)

We must also keep in mind that, in seawater, a concentration of $1.00 \text{ mg/L} \neq 1.00 \text{ ppm}$ since the density of seawater is 1.035 kg/L. Hence, in theory:

$$1.00 \text{ mg/L seawater} = 1.00 \text{ mg/L} \times 1 \text{ L}/1.035 \text{ kg}$$
$$= 0.966 \text{ mg/kg or } 0.966 \text{ ppm}$$

In reality, most reported concentrations of contaminants in seawater do not take the density of seawater into consideration, probably due to the very minor actual difference and the complexity that would arise when sampling estuarine environments, where salinity varies in time and space.

2.6 Water Sampling and Analysis

Water generally contains low levels of contaminants and for this reason, small errors caused by poor sampling and/or analytical processes can cause large relative impacts on the final measured concentrations and compromise correct interpretations. ◻ Table 2.3 provides some details about procedures and analytical tools for different types of water analyses. Due to the behaviour of different classes of chemical contaminants, it is usual practice for samples to be collected in different types of containers so that the material the containers are made of doesn't interact with the sample. For example, some parameters such as chlorophyll *a* will photodegrade, and therefore, need to be protected from light on collection (usually by using opaque containers or wrapping containers in foil). Other parameters will require pretreatment such as filtration and/or acidification at the time of collection in order to stabilise the sample prior to storage. Containers used in sample collection and laboratory analysis should be acid washed prior to use for later metal analysis procedures, and ethanol washed prior to use for later organic chemical analysis.

2.6.1 Surface Water

Sampling water at the water surface is fairly easy but a few key points should be remembered:

– Containers for sampling should be rinsed three times with water that is to be collected.
– For dissolved contaminants, water should be collected just below the surface and not specifically from the water–air interface.
– The bottle should remain capped until it is fully submerged, the cap can then be removed, the bottle filled, and the cap replaced under the water. Exceptions to this would be if the sampling effort was specifically focused on floating material (i.e. not dissolved) such as oils or microplastics.

Some analyses can be completed in situ using various probes and detectors including pH, dissolved oxygen, temperature, turbidity, conductivity, redoxpotential, total dissolved solids, chlorophyll *a,* and pressure. **Data loggers** can be deployed for long time periods for continuous or high-frequency analysis. The data is either stored within the instrument until retrieval and downloading or is delivered in real time using phone or satellite systems. More advanced **field equipment** enables measurements of parameters including labile ions and trace metals, but the detection limits can be much higher than those available in laboratory analyses, and therefore, the data may only be useful for broad screening. Deployed equipment can be damaged by vandalism, biofouling, rough seas, floods, and bad weather; therefore, systems need to be in place to minimise such risks.

Sampling the interface of the water and atmosphere is also important for some studies. Floating material such as debris, nanoplastics, and films or slicks

Table 2.3 Approaches for the analyses of some contaminants in seawater, techniques, sample collection and storage. Created by K. Summer

Contaminant/analyses class	Specific analyses	Typical instrumentation/method (LOD)	Sample volume required	Sample container	Sample storage	Maximum storage time
Metals and metalloids	Total, dissolved	ICP–MS (0.5–5 µg/L; metal dependent)	10 mL	Polycarbonate	0.45 µm filtration; acidify with conc. HNO_3; R	6 mo (28 d for Hg)
	Acid extractable	ICP–MS (0.5–5 µg/L; metal dependent)	10 mL	Polycarbonate	Acidify with conc. HNO_3; R	6 mo (28 d for Hg)
	Speciation	HPLC–ICP-MS, GC–MS (low ng/L)	10 mL–1 L; compound/method dependent	Polycarbonate for metals/metalloids; Glass for organometallics	R; may require dilution/further preparation	14 d
Organic contaminants	Pesticides	HPLC, HPLC-FLD (0.2–10 µg/L–compound/method dependent)	Up to 1 L; compound/method dependent	Glass solvent-rinsed amber bottles; PTFE-lined caps and septa; fill to neck	Preserve (e.g., ascorbic acid, $Na_2S_2O_3$) if necessary (dependant on sample redox conditions); LS; R	28 d
	Hydrocarbons (e.g. PAHs, TPHs, BTEX, oil/grease)	HPLC–ICP-MS + UVD and/or FLD; GC–MS + FID (0.5–2 µg/L); HEGM (2 mg/L)	2–40 mL (PAH, BTEX, TPH C6-C9); 1L (TPH C10-C36); 500 mL oil/grease	G solvent-rinsed amber bottle; PTFE-lined caps and septa; do not pre-rinse with sample	Acidify with HCl/H_2SO_4 if analysis delayed; LS; R	7 d
	Surfactants	HPLC–ICP-MS (0.1 mg/L)	100 mL	Glass	R; avoid foam formation	2 d
Pharmaceuticals	Specific drugs	HPLC–ICP-MS (1–2000 ng/L; depending on compound)	500 mL	Glass PTFE-lined caps and septa;	Immediately dechlorinate using 50 mg ascorbic acid/L sample and add 1 g NaN_3/L sample as biocide; R; LS	28 d
Nutrients/trophic status	Nitrogen and phosphorous (total)	Persulfate-FIA (2 µg/L)	50 mL	Acid-washed P/G vials	Acidify with H_2SO_4/HCl; R	28 d
	Chlorophyll-a	HPLC; EF (5 µg/L)	500 mL	Opaque P/G container; filter to remove algae	Cool/R (2 d) or F (28 d); LS	2 d, 28 d
	BOD	DO meter, incubation bottles	500 mL	Clean glass bottles or disposable plastic BOD bottles with stoppers	LS during incubation period; cool	24 h
Nano/micro-particles	Nano-materials (1–100 nm)	HPLC–ICP-MS; variable dependent on type	1 L or potentially greater volumes, then filter	Plastic (inorganic) or Glass (organic constituents)	LS; depends on particle type	Depends on particle type
	Micro-plastics (<5 mm)	WPO-DS-GA; ATR-FTIR; (0.1 g/L)	1 L	Glass; whole sample or surface net capture (mesh size depending on desired size fraction), record transect length	Rinse from net/preserve with 70% ethanol; Cool; LS	Depends on particle type

(continued)

2

Table 2.3 (continued)

Contaminant/ analyses class	Specific analyses	Typical instrumentation/ method (LOD)	Sample volume required	Sample container	Sample storage	Maximum storage time
Physicochemical analyses	e.g. DO, pH, EC, temperature, turbidity	Relevant calibrated meters and probes (single/multi-parameter)	20–500 mL (parameter dependent)	Plastic or lass; analyses generally performed on-site	Cool	6 h–28 d, depending on parameter
	TOC/DOC, TSS	Oven-dried TSS	50 mL; 250 mL (depends on clarity)	Plastic or glass; filter to 0.45 µm for DOC	R; LS	28 d
	Alkalinity	HCl/H_2SO_4 titration (1 mg/L)	100 mL	Plastic or glass	R	14 d
Gases	e.g. CO_2, CH_4	CRDS			Generally analysed on-site	
Radionuclides	e.g. ^{222}Rn, ^{131}I and stable isotopes ^{12}C, ^{13}C	HPGe; GFPC (1 – 100 pCi/L)	20 mL – 18 L (depending on analyses)	Plastic or glass	Acidify with H_2SO_4/HCl; R	6 mo (8 d ^{222}Rn, 14d ^{131}I)
Pathogens	e.g. *E. coli*	Cell culture, microscopy (cfu/100 mL); HPLC–ICP–MS	100 mL	Plastic or glass	R	96 h

Sources and examples: Batley and Gardener (1977); SA and SNZ (1998); Takeuchi et al. (2000); Rovedatti et al. (2001); US EPA (2001); Anderson et al. (2002); Schultz et al. (2003); Niyogi and Wood (2004); Smith (2006); Becker (2008); Hassellöv et al. (2008); Gattuso et al. (2010); Hirai et al. (2011); Buesseler et al. (2012); Kroon et al. (2012); Moynihan et al. (2012); Chen et al. (2013); Maher et al. (2015); Masura et al. (2015); Downs et al. (2016); Viršek et al. (2016); Baird et al. (2017); Rosentreter et al. (2017); Reichelt-Brushett et al. (2017)

F = freeze, R = refrigerate (transport on ice, store < 4 °C), LS = light sensitive (store in dark/foil wrapped), ATR-FTIR: attenuated total reflectance fourier-transform infrared spectroscopy; BOD: biochemical oxygen demand; BTEX: benzene, toluene, ethylbenzene and xylene; CRDS: cavity ring down spectroscopy; EDTA: ethylenediaminetetraacetic acid; EF: acetone-HCl pigment extraction and fluorometric determination; FID: flame ionisation detector; FLD: fluorescence detector; GFPC: gas flow proportional chamber; HEGM: hexane extractable gravimetric method; HPLC-ICP-MS: high performance liquid chromatography coupled with ICP-MS; partition of compounds based on transfer rate differences between stationary phase (column) and mobile phase (gas/liquid); HPGe: high purity germanium detection; ICP-MS: inductively coupled mass spectrometry; element identification and quantification based on atomic mass; LLE-GC-HPLC: liquid–liquid extraction (chloroform) coupled with gas chromatography or high performance liquid chromatography; LOD: limit of detection; Persulfate-FIA: persulfate digestion with flow injection analyser; PFOA/PFOS: perfluorooctanoic acid/perfluorooctane sulfonate; PTFE: polytetrafluoroethylene, TOC: total organic carbon; TPHs: total petroleum hydrocarbons; UVD: ultra-violet detector; WPO-DS-GA: wet peroxide oxidation of organic material, density separation, gravimetric analysis

of oil will be present at this boundary layer. Sampling is different for solid and liquid materials, and sampling procedures that involve sieving or filtration need to consider the effort per unit area of sea surface sampled. Adsorption discs can be used to collect contaminants from liquid samples, but some techniques can be weather dependent (e.g. as oils degrade they can become more solid than liquid). Measuring gas fluxes between the water and air requires different measurement and quantification approaches.

skin bottles or similar vessels (e.g. Knudsen, Nansen, and Rosette Friedinger samplers) are used which are deployed open, a messenger is sent to shut and seal the container, and a volume of water is bought to the surface in the sealed bottle for sampling (Mudroch and Macknight 1994). SCUBA divers can take samples at depths up to about 30 m. Sample collections from deeper than about 100 m are best achieved by robotic sampling devices attached to powered undersea vessels. ◘ Figure 2.3 provides some examples of different water sampling devices.

2.6.2 Water from Depth

There is a range of equipment available for collecting water samples at **depth** and the best method will depend on the depth required and the vessel available. Not surprisingly, generally the deeper the sample the costlier the sample collection process. For sampling down to a depth of about 10 m depth, a weighted tube can be deployed and a pump used to draw water up from depth with the sample being taken after a calibrated pumping period, depending on the depth being sampled. Beyond this, Ni-

2.6.3 Pore Water and Groundwater

Pore water or **groundwater** is the water that occupies the space between sediment or soil particles, making up about 5% of the volume of surface sediments (e.g. Presley et al. 1980). It can move through sediments and interacts with the contaminant load in the sediment; it is often a place of anoxic and reducing physicochemical conditions. These parameters influence contaminant behaviour, and some contaminants that would normally bind to sediment particles mobilise into the

◘ **Figure 2.3** Underwater sampling devices **a** a Niskin bottle rosette (*Photo*: Hanness Grobe. Creative Commons: CC-BY-SA-2.5), **b** Lowering a Niskin bottle (*Photo*: Hanness Grobe. Creative Commons: CC-BY-SA-2.5), **c** remotely operated underwater vehicle (*Source to Photo*: Mountains in the Sea Research Team; the IFE Crew; and NOAA/OAR/OER. NOAA Photo Library, Flickr), **d** a tethered management system atop a remotely operated vehicle (*Photo*: Gulf of Mexico Deep Sea Habitats Expedition/NOAA/OAR/OER CC By 2.0)

pore water. As a result, pore water can be highly contaminated and an important transport route for contaminants (Chapman et al. 1998; Simpson et al. 2016). Characteristics of the sediment particles and macro-biological structures, such as plant roots and burrows of benthic organisms, can increase the reactive surface area of the sediment and the pore water volume. Benthic organisms can be exposed to pore water, however, many burrowing species have well-aerated burrows that maintain a micro-layer of oxic sediment within their habitat which acts as a barrier to the anoxic sediment chemistry.

2.7 Sediment Sampling and Analysis

Adsorption of contaminants onto sediment surfaces plays an important role in the removal of contaminants from water. The capacity for contaminants to bind to surfaces will depend on the size, composition, and abundance of the particles, concentration of other ions in the solution, the type of charge associated with the contaminant, hydrophobicity, and the pH of the solution. Particles with bound contaminants eventually settle in low-energy environments and these **depositional areas** may have enriched contaminant loads. Once particles are settled and become sediments, the surrounding pore water chemistry may change and influence the adsorption/desorption behaviour. Adsorption and **desorption** are important mechanisms that influence the solubility, mobility, and dispersion of contaminants. The way contaminants behave is dynamic and influenced by many factors, some of which are highlighted in ◘ Table 2.4.

There is much discussion in the scientific literature about how to treat sediment samples and the effects that such treatments have on analytical results (e.g. Ajayi and Vanloon 1989; Markert 2008; Simpson and Batley 2016; Csuros 2018) (◘ Table 2.5). As with water samples, sediment samples destined for organic contaminant analysis should not come into contact with plastics and instead should be stored in aluminium foil or amber glass containers. Sediment samples for metal analysis can be stored in clean plastic bags or containers. An additional effort may be required to minimise geochemical changes in sediment resulting from oxidation processes in the newly-exposed sediments, especially where sequential or partial extraction analyses are to be performed. When metal speciation and/or toxicity studies are important, samples should be collected and immediately stored in nitrogen-sealed bags prior to analysis. During sample processing, the use of nitrogen sparring or a nitrogen atmosphere glove box may be justified to avoid sample oxidation. The care given to avoiding sediment oxidation will depend on the analyte/s of interest. ◘ Table 2.6 provides a snapshot of general techniques for sediment sampling and processing.

2.7.1 Surface Sediments

The top 15 cm of sediment is the primary area of sediment and water interaction and biological activity. Such interactions can also occur deeper in a sediment profile as a result of ground water movement, deeply-burrowing organisms, and natural (e.g. extreme weather) and artificial (e.g. dredging) disturbances. Surficial sediments are generally of most interest in sampling sediment contamination and sampling methods are designed to collect these surficial sediments. There are various types of grab samplers available for sediment collection. The **Van Veen grab** sampler (◘ Figure 2.4) is arguably the most commonly used sediment-sampling device; other options include: the Birge-Ekman Sampler, Ponar Grab Sampler, Smith-Mcintyre Grab Sampler, and Petersen Grab Sampler, all with various benefits and applications (Mudroch and Macknight 1994; Simpson et al. 2016). They all work on a similar principle: the grab sampler is lowered slowly through the water (either by hand or hydraulic winch) until landing on the sediment surface. The release of weight tension on the device from landing triggers the release of the pin that allows the jaws to close containing a sediment sample. Expert SCUBA divers are able to minimise disturbance whilst collecting surface sediment samples which can be particularly useful in studies of the sediment–water interface (Mudroch and Macknight 1994). Sediment sampling in extremely deep waters requires mechanised and usually remotely operated equipment.

2.7.2 Sediment Cores

Sediment cores taken from low-energy depositional sites can provide a wealth of historical information about contaminant loadings over time, and analyses can determine valuable pre-contamination reference points. The inclusion of age dating such as carbon dating can also provide a chronology and historical time series (Reichelt-Brushett et al. 2017). Such a time series may be linked to historical events that have occurred at the sites and locations of interest. Finding suitable sites that are not constricted by habitat types (e.g. mangrove roots, biogenic solids such as coral reefs), physical impediments to sampling (e.g. river stones), or impacted by disturbance events can be challenging. It is best to target low-energy environments with a sediment accumulation rate of mm to cm per year; a high rate of sediment accumulation limits the duration of the deposition history within a given length of core.

Similar to grab samplers, there is a range of core sampling devices such as gravity corers, box corers, piston corers, vibra corers, and boomerang corers (Mudroch and Macknight 1994; Batley and Simpson 2016), and as the name suggests, they are designed to retrieve

□ Table 2.4   Some examples of how sediment characteristics can influence contaminant loads, and measures used to mitigate errors in data interpretation

Physicochemical parameter	Influence on contaminant behaviour	Considerations in sample processing and interpretation
Mineralogy	Adsorption–desorption potential (also related to cation exchange capacity); influences contaminant storage/mobility	Complete X-ray diffraction and X-ray fluorescence analysis to determine mineralogy
Grain size	Finer-sized particles have a greater surface area to volume ratio, and therefore a higher capacity for surface adsorption and chemical reactions; grain size is also related to particle (and bound contaminant) distribution; negative charges dominate clay surfaces to which positively charged cations in solution are readily adsorbed	Normalisation[a] of contaminants with grain size fraction; filter water samples according to operationally defined processes to clarify grain size of interest
Colloids particles	Negative charges on clays are much greater than positive charges, and positively charged cations in solution are adsorbed onto clays	Filter water samples; use operationally defined processes to clarify grain size of interest
Salinity gradients	Common in estuaries: a steep ionic gradient destabilises fine suspensions of colloidal material and causes the suspension suspended particles to flocculate, carrying with it them the bound contaminants	Measure salinity at sampling sites; note variability between wet and dry seasons
pH and Redox	Both parameters influence the mobility/immobility of contaminants (particularly metals) in sediment–water systems	Measure pH and redox to assist with interpretation and risk assessment
Total organic carbon (TOC)	Humic substances in sediment TOC pools strongly bind contaminants and influence the partitioning of organic and inorganic contaminants within sediment matrices; major determinant of bioavailability	Normalisation of contaminants with TOC content; measure when TOC exceeds 1%

[a] for the purpose of comparison between sites

■ Table 2.5 Treatment methods for sediment samples and purposes for doing so

Treatment	Procedure	Purpose	Example of use
Homogenisation	Use large pre-cleaned glass bowl; debris documented and removed (e.g., shells, wood, stones, seagrass); should be performed quickly and efficiently to reduce sample oxidation	Obtain representative samples	ANZG (2018)
Compositing	Sub-sample combination; use equipment made of inert/non-contaminating materials; process within 24 h of collection; multiple replicate samples should be taken as opposed to compositing where sediment disturbance needs to be minimised	Determine broad site characteristics; if large sample volumes required; depends on sediment heterogeneity, desired analytical resolution/study objectives	
Sieving	Dry press-sieving is preferred using a 2 mm mesh sieve; not recommended as can substantially alter sample physicochemical characteristics (use forceps as an alternative for small sample volumes)	Remove organisms and debris	ANZG (2018)
Pore-water collection	In situ (peeper chambers, suction); laboratory methods (centrifugation, pressurisation, suction); require 1–2 kg sediment sample or 500 mL pore water; analyse as soon as possible as chemistry will change substantially	Expected to be in equilibrium with sediment; understand partitioning of contaminants within sediment matrix; may reflect groundwater; toxicity testing	Pagano et al. (2017)
Preparation of elutriates	1:4 sediment:water; shaken for 1 h prior to centrifugation and analyses	Evaluate aqueous extraction of suspended sediment to assess effects of contaminant resuspension during disturbance events (e.g. by dredging, bioturbation, storms)	Lesueur et al. (2015)
Spiking	Wet methods are recommended over dry; jar rolling (large vol, 2 h) hand mixing (small vol); ensure homogenous mixing (confirm by replication) and sufficient equilibration time (2 h–28 d); measure physicochemical parameters; determine moisture content to normalise dry weight concentration	Chemical addition to determine recovery rates (QA/QC); toxicity testing	Simpson et al. (2004)

Table preparation assisted by K. Summer and P. Howe
Sources: ASTM (2000), US EPA (2001), Simpson and Batley (2016)

Table 2.6 Approaches for the analyses of contaminants in sediments including sample collection and storage procedures

Contaminant/analyses class	Specific analyses	Recommended preparation and instrumentation (LOD)	Collection container, amount and storage[a–c]	Preferred maximum storage time
Metals and metalloids	Total/acid extractable	HNO₃/HCl acid digestion at 85 °C for 1 h; detection by ICP–MS (0.2–1 mg/kg)	P; 20 g; F	6 mo (6 wk for Hg)
	Operationally defined bio-available	1.0 M HCl acid extraction at 25 °C for 1 h; detection by ICP–MS (0.2–1 mg/kg)	P; 20 g; F	6 mo (6 wk for Hg)
	Speciation	Extraction prior to analysis by HPLC–ICP–MS (0.2–1 mg/kg)	P (metal/metalloid) G (organometallics); 20 g; F	6 mo (6 wk for Hg)
Non-metallic contaminants	Acid-volatile sulfides (AVS)	PT; spectrophotometry	Allow 10% headspace; F	28 d
	Ammonia	ASE	R	28 d
Organic contaminants	e.g., pesticides, hydrocarbons, surfactants, phenols, cyanides, pharmaceuticals, and PCPs	Extraction[a] prior to analysis generally by HPLC–ICP–MS (µg–mg/kg)	250 g–1 kg; G amber or foil wrapped with PTFE- lined caps/septa; F	7 d (before extraction), 30 d (after extraction)
Nutrients and organic material	Nitrogen and phosphorous (total)	Extraction[a] followed by persulfate-FIA (mg/kg)	P/G; R; 20 g	28 d
	Stable C, N isotope ratios	Elemental analyser-IRMS	P/G; F	28 d
	TOC	Wet oxidation-titration; combustion-IR; GC-FID (µg/kg)	P/G; R; 20 g; remove larger organic debris (>2 mm)	28 d
Nano/micro-particles	Nano-materials (1–100 nm)	Extraction[a] followed by HPLC–ICP–MS; variable dependent on type (see)	P (inorganic) or G (organic constituents)	6–12 mo
	Micro-plastics (<5 mm)	WPO-DS-GA; microscopy	G; sieved to remove > 5 mm plastics	Not defined
Physicochemical analyses	Particle size analysis/texture, bulk density/porosity	LDA; may be impregnated with epoxy or polyester resins	P/G; 300 g	Not defined

(continued)

Table 2.6 (continued)

Contaminant/analyses class	Specific analyses	Recommended preparation and instrumentation (LOD)	Collection container, amount and storage[a–c]	Preferred maximum storage time
	pH and redox potential	pH/ORP electrode	Generally in situ	
	Water content	Oven dried at 40 °C for 24–48 h; wet mass-dry mass (g/cm^3)		28 d
	Radionuclides (e.g., ^{7}B, ^{210}Pb, ^{222}Rn) and stable isotopes	HPGe; GFPC; IRMS (1–100 pCi/L)	P/G; 50 g	Variable
Biological analyses	Pathogens	Cell culture; microscopy; DNA extraction-bioinformatics; variable depending on the organism	P/G; R/F	96 h-28 d

Table preparation assisted by K. Summer and P. Howe

Sources and examples: Ankley et al. (1990), Chapman et al. (1998), ASTM (2000), Petrović et al. (2001), US EPA (2001), Hassellöv et al. (2008), Bainbridge et al. (2012), Zhang (2014), O'Reilly et al. (2015), Koziorowska et al. (2016), Simpsor and Batley (2016), Brady and Weil (2016), Al-Mur et al. (2017), Nascimento et al. (2017), Reichelt-Brushett et al. (2017), Liu et al. (2018)

P = plastic (polyethylene/polypropylene), G = glass; F = freeze, R = refrigerate (transport on ice, store < 4 °C), LS = light sensitive (store in dark/foil wrapped); [a]extraction methods may include Soxhlet, sonication, centrifugation, supercritical fluid, accelerated solvent, and/or microwave-assisted extraction. ASE: ammonia-selective electrode; GC-FID: gas chromatography-flame ionisation detector; GFPC: gas flow proportional chamber; HPGe: high purity germanium detection; HPLC–ICP–MS: high performance liquid chromatography coupled with ICP–MS; ICP–MS: inductively coupled mass spectrometry; IR: infrared detection; IRMS: isotope ratio mass spectrometry; LDA: laser diffraction analysis; NMR: nuclear magnetic resonance; PCPs: personal care products; Persulfate-FIA: persulfate digestion with flow injection analyser WPO-DS-GA: wet peroxide oxidation of organic material, density separation, gravimetric analysis

2

◧ **Figure 2.4** A Van Veen grab sampler shown open. *Photo*: A. Reichelt-Brushett

an intact sediment core sample. There is normally a cutting head on the end that is pushed into the sediment, the core barrel which can be of various lengths, a weighted collar, and a sealing mechanism. Shallow-water, hand-operated corers can easily be made with equipment from local hardware stores (care should be taken to avoid material that could be a potential contamination source of the analytes of interest). There is a range of suitable mechanised deployment options for mid-range water depths and SCUBA divers can collect sediment cores to a depth of around 30 m. For deeper waters of around 50–80 m, sophisticated oceanographic sampling equipment is required.

Coring devices vary in width and capacity to successfully sample sediments with differing physical features such as variable grain size. A wide core will provide more sediment for analyses at each depth interval but will be heavy and harder to manage than a narrow core. Deeper sediments (> 50 cm from the sediment–water interface) usually have less water content than shallower sediments which increases the sediment-to-water ratio. Specialised equipment (e.g.

penetrometers, acoustic surveys) may be used to understand specific physical sediment characteristics at different sites, and hence aid in the sampling design (Mudroch and Macknight 1994; Simpson and Batley 2016).

2.7.3 Suspended Particulate Matter (SPM)

There is interest in measuring contaminant loads associated with SPM. The interest lies in the particulate matter as an exposure route to biota, and in understanding contaminant transport, dispersion, and relocation through ecosystems (Mudroch and Macknight 1994; Simpson and Kumar 2016). As noted previously, suspended sediments are sites of adsorption for contaminants and therefore influence their cycling through the environment and commonly make up the bulk of SPM. Suspended particles can range in size from colloids (< 0.05 µm) to particles > 2 mm (Mudroch and Macknight 1994). There is considerable research delineating what is truly particulate and what is dissolved. Commonly, but arguably, operationally defined suspended particulates are those that are captured on a 0.45-µm filter and the filtrate is the dissolved fraction. The dissolved fraction, < 0.45 µm, can be further classified (◧ Table 2.7). Suspended sediments can be collected in various ways such as grab sampling, pump samplers, sediment traps, and integrating samplers (Batley and Simpson 2016). Water sampling devices are suitable for collecting SPM but samples require filtration on collection. Sediment traps and settling containers can also be deployed in situ, but their success, and our ability to make direct comparisons between sites, will depend on local turbulence, current, and tide conditions/interactions.

Complexities with Estuarine Waters
Estuaries are a mixing zone between seawater and fresh water. Within the mixing zone, often referred to as a **salt water wedge**, a steep ionic gradient destabi-

◧ **Table 2.7** Operationally defined size classes and the range of chemical forms potentially present within each class

Chemical form	Size class
Dissolved	< 1 nm
'Dissolved' organic chelates e.g. weak binding ligands	1 nm–0.2 µm
Biogenic and organic colloids and particles e.g. detritus, faecal pellets, and humic and Fulvic substances	0.05 µm–0.45 µm
Inorganic colloids and particulates e.g. clay platelets detrital alumno-silicates	0.01 µm–0.45 µm
Adapted from von der Heyden and Roychoudhury (2015)	

lises fine suspensions of colloidal material and causes the suspended particles to **flocculate**, carrying with it the bound contaminants. A pH gradient between the fresh and marine water may also exist. Sedimentation in an estuary is not only controlled by flow rates, but is also electrolytically driven by divalent-cations (commonly Mg^{2+}, Fe^{2+}, and Ca^{2+}) bridging between fine (negatively charged) particles causing flocculation. **Sediment suspensions** may also be continually reworked by the physical effects of tidal currents and wind action on the water surface to produce characteristic turbidity, known as the **turbidity maximum**. The region of turbidity maximum is exceptional in regard to the many chemical reactions involved in shifting phase between dissolved and particulate forms.

2.8 Biota Sampling

How Biotaorganisms interact with the abiotic environment influences contaminant **exposure and uptake pathways**. Uptake of contaminants by organisms also depends on numerous physicochemical factors such as chemical speciation, partitioning, and degradability (Connell et al. 1999; Maher et al. 2016). Biological variables (e.g. species, habitat, physiology, feeding habits, age, etc.) can play major roles in uptake, and environmental factors such as season may also alter the distribution and availability of contaminants (Connell et al. 1999; Maher et al. 2016). What is clear is that an understanding of the biological and ecological characteristics of a given organism is essential to understand contaminant loads in that organism. The following definitions outline the differences between bioaccumulation, bioconcentration, and biomagnification (see also ► Chapter 7):

Bioconcentration: The process whereby chemicals enter aquatic organisms through the gills or epithelial tissue directly from the water (or surrounding environmental medium) and become more concentrated in the organism than in the surrounding environmental medium (water, soil, etc.).

Bioaccumulation: Chemical uptake by an organism, attributable to both bioconcentration and dietary accumulation. Bioaccumulation is related to organism-specific rates of uptake, metabolism, and elimination, and occurs when a substance is absorbed at a faster rate than it is lost.

Biomagnification: The process whereby tissue concentrations of a bioaccumulated chemical increase with successively higher levels in a food chain (at least two trophic levels) (► Chapter 7). This phenomenon is rare; however, those specific contaminants that do biomagnify pose serious environmental problems, and consequently, receive great publicity. Contaminants known to biomagnify include:

- chlorinated hydrocarbons;
- insecticides (e.g. DDT group and cyclodienes—dieldrin);
- non-insecticides (e.g. PCBs—hydraulic fluids, heat-transfer fluids); and
- some organometallic compounds (e.g. methylmercury [MeHg], tributyltin [TBT]).

These contaminants characteristically have a high lipid solubility (i.e. high K_{ow} value; see ► Chapter 7 for further explanation), and therefore, a strong affinity to accumulate in biological tissue with high-fat content (including adipose tissue/blubber, white muscle, and some organs). They are stable and persistent within the organism, and accumulation can lead to disease and mortality in higher order predators.

Bioavailability is another common term that refers to the fraction of an element (or compound) that is available to be taken up by an organism. The entry pathway of a contaminant into an organism (e.g. via water, sediment, food, etc.) will also influence the bioavailability. Metals that are weakly bound to sediment particles are bioavailable. However, some contaminants are very strongly adsorbed to sediment particles or bound within the lattice structure of the particle: they will not be reactive within the organism and will likely pass through the organism with the particle.

2.8.1 Tissue Sampling

Sampling of organisms to determine contaminant loads in tissues involves firstly identifying and collecting representative species found at the range of sites in a sampling program. Consideration should be given to the need for depuration of gut contents, particularly if the organism is going to be acid digested whole for analysis. Often specific tissues will be dissected from the organism for separate analysis, typically; tissue, liver, gill, and gonads, but specifics will depend on the research question. In the dissection process, it is important to develop a procedure that avoids contamination of the tissue samples, even between tissue types. To minimise the transfer of contaminants between different tissue types during storage, dissections are best done at the time of sampling and prior to freezing and storage. ▢ Table 2.8 provides some guidelines and considerations for tissue sampling and analysis. Sample processing and analytical procedures are similar to those used for sediment.

2

□ Table 2.8 Analyses of marine contaminants: general considerations, rationale, and requirements for biota samples

Traits	Considerations	Example references
Biology/ecology		
Organism type	Depends on resources and research question (e.g. intensive study, ongoing biomonitoring programs, toxicity testing, community/ecosystem, contaminant/s of interest, human consumption); organism should be abundant, socially and/or ecologically important, representative, and biology/taxonomy well-understood	US EPA (2000), (2002), Lu (2005), Bricker et al. (2014)
Organism size, age, life stage etc.	Related to organism sensitivity, rates of BC, BA, BM; different uptake, metabolism and elimination pathways depending on developmental stage and physiology; should be standardised for comparability	Walker et al. (2012)
Biotic interactions	Consider tropic level and predator–prey relationships; BM- increased tissue concentrations with trophic level by virtue of contaminant persistence (maximum BM in apex predators)	Kelly et al. (2009)
Abiotic interactions: habitat, ecological niche	Consider pelagic, benthic, sediment-dwelling; related to contaminant exposure and uptake pathways; filter/deposit feeding enhances contaminant exposure	Walker et al. (2012)
Behaviour, distribution and life cycle traits	Consider sessile, sedentary, mobile characteristics, migration, reproduction and influences on exposures, sensitivity, contaminant transfer to offspring	Russell et al. (1999)
Sample type	Contaminant concentrations vary between different tissue types and organs, living and non-living tissues, symbiont-host organisms	Reichelt-Brushett and McOrist (2003), Maher et al. (2016)
Live/dead organism	Tissue sample: live- biopsy (non-lethal; limitations to sample type and quantity), dead- necropsy; determines equipment and procedures	Cagnazzi et al. (2019)
Permits and approvals	Ensure necessary permits and ethics approval are obtained prior to sampling	NHMRC (2013)
Equipment and sample collection	Analytical grade/acid-washed containers P/G (inorganics) (zip-lock bags/containers) or G or foil (organics); stainless steel HHI, biopsy needle or gun-dart system, PPE; equipment cleaned of adhering material and washed with deionised water and methanol between samples (new scalpel blades); collection devices and equipment dependent on sample type- may require additional resources (e.g. boat, personnel, aquaria)	US EPA (1994), (2000)
Sample storage and preparation	Dissection and cleaning using ambient seawater or deionised water; store on ice in field and freeze; if analyses delayed >24 h store in liquid nitrogen whilst in field and freeze -80 °C; preparation and extraction for specific analyses similar to □ Table 2.3 (sediments); do not freeze–thaw-freeze	US EPA (1994), (2000)
Whole organism	Small organisms (e.g. shellfish, anemones, larvae); often used in toxicity testing; *in situ* biomonitoring (e.g., shellfish); biochemical analyses (e.g. enzyme activity)	US EPA (1994), (2000)
Tissue analysis		
Muscle, fat/blubber	Analyses of inorganic and organic contaminants	Cagnazzi et al. (2019), Gilbert et al. (2015a, 2015b)
Gills	Primary sites of contaminant contact, uptake, deposition, gas/ion exchange; identification of ultrastructural changes caused by contaminants	Sweidan et al. (2015)

(continued)

◼ **Table 2.8** (continued)

Traits	Considerations	Example references
Organs	Animal organs e.g. liver, kidney; analyses of lipid-soluble organic contaminants; understanding contaminant compartmentalisation, metabolism, secondary metabolites and excretion; sample from same portion of organ (e.g. top left lobe of liver) as contaminant distribution varies	US EPA (1994)
Fluids	e.g. urine- analyses of aqueous contaminants, metabolism and excretion; e.g. blood- lipid-soluble contaminant storage, assessment of contaminant effects on endocrine and immune systems	Nomiyama et al. (2014)
Reproductive organs/tissues, milk	e.g. gonads, eggs; understand maternal contaminant transfer and genotoxicity	Russell et al. (1999)
Other structures	e.g. hair, feathers, skin; indicators of whole-body metal burden	Finger et al. (2015)
Microbiology	e.g. bacteria, viruses, protozoa; pathogenic organisms and changes to beneficial microbial communities associated with toxicity and disease	Kamp et al. (2015), Gissi et al. (2019)
DNA	Isolated from frozen sample; genetic lesions indicate toxicity	Downs et al. (2016)

Table preparation assisted by K. Summer

BC: bioconcentration (concentration in organism via surrounding environmental media), BA: bioaccumulation (BC + dietary accumulation), BM: biomagnification (increased concentration with successive trophic levels); P = plastic (polyethylene/polypropylene), G = glass, acid washed and rinsed using deionised water and solvent, dried and sealed between cleaning and collection; In general, the shorter the time that elapses between sample collection and analysis, the more reliable will be the analytical results. HHI: stainless steel hand-held implements (e.g. sterile knives, scissors; dissecting scalpel, forceps, needles, and trays), MeHg: methylmercury, OCs: organochlorine pesticides, PCBs: polychlorinated biphenyls, PPE: personal protective equipment, including non-vinyl gloves

2.8.2 Biomonitors

The rate of uptake and depuration of a chemical will influence its toxicity, and all **uptake pathways** (through water, food, sediment, etc.) contribute to chemical concentrations in an organism. If we understand **chemical uptake rates,** uptake pathways, and depuration rates of a contaminant by an organism, we can potentially use living organisms as monitors of the environment. This use of organisms in environmental assessment is termed biomonitoring and is an established and growing field of research in ecosystem risk assessment. Biomonitoring studies provide a longer-term integration of environmental contaminants compared to single time and point sampling of abiotic compartments. Biomonitoring studies may include chemical loads in organism tissues, changes in biochemical, physiological, morphological, or behavioural aspects of organisms, as well as ecological aspects such as species diversity and abundance (Connell et al. 1999).

Biomonitoring studies provide an integrated assessment of conditions over time, whereas water samples are usually a snapshot in time.

Requisites for organisms used in bioaccumulation studies:

- easy to identify and preferably established knowledge of biology and ecology;
- geographically widespread;
- not highly specialised in habitat;
- wide tolerance to environmental conditions (e.g. temperature, O_2, salinity);
- common and abundant;
- sedentary or relatively immobile;
- relatively large in size;
- long-lived (ageing methods established);
- hardy animal in the lab (for controlled experiments); and
- high concentration factor for the pollutant under study.

Organisms within a survey should exhibit the same correlation between the pollutant level and that of the surrounding medium. However, before data can be interpreted, the following questions need to be addressed:

- How rapidly is the contaminant taken up?
- How quickly is it lost/depurated?

- Can the organism exercise control over uptake and/or loss?
- Is the pollutant differentially distributed within the organism (e.g. flesh vs. liver for fish, flesh vs. carapace for prawns)?

Biomarkers should not be mixed up with biomonitors. Biomarkers are endpoints that can be used to define a potential effect of contaminants on organisms in the environment, and are employed in laboratory-based ecotoxicological testing (Connell et al. 1999; Taylor and Maher 2016). Details on ecotoxicological studies are provided in ▶ Chapter 3.

2.8.3 Collecting Pelagic Species

Various technique can be used to collect pelagic biota with equipment often targeting a particular size class. Importantly, the same approaches should be used across all sites. Field staff need to have suitable taxonomic knowledge to ensure the correct species are collected. Plankton tows are suitable for small species including both plankton and nekton communities. There are many regular fishing devices, such as traps, rods, and nets, which can be purchased off the shelf for use or adapted for sampling. You may also find and capture some species using snorkel or SCUBA. Consider the potential benefits of selecting a range of species that comprise of trophic levels.

2.8.4 Collecting Benthic Species

Sediment grab samplers (◘ Figure 2.4) and pipe dredge samplers can be used to collect benthic species (see also Maher et al. 2016). Sediments may need to be sieved to retrieve specimens and samples should be rinsed (and dissected if necessary) before storage. You may also locate and collect specimens by snorkelling or SCUBA diving or even on foot at shorelines and in intertidal areas.

Box 2.1: NOAA Mussel Watch Program, United States of America
Dr. Pelli Howe, Environmental Scientist.
Since commencement in 1986, the United States of America (USA) National Oceanic and Atmospheric Administration (NOAA) Mussel Watch Program has become one of the world's longest, and arguably most successful, continuous ecosystem monitoring efforts. It was designed to provide data on the spatial distribution and temporal trends of contaminants in coastal marine and estuarine environments of the United States of America, including Alaska, Puerto Rico, and Hawaii (O'Connor, 1998). The program is based on annual collections of resident mussels and oysters and comprises almost 300 monitoring sites between 10 and 100 km apart (◘ Figure 2.5). Tissue analyses of over 140 organic and inorganic

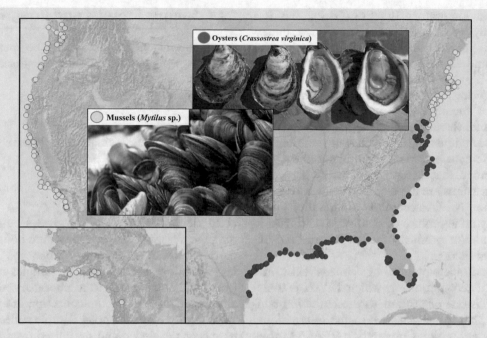

◘ Figure 2.5 ► Box 2.1: Distribution of marine bivalve species and sampling locations comprising the Mussel Watch Program throughout coastal and estuarine environments of the United States. Adapted from Kimbrough et al. (2008) by P. Howe

compounds and a rigorous QA/QC process is employed among analytical laboratories to ensure data accuracy and comparability (Kimbrough et al. 2008). Samples are also stored to enable retrospective analyses for new and emerging contaminants of concern.

Bivalves are commonly used as indicators of water and sediment quality as they readily bioconcentrate and bioaccumulate contaminants by virtue of filter feeding (Melwani et al. 2014). They are sedentary and therefore representative of a given location, widely distributed, and have stable populations that can withstand repeated sampling (Farrington et al. 2016). Many are also commercially important seafood species. Mussel Watch data is useful for characterising the impact of marine contaminants at local, regional, and national scales and helping to understand current and emerging contaminants, potential food safety risks, and can be used to evaluate the efficacy of management and remediation strategies (Kimbrough et al. 2008).

Such long-term and extensive sampling efforts are costly and difficult to sustain. The success of the Mussel Watch Program can be attributed to collaboration between national and state governments, regional and local groups, private sector partners, and citizen scientists who aid in collection (Kimbrough et al. 2008). One justified criticism is that not all contaminants are readily bioconcentrated/bioaccumulated (e.g. those with high water-solubility) and are therefore not addressed by the program (Farrington et al. 2016). Furthermore, large scale geographic comparisons are constrained by differences in target species and environmental factors which may influence bioaccumulation and bioconcentration rates (Farrington et al. 2016). Nonetheless, the program provides valuable information for decision making. See Kimbrough et al. (2008) for further information and data summaries.

2.9 Quality Assurance and Quality Control

2.9.1 NATA Registration and Other Global Systems

To standardise quality assurance and quality control, a diverse range of tools can be applied (e.g. calibration services, inspection organisations, certified reference materials, and providers of proficiency testing). Companies provide these services and products to help ensure that users have confidence in their results. Labora-

tory accrediting organisations conduct regular audits of member laboratories to ensure that a high standard is maintained. Accreditation is a valuable tool for effective policy making, it contributes to maintaining fair markets, and improves regulation and governance in diverse areas including; food production, environmental protection, healthcare, construction, and waste management.

In Australia, the National Association of Testing Authorities, Australia (NATA) performs this role, and monitors compliance with the Organisation of Economic and Cooperation and Development OECD principles of Good Laboratory Practise (GLP). This is a public company which has a memorandum of understanding with Federal and State Governments, who recommend (and in many cases enforce) the engagement of NATA-accredited organisations for certain services. Acquiring and maintaining NATA-accreditation requires organisations that provide analytical services to establish and uphold high quality, strictly standardised techniques, analyses, instrumentation, use of certified reference materials, data management, and consistently prove that exceptional quality assurance and control measures are in place and being used effectively. Therefore, NATA-accredited organisations offer the community an assurance of confidence and trust in their services and/or products, facilitate trade, and improve tendering success.

International accreditation agreements exist and provide mutual recognition, to which NATA is a signatory. NATA is one of around 100 accreditation bodies worldwide that are signatories to the International Laboratory Accreditation Cooperation (ILAC) Mutual Recognition Agreement (MRA). This minimises trade barriers between accredited organisations. The ILAC is for accreditation bodies that involve calibration, testing (environment and medical), inspection, and proficiency testing providers. The ILAC offers independent evaluation of conformity to recognised standards, and works closely with regional co-operation bodies such as the Asia–Pacific Accreditation Cooperation (APAC), European Accreditation (EA), Inter-America Accreditation Cooperation IAAC in the Americas, the African Accreditation Cooperation (AFRAC), the Arab Accreditation Cooperation (ARAC), and the South African Development Community Cooperation in Accreditation (SADCA). In 2018, more than 10,500 inspection bodies and almost 76,500 laboratories were accredited by ILAC MRA signatories.

2.9.2 Chain of Custody

Reliable reporting of analytical results requires a great deal of care in handling samples and is an important requirement for accreditation. A **chain of custody** accurately documents the movement of samples through an organisation, from collection and submission for analyses, transfer between sections of an organisation, preparation, analysis, and storage of samples, and reporting of results. Strict practices ensure that samples are not mislaid or mislabelled (or not labelled), that holding times are appropriate, and enable samples to be easily located at all times. As you can imagine, an unlabelled or mislabelled sample container with a nondescript sample in a busy laboratory is extremely problematic. Each stage of the chain of custody requires appropriate sample storage and handling.

2.9.3 Sample Storage and Integrity

Some parameters should ideally be measured in the field as they are likely to change during collection, transport, and storage, including:
- temperature;
- pH;
- conductivity;
- redox (reduction/oxidation potential);
- dissolved oxygen;
- turbidity; and
- chloride.

Field measurements may not always be possible for a range of reasons, and nor may they be ideal, as much more accurate measurements may be available using laboratory instruments. If samples are to be collected and transported to a laboratory, the techniques and materials used for the collection are important for the integrity of the sample. Sample collection methods have been developed and optimised over many years of research to minimise potential changes during transport (see ◨ Tables 2.1, 2.5, and 2.7). The collection method should be carefully chosen with consideration of the target analyte/s and the overall objectives of the sampling. These must be understood and strictly adhered to by all involved.

Strict protocols exist for accredited analytical laboratories to ensure that sample handling and storage maintain sample integrity. Obviously, appropriate handling and storage differ for different types of samples, and also depend upon the type of analyses that will be conducted. General protocols for samples include:
- ensuring that sample containers cannot break or leak and cause cross-contamination;
- filtering in the field is mandatory for many tests;
- minimising the exposure of samples to air is critical for many tests;
- most samples should be chilled to $< 4\,°C$ or $< 6\,°C$ (depending on the analyte/s of interest) on collection. There are also cases where samples need to be stored at -20 °C or -80 °C;
- some samples need to be treated with preservation chemicals on collection; and
- strict adherence to established recommended holding times (RHTs) for different samples and for different analyses.

2.9.4 Step to Ensure Analytical Certainty

Certified Reference Materials
As mentioned previously, manufacturers of certified reference materials require accreditation, and so must maintain an exceptionally high and consistent standard. They guarantee their product to the analytical lab-

oratories that use them to maintain their own QA/QC. Reference materials of a known substance and concentration provide a **reference sample** and are prepared in the sample way as all other samples. The analytical results of reference materials are used to ensure that the sample preparation and analyses have been conducted correctly. For example, certified reference sediment is used alongside sediment samples for analyses of trace elements by inductively coupled plasma mass spectrometry (ICP–MS); the concentrations in the reference sediment are known (and are certified by the manufacturer), and so any significant deviation (e.g. $\pm 10\%$) from the expected concentrations will alert the operator to a potential problem.

The ILAC defines five types of reference materials:
- physicochemical reference substances;
- matrix reference materials;
- pure substances;
- standard solutions and gas mixtures; and
- artefacts or objects.

Spiked Additions
Another means of maintaining QA/QC is through spiked additions of known substances at known concentrations to randomly chosen samples. This is a method used to ensure that analytical instruments are calibrated properly and are functioning normally. Accredited laboratories will, for example, add a known volume of a known solution to a duplicate of a sample that is being analysed normally. The software is then programmed to recognise the appropriate increase in concentration in the spiked sample compared to the unspiked sample and will flag unexpected problem results. The operator must then identify the problem, which may not always be easy. The first step is to carefully spike and analyse the sample again in order to discount human error in the preparation of the spiked sample (e.g. incorrect ratio volume of spike solution to sample). Next, the spiked solution should be carefully remade to discount this as the source of the error. If the results still seem erroneous, it may be that the sample contains substances that interact with the spiked solution and interfere with the analysis. As a final check to discount a non-instrument-related problem, a different sample should be spiked. If none of these steps resolve the discrepancy between the concentration in the spiked addition and the concentration measured by the instrument, then there is a problem with the instrument and/or the calibration, and all sample analyses are ceased until this is resolved. Spiked additions are generally used every 10 or so samples to enable reasonably immediate identification of possible errors in the data. For the same reason, samples are analysed in duplicate at a similar frequency. If the result of a spiked or duplicated sample is inconsistent with the expected result, all the samples that have been ana-

lysed since the last correct measurement need to be re-analysed.

2.9.5 Detection Limits

Analytical laboratories have minimum detection limits, which generally depend on the instrumentation that is available. These are usually reported as the limit of detection (LOD), or the limit of reporting (LOR). LOR is a value below which the laboratory cannot confirm the repeatability of the result. Although laboratory instruments and equipment also have upper limits to the concentrations that they can reliably measure (and which may be governed by the calibration), if samples present very high concentrations they can be diluted prior to analyses. Dilutions up to 10,000 times may need to be done on some samples to enable analyses, but this introduces potential errors. The operator needs to be aware of the limitations of the instrumentation, for example, analysing samples with very high concentrations of some substances can severely contaminate the instrument as well as all other samples that are analysed subsequently.

2.9.6 Dealing with Difficult Samples

Seawater has a very different chemical composition than fresh water ($\blacktriangleright$ Chapter 1), and for many types of analyses, an entirely different method is required. Analyses of seawater samples by ICP–MS, for example, are complicated by the usually very low concentrations of trace elements, and by the very high salt content, which can cause complex interferences. These interference problems have mostly been overcome by advancements in instrument design. There may also be different analytes in the same sample that require different pre-treatment and instrument operations. Nutrient analyses of soils and sediments may be required for total nitrogen and total phosphorous, and also for relevant nitrogen species (e.g. nitrate, nitrate, ammonia) and phosphorous (e.g. orthophosphate) species. Analyses of total nutrients require pre-treatment via chemicals and autoclave digestion, whereas analyses of nutrient species do not. Hence, the different sample matrix requires the instrument to be calibrated and operated differently for total nitrogen and phosphorus as compared to their species.

2.9.7 Dealing with Novel Contaminants

The rapidly increasing number and quantity of new and emerging contaminants are challenging for analytical scientists. It is difficult to develop reliable and accurate testing methods at the same pace as new contaminants are being identified, although efforts continue (e.g. Liu et al. 2015a, 2015b) (▶ Chapter 13). Without analytical methods to identify the presence and concentration of novel contaminants, their toxicity and environmental behaviour cannot be investigated and their effects on the environment will remain unknown.

2.10 Identifying Contamination

As discussed in ▶ Chapter 1, a contaminant is a substance that is present in unnaturally high concentrations (i.e. above background concentrations). For all synthetic substances, this is any measured concentration, whilst for naturally occurring substances, the background concentration will vary with location. In aquatic systems, some substances are far more toxic in seawater than in fresh water, and vice versa, and the levels of contamination that may have negative effects on organisms and ecosystems differ greatly. Many substances are naturally present in different types of ecosystems at very different concentrations. The same concentrations of nutrients, for example, could indicate contamination in oligotrophic marine waters (e.g. coral reefs), but might be a normal background concentration in an estuaries.

2.10.1 Determining Background Concentrations

By definition, the assessment of contamination depends on **knowledge of the normal**, or **background concentrations**. Whilst most synthetic substances have no normal background concentration, there are exceptions to this with contaminants (such as DDT) that are now distributed globally.

In Australia state-based environmental protection agencies describe background concentrations as being natural or ambient:

- **Natural**—the amount of a chemical substance that is naturally occurring and is derived/originated from natural processes (e.g. erosion and dissolution of minerals), and is related to specific human activities or sources. The concentrations will depend on a wide range of factors such as the geology, geography, topography, and biological and chemical characteristics of the receiving environments.
- **Ambient**—the concentration of a chemical substance that is representative of the surrounding area and is not from a single source (e.g. widespread diffusion, historical activities). If the determination of the natural background concentration is not possible due to long-term human impacts, the ambient background concentration still provides a means of identifying increases due to future inputs.

In the absence of background concentration data, reference sites are often used. Appropriate **reference sites** must have similar physicochemical conditions and not be impacted. Samples from reference sites should be taken at the same time as samples from the study site, handled in the same manner, and preferably analysed by the same laboratory. Samples collected for the purpose of determining background concentrations should not be combined. Reference sites are selected on a case-by-case basis. Careful collection of samples for determining background concentrations is critical, since this information will be used to identify contamination (e.g. Crommentuijn et al. 2000).

Water
Sampling to determine background concentrations in water requires the following considerations:
- **samples should be taken upstream or up current (if relevant)**—awareness of conditions such as inflow, outflow, and currents;
- **recent rainfall or extended dry periods**—storm events may result in short-lived changes in water quality that may not be representative of background concentrations, and concentrations of potential contaminants may also not be representative of typical background concentrations following unusually dry periods; and
- **water quality parameters** (e.g. pH, hardness, conductivity, temperature, suspended and total dissolved solids, and dissolved oxygen)—should be measured prior to and during sampling to provide assurance of the stability of the system at the time of sampling.

Sediment
Sampling for the purpose of determining background concentrations in soils and sediments requires the following considerations:
- the samples should be taken from relatively undisturbed sites at a higher elevation and upwind of the study site;
- sediments should have similar lithology as the study area;
- samples should have no odour or staining; and
- reference sites should be geographically, chemically, physically, and biologically similar to the impact site.

Activities or events that result in sediment deposition and their frequency and intensity need to be understood to help interpret geochemical signatures of contaminants in sediments and sediment cores. For example, inputs into estuaries that are derived from broad-scale agriculture activities in river catchments may have a deposition record linked to pulse rainfall events.

A range of information is required to fully characterise an area of interest and identify when measured concentrations exceed normal background levels and these may vary considerably throughout a study site. By using approaches like sediment normalisation to determine regional geochemical baselines, a more integrated understanding of contaminant concentrations over space and time can be achieved.

Biota

It may be important to determine background concentrations of contaminants in Biota. This is particularly relevant for biomonitoring, whereby it is necessary to understand the naturally occurring concentrations of potential contaminants in organisms to enable the identification of an increased body burden. Oysters, for example, are often used as biomonitors to measure environmental concentrations of metals. As the normal concentrations of metals in oysters tissue may vary depending on geological location, it is important that background concentrations are determined in areas where oysters are being used as biomonitors (e.g. Scanes and Roach, 1999).

2.10.2 Normalising Techniques

It is common to normalise the total contaminant concentration to the organic carbon content in sediments for hydrophobic organic compounds (Simpson and Batley, 2016). It is a useful approach to establish differences in organism exposure because the organic carbon is important in establishing the equilibrium between the solid and liquid phases of sediments (Di Toro et al. 1991) and takes account of the relative partitioning between pore water, organic carbon, and Biota (Simpson and Batley, 2016). Contaminants may also be normalised to the sediment grain size either by analysis of selective sediment grain size fractions or by completing post-extraction normalisation procedures (e.g. Birch and Snowdon, 2004). It is also a standard procedure to remove sediment particles > 2 mm as part of sample processing prior to analysis (ANZG, 2018). For Biota sampling, hydrophobic organic compound concentrations may also be normalised to lipid content. Consideration should be given to normalisation procedures prior to sampling but also once early results are available. It is helpful to ensure that some portion of the sample remains intact and well stored in case retrospective sampling needs to be undertaken.

2.10.3 Understanding Degradation

Following the distribution of contaminants in the environment, some compounds (mainly organic compounds) degrade over time. As the compounds degrade or break down, they generally become less toxic, but there are cases where the degradation products are actually more toxic than the original compound. The herbicide diuron (1,1-dimethyl, 3-(3',4'-dichlorophenyl) urea) is used in agriculture for weed control around water bodies and is a component of marine antifouling paints. There is evidence that most of the degradation products of diuron are much more toxic than the parent molecule (Tixier et al. 2001; Giacomazzi and Cochet, 2004). For further details on degradation, see Chapters 7 and 8.

2.10.4 Using Guideline Values

Water, Biota, and sediment quality guidelines or trigger values provide a tool for assessing whether a given chemical or physical stressor is likely to cause unacceptable harm to a specific community value (e.g. human or agricultural health, recreation, or ecological protection). As environmental conditions are dynamic and infinitely variable, site-specific guidelines are always the most appropriate, but are not always available, in which case non-site-specific values are used (ANZG, 2018).

Water Quality Criteria (WQC) are defined and expressed differently throughout the world. Australia, New Zealand, and the United States, and more recently China and South Africa, use the term WQC or Water Quality Guideline (WQG). The first WQC was published in America in 1968 by the National Technical Advisory Committee (NTAC) and has been continually revised by the United States Environmental Protection Agency (US EPA) since 1972. The most recent WQC published by the US EPA lists 120 priority pollutants, 43 nonpriority pollutants, and 23 pollutants with organoleptic (i.e. taste and odour) effects (US EPA, 2009). The US EPA WQC provides:

- **criteria maximum concentrations (CMCs)**, which are the estimated highest concentrations of a substance in surface water to which brief exposure is not predicted to cause unacceptable effects on aquatic ecosystems; and
- **criteria continuous concentrations (CCCs)**, which are estimations of the concentrations to which an aquatic ecosystem can be indefinitely exposed without experiencing unacceptable effects.

However, for the majority of these pollutants, only concentrations that are relevant to human health (e.g. by exposure or ingestion) are provided.

Environmental risk limits (ERLs) are used to derive environmental quality standards (EQS) in several countries (e.g. the Netherlands; Crommentuijn et al. 2000). ERLs include negligible concentrations, maximum per-

2

missible concentrations, maximum concentrations that are considered acceptable for ecosystem protection, serious risk concentrations, and maximum permissible concentrations.

Guideline values for aquatic ecosystems may be based on reference-site data, field-effects data, or laboratory-effects data. Increasingly, importance is placed on using multiple lines of evidence (i.e. a combination of at least two of these data types) (See Chapters 3 and 7 for further details on how guideline values are developed).

2.10.5 Development of Guidelines for New and Emerging Contaminants

For the benefit of human and ecological health, in recent years a lot of research effort has focussed on new and emerging contaminants (▶ Chapter 13). This is critically important as the number and quantity of new chemicals are increasing rapidly. You may be aware of numerous examples of substances that were not known to be toxic until well after they caused devastating effects, such as DDT and the Minamata mercury poisoning event. This point should be considered when determining the amount of research effort that should be directed towards developing methods for measuring and assessing the toxicity of new and emerging contaminants.

Water and sediment quality guidelines and trigger values provide tools for assessing potential impacts of a wide range of contaminants, and arriving at meaningful estimates of these values takes dedicated time and effort from people with a wide range of expertise. As mentioned earlier, there are no analytical techniques for many emerging contaminants, and this must be resolved before environmental toxicologists, analysts, and policymakers can even begin to estimate the risks of new and emerging contaminants.

2.11 Summary

The first step in understanding marine pollution is understanding the contaminants in the environment. Before we can investigate the effects or potential effects of contaminants, we need to be able to measure them, which requires sampling. The sampling design (e.g. location(s), sites (e.g. reference sites, number of sites), sample number/frequency, sampling methods) will depend on the **purpose of the research**. The quality of the sampling will become irrelevant if the sampling is not designed in a way that can provide meaningful data. After the research purpose has been clearly defined, the sampling program can be designed.

The **accuracy**, and therefore, relevance, of research findings involving environmental samples depends on several consecutive processes. If any one of these processes is not carried out correctly, it can jeopardise the results, regardless of how well every other process was conducted. The importance of maintaining the **integrity** of samples begins with sampling, transport, and storage, prior to analyses. The accurate analysis depends on correct sample **handling**, processing and pre-treatment, preparation for analysis, operation of instrumentation, instrumentation accuracy/calibration, and finally data **analyses**, **interpretation**, and **reporting**. Most of these stages have several elements.

Accurate measurement of contaminants in environmental samples, alongside an understanding of the effects of such contaminants, allows, finally, for an ability to identify the presence and degree of contamination, and provide guidance for policy and regulation to protect environmental and human health. This entire process obviously requires a lot of **time** and **effort** from a wide range of highly **trained people** with a wide range of **expertise**, and even more so in the case of new and emerging contaminants.

2.12 Study Questions and Activities

1. Imagine you were designing a sampling program to assess the extent of contamination from a toxic chemical spill. Try and list in order the steps you would take to develop the program. It is expected that you would include at least 10 steps.
2. Using ◘ Table 2.6, determine how much sediment sample that would be required to collect from each site to complete the following analyses and explain how you would store your samples on collection. Analyses to be completed: total trace metals, TOC, pesticides, and grain size.
3. You are about to embark on a field trip to sample sediments at 27 sites for metals and pesticides, you will also be collecting physicochemical water quality data at each site and will need to determine sediment grain size and TOC for each sample. Using the guiding principles described in this chapter, create a checklist of all your sample collection equipment. Assume that there are no shops nearby and you need to be 100% self-contained (i.e. don't forget plenty of permanent markers and all those other minor, but essential, items).

References

Ajayi SO, Vanloon GW (1989) Studies on redistribution during the analytical fractionation of metals in sediments. Sci Total Environ 87–88:171–187

Al-Mur BA, Quicksall AN, Kaste JM (2017) Determination of sedimentation, diffusion, and mixing rates in coastal sediments of the eastern Red Sea via natural and anthropogenic fallout radionuclides. Mar Pollut Bull 122(1–2):456–463

Anderson DM, Glibert PM, Burkholder JM (2002) Harmful algal blooms and eutrophication: nutrient sources, composition, and consequences. Estuaries 25(4B):704–726

Ankley GT, Katko A, Arthur JW (1990) Identification of ammonia as an important sediment-associated toxicant in the lower Fox River and Green Bay, Wisconsin. Environ Toxicol Chem 9(3):313–322

ANZG (Australian and New Zealand Guidelines) (2018) Guidelines for fresh and marine water qality, toxicant default guideline values for sediment quality. Available at: ▶ https://www.waterquality.gov.au/anz-guidelines/guideline-values/default/sediment-quality-toxicants. Accesssed 10 Dec 2021

ASTM (American Society for Testing Materials) (2000) Standard guide for collection, storage, characterisation, and manipulation of sediments for toxicological testing. E 1391–94. ASTM, PA, USA, p 95

Bainbridge ZT, Wolanski E, Álvarez-Romero JG, Lewis SE, Brodie JE (2012) Fine sediment and nutrient dynamics related to particle size and floc formation in a Burdekin River flood plume, Australia. Mar Pollut Bull 65(4–9):236–248

Baird RB, Eaton AD, Rice EW (eds) (2017) Standard methods for the examination of water and wastewater, 23rd edn. Prepared and published jointly by APHA (American Public Health Association), AWWA (American Water Works Association) and WEF (Water Environment Federation), p 1504

Batley GE, Gardner D (1977) Sampling and storage of natural waters for trace metal analysis. Water Res 11(9):745–756

Batley GE, Simpson GE (2016) Sediment sampling, sample preparation and general analysis. In: Simpson S, Batley G (eds) Sediment quality assessment—a practical guide, 2nd edn. CSIRO Press, Clayton South, pp 15–46

Becker JS (ed) (2008) Inorganic plasma mass spectrometry: principles and applications. Wiley, Julich, p 514

Birch GF, Snowdon RT (2004) The use of size-normalisation techniques in interpretation of soil contaminant distributions. Water Air Soil Pollut 157:1–12

Brady NC, Weil RR (2016) The nature and properties of soils. 15th edn. Pearson Education Inc., New Jersey, p 912

Bricker S, Lauenstein G, Maruya K (2014) NOAA's mussel watch program: incorporating contaminants of emerging concern (CECs) into a long-term monitoring program. Mar Pollut Bull 81(2):289–290

Buesseler KO, Jayne SR, Fisher NS, Rypina II, Baumann H, Baumann Z, Breier CF, Douglass EM, George J, Macdonald AM, Miyamoto H, Nishikawa J, Pike SM, Yoshida S (2012) Fukushima-derived radionuclides in the ocean and biota off Japan. Proc Natl Acad Sci USA 109(16):5984–5988

Cagnazzi D, Broadhurst MK, Reichelt-Brushett A (2019) Metal contamination among endangered, threatened and protected marine vertebrates off south-eastern Australia. Ecol Ind 107:105658

Chapman PM, Wang F, Janssen C, Persoone G, Allen HE (1998) Ecotoxicology of metals in aquatic sediments: binding and release, bioavailability, risk assessment, and remediation. Can J Fish Aquat Sci 55(10):2221–2243

Chen X, Han C, Cheng H, Wang Y, Liu J, Xu Z, Hu L (2013) Rapid speciation analysis of mercury in seawater and marine fish by cation exchange chromatography hyphenated with inductively coupled plasma mass spectrometry. J Chromatogr A 1314:86–93

Connell D, Lam P, Richardson B, Wu R (1999) Introduction to ecotoxicology. Blackwell Science Ltd., Cornwall, p 180

Crommentuijn T, Sijm D, De Bruijn J, Van den Hoop MAGT, Van Leeuwen K, Van de Plassche E (2000) Maximum permissible and negligible concentrations for metals and metalloids in the Netherlands, taking into account background concentrations. J Environ Manage 60(2):121–143

Csuros M (2018) Environmental sampling and analysis for technicians. CRC Press, Boca Raton, p 560

Di Toro DM, Zarba CS, Hansen DJ, Berry WJ, Swartz RC, Cowan CE, Pavlou SP, Allen HE, Thomas NA, Paquin PR (1991) Technical basis for establishing sediment quality criteria for nonionic organic chemicals by using equilibrium partitioning. Environ Toxicol Chem 10:1541–1583

Downs CA, Kramarsky-Winter E, Segal R, Fauth J, Knutson S, Bronstein O, Ciner FR, Jeger R, Lichtenfeld Y, Woodley CM, Pennington P, Cadenas K, Kushmaro A, Loya Y (2016) Toxicopathological effects of the sunscreen UV filter, oxybenzone (benzophenone-3), on coral planulae and cultured primary cells and its environmental contamination in Hawaii and the U.S. Virgin Islands. Arch Environ Contam Toxicol 70(2):265–288

Farrington JW, Tripp BW, Tanabe S, Subramanian A, Sericano JL, Wade TL, Knap AH (2016) Edward D. Goldberg's proposal of "the mussel watch": reflections after 40 years. Mar Pollut Bull 110(1):501–510

Finger A, Lavers JL, Dann P, Nugegoda D, Orbell JD, Robertson B, Scarpaci C (2015) The little penguin (*Eudyptula minor*) as an indicator of coastal trace metal pollution. Environ Pollut 205:365–377

Gattuso J-P, Lee K, Rost B, Schulz K (2010) Approaches and tools to manipulate the carbonate chemistry. In: Riebesell U, Fabry V, Hansson L, Gattuso J-P (eds) Guide to best practices for ocean acidification research and data reporting. Publications Office of the European Union, Luxembourg, pp 41–52

Giacomazzi S, Cochet N (2004) Environmental impact of diuron transformation: a review. Chemosphere 56(11):1021–1032

Gilbert JM, Reichelt-Brushett AJ, Butcher PA, McGrath SP, Peddemors VP, Bowling AC, Christidis L (2015a) Metal and metalloid concentrations in the tissues of dusky *Carcharhinus obscurus*, sandbar *C. plumbeus* and great white *Carcharodon carcharias* sharks from south-eastern Australian waters, and implications for human consumption. Mar Pollut Bull 92:186–194

Gilbert J, Badual C, Reichelt-Brushett AJ, Butcher P, McGrath S, Peddemores VM, Mueller J, Christidis L (2015b) Bioaccumulation of PCBs in liver tissue of dusky *Carcharhinus obscurus*, sandbar *C. plumbeus* and white *Carcharodon carcharias* sharks from south-eastern Australian waters. Mar Pollut Bull 101:908–913

Gissi F, Reichelt-Brushett A, Chariton A, Stauber J, Greenfield P, Humphrey C, Salmon M, Stephenson S, Creswell T, Jolley D (2019) The effect of dissolved nickel and copper on the adult coral *Acropora muricata* and its microbiome. Environ Pollut 250:792–806

Hassellöv M, Readman JW, Ranville JF, Tiede K (2008) Nanoparticle analysis and characterization methodologies in environmental risk assessment of engineered nanoparticles. Ecotoxicology 17(5):344–361

Hirai H, Takada H, Ogata Y, Yamashita R, Mizukawa K, Saha M, Kwan C, Moore C, Gray H, Laursen D, Zettler ER, Farrington JW, Rcddy CM, Peacock EE, Ward MW (2011) Organic micropollutants in marine plastics debris from the open ocean and remote and urban beaches. Mar Pollut Bull 62(8):1683–1692

Kahru A, Dubourguier HC (2010) From ecotoxicology to nanoecotoxicology. Toxicology 269(2–3):105–119

Kamp A, Høgslund S, Risgaard-Petersen N, Stief P (2015) Nitrate storage and dissimilatory nitrate reduction by eukaryotic microbes. Front Microbiol 6

Kelly BC, Ikonomou MG, Blair JD, Surridge B, Hoover D, Grace R, Gobas FAPC (2009) Perfluoroalkyl contaminants in an arctic marine food web: trophic magnification and wildlife exposure. Environ Sci Technol 43(11):4037–4043

2

Kimbrough KL, Johnson WE, Lauenstein GG, Chritensen JD, Apeti DA (2008) An assessment of two decades of contaminant monitoring in the Nation's coastal zone. NOAA Technical Memorandum NOS NCCOS 74, p 105. Available: ► https://repository.library.noaa.gov/view/noaa/2499. Accessed 10 Dec 2021

Koziorowska K, Kuliński K, Pempkowiak J (2016) Sedimentary organic matter in two Spitsbergen fjords: terrestrial and marine contributions based on carbon and nitrogen contents and stable isotopes composition. Cont Shelf Res 113:38–46

Kroon FJ, Kuhnert PM, Henderson BL, Wilkinson SN, Kinsey-Henderson A, Abbott B, Brodie JE, Turner RDR (2012) River loads of suspended solids, nitrogen, phosphorus and herbicides delivered to the Great Barrier Reef lagoon. Mar Pollut Bull 65(4–9):167–181

Lesueur T, Boulangé-Lecomte C, Restoux G, Deloffre J, Xuereb B, Le Menach K, Budzinski H, Petrucciani N, Marie S, Petit F, Forget-Leray J (2015) Toxicity of sediment-bound pollutants in the Seine estuary, France, using a *Eurytemora affinis* larval bioassay. Ecotoxicol Environ Saf 1(13):169–175

Liu AF, Tian Y, Yin NY, Yu M, Qu GB, Shi JB, Du YG, Jiang GB (2015a) Characterization of three tetrabromobisphenol-S derivatives in mollusks from Chinese Bohai Sea: a strategy for novel brominated contaminants identification. Sci Rep 5:11741

Liu A, Qu G, Zhang C, Gao Y, Shi J, Du Y, Jiang G (2015b) Identification of two novel brominated contaminants in water samples by ultra-high performance liquid chromatography-Orbitrap Fusion Tribrid mass spectrometer. J Chromatogr A 1377:92–99

Liu S, Jiang Z, Deng Y, Wu Y, Zhang J, Zhao C, Huang D, Huang X, Trevathan-Tackett SM (2018) Effects of nutrient loading on sediment bacterial and pathogen communities within seagrass meadows. Microbiology Open 7(5):e00600

Lu L (2005) The relationship between soft-bottom macrobenthic communities and environmental variables in Singaporean waters. Mar Pollut Bull 51(8–12):1034–1040

Maher WA, Ellwood MJ, Krikowa F, Raber G, Foster S (2015) Measurement of arsenic species in environmental, biological fluids and food samples by HPLC-ICPMS and HPLC-HG-AFS. J Anal at Spectrom 30(10):2129–2183

Maher WA, Taylor AM, Batley GE, Simpson SL (2016) Bioaccumulation. In: Simpson S, Batley G (eds) Sediment quality assessment—a practical guide, 2nd edn. CSIRO Press, Clayton South, pp 123–156

Markert B (ed) (2008) Environmental sampling for trace analysis. Wiley, Weinheim, p 559

Masura J, Baker J, Foster G, Arthur C, Herring C (2015) Laboratory methods for the analysis of microplastics in the marine environment: recommendations for quantifying synthetic particles in waters and sediments. NOAA Technical Memorandum NOS-OR&R-48. National Oceanic and Atmospheric Administration (NOAA), USA, p 31

Melwani AR, Gregorio D, Jin Y, Stephenson M, Ichikawa G, Siegel E, Crane D, Lauenstein G, Davis JA (2014) Mussel watch update: long-term trends in selected contaminants from coastal California, 1977–2010. Mar Pollut Bull 81(2):291–302

Moynihan MA, Baker DM, Mmochi AJ (2012) Isotopic and microbial indicators of sewage pollution from Stone Town, Zanzibar, Tanzania. Mar Pollut Bull 64(7):1348–1355

Mudroch A, Macknight SD (1994) Handbook of techniques for aquatic sediments sampling, 2nd edn. CRC, Boca Raton, p 256

Nascimento RA, de Almeida M, Escobar NCF, Ferreira SLC, Mortatti J, Queiroz AFS (2017) Sources and distribution of polycyclic aromatic hydrocarbons (PAHs) and organic matter in surface sediments of an estuary under petroleum activity influence, Todos Santos Bay, Brazil. Mar Pollut Bull 119(2):223–230

NHMRC (National Health and Medical Research Council) (2013) Australian code for the care and use of animals for scientific purposes, 8th edn. NHMRC, Canberra. Available at: ► https://www.nhmrc.gov.au/about-us/publications/australian-code-care-and-use-animals-scientific-purposes. Accessed 23 Feb 2022

Niyogi S, Wood CM (2004) Biotic ligand model, a flexible tool for developing site-specific water quality guidelines for metals. Environ Sci Technol 38(23):6177–6192

Nomiyama K, Kanbara C, Ochiai M, Eguchi A, Mizukawa H, Isobe T, Matsuishi T, Yamada TK, Tanabe S (2014) Halogenated phenolic contaminants in the blood of marine mammals from Japanese coastal waters. Mar Environ Res 93:15–22

O'Connor TP (1998) Mussel watch results from 1986 to 1996. Mar Pollut Bull 37(1):14–19

O'Reilly C, Santos IR, Cyronak T, McMahon A, Maher DT (2015) Nitrous oxide and methane dynamics in a coral reef lagoon driven by pore water exchange: insights from automated high-frequency observations. Geophys Res Lett 42(8):2885–2892

Pagano G, Thomas P, Guida M, Palumbo A, Romano G, Oral R, Trifuoggi M (2017) Sea urchin bioassays in toxicity testing: II. sediment evaluation. Expert Opin Environ Biol 6(1):1000141

Petrović M, Eljarrat E, López de Alda MJ, Barceló D (2001) Analysis and environmental levels of endocrine-disrupting compounds in freshwater sediments. TrAC Trends Anal Chem 20(11):637–648

Presley BJ, Trefry JH, Shokes RF (1980) Heavy metal inputs to Mississippi Delta sediments—a historical view. Water Air Soil Pollut 13(4):481–494

Reichelt-Brushett AJ, McOrist G (2003) Trace metals in the living and nonliving components of scleractinian corals. Mar Pollut Bull 46(12):1573–1582

Reichelt-Brushett A, Clark M, Birch GF (2017) Physical and chemical factors to consider when studying historical contamination and pollution in estuaries. In: Weckström K, Sauunders K, Gell P (eds) Applications of paleoenvironmental techniques in estuarine studies. Springer, Netherlands, pp 239–276

Rosentreter JA, Maher DT, Ho DT, Call M, Barr JG, Eyre BD (2017) Spatial and temporal variability of CO_2 and CH_4 gas transfer velocities and quantification of the CH_4 microbubble flux in mangrove dominated estuaries. Limnol Oceanogr 62(2):561–578

Rovedatti MG, Castañé PM, Topalián ML, Salibián A (2001) Monitoring of organochlorine and organophosphorus pesticides in the water of the Reconquista River (Buenos Aires, Argentina). Water Res 35(14):3457–3461

Russell RW, Gobas FAPC, Haffner GD (1999) Maternal transfer and in ovo exposure of organochlorines in oviparous organisms: a model and field verification. Environ Sci Technol 33(3):416–420

SA and SNZ (Joint Standards Australia/Standards New Zealand Committee) (1998) Australian/New Zealand Standard 5667.1.1998: water quality sampling. Sydney, Australia/Wellington, New Zealand. Available at ► https://www.saiglobal.com/pdftemp/previews/osh/as/as5000/5600/56671.pdf. Accessed 10 Dec 2021

Scanes PR, Roach AC (1999) Determining natural "background" concentrations of trace metals in oysters from New South Wales, Australia. Environ Pollut 105:437–446

Schultz MM, Barofsky DF, Field JA (2003) Fluorinated alkyl surfactants. Environ Eng Sci 20(5):487–501

Simpson SL, Kumar A (2016) Sediment ecotoxicology. In: Simpson S, Batley G (eds) Sediment quality assessment—a practical guide, 2nd edn. CSIRO Press, Clayton South, pp 47–76

Simpson SL, Angel BM, Jolley DF (2004) Metal equilibration in laboratory-contaminated (spiked) sediments used for the development of whole-sediment toxicity tests. Chemosphere 54(5):597–609

Simpson SL, Batley GE (eds) (2016) Sediment quality assessment—a practical guide, 2nd edn. CSIRO Press, Clayton South, p 346

Simpson SL, Batley GE, Maher WA (2016) Chemistry of sediment contamination. In: Simpson S, Batley G (eds) Sediment quality assessment—a practical guide. 2nd edn. CSIRO Press, Clayton South, pp 47–76

Smith VH (2006) Responses of estuarine and coastal marine phytoplankton to nitrogen and phosphorus enrichment. Limnol Oceanogr 51(1):377–384

Sweidan AH, El-Bendary N, Hegazy OM, Hassanien AE, Snasel V (2015) Water pollution detection system based on fish gills as a biomarker. Procedia Comput Sci 65:601–611

Takeuchi M, Mizuishi K, Hobo T (2000) Determination of organotin compounds in environmental samples. Anal Sci 16(4):349–359

Taylor A, Maher WA (2016) Biomarkers. In: Simpson S, Batley G (eds) Sediment quality assessment—a practical guide, 2nd edn. CSIRO Press, Clayton South, pp 157–194

Tixier C, Sancelme M, Sancelme M, Bonnemoy F, Cuer A, Veschambre H (2001) Degradation products of a phenylurea herbicide, diuron: synthesis, ecotoxicity, and biotransformation. Environ Toxicol Chem 20(7):1381–1389

US EPA (United States Environment Protection Agency) (1994) Recommended guidelines for sampling marine mammal tissue for chemical analyses in Puget Sound. US EPA and Puget Sound Water Quality Authority, Seattle, p 71. Available at: ▶ https://www.cascadiaresearch.org/files/publications/sampling.pdf. Accessed 10 Dec 2021

US EPA (United States Environment Protection Agency) (2000) Guidance for assessing chemical contaminant data for use in fish advisories. Volume 1: fish sampling and analysis, 3rd edn. US EPA, Office of Science and Technology, Office of Water, Washington DC, p 485. Available at: ▶ https://www.epa.gov/sites/default/files/2015-06/documents/volume1.pdf. Accessed 10 Dec 2021

US EPA (United States Environment Protection Agency) (2001) Methods for collection, storage and manipulation of sediments for chemical and toxicological analyses: technical manual. EPA-823-B-01-002. US EPA, Office of Water, Office of Science and Technology, Washington DC, p 208. Available at: ▶ https://www.epa.gov/sites/default/files/2015-09/documents/collectionmanual.pdf. Accessed 10 Dec 2021

US EPA (United States Environment Protection Agency) (2002) Methods for measuring the acute toxicity of effluents and receiving waters to freshwater and marine organisms, 5th edn. US EPA, Washington DC, p 275. Available at: ▶ https://www.epa.gov/sites/default/files/2015-08/documents/acute-freshwater-and-marine-wet-manual_2002.pdf. Accessed 10 Dec 2021

US EPA (United States Environmental Protection Agency) (2009) National recommended water quality criteria. Office of Water and Office of Science and Technology, Washington, DC, p 23. Available at: ▶ https://www.epa.gov/wqc/national-recommended-water-quality-criteria-aquatic-life-criteria-table. Accessed 10 Dec 2021

Viršek MK, Palatinus A, Koren Š, Peterlin M, Horvat P, Kržan A (2016) Protocol for microplastics sampling on the sea surface and sample analysis. J vis Exp 118:55161

von der Heyden BP, Roychoudhury AN (2015) A review of colloidal iron partitioning and distribution in the open ocean. Mar Chem 177:1–11

Walker CH, Sibly RM, Hopkin SP, Peakall DB (2012) Principles of ecotoxicology, 4th edn. CRC, Boca Raton, p 328

Zhang C (2014) Effects of sediment geochemical properties on heavy metal bioavailability. Environ Int 73:270–281

Assessing Organism and Community Responses

Amanda Reichelt-Brushett, Pelli L. Howe, Anthony A. Chariton and Michael St. J. Warne

Contents

3.1 **Introduction** – 54

3.2 **Ecotoxicology** – 54
3.2.1 General Principles of Ecotoxicology – 56
3.2.2 Factors Influencing Toxicity – 56
3.2.3 Considerations for Planning Ecotoxicology Experiments – 57
3.2.4 Selecting Species for Toxicity Testing – 59

3.3 **Current Status of Marine Ecotoxicology** – 63
3.3.1 Temperate Marine Ecotoxicology – 63
3.3.2 Polar Marine Ecotoxicology – 64
3.3.3 Tropical Marine Ecotoxicology – 64

3.4 **Using Ecotoxicological Data to Set Guideline Values** – 65
3.4.1 Deriving Limits – 65

3.5 **Limitations of Species Toxicity Studies** – 67

3.6 **Assessing Responses from Organisms at the Community Level** – 67
3.6.1 In situ Studies – 67
3.6.2 Experimental In situ Studies – 69
3.6.3 Laboratory Studies – 70

3.7 **Summary** – 71

3.8 **Study Questions and Activities** – 71

© The Author(s) 2023
A. Reichelt-Brushett (ed.), *Marine Pollution—Monitoring, Management and Mitigation*,
Springer Textbooks in Earth Sciences, Geography and Environment,
https://doi.org/10.1007/978-3-031-10127-4_3

Acronyms and Abbreviations

ANOVA	Analysis of variance
AF	Assessment factor(s)
EC50	Concentration of a toxicant that causes a measured negative effect to 50% of a test population
EC10	Concentration of a toxicant that causes a measured negative effect to 10% of a test population
EDA	Effects-directed analysis
HC1	Harmful concentration for 1% of species. Equivalent to the PC99
HC5	Harmful concentration for 5% of species. Equivalent to the PC95
LC50	Concentration of a toxicant that causes a 50% mortality to a test population
LOE	Line of evidence
NOEC	No observed effect concentration
PC99	The protective concentration for 99% of species
PC95	The protective concentration for 95% of species
POPs	Persistent organic pollutants
QSAR	Quantitative structure–activity relationship
SF	Safety factor(s)
SSD	Species sensitivity distribution
TIE	Toxicity identification evaluation
US EPA	United States Environmental Protection Agency
WET	Whole effluent toxicity test
WOE	Weight of evidence

3.1 Introduction

Many of the chemicals in the environment are naturally derived from compounds in plants, petroleum oils, or minerals in rocks. However, their chemical composition, concentration, and distribution through the environment have been altered by humans, usually as a result of an economic incentive (e.g. mining). Other chemicals are synthetic, produced in laboratories, and manufactured for specific uses. These manufactured chemicals are known as xenobiotics and include some fertilisers, pesticides, dyes, manufactured petroleum products, personal care products, and pharmaceuticals. What is common to all natural and Synthetic chemicals is that they are potentially toxic and likely to have come from a small geographic area or a limited number of sources. The chemicals are then redistributed in the environment through natural and anthropogenic activities, where organisms can intentionally or unintentionally be exposed to them. Some exposed species, and indeed some individuals, will be more sensitive than others, which can lead to adverse effects at the population level. When sensitive species are keystone or foundation species for a particular ecosystem, or enough species are affected, this can alter the structure and function of the exposed communities, having flow-on effects at the ecosystem level (◘ Figure 3.1).

3.2 Ecotoxicology

Toxicity testing of organisms has been developing since the 1940s (Cairns Jr and Niederlehner 1994) because of the need to understand the effects of chemicals on organisms. Its application in environmental monitoring has grown rapidly (e.g. Auffan et al. 2014). In fact, the term **ecotoxicology**, which the field of study is now referred to as, was first used in 1969 by René Truhaut, defining it

» *as a science describing the toxic effects of various agents on living organisms, especially on populations and communities within ecosystems*

Ecotoxicology is a multidisciplinary science that combines chemistry, biology, ecology, pharmacology, epidemiology, and of course toxicology. It seeks to understand and predict the effects of chemicals on organisms and ecosystems and is constantly evolving as a discipline area (Sánchez-Bayo et al. 2011). Pollution studies use ecotoxicology as a tool to document the effects of pollutants at known concentrations on living organisms (Phillips 1977; Chapman and Long 1983) and to supplement conventional pollutant concentration data.

Ecotoxicology is used in a multiple **lines of evidence** (**LOE**) approach to risk assessment. This means that you use more than one source of information to support and understand the risk. Ecotoxicological experiments are most often conducted in laboratories under controlled conditions, and this chapter provides a guide

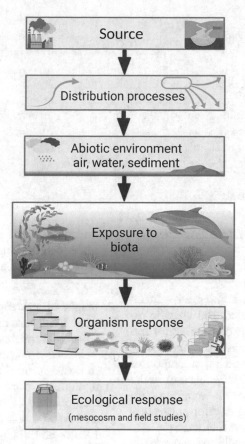

Source

Distribution processes

Abiotic environment
air, water, sediment

Exposure to
biota

Organism response

Ecological response
(mesocosm and field studies)

◘ Figure 3.1 Transport of chemicals into biological systems. Chemicals come from a source and are distributed through the abiotic environment by air, water, and soil/sediment movement. Through this process, organisms are exposed to the chemicals and they too become part of the distribution process. Scientists use ecotoxicological experiments (see also ◘ Figure 3.3) to test the effects of exposure at the organism level and also at the ecosystems level using mesocosms and other multispecies assessments. *Image*: A. Reichelt-Brushett with Biorender.com

to how these experiments are performed. While the resulting information is limited by its lack of relevance to conditions in the environment, it provides important standard approaches for comparative assessment to help understand the relative sensitivity of different species and the relative toxicity of chemicals. Standard approaches to toxicity testing are explained in ► Section 3.2.3. Non-laboratory approaches for assessment under more relevant environmental conditions are discussed in ► Section 3.6.1.

The expansion of ecotoxicology to develop **species sensitivity distribution (SSD)** using toxicity data for a range of species and taxonomic groups (for detail see ► Section 3.4) increases the ecological relevance of laboratory-based toxicity results. The greater the number and diversity of species used in an SSD the greater the confidence in predicting concentrations that should protect any chosen percentage of species (see ► Section 3.4). Another means of gaining a better understanding of ecological interactions through ecotoxicological assessment is by using microcosm or mesocosm level studies in the field or laboratory (► Box 3.1).

Box 3.1. Microcosms and Mesocosm Studies in Ecotoxicology

There are many benefits of assessing toxicity using highly controlled single-species laboratory toxicity tests. Strictly controlled test conditions (e.g. temperature, photoperiod, contaminant dispersion, concentration, etc.) isolate a chosen contaminant (or contaminants) as the cause of any toxic effects. However, ecotoxicological experiments using enclosed experimental ecosystems (microcosms and mesocosm) provide considerably more ecologically relevant understanding of contaminant effects in the environment. They can also be used as a line of evidence along with single-species tests.

Microcosms are similar to standard toxicity tests, being generally conducted in a laboratory in small experimental vessels. The main difference is that microcosm studies involve exposing numerous interacting species to a contaminant, rather than a single species. This provides insight into contaminant effects in an environment with a much higher (and more realistic) level of biological organisation and interaction. For example, a species may not be directly affected by a contaminant but suffer effects from ingesting organisms which have absorbed the contaminant. Or, a species may not be directly affected, but its population may be decimated by a predator that has lost another food supply (i.e. the population of another prey species has been affected by the contaminant).

mesocosm are similar to microcosms but are usually (though not always) conducted in the field and are much larger. mesocosm experiments incorporate natural abiotic effects on the toxicity of a contaminant (i.e. contaminants are exposed to the elements). mesocosm are as close as scientists can get to "replicating" the effects of contaminants in natural ecosystems without intentionally distributing contaminants into the environment. However, they are generally constrained by costs and logistics, limiting the number of replicates and interactions, and thus, the statistical confidence.

3.2.1 General Principles of Ecotoxicology

Ecotoxicologists are guided by four general principles, including the following:

- **You can only find what you are looking for**—toxicity tests can only provide targeted and specific information. For example, only the contaminants and responses of interest are measured, although other contaminants may be present and other responses may be occurring. This is particularly relevant in the context of the explosion of new and emerging contaminants, as well as the need to develop more sensitive and ecologically relevant test methods.
- **The dose determines the poison**—sediment and aquatic toxicity tests are only able to arrive at an effect concentration, rather than a dose. The consumption (dose) of a contaminant can generally not be quantified due to the multiple exposure routes including ingestion (in food, water, particulates, and sediment) and direct absorption from water or sediment. The influence of different types of exposure differs depending on the organism (e.g. behaviour, physiology, life stage, etc.), the contaminant (e.g. different substances may dissolve, bind to sediments or suspended particles, etc.), and the environmental conditions (e.g. physiochemistry, hydrological processes, etc.). For example, in a given environment, different species (or life stages of the same species) may be exposed to vastly different concentrations and types of contaminants due to their different feeding behaviours, preferred food sources, and detoxification and depuration abilities. Bioavailability is very important to consider here, since a high aqueous concentration does not determine toxicity if a substance is not bioavailable.
- **Toxicity can only be measured by living material** (e.g. organisms, cell lines, or enzymes). However, models such as quantitative structure–activity relationships (QSARs) and quantitative activity-activity relationships (QAARs) can be useful to predict potential toxicological effects. In QSARs, structural characteristics of chemicals are used to predict the toxicity (activity) of chemicals without toxicity data.

In contrast, in QAARs, the toxicity (activity) of chemicals to one organism is used to predict the toxicity to another species.

Weight of Evidence

A **weight of evidence (WOE)** approach, which incorporates LOE has long been used in legal systems. It is a broad term that simply means that several pieces of evidence are considered together, rather than basing a decision on a single piece of evidence (Chapman et al. 2002). Court decisions should use a WOE approach, instead of relying on a single LOE, to provide **proof beyond a reasonable doubt**. According to the Federal Rules of Evidence in the USA (Annas 1994, 1999), court judges must consider the following:

- whether hypotheses used in experimental studies are testable or falsifiable;
- whether the relevant techniques/theories have been peer-reviewed and published;
- the potential rate of error in the methods; and
- if there is a general acceptance of the theory or method (similar to the point above).

More recently, a WOE approach has been used in the context of ecotoxicological risk assessments and is determined by multiple LOE. The LOE may include experimental (e.g. toxicity tests, contaminant characteristics) and observational (e.g. field assessments, physiological biomarkers) data, with each having its quality (e.g. were appropriate methods and analyses used) and extent (e.g. short/long term, local/regional/global scale) assessed. Qualitative LOE may also be considered in the WOE (e.g. best professional judgement) to help arrive at a prediction of the ecological risks based on the various LOE (e.g. Suter II 2016).

3.2.2 Factors Influencing Toxicity

The concentration of a chemical is one of the more obvious factors that will affect its toxicity. You may have heard the historic quote from Paracelsus (1493–1541), who expressed:

> *All substances are poisons; there is none that is not a poison. The right dose differentiates a poison and a remedy.*

However, many other environmental, physiological, biological, and chemical factors influence the toxicity of a chemical (e.g. Chariton et al. 2010a; de Almeida Rodrigues et al. 2022). More specifically:

— **Temperature**—the toxicity of many substances differs with temperature, however, there is no general rule as to whether a higher or lower temperature will elicit a higher toxicity. For example, some metals are more toxic in tropical marine ecosystems than in temperate or polar regions, whereas for other metals this is reversed (Chapman et al. 2006). This makes the extrapolation of toxicity data from different climatic regions unreliable. This is one reason that some countries such as Canada and the USA have requirements to use toxicity data for endemic species. This approach is possible for North American and European countries because most of the test organisms used in toxicity tests originated from those regions. However, in other countries with smaller populations, different climates, and/or different ecosystems and species, this approach is not logistically possible, and instead toxicity data generated using non-endemic species must be relied upon.

— **pH**—affects solubility, and therefore bioavailability, of contaminants such as metals. Generally, a decrease in pH increases the toxicity of metals as they become more bioavailable. Since marine waters are buffered, they resist changes in pH (to a point).

— **Salinity**—can decrease the toxicity of chemicals by decreasing their aqueous solubility or changing their chemical form, although this depends on the type and concentration of other competing substances such as dissolved organic compounds (e.g. Hall and Anderson 2008).

— **Suspended sediment**—may increase toxicity (for example by providing a surface for contaminants to bind to, allowing uptake of contaminants by some organisms), but can also decrease toxicity (e.g. by decreasing the bioavailability of hydrophobic chemicals to aquatic organisms). This largely depends on the type of contaminant and the type of sediment. Suspended sediments may also act as a stressor in their own right.

— **Dissolved organic carbon**—may decrease toxicity by decreasing a chemical's aqueous solubility, and its ability to pass through membranes or bioavailability.

— **Previous exposure/resistance**—populations or strains of species may develop resistance to toxicants, so that considerable differences in sensitivity may exist within one species. For instance, species in areas with naturally elevated concentrations of metals may become more tolerant to those metals.

— **Organism life stage**—generally the early and oldest life stages are the most sensitive to the harmful effects of chemicals.

— **Duration of exposure**—generally a longer period of exposure will result in adverse effects at lower concentrations of a toxicant than the same effects measured after a short period of exposure.

3.2.3 Considerations for Planning Ecotoxicology Experiments

Experimental Procedures

As with all scientific experiments, ecotoxicology requires strict adherence to protocols including replication, quality control, statistical analyses, and interpretation of results. The assessment of the toxicity of a chemical usually starts with a **range finder** test. The purpose of these tests is to determine a broad range of concentrations of the chemical that include no effect through to 100% effect. For example, a range finder test might include concentrations of a test chemical including 0 (control), 0.001, 0.01, 0.1, and 1.0 mg/L. It is important that the amount or concentration of the chemical is measured using appropriate analytical equipment. If they are not measured, they are defined as nominal concentrations. Generally, only studies that have measured concentrations are publishable in scientific literature. Once the relevant EC50 (concentration of a toxicant that causes a measured negative effect to 50% of a test population) or LC50 (concentration of a toxicant that causes 50% mortality of a test population) has been determined from the range finder test, a **definitive test** is completed using concentrations in a much tighter range around the concentrations which induced the predetermined toxicological response (endpoint). ◘ Figure 3.2 shows the basic approach to replication in a static system and highlights that each replicate is wholly independent of the others, with the intention of avoiding pseudo-replication (e.g. many organisms in a single container for each test concentration). It is important to record the key physicochemical parameters of the media used in toxicity tests (e.g. dissolved oxygen content, temperature, pH, and water hardness) throughout the duration of the test. This information is important because these parameters may influence the speciation or behaviour and subsequent toxicity of the chemical in question or exert toxic effects in their own right. It is undesirable to have an additional stressor of declining water quality in combination with the chemical stressor of interest, unless the experiment is specifically designed to test the effect of multiple stressors.

Laboratory toxicity tests may be static (no renewal of the test solution), semi-static or static-renewal (renewal of the test solutions at set times, e.g. every 24 or 48 h for the duration of the test), or flow through (constant renewal of test solution for the duration of the

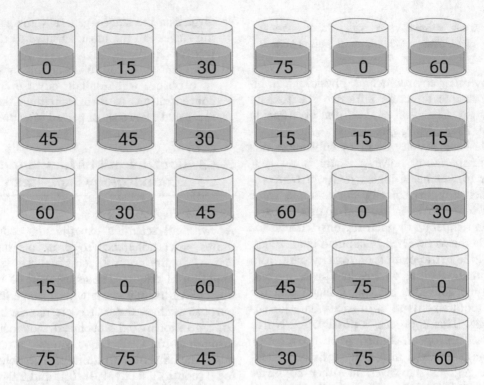

Figure 3.2 A typical experimental design for a replicated definitive ecotoxicological test. There are six treatments including a control (the concentrations, µg/L, are the numbers in the test containers), each with five replicates. A fixed number of test organisms are placed in each test chamber. The replicates of each concentration are randomly located in the testing room or incubator. The biological effect will be determined at set time intervals such as 0, 2, 24, 48, 72, and 96 h. These data will then be statistically analysed and the concentrations that cause biological effects of a certain magnitude are determined. *Image*: A. Reichelt-Brushett using Biorender.com

test). **Static systems** require simple equipment and are cost-efficient, however, they have the potential to provide inaccurate results due to changes in the concentration of contaminants during the test period, which may be absorbed and metabolised by test organisms, or be volatilised, degraded, and/or adsorbed onto the test container. The advantage of **semi-static** and **flow-through** test systems is that they maintain much more consistent chemical concentrations and minimise the accumulation of food, faeces, waste products (e.g. ammonia), algae, etc. in the test containers, which may influence the results. Flow-through tests, if maintained properly, provide the most consistent chemical concentrations. Another type of toxicity test is termed **pulse-exposure**, which, as the term suggests, exposes organisms to pulses of contaminant loads. This approach mimics an exposure regime that you might expect to see from rainfall and runoff events.

Test Endpoints
Ideally, standardised toxicity tests should have a well-defined, easily quantifiable endpoint (biological effect) that does not require a particularly high level of expertise to measure and interpret. Importantly, the duration of exposure needs to be considered and there are two main terms used to describe this. **Acute** toxicity

tests are short-term tests that measure the lethal or sub-lethal effects of exposure to relatively high concentrations of chemicals. The duration which is considered short-term depends on the lifespan of the test organism. For example, durations of up to 96 h tests are commonly considered acute; however, many microorganisms (e.g. algae) double their cell number several times within this time frame, and hence 96 h is a chronic or longer-term exposure period for those species. **Chronic** toxicity tests are longer term and usually encompass a large period of the life cycle of the test organisms—typically of greater than 10% of the organism's lifespan (Newman 2010). Endpoints include both sub-lethal effects (e.g. reproduction, growth, population growth rate, and immobilisation) and lethal effects.

Lethality is the most basic test endpoint, whereby test organisms are determined as either dead or alive after a given exposure time, and the median lethal concentration (i.e. LC50) is calculated from the test data. While lethal endpoints are relevant to fish kills or exposure to high chemical concentrations, they are not ideal for deriving toxicant limits designed to protect the form and function of ecosystems (refer to ▶ Section 3.4).

Sub-lethal toxicity endpoints assess the effect of contaminants on a particular life stage and may be shorter (e.g. 1 h sea urchin fertilisation) or longer (e.g.

7-day fish growth) than acute lethal tests. Tests using these endpoints allow estimations of effect concentrations (EC values, e.g. EC50) rather than lethal concentrations (LC values). Traditional sub-lethal endpoints used for marine species include the following:

- growth (e.g. algae, juvenile and adult fish);
- germination (e.g. algae);
- fertilisation (e.g. sea urchin);
- early life-stage development (e.g. larval development in oysters, mussels, scallops); and
- behaviour (e.g. fish imbalance).

Considerable research effort has been (and continues to be) directed towards developing and standardising new toxicity tests with ecologically relevant sub-lethal endpoints. Examples of well-developed new test endpoints include the following:

- behaviour (e.g. coral larvae motility [Reichelt-Brushett and Harrison, 2004] and motor behaviour in fish [Harayashiki et al. 2019]);
- physiology (e.g. heart rate, neurotoxicity, and production of reactive chemical species);
- development (e.g. species-specific larval/juvenile development, larval malformation rate, heart rate, spontaneous movements, tail length, enzyme activities,biomarker genes and plant root elongation [Howe et al. 2014, Zhu et al. 2014, Rodriguez-Ruiz et al. 2014, van Dam et al. 2016])
- reproduction population growth (e.g. asexual reproduction rate and algal biomass change); and
- photosynthesis (e.g. in algae, plants, and symbiotic organisms such as corals).

Sediment Toxicity
Sediments are repositories for many contaminants, and for some contaminants, particularly hydrophobic contaminants, concentrations may be orders of magnitude higher in sediments than in the overlying waters. The metals and organic contaminants in sediment that are of most interest to ecotoxicologists are those that are available for uptake (i.e. they are bioavailable) by organisms exposed to sediments and/or sediment pore water. The presence of pollutants in aquatic sediments may cause toxic responses from benthic (sediment dwelling) organisms and bottom-feeding animals (e.g. prawns, some fish), and suspended sediments interfere with filter-feeding species such as bivalve molluscs. Most of the toxic effects result from toxicants dissolved in the interstitial water of the sediment, since animal gills are the prime sites of toxic action, although toxicants can be bioaccumulated from food and sediment ingestion. Additionally, organism interaction with sediments via feeding on the sediment, burrowing, and bioturbation may also change the local physicochemical conditions and alter the availability and/or toxicity of the contaminants (e.g. pH change through digestive acids and organic complexation through mucus secretion) (McCon-

chie and Lawrence 1991; Han et al. 1996; Luoma 1996; Reichelt-Brushett and McOrist 2003). Sediment toxicity assessment is challenging because it is very hard to define the exposure/dose, and different sediment types influence the bioavailability of the contaminant and this availability will vary between the various compounds or complexes being tested (Chariton et al. 2010a). Some studies have investigated the status of tropical and temperate sediment toxicity although testing (e.g. Adams and Stauber 2008) and concluded that further tests for ecologically relevant species need to be developed. This is an ongoing field of research although there are some standard sediment toxicity test procedures established by the United States Environmental Protection Agency (US EPA) and Organisation for Economic Co-operation and Development (OECD) (◘ Table 3.1).

3.2.4 Selecting Species for Toxicity Testing

Traditional Species
The majority of available standard toxicity test species are freshwater temperate species. This is because ecotoxicological work has traditionally been conducted in temperate regions in the Northern Hemisphere and because contamination of freshwater ecosystems has been recognised for longer than marine contamination. It is valuable to determine the toxicity of a number of different taxonomic groups to help represent the ecosystem composition and water and sediment exposure. ◘ Figure 3.3 shows a range of taxonomic groups that are commonly used in toxicity tests. Standard test methods have been developed throughout the world for different species that represent these taxonomic groups.

Novel Toxicity Test Species
A lot of research effort has been directed towards developing toxicity test methods for novel species, particularly **keystone or** foundation **species** of specific ecosystems, to increase the ecological relevance of results. This is particularly relevant for tropical marine species, which are under-represented in standard toxicological testing (e.g. van Dam et al. 2008). The following criteria are usually considered in species selection:

- suitability for culturing in laboratory conditions (e.g. tolerant of handling and laboratory culturing conditions, not particularly large);
- high reproduction rate and easily induced reproduction;
- ecological relevance (e.g. wide-ranging, ecologically relevant, representative species); and
- quantifiable toxicological responses.

Sometimes a species or taxonomic group will not meet all these requirements, and where keystone or foundation species for ecosystems are concerned, considera-

◘ Table 3.1 Examples of whole sediment toxicity tests for marine and estuarine species

Type of organism	Species	Temperate/Tropical	Test endpoint	Acute/Chronic
Bacterium	*Vibrio fischeri*	Temperate	15-min luminescence	Acute
Microalga	*Entomoneis punctulata*	Temperate	72-h growth	Chronic
Amphipod	*Melita plumulosa*	Temperate	10-d survival	Acute
			28–42 d reproduction	Chronic
	Grandidierella japonica	Temperate	10-d survival 28-d growth	Chronic
	Corophium cola	Temperate	10-d survival and emergence	Acute
			14-d growth	Chronic
	Corophium insidiosum	Temperate	10-d survival	Acute
Crab	*Diogenes* sp.	Tropical	10-d survival	Acute
Bivalve	*Tellina deltoidalis*	Temperate	10-d survival	Acute
	Paphies elongate	Temperate	10-d survival	Acute
			28-d growth	Chronic
	Donax cuneate/Donax columbellia	Temperate/Tropical	10-d survival	Acute
Polychaete worm	*Australonereis ehlersi*	Temperate	10-d survival	Acute
	Ceratonereis aequisetis	Temperate	10-d survival	Acute

Examples of standard test species

vertebrates	Atlantic silverside -*Menidia menidia* Sheepshead minnow -*Cyprinodon variegatus*
echinoderms	Red sea urchin -*Heliocidaris turbuculata* Purple sea urchin -*Paracentrotus lividus*
crustaceans	Tiger prawn -*Penaus monodon* Amphipods -*Corophium* spp. -*Gammarus* spp.
molluscs	Blue mussel -*Mytilus edulis*
cnidarians	Scleractinian coral -*Acropora* spp. Anemone -*Exaiptasia pallida*
plants	Neptune's necklace -*Hormosira banksii* Kelp -E*klonia radiata* Algae -*Nitzchia closterium* -*Isochrysis galbana*

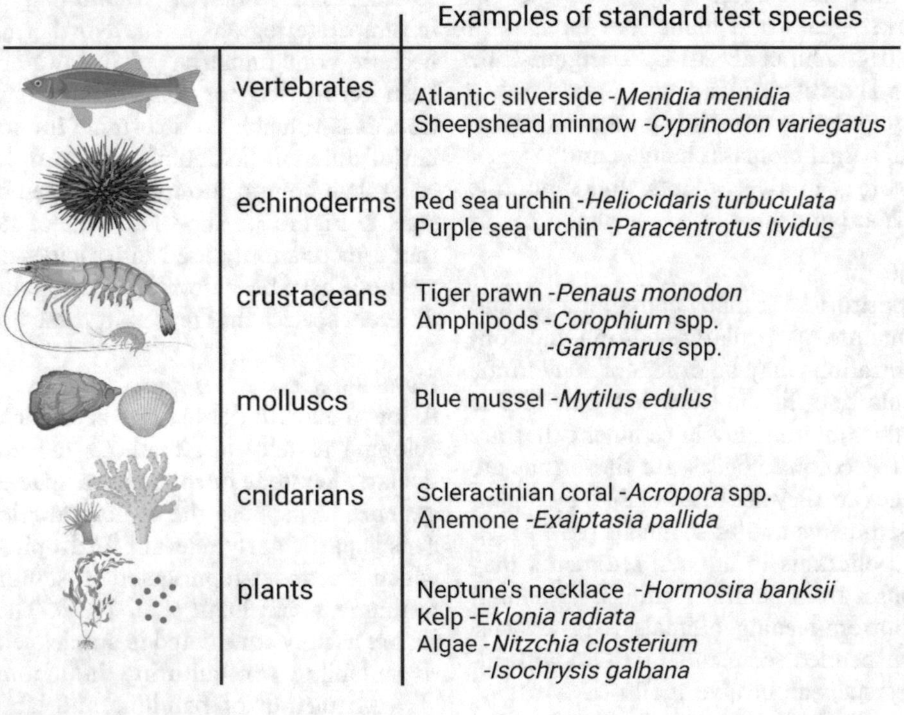

◘ Figure 3.3 Examples of some standard taxonomic groups used in marine toxicity test species. *Image*: A. Reichelt-Brushett using Biorender.com

ble laboratory infrastructure and experimental design may need to be developed. Reef-building scleractinian corals are an example of foundation species that require intensive animal husbandry to maintain in aquarium conditions for ecotoxicology testing. Since most scleractinian corals are broadcast spawning and fertilisation occurs in the water, followed by metamorphosis (◘ Figure 3.4), reproduction is considered a particularly sensitive stage of development to chemical exposure (e.g. Reichelt-Brushett and Harrison 2004;

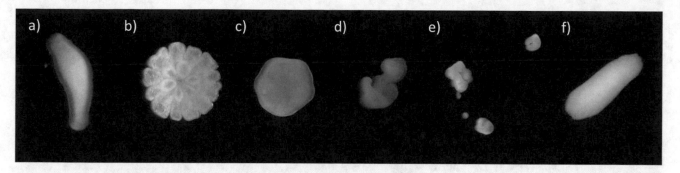

Figure 3.4 Examples of normal and abnormal larvae and recruits observed during larval *Acropora millepora* assays with exposure to total aromatic hydrocarbons (TAH). Morphologies observed included: **a** normal-sized planula larva (0–100 μg/L TAH), **b** fully metamorphosed recruit (0–100 μg/L TAH), **c** early-stage metamorphosed recruit (10–500 μg/L TAH), **d** severely deformed larvae undergoing fragmentation (10–500 μg/L TAH), **e** swimming larval fragments and deformed larvae undergoing fragmentation (10–500 μg/L TAH), and **f** larva-shaped mass of dead cells (>350 μg/L TAH). Examples extracted from photographs obtained using a Leica MS5 dissecting microscope with a 5.1 MP camera calibrated using the ToupView software. *Source*: Nordborg et al. 2021 with permission

Reichelt-Brushett and Hudspith 2016; Nordborg et al. 2021). External fertilisation is quite common among marine invertebrates, which results in gametes being directly exposed to chemicals in the water.

Animal Ethics Considerations

In 1959, the publication of the seminal book *The Principles of Humane Experimental Technique* by Russell and Burch encouraged scientific researchers using animals to "*remove the inhumanity*" of animal research by considering the **three Rs**—reduction (reduce the number of animals needed to obtain a given data set by controlling variability and optimising the design and analysis, so as to avoid repeating tests), refinement (techniques to minimise suffering), and replacement of animal use (use of non-animal alternatives wherever possible) (Russell and Burch 1959). These concepts aimed to minimise the unnecessary suffering of animals. In many countries, animal ethics approval must be acquired for research using animals although the definition of an **animal** may vary with jurisdiction. For example, in the Australian Code for the Care and Use of Animals for Scientific Purposes (NHMRC 2013), animals are defined as

» "*any live non-human vertebrate (that is,* fish, amphibians, reptiles, birds *and* mammals *encompassing domestic animals, purpose-bred animals, livestock, wildlife) and cephalopods*" [Box 3.2].

Box 3.2: Global Horizon Scanning Project

The Global Horizon Scanning Research Prioritization Project was launched by the Society of Environmental Toxicology and Chemistry (SETAC) World Council. The purpose of the project was to identify research needs that are geographically specific and improve our understanding of the effects of different types of stressors on environmental sustainability (see ► https://globe.setac.org/ghsp-2017-recap/). Participants involved in the global research were asked to consider the following aspects when proposing their priority research needs:

- Does the research address important knowledge gaps?
- Can the research questions be answered by the implementation of a realistic research design which will enable the arrival at a factual answer that is not dependent on value judgements?
- Does it cover a temporal and spatial scale that could realistically be addressed by a research team?
- For research questions regarding impacts and interventions, does it contain a subject, intervention, and a quantifiable outcome?

Examples of proposed priority research needs in the Australasian region include the following:

"*How can we identify and examine the environmental fate and toxicity of ingredients other than the stated 'active' components in commercial formulations individually and in chemical mixtures?*" (Gaw et al. 2019 p. 74).

"*How do we advance ecotoxicology testing to be more relevant to ecological systems?*" (Gaw et al. 2019 p. 76).

Other proposed priorities for Australasia included the following:

- improving predictive risk assessment tools relevant to environmental exposure and toxicology;
- reducing and replacing animal testing;

- development of non-target analytical screening methods to identify priority contaminants in ecosystems which are exposed to complex mixtures;
- effects of multiple stressors;
- vulnerability of regional flora and fauna;
- improved management of ecosystems that are unique;
- stress from global trends (e.g. urbanisation, deforestation); and
- climate change related stress.

Priority research areas have also been identified in Europe (van den Brink et al. 2018), Latin America (Furley et al. 2018), and North America (Fairbrother et al. 2019).

Toxicity Identification Evaluation Analysis

Identification of sources of toxicity in sediment, water, or effluent samples provides information that can be used to develop methods to treat and reduce their toxicity, characterise priority substances in contaminated sites to guide remediation, identify the **active stressors** in an environmental sample to facilitate relevant ecological risk assessment, diagnose stressors that are impairing ecosystem function in watersheds and develop management strategies and policies to reduce the concentration of the stressors, and identify emerging contaminants (Burgess et al. 2013).

The **toxicity identification evaluations (TIEs)** framework was developed by the United States Environment Protection Agency (US EPA). This framework combines toxicity testing, physical and chemical separation procedures, and chemical analysis to identify and quantify toxicants in samples when the sample is toxic and the cause(s) of the toxicity is unknown (e.g. in complex mixtures such as effluents or environmental samples) (◘ Figure 3.5). The resulting knowledge may enable the mitigation of the toxic component(s), for example by targeted treatment (i.e. removal or reduction of the source of toxicity).

In a TIE, potential sources of toxicity are systematically removed by treating the sample, and the remaining sample is re-tested to determine if its toxicity has decreased or remained the same. A decrease in toxicity indicates that the type of chemicals removed by the treatment is the cause of, or contributes to, the toxicity of the original sample. Potential sources of toxicity are removed by physical or chemical treatments (e.g. pH adjustment to remove acids or bases, aeration to remove volatile chemicals, filtering to remove particulates, passing through a cation-exchange column to remove cations, and passing through a C18 column to remove hydrophobic organic chemicals). Chemical separation and identification techniques are then used to identify the chemical or chemicals contributing to the toxicity based on the earlier results. Solutions of the identified chemicals at their concentrations in the original sample are then created and their toxicity determined. If they result in the same toxicity as the original sample, then the chemicals causing the toxicity have been identified. If the toxicity of the solutions is not as great as the original sample, then further TIE work is needed to identify other toxicants. Specialised techniques for TIEs need to be developed for individ-

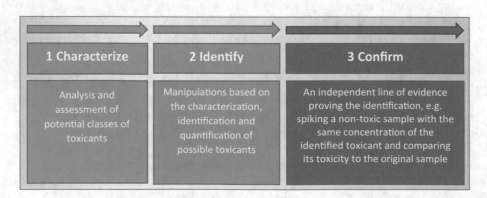

◘ **Figure 3.5** Summary of a toxicity identification, evaluation (TIE) process for sediments, effluents, and receiving waters. *Image*: A. Reichelt-Brushett and M St. J. Warne

◘ Table 3.2 Primary differences between toxicity identification evaluations (TIEs) and effects-directed assessments (EDAs). Adapted from Burgess et al. 2013

Parameter	TIEs (in-vivo)	EDAs (in-vitro)
Toxicological endpoint	Whole organism: e.g. survival, reproduction, etc.	Genotoxicity, endocrine disruption, and mutagenicity
Targeted toxicants	All toxicants	Organic toxicants
Bioavailability	Is considered	Not considered, may be a source of inaccuracy due to how the compounds are extracted from the sample
Form of sample	Whole water, interstitial water, and sediment samples	Organic solvent extracts of sediment, water, interstitial water, biota, technical mixtures, and consumer products
Chemical analysis	Usually targeted analysis for suspected toxicants	Commonly non-targeted analyses and elucidation of structure
Specificity of toxicant identification	High for groups of contaminants and moderate for individual toxicants	High for individual toxicants
Relevance to natural exposure conditions	Primary goal of TIEs	Secondary goal of EDAs

ual contaminants and their degradation/transformation products, and hence much research effort needs to be directed towards developing techniques for isolating the effects of new and emerging contaminants (e.g. Dévier et al. 2011). Despite recent advances, it is not always possible to identify all the causes of toxicity in a complex sample.

Effects-Directed Analysis
A second tool used to identify chemicals causing toxicological effects in the environment is **effects-directed analysis (EDA)**. EDA is an approach used to reduce the complexity of possible or actual toxicity while limiting the chance of overlooking significant chemicals that contribute to risks and effects (Brack et al. 2016). The general approach is to test the biological activity of a sample using responses from sub-cellular systems or whole organisms; samples are then fractionated (separated) and analysed to quantify and characterise the toxic components. This fractionation and effects assessment can be repeated to eliminate fractions that are not biologically active, enabling the isolation and identification of the toxic components (Brack et al. 2016). This method has some fundamental differences to TIEs and should be seen as complementary, rather than being interchangeable with TIEs (◘ Table 3.2). Although there are many advantages of EDAs over TIEs, EDAs have some important limitations (e.g. only organic chemicals can be assessed and their bioavailability is not considered). Also, care must be taken when interpreting EDAs, as the techniques used to extract the toxicants may alter their bioavailability compared to natural conditions and so overestimate their toxicity (Burgess et al. 2013).

3.3 Current Status of Marine Ecotoxicology

The vast majority of aquatic ecotoxicological data is for freshwater species because humans have been aware of contamination of fresh water for much longer. We have had a much greater investment in fresh water, and hence pollution of freshwater ecosystems has historically been more relevant to us and more noticeable. As discussed in ► Chapter 1, it is only relatively recently in human history that we have become aware of the effects of pollution in marine ecosystems, and the vast majority of marine data is for temperate, Northern Hemisphere species, because that is where most ecotoxicology has been conducted (Lacher and Goldstein 1997).

3.3.1 Temperate Marine Ecotoxicology

As you can see from ◘ Table 3.1, most species used in sediment toxicity testing are temperate, and this is the same for aquatic toxicity tests. Because a lot of toxicity data exists for temperate marine species, and ecotoxicological risk assessment is much more relevant when a larger amount of data are available, considerable effort has been directed towards understanding whether temperate data can be applied for the ecosystem protection in other climatic regions. Research has illustrated that there are no predictable patterns in toxicity between temperate and tropical, or temperate and polar species (Chapman et al. 2006; Wang et al. 2014). Rather, it is evident that the relative toxicity depends on the contaminant (i.e. some metals are more toxic to tropical species than temperate species, and vice versa) (Kwok et al. 2007).

3

3.3.2 Polar Marine Ecotoxicology

Despite the remoteness and isolation of polar regions from the centres of anthropogenic activity, contamination is increasingly being identified in these regions, including in deep ocean sediments (e.g. Isla et al. 2018), benthic organisms, and in the tissues of organisms high in the food chain (e.g. polar bears and other mammals, large seabirds [e.g. Eckbo et al. 2019], and sharks [Ademollo et al. 2018]). While there are a few isolated point sources of contaminants (e.g. sewage and other waste from research stations, fuel, and oil), the primary concern and challenge is that contaminants are being transported to polar regions in ocean currents, in the atmosphere, (refer to ▶ Chapter 7 for more detail), and by trophic transfer. These dispersed contaminants include a wide range of persistent organic pollutants (POPs) (see Chapters 7 and 8), plastics (including micro- and nano-plastics, microfibres, and their degradation products) (e.g. Mishra et al. 2021) (see ▶ Chapter 9), and pesticides (see ▶ Chapter 7).

Unfortunately, to date, ecotoxicological risk assessments for polar environments are constrained by the very limited amount of regionally relevant toxicity data. Consequently, they are mostly derived from extrapolations of temperate and tropical toxicity data. There is a valid argument that extrapolation of data from other regions is better than no data at all; however, taxonomic compositions, chemical toxicity, and organism physiology are extremely different in the consistently low temperatures experienced in polar regions (e.g. Kefford et al. 2019). Obtaining the necessary toxicological data to enable the development of relevant water quality guidelines for these ecosystems is currently the subject of dedicated research effort (e.g. King et al. 2006; Gissi et al. 2015; Alexander et al. 2017; Koppel et al. 2017; Kefford et al. 2019; van Dorst et al. 2020).

Generally, polar species are more sensitive to long-term exposure to contaminants than tropical or temperate species. Chapman and Riddle (2005) suggest the following possible reasons for this:

- many species have relatively long lifespans and long development times (and so have a long time to accumulate contaminants);
- many species are relatively large, exhibiting gigantism, which may influence the response time (and so have a slower uptake of contaminants due to the low surface-area-to-volume ratio);
- slower metabolic rates and slow uptake kinetics (resulting in slower accumulation of contaminants, but also slower detoxification/depuration);
- less energy consumed (so less energy is available for detoxification/depuration); and/or
- high lipid content (so accumulate higher concentrations of lipophilic contaminants).

Some of these factors also affect the way that toxicity tests need to be conducted. For example, toxicity tests may need to be continued for longer time periods to account for slower metabolism and slower transition through different life stages, and different endpoints or assessments may be needed for lipophilic substances (Chapman and Riddle 2005).

3.3.3 Tropical Marine Ecotoxicology

Tropical marine ecosystems have a very different taxonomic composition, biodiversity, and physiology of organisms compared to temperate marine ecosystems. Some tropical marine ecosystems have extremely high levels of biological complexity, organisation, and diversity. For example, the tropical area known as the **Coral Triangle** (a marine region that spans parts of Indonesia, Malaysia, Papua New Guinea, the Philippines, the Solomon Islands, and Timor-Leste) is recognised as a global hotspot of biodiversity for corals and reef fishes (Allen 2007). Hence, there are more species that are susceptible to being exposed to contaminants, as well as more complex ecological interactions which further complicates risk assessment. Many tropical marine waters are oligotrophic (see ▶ Chapter 4), which provides less opportunities for contaminants to form complexes and so can result in higher bioavailability. Physiologically, organisms generally have a higher metabolism in warmer temperatures, which can increase either or both uptake and detoxification of contaminants. Degradation (biological and abiotic) might be expected to be enhanced in tropical marine systems compared to temperate and polar marine systems; however, research by Mercurio et al. (2015) found that five herbicides had half-lives of greater than 1 year in tropical marine water compared to earlier studies in temperate laboratories that reported half-lives of months.

Ecotoxicology in tropical marine environments is limited and there is a dearth of data on the dose–response characterisations of pollutants, particularly for early life stages. Lacher and Goldstein (1997) discussed the rapid increase in agricultural, urban, and industrial development in tropical regions. Peters et al. (1997) stressed that managers of tropical marine ecosystems have few tools to aid in decision-making and policy implementation and presented **conceptual models** as a future tool for the problem formulation phase of ecological risk assessment. Measurable responses to stressors, such as the concentrations of chemicals (i.e. ecotoxicological studies), are used within these models and are pertinent to the decisions that may be made to protect the environment (Peters et al. 1997). Since the study by Peters et al. (1997), some progress has been made in developing an understanding of the impacts of trace metals on tropical species (Chapman et al. 2006). However, there is a paucity of fully developed regionally relevant marine toxicity testing methods for tropical marine systems (e.g. van Dam et al. 2008). Fortunately, research effort is growing in tropical marine ecotoxicology.

3.4 Using Ecotoxicological Data to Set Guideline Values

As the preceding text has shown, chemicals, if present at sufficiently high concentrations, can cause a diverse range of harmful effects. Largely as the result of some particularly disturbing pollution events in the United States of America, the United States Environmental Protection Agency (US EPA) started to develop maximum concentrations of chemicals in the water that are **safe** or provide a high degree of protection to aquatic ecosystems. Subsequently, numerous countries, states, and provinces have developed similar limits for chemicals in water, soil, sediment, and animal tissue in order to protect ecosystems. These limits are called guidelines, criteria, standards, or objectives, depending on the legal framework of the jurisdiction developing the limits. Although these terms are often used interchangeably, they have different meanings. Criteria and standards generally have some legal standing and if they are exceeded, this can lead to prosecution in courts of law. Guidelines do not have any legal standing but rather provide guidance on what is a safe concentration. Typically, if environmental concentrations are greater than a guideline concentration, then further work is required. This work can take several forms such as the development of management actions to decrease the concentration, amount, or type of chemicals released, or investigations to determine if the guideline is appropriate or if there are special conditions at the site that may increase or decrease the degree of protection provided. Criteria, standards, and guidelines are all based on the available scientific information, but scientific information is only one of the multiple factors that may be considered in deriving objectives. Other potentially relevant factors include costs and benefits, commercial considerations, and religious and cultural values. Objectives, criteria, standards, and guidelines are limits that reflect the management goals for a particular part of the ecosystem. For the remainder of this section, the term **limit** will be used generically to mean guidelines, objectives, criteria, and standards.

3.4.1 Deriving Limits

There are three main methods for deriving limits: background concentrations, assessment or safety factors (AF or SF), and SSDs.

Background Concentration Method
This method determines a fixed percentile (e.g. the median or 90th percentile) of the **background concentration** of a chemical and adopts that as the limit. While this is conceptually straightforward, it is often quite complex to obtain background concentrations, particularly in areas with a long history of human activity (e.g. in-shore regions near major urban developments). However, while they may not be relevant for particular sites, publications or databases of background concentrations are often available.

Assessment Factor Method
The **assessment factor** (**AF**) method requires a literature search for available data on the responses of marine organisms to toxicants. The data are then screened and assessed for quality, and inappropriate and/or low-quality data are removed. Then the lowest toxicity value is identified and divided by an AF to derive the limit. The magnitude of the AF depends on the amount and type of toxicity data that are available (◻ Table 3.3). Basically, an AF of 10 is applied to account for the following: a lack of data, the difference between toxicity values from acute (short-term) and chronic (long-term) exposures, and differences in toxicity data from laboratory-based and field-based experiments (◻ Table 3.3). This method is easy to understand, and the resulting limit will prevent any of the toxic effects reported in the literature, but it can lead to very low limits.

A key criticism of the AF approach is that there is little scientific justification for the magnitude of the AFs. A crit-

◻ **Table 3.3** Assessment factors applied to the minimum toxicity value depend on the type and amount of toxicity data available

Type of toxicity data	Assessment factor	Type of extrapolation
Chronic NOEC[a,b]	10	Field to laboratory
Acute EC50 or LC50[a]	100 (10 × 10)	Field to laboratory and acute to chronic
Acute EC50 or LC50 for 1 or 2 species	1000 (10 × 10 × 10)	Field to laboratory and acute to chronic and few to many

[a]Data are available for at least one species of algae, a crustacean, and a fish (OECD 1992). [b] No observed effect concentration is the highest concentration used in a toxicity test that does not cause a statistically significant effect compared to the control

ical assessment of the strengths and weaknesses of the AF methods is provided in Warne (1998). Experimentally determined AFs termed acute to chronic ratios (ACRs) have been developed to convert acute toxicity data to chronic data. However, while these are better than the above default AF of 10, they also have limitations (Warne 1998).

Species Sensitivity Distribution Method

The newer and currently preferred method for deriving limits is the SSD approach. This approach was developed in 1985 by Stephan and colleagues (Stephan et al. 1985) and has subsequently been extensively improved. All these SSD methods require a thorough search of the literature, followed by screening and assessing the quality of the toxicity data. The data that pass the screening and quality assurance process are then manipulated to obtain a single value to represent each species for which are data available (e.g. Saili et al. 2021). The data are ordered from highest to lowest toxicity (i.e. lowest to highest concentration at which toxic effects occur) and then given a ranking increasing from one. The cumulative frequency (a percent value) for each species is then calculated by:

$$\text{Cumulative frequency} = \text{rank}/(n+1) \times 100.$$

where n is the number of species for which toxicity data are available (i.e. the highest rank number). Thus, if there are toxicity data for 10 species, the cumulative frequency values for the first three species would be 9.09% ($1/11 \times 100$), 18.18% ($2/11 \times 100$), and 27.27% ($3/11 \times 100$). The cumulative frequency values for each species are then plotted against the toxicity value representing the species and a statistical distribution is fitted to the data. Once the statistical distribution that best fits the data has been identified, that distribution is used to calculate the concentration that corresponds to protecting any selected percentage of species (conversely, the concentration that will permit a certain percentage of species to experience adverse effects). An example SSD for a hypothetical toxicant to marine species is presented in ◘ Figure 3.6.

The usual percentages of species selected to be protected for toxicant limits are 99 and 95%, and the usual limits of species that are permitted to be harmed are 1 and 5%. The concentrations that correspond to these levels of protection are termed the protective concentrations for 99 and 95% of species (i.e. PC99 and PC95, respectively) or the harmful concentrations for 1 and 5% of species (i.e. HC1 and HC5, respectively). While these are the most commonly used levels of protection, it is possible to calculate the concentration that corresponds to any percentage of species desired to be protected or harmed. Examples of the protocols for deriving limits are those used in Canada (CCME 2007) and Australia and New Zealand (Warne et al. 2018), and the software packages used to generate SSDs include ssdtools (Thorley and Schwarz (2018) and Burrlioz (CSIRO 2016). A critical assessment of the strengths and weaknesses of the SSD approach and the validity of its assumptions is presented in Warne (1998).

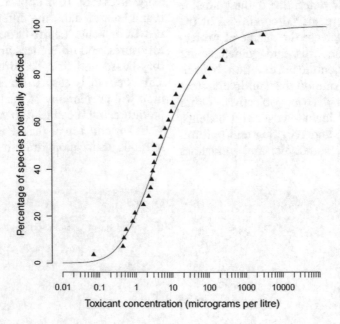

◘ **Figure 3.6** A cumulative frequency plot of the sensitivity of marine species (a species sensitivity distribution [SSD]) to a hypothetical toxicant. Each black triangle represents the concentration of the toxicant at which toxic effects commence for a species. *Image*: M St. J. Warne, output generated by Burrlioz V2

3.5 Limitations of Species Toxicity Studies

There are many limitations with toxicity studies and these are principally related to the fact that they are conducted in laboratories. Some of the limitations are as follows:

- The experimental conditions are highly standardised and controlled to minimise variation (i.e. not like the real world). For example, toxicity tests try to maintain the concentration of the test chemical for the duration of the test. This makes calculations of the toxicity simpler, but organisms in the environment are exposed to concentrations that change over time.
- The experimental conditions are usually optimal for the species, whereas that is often not the case in the environment and could lead to an underestimation of toxicity.
- The species used in toxicity tests have often been chosen because of their ease of being cultured in the laboratory or other pragmatic considerations. Such organisms may not be the most sensitive to chemicals. Rare and endangered species are seldom used in toxicity tests—yet these might be organisms that are important to protect.
- Toxicity test methods have only been developed for some species and are therefore biassed with many important organism types not being included, or there is a marked bias in the proportion of organism types with toxicity data.
- The duration of short-term (acute) tests is based on pragmatic considerations rather than biological reasons. For example, many acute tests are of 96 h duration to permit a toxicity test to be established and completed in a working week.
- Most toxicity tests only expose individuals of a single species and can therefore only measure the direct effects of chemicals on the test organism. Whereas, in the real-world, multiple species will be simultaneously exposed and both direct and indirect effects of chemicals on the test organisms can occur.
- Most toxicity tests only expose the test organism to a single chemical, whereas in reality, organisms are usually exposed to mixtures of chemicals (e.g. Warne et al. 2020). This can lead to underestimation or overestimation of the harmful effects caused by chemicals.

3.6 Assessing Responses from Organisms at the Community Level

So far in this chapter, we have focussed on single-species ecotoxicological assays, with these generally based on the exposure to one or a small number of toxicants under stable laboratory conditions. Ideally, these are designed to provide guidance about the exposure and the concentrations of contaminants at which toxicity commences which can then be used to derive chemical limits and protect marine environments. However, as discussed above (▶ Section 3.5), these approaches are not without limitations, and extrapolating laboratory-derived predictions about the adverse effects of contaminants at specific concentrations is fraught with ambiguity. This is not only because organisms are generally exposed to multiple stressors in the field (see ▶ Chapter 14) but also because environmental protection focuses on communities rather than individual species.

Even without pollutants, marine communities are complex and dynamic systems. Species migrate and immigrate, and interact with each other, sometimes favourably (e.g. symbiotic or mutualistic relationships), other times, less favourably (e.g. predation, parasitism, disease, and competitive displacement). Overlaying these complex interactions are a myriad of abiotic conditions (e.g. seasonality, substrate differences, temperature, depth, salinity, etc.). In many cases, these variables can also be stressors, albeit natural stressors. For example, while many organisms are accustomed to residing in relatively stable marine waters, living in estuaries is far more challenging, with marked tidal changes in salinity, temperature, pH, etc., and only a relatively small number of species are physiologically equipped to deal with such conditions. **Marine communities change over space and time**, and the challenge for scientists is being able to distinguish the effects of any contaminants over the natural variation. This will assist in determining whether the ecological impacts of the contaminants are significant and, therefore, whether the contamination is deemed pollution. Ideally, the assessment would identify which contaminant(s) are driving any observed changes. Here, we will discuss three approaches for examining the effects of contaminants on marine communities: in situ (field) surveys; experimental in situ studies; and community-level laboratory studies.

3.6.1 In situ Studies

Logically, the most common approach for examining the potential effects of contaminants on marine communities is by **in situ surveys**, also known as **field studies**. Given that marine ecosystems can encompass many different types of environments (e.g. seawalls, pelagic, coral, soft-substrate, intertidal, abyssal, etc.), each with its own range of communities, it is imperative to first establish which communities or assemblages (group of taxonomically related species, e.g. fish) should be targeted. This decision should be driven by a number of factors, including the following:

- whether a particular community or assemblage has a high conservation, socio-economic, or other values (e.g. key diet species of local people);
- the type of contaminant and its primary exposure pathway;
- accessibility, time, and cost to collect and process samples;
- relevant taxonomic and ecological expertise; and
- whether sufficient and representative community-level samples can be obtained.

Hence, it is essential that the targeted community is of ecological and ecotoxicological relevance to the potential pollutant(s).

While the types of marine communities captured in ecotoxicological studies are highly varied and can include rocky tidal platform communities, fish, and even the microbiome associated with particular host species (e.g. sponges [Glasl et al. 2017]), the most common approach is to examine the **macrobenthic communities** associated with soft-bottom sediments. This includes taxa such as polychaetes, amphipods, bivalves, and gastropods. This is because, firstly, sediments often contain far greater concentrations of contaminants than the water column and, thus, can pose a significant risk to the whole ecosystem. Secondly, macrobenthos interact with the sediment via consumption or residing within it, and therefore may experience multiple exposure pathways. Thirdly, macrobenthos are numerous and diverse, and therefore not only capture a wide variety of life strategies and sensitivities, but also are generally in numbers sufficient for robust statistical analysis. Fourthly, because of their size and historic use, they tend to be relatively easy to identify at the family level of taxonomic rank and higher. Fifthly, macrobenthic invertebrates are also generally relatively sessile, and therefore, their composition reflects the condition of the environment they were sampled in. Finally, and importantly, benthic communities are a crucial part of near-shore food webs, and consequently, changes in their composition may have cascading effects on other components of the system (e.g. fish) (Antrill and Depledge 1997; Fleeger et al. 2003).

While macrobenthic communities are typically the focus of in situ field studies, it is important to reiterate that the choice of targeted community in any in situ study is dependent on several factors and is by no means limited to macrobenthic invertebrates. In fact, there is an increasing trend to use DNA-based approaches such as metabarcoding to capture a far wider range of taxa than can be obtained using traditional means (Chariton et al. 2010b; Cordier et al. 2020; DiBattista et al. 2020).

Broadly speaking, two different approaches are most commonly used in in situ community surveys: reference/condition and gradient (Quinn and Keough 2002; Chariton et al. 2016). The first approach is a comparison of reference and condition sites. For example, the composition of the targeted communities from several relatively unmodified reference sites is compared to those from several sites exposed to the contaminant(s) of interest (e.g. sites with elevated copper derived from mine tailings). It is emphasised that **multiple sites** are required for each treatment, with this approach being founded on a **factorial design**, like those where an Analysis of Variance (ANOVA) can be applied. One of the biggest challenges of this approach is finding replicated reference locations, which is becoming increasingly difficult with the increasing loss of natural marine habitats. Careful consideration is required when choosing **reference sites**, as it is possible that other variables (e.g. grain size, seagrass cover, and unmeasured contaminants), and not the stressor of interest, may be driving any potential differences between the reference and impacted sites. This issue was highlighted in a comprehensive survey of North Carolina estuaries (Hyland et al. 2000) where the authors found that benthic assemblages were impaired in approximately one-quarter of sites (27%) even though no significant concentrations of contaminants were observed. This suggests that in these cases the communities were either being modified by natural variables, unmeasured contaminants, or a combination of the two.

In contrast to the factorial design which underpins the reference/condition sites approach, **gradient studies** aim to detect variability along a dominant pollution gradient. For example, sites are sampled at increasing distances from the deposition point for deep-sea tailings. Indeed, the approach can be used to capture multiple gradients, both natural and anthropogenic, with an increasing number of statistical tools becoming available which enable scientists to identify the proportion of variation in the community data which can be explained by the measured contaminants and other environmental variables (Chariton et al. 2010b). Gradient studies can be expensive and time-consuming, requiring sufficient environmental data (e.g. metals, pesticides, and natural stressors) and community data to capture the correlative patterns which underpin the gradient(s). However, if designed and implemented properly, they can provide key insights into how communities may be being shaped by both natural and anthropogenic variables, as well as their interactions. Preliminary or investigative studies that provide a reasonable understanding of the factors that may be at play will help in designing gradient studies.

At first glance, it would be logical to assume that either in situ approaches would result in determining whether a contaminant is causing the observed negative impairments to the marine community. However, this is not the case, as in situ studies are correlative and consequently cannot be used to state causality. For example, even if there was a very strong correlation between sediment copper concentrations and benthic diatom communities, there may be other reasons for the observed trend, such as another unmeasured contaminant or natural variation. What in situ studies can tell us is that there is evidence that the contaminant is causing an adverse effect and is acting as a pollutant. Very much like a legal court case where no single LOE can be used to make a verdict, additional LOEs are required to state with a high level of certainty that the contaminant is causing an effect. This evidence may be obtained from ecotoxicological data such as described earlier in this chapter, bioaccumulation and biomarker studies, or via the use of manipulative experiments.

To reiterate, in situ studies are an important tool for helping to determine whether contaminants may be negatively impacting a marine community and therefore causing pollution. However, they cannot determine causality, and thus their findings must be used within the context that they are a correlative LOE. Additional information on the fundamental designs and analyses associated with in situ studies can be found in Underwood (1994), Quinn and Keough (2002), and Chariton et al. (2016).

3.6.2 Experimental In situ Studies

In order to validate correlative studies and to increase our understanding of how contaminants affect marine communities, the testing of models founded on cause and effect is essential. One way to do this is via in situ experiments. In **marine community ecotoxicological studies,** the two most common approaches are spike and translocation studies. **Spiked studies** are predominately sediment-based experiments that involve dosing a sediment with the contaminant of interest. Non-dosed (control) and dosed sediments are then transferred into containers and placed into the substrate unimpacted site(s) (e.g. Lu and Wu 2006; Birrer et al. 2018). The containers then remain in the sediment for sufficient time for them to be recolonized by the native biota, and comparisons between the compositions of the recolonized communities are used to determine whether the spiked sediment altered the species composition, and if so, at what concentrations effects were observed.

One of the challenges of spiked studies is ensuring that the contaminant remains bound to the sediment and does not alter the physicochemical properties of the sediments, minimising any differences between the controls and spiked sediments, apart from the toxicant of interest. Other additional limitations associated with this approach are that the endpoint is based on recolonized communities, and these may behave differently to established fauna; this approach is also limited to contaminants that bind to sediments, as hydrophilic contaminants will be released into the water column and be no longer present in the sediment.

Importantly, considerable resources are required to perform such studies, especially when you consider the quantity of sediment that may be required to fill, for example, 50×5-L containers. Furthermore, upscaling the approach to capture multiple contaminants at multiple concentrations is challenging. Imagine if you needed five containers of each treatment (replicates). If you just had a control and one contaminant at one concentration, 10 containers would be required. If you had one control and four concentrations (contaminant A, e.g. copper) this would require 25 containers (◘ Figure 3.7a). If you added another stressor (contaminant B, e.g. endosulfan) even at one concentration, you would need 5 control containers, 20 containers of contaminant A, 5 containers of contaminant B by itself, and 20 containers to capture each concentration of contaminant A plus contaminant B. That's 50×5-L containers (◘ Figure 3.7b), and as sediment often weighs around 2 kg/L, that is roughly 500 kg of sediment which needs to be manipulated. As you can see, if you wanted multiple concentrations of contaminant B, the experiment would get rather big and complex, not only requiring a lot of sediment to be spiked, but also an extraordinary large amount of effort to install, recover, and process the samples. Consequently, spiked studies are generally restricted to a limited number of treatments.

Translocation studies are similar to spiked studies in that they involve placing containers of sediments into a reference site; however, in this case, the sediments are sourced from the locations sampled in the field study. The aim is not specifically to identify if a specific contaminant is causing an effect, but rather to test whether there is something about the sediments per se which is causing the effect. That is, translocation studies are designed to remove the effect of location by translocating all the sediments to a single location, enabling a direct comparison between sediments obtained from multiple locations. One of the challenges of this process is keeping the sediments intact, including their contaminants, while simultaneously removing the biota. This can be done either by freezing or anoxia (Chariton et al. 2011; O'Brien and Keough 2013) and is essential, given that recolonisation is the endpoint and all sediments must start with the same de-faunated state (i.e. no organisms).

While by no means routine, both spike and translocation studies can provide an additional line of community-level information to complement in situ field studies and laboratory-based toxicity tests. Spiking experiments aim to provide experimental evidence of whether a specific contaminant has the capacity to alter community composition, as well as some insight about at what concentration this may occur. Translo-

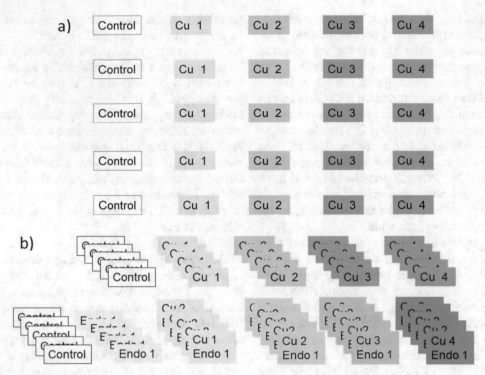

Now consider multiple concentrations of endosulfan and combining each of these with multiple copper concentrations

■ **Figure 3.7**   Representative experimental designs for in situ spiked sediment tests **a** for a single contaminant copper (Cu) and **b** for two contaminants, copper at four concentrations and one concentration of endosulfan (Endo). *Image*: A. Reichelt-Brushett

cation studies, on the other hand, remove the potential confounding influence of location. Both approaches are resource-intensive and thereby place constraints on the experimental design, often limiting their statistical power (Chariton et al. 2011).

3.6.3 Laboratory Studies

While in situ experiments can provide community-level responses under environmentally relevant conditions, they are not without limitations. Most notably, they are very much sediment focussed and not easily amendable to hydrophilic chemicals, chemicals with short-half-lives (e.g. some herbicides), or for exploring the toxicity of contaminants within the water column. In such cases, it may be more appropriate to expose whole communities to a contaminant of interest under laboratory conditions (e.g. in replicated aquaria). Under such conditions, the physicochemical properties of the water column as well as the concentration of the stressor can be controlled. Furthermore, the overlying waters can be continually renewed, ensuring that metabolic waste such as ammonia is removed and not impairing the health of the exposed communities.

The power of **laboratory community assays** was demonstrated by Gissi et al. (2019) who exposed the coral *Acropora muricata* to a range of dissolved nickel

and copper doses. While the coral itself is a species, each fragment contains its own microbial assemblages. Consequently, in this study, the authors were able to gain experimental laboratory species-level data from the host (*A. muricata*) as well as community-level information by examining the host's external microbiomes. In the case of Gissi et al. (2019), the corals were wild-caught and allowed to acclimate for many weeks prior to exposure to the copper and nickel. While this helps ensure that the microbial communities are similar across all individuals, it does not infer that the microbial communities are the same as those which naturally reside on the corals at the site of collection, with other studies showing that marked differences in community structure can occur when transferring communities from the field to the laboratory (Ho et al. 2013; Chariton et al. 2014).

In a novel study by Ho et al. (2013), the authors examined the effects of the antibacterial agent triclosan on marine meiobenthic and macrobenthic communities. Their approach was to collect whole communities, including their residing sediment, and allow the whole sediment communities to acclimatise within a facility under a continual flow system that also supplied food. Instead of dosing the sediments, the authors placed a layer of the toxicant in a slurry on top of the community, and then 2 weeks later applied a clean sediment on top of this. The authors hypothesised that those animals which were alive would migrate through the clean sediment enabling them to be sampled, thereby providing a different community-level endpoint to the recolonized fauna obtained in in situ spiked and translocation experiments.

As in the case of in situ field experiments, laboratory-based community experiments can be logistically challenging and require significant resources and expertise. As a rule of thumb, replication is generally kept to a minimum, and designs incorporating the interactions between multiple stressors are generally avoided.

The data generated by both in situ field and laboratory-based community experiments can be used in SSDs to derive limits. The data could be used by itself or by combining it with more traditional single-species laboratory-based data (e.g. Leung et al. 2005).

3.7 Summary

Obtaining a comprehensive understanding of the effects of contaminants on marine organisms and communities requires a combination of both laboratory and field studies. Ecotoxicology is one of the LOEs that can be used in risk assessment. Experiments are most often conducted in laboratories under controlled conditions. As with all studies, ecotoxicology requires strict adherence to protocols including replication, quality control, statistical analyses, and interpretation of results. Physicochemical conditions need to be standardised throughout experiments and measured to ensure experimental conditions are suitable for organism survival. While this information is limited by its lack of relevance to changing conditions in the environment, it provides some important criteria for comparative assessment and is used in a multiple LOEs approach to develop guideline values for water and sediment quality. Studies of the effects of contaminants on community structure and function are more often based in situ. Due to the complex and dynamic nature of marine communities, thought must be given to the many variables that will influence toxicity.

3.8 Study Questions and Activities

1. What are the benefits of using sub-lethal endpoints to assess toxicity, as opposed to lethal endpoints?
2. What considerations must be given to chronic toxicity test procedures?
3. Describe what a species sensitivity distribution curve is and how it is used.
4. Explain why range finder experiments are used in ecotoxicology.-

5. Design a laboratory toxicity test to assess the effects of the pesticide imidacloprid on a marine species. Consider the species of interest to you, the experimental design, the duration of the exposure, what endpoint you will use, the concentration range you will use, how and when you will measure the test conditions (including the imidacloprid concentrations), and how you will interpret the results.
6. What are the advantages and disadvantages of in situ experiments?

References

Adams MS, Stauber JL (2008) Marine whole sediment toxicity tests for use in temperate and tropical Australian environments: current status. Australasian J Ecotoxicol 14(2/3):155–167

Ademollo N, Patrolecco L, Rauseo J, Nielsen J, Corsolini S (2018) Bioaccumulation of nonylphenols and bisphenol A in the Greenland shark *Somniosus microcephalus* from the Greenland seawaters. Microchem J 136:106–112

Alexander F, King CK, Reichelt-Brushett AJ, Harrison PL (2017) Fuel oil and dispersant toxicity to the Antarctic sea urchin (*Sterechinus neumayeri*). Environ Toxicol Chem 36(6):1563–1571

Allen GR (2007) Conservation hotspots of biodiversity and endemism for Indo-Pacific coral reef fishes. Aquat Conserv Mar Freshwat Ecosyst 18(5):541–556

Annas GJ (1994) Scientific evidence in the courtroom—the death of the Frye rule. N Engl J Med 330(14):1018–1021

Annas GJ (1999) Burden of proof: judging science and protecting public health in (and out of) the courtroom. Am J Public Health 89(4):490–493

Antrill MJ, Depledge MH (1997) Community and population indicators of ecosystem health: targeting links between levels of biological organisation. Aquat Toxicol 38:183–197

Auffan M, Tella M, Santaella C, Brousset L, Paillès C, Barakat M, Espinasse B, Artells E, Issartel J, Masion A, Rose J (2014) An adaptable mesocosm platform for performing integrated assessments of nanomaterial risk in complex environmental systems. Sci Rep 4:5608

Birrer S, Dafforn K, Simpson SL, Kelaher BP, Potts J, Scanes P, Johnston E (2018) Interactive effects of multiple stressors revealed by sequencing total (DNA) and active (RNA) components of experimental sediment microbial communities. Sci Total Environ 637–638:1383–1394

Brack W, Ait-Aissa S, Burgess RM, Busch W, Creusot N, Di Paolo C, Escher BI, Hewitt LM, Hilscherova K, Hollender J, Hollert H, Jonker W, Kool J, Lamoree M, Muschket M, Neumann S, Rostkowski P, Ruttkies C, Schollee J, Schymanski EL, Schulze T, Seiler TB, Tindall AJ, De Aragão Umbuzeiro G, Vrana B, Krauss M (2016) Effect-directed analysis supporting monitoring of aquatic environments—an in-depth overview. Sci Total Environ 544:1073–1118

Burgess RM, Ho KT, Brack W, Lamoree M (2013) Effects-directed analysis (EDA) and toxicity identification evaluation (TIE): complementary but different approaches for diagnosing causes of environmental toxicity. Environ Toxicol Chem 32(9):1935–1945

Cairns J Jr, Niederlehner BR (1994) Estimating the effects of toxicants on ecosystem services. Environ Health Perspect 102(11):936–939

CCME (Canadian Council of Ministers of the Environment) (2007) A protocol for the derivation of water quality guidelines for the protection of aquatic life 2007. Available at: ► https://www.ccme.

ca/files/Resources/supporting_scientific_documents/protocol_aql_2007e.pdf [Accessed: 20 Oct 2020]

Chapman PM, Long ER (1983) The use of bioassays as part of a comprehensive approach to marine pollution assessment. Mar Pollut Bull 14(3):81–84

Chapman PM, Riddle MJ (2005) Toxic effects of contaminants in polar marine environments. Environ Sci Technol 39(9):200A-207A

Chapman PM, McDonald BG, Lawrence GS (2002) Weight of-evidence issues and frameworks for sediment quality (and other) assessments. Hum Ecol Risk Assess 8(7):1489–1515

Chapman PM, McDonald BG, Kickham PE, McKinnon S (2006) Global geographic differences in marine metals toxicity. Mar Pollut Bull 52(9):1081–1084

Chariton AA, Roach AC, Simpson SL, Batley GE (2010a) Influence of the choice of physical and chemistry variables on interpreting patterns of sediment contaminants and their relationships with estuarine macrobenthic communities. Mar Freshw Res 61:1109–1122

Chariton AA, Court LN, Hartley DM, Colloff MJ, Hardy CM (2010b) Ecological assessment of estuarine sediments by pyrosequencing eukaryotic ribosomal DNA. Front Ecol Environ 8:233–238

Chariton AA, Maher WA, Roach AC (2011) Recolonisation of translocated metal-contaminated sediments by estuarine macrobenthic assemblages. Ecotoxicology 20:706–718

Chariton AA, Ho KT, Proestou D, Bik H, Simpson SL, Portis LM, Cantwell MG, Baguley JG, Burgess RM, Pelletier MM, Perron M, Gunsch C, Matthews RA (2014) A molecular-based approach for examining responses of eukaryotes in microcosms to contaminant-spiked estuarine sediments. Environ Toxicol Chem 33(2):359–369

Chariton A, Baird DJ, Pettigrove V (2016) Ecological assessment. In: Simpson S, Batley G (eds) Handbook for sediment quality assessment, 2nd edn. CSIRO, Clayton South, pp 151–189

Connell D, Lam P, Richardson B, Wu R (1999) Introduction to ecotoxicology. Cornwell, Blackwell Science Ltd., p 170

Cordier T, Alonso-Sáez L, Apothéloz-Perret-Gentil L, Aylagas E, Bohan DA, Bouchez A, Chariton A, Creer S, Frühe L, Keck F, Keeley N, Laroche O, Leese F, Pochon X, Stoeck T, Pawlowski J, Lanzén A (2020) Ecosystems monitoring powered by environmental genomics: a review of current strategies with an implementation roadmap. Mol Ecol 30:2937–2958

CSIRO (Commonwealth Scientific and Industrial Research Organisation) (2016) Burrlioz 2.0 Statistical software package to generate trigger values for local conditions within Australia. CSIRO (► http://www.csiro.au). Available at: ► https://research.csiro.au/software/burrlioz/ [Accessed 27 Sept 2020]

de Almeida Rodrigues P, Ferrari RG, Kato LS, Hauser-Davis RA, Conte-Junior CA (2022) A systematic review on metal dynamics and marine toxicity risk assessment using crustaceans as bioindicators. Biol Trace Elem Res 200:881–903

Dévier MH, Mazellier P, Ait-Aissa S, Budzinski H (2011) New challenges in environmental analytical chemistry: identification of toxic compounds in complex mixtures. C R Chim 14(7–8):766–779

DiBattista JD, Reimer JD, Stat M, Masucci GD, Biondi P, De Brauwer M, Wilkinson SP, Chariton A, Bunce M (2020) Environmental DNA can act as a biodiversity barometer of anthropogenic pressures in coastal ecosystems. Sci Rep 10:8365

Eckbo N, Le Bohec C, Planas-Bielsa V, Warner NA, Schull Q, Herzke D, Zahn S, Haarr A, Gabrielsen GW, Borgå K (2019) Individual variability in contaminants and physiological status in a resident Arctic seabird species. Environ Pollut 249:191–199

Fairbrother A, Muir D, Solomon KR, Ankley GT, Rudd MA, Boxall ABA, Apell JN, Armbrust KL, Blalock BJ, Bowman SR, Campbell LM, Cobb GP, Connors KA, Dreier DA, Evans MS,

Henry CJ, Hoke RA, Houde M, Klaine SJ, Klaper RD, Kullik SA, Lanno RP, Meyer C, Ottinger MA, Oziolor E, Petersen EJ, Poynton HC, Rice PJ, Rodriguez-Fuentes G, Samel A, Shaw JR, Steevens JA, Verslycke TA, Vidal-Dorsch DE, Weir SM, Wilson P, Brooks BW (2019) Towards sustainable environmental quality: priority research questions for North America. Environ Toxicol Chem 38(8):1606–1624

Fleeger JW, Carman KR, Nisbel RM (2003) Indirect effects of contaminants in aquatic ecosystems. Sci Total Environ 317:207–233

Furley TH, Brodeur J, Silva de Assis HC, Carriquiriborde P, Chagas KR, Corrales J, Denadai M, Fuchs J, Mascarenhas R, Miglioranza KS, Miguez Carames DM (2018) Toward sustainable environmental quality: identifying priority research questions for Latin America. Integr Environ Assess Manag 14(3):344–357

Gaw S, Harford A, Pettigrove V, Sevicke-Jones G, Manning T, Ataria J, Dafforn KA, Leusch F, Moggridge B, Cameron M, Chapman J (2019) Towards sustainable environmental quality: priority research questions for the Australasian region of Oceania. Integr Environ Assess Manag 15(6):917–935

Gissi F, Adams MS, King CK, Jolley DF (2015) A robust bioassay to assess the toxicity of metals to the Antarctic marine microalga *Phaeocystis antarctica*. Environ Toxicol Chem 34(7):1578–1587

Gissi F, Reichelt-Brushett AJ, Chariton AA, Stauber JL, Greenfield P, Humphrey C, Salmon M, Stephenson SA, Cresswell T, Jolley DF (2019) The effect of dissolved nickel and copper on the adult coral *Acropora muricata* and its microbiome. Environ Pollut 250:792–806

Glasl B, Webster NS, Bourne DG (2017) Microbial indicators as a diagnostic tool for assessing water quality and climate stress in coral reef ecosystems. Mar Biol 164:91

Hall L, Anderson D (2008) The influence of salinity on the toxicity of various classes of chemicals to aquatic biota. Crit Rev Toxicol 25(4):281–346

Han B-C, Jeng W-L, Hung T-C, Wen M-Y (1996) Relationship between copper speciation in sediments and bioaccumulation by marine bivalves of Taiwan. Environ Pollut 91:35–39

Harayashiki CAY, Reichelt-Brushett A, Benkendorff K (2019) Behavioural and brain biomarker responses in yellowfin bream (*Acanthopagrus australis*) after inorganic mercury ingestion. Mar Environ Res 144:62–71

Ho KT, Chariton AA, Portis LM, Proestou D, Cantwell MG, Baguley JG, Burgess RM, Simpson S, Pelletier MC, Perron MM, Gunsch CK, Bik HM, Katz D, Kamikawa A (2013) Use of a novel sediment exposure to determine the effects of triclosan on estuarine benthic communities. Environ Toxicol Chem 32:384–392

Howe P, Reichelt-Brushett AJ, Clark MW (2014) Effects of Cd Co, Cu, Ni, and Zn on the asexual reproduction and early development of the tropical sea anemone *Aiptasia pulchella*. Ecotoxicology 23:1593–1606

Hyland JL, Balthis WL, Hackney CT, Posey M (2000) Sediment quality of North Carolina estuaries: an integrative assessment of sediment contamination, toxicity, and condition of benthic fauna. J Aquat Ecosyst Stress Recover 8:107–124

Isla E, Pérez-Albaladejo E, Porte C (2018) Toxic anthropogenic signature in Antarctic continental shelf and deep sea sediments. Sci Rep 8(1):9154

Kefford B, King CK, Wasley J, Riddle MJ, Nugegoda D (2019) Sensitivity of a large and representative sample of Antarctic marine invertebrates to metals. Environ Toxicol Chem 38(7):1560–1568

King CK, Gale SA, Stauber JL (2006) Acute toxicity and bioaccumulation of aqueous and sediment-bound metals in the estuarine amphipod *Melita plumulosa*. Environ Toxicol 21(5):489–504

Koppel DJ, Gissi F, Adams MS, King CK, Jolley DF (2017) Chronic toxicity of five metals to the polar marine microalga *Cryothecomonas armigera* –application of a new bioassay. Environ Pollut 228:211–221

Kwok KWH, Leung KMY, Chu VKH, Lam PKS, Morritt D, Maltby L, Brock TCM, van den Brink PJ, Warne MStJ, Crane M (2007) Comparison of tropical and temperate freshwater species sensitivities to chemicals: implications for deriving safe extrapolation factors. Integr Environ Assess Manag 3(1):49–67

Lacher TE Jr, Goldstein MI (1997) Tropical ecotoxicology: status and needs. Environ Toxicol Chem: Int J 16(1):100–111

Leung KMY, Gray JS, Li WK, Lui GCS, Wang Y, Lam PKS (2005) Deriving sediment quality guidelines from field-based species sensitivity distributions. Environ Sci Technol 39:5148–5156

Lu L, Wu RSS (2006) A field experimental study on recolonization and succession of macrobenthic infauna in defaunated sediment contaminated with petroleum hydrocarbons. Estuar Coast Shelf Sci 68:627–634

Luoma SN (1996) The developing framework of marine ecotoxicology: pollutants as a variable in marine ecosystems. J Exp Mar Biol Ecol 200:29–55

McConchie DM, Lawrence IM (1991) The origin of high cadmium loads to some bivalve molluscs from Shark Bay, Western Australia: a new mechanism for cadmium uptake by filter feeding organisms. Arch Environ Contam Toxicol 21:1–8

Mercurio P, Mueller JF, Eagleshan G, Negri AP (2015) Herbicide persistence in seawater simulation experiments. PLoS ONE 10(8):e0136391

Mishra AK, Singh J, Mishra PP (2021) Microplastics in polar regions: an early warning to the world's pristine ecosystem. Sci Total Environ 784:147149

Newman MC (2010) Fundamentals of ecotoxicology. CRC Press, Boca Raton, pp 247–272

NHMRC (National Health and Medical Research Council) (2013) Australian Code for the Care and Use of Animals for Scientific Purposes, 8th Edition. Canberra, National Health and Medical Research Council, p 99

Nordborg FM, Brinkman DL, Ricardo GF, Augustí S, Negri AP (2021) Comparative sensitivity of the early life stages of a coral to heavy fuel oil and UV radiation. Sci Total Environ 781:146676

O'Brien AL, Keough MJ (2013) Detecting benthic community responses to pollution in estuaries: a field mesocosm approach. Environ Pollut 175:45–55

OECD (Organisation for Economic Co-operation and Development) (1992) Report of the OECD workshop on extrapolation of laboratory aquatic toxicity data to the real environment. OECD Environment Monographs 59, Paris, OECD, p 43. Available at: ► https://www.oecd.org/chemicalsafety/testing/34528236.pdf [Accessed 24 February 2022]

Peters EC, Gassman NJ, Firman JC, Richmond RH, Power EA (1997) Ecotoxicology of tropical marine ecosystems. Environ Toxicol Chem 16(1):12–40

Phillips DJH (1977) The use of biological indicator organisms to monitor trace metal pollution in marine and estuarine environments -a review. Environ Pollut 13:281–317

Quinn GP, Keough MJ (2002) Experimental design and data analysis for biologists. Cambridge, Cambridge University Press, p 537

Reichelt-Brushett A, Harrison P (2004) Development of a sub-lethal test to determine the effects of copper and lead on scleractinian coral larvae. Arch Environ Contam Toxicol 47(1):40–55

Reichelt-Brushett A, Hudspith M (2016) The effects of metals of emerging concern on the fertilization success of gametes of the tropical scleractinian coral *Platygyra daedalea*. Chemosphere 150:398–406

Reichelt-Brushett AJ, McOrist G (2003) Trace metals in the living and nonliving components of scleractinian corals. Mar Pollut Bull 46(12):1573–1582

Rodriguez-Ruiz A, Asensio V, Zaldibar B, Soto M, Marigómez I (2014) Toxicity assessment through multiple endpoint bioassays in soils posing environmental risk according to regulatory screening values. Environ Sci Pollut Res 21(16):9689–9708

Russell WMS, Burch RL (1959) The principles of Humane experimental technique. London, Methuen, p 238

Saili KS, Cardwell AS, Stubblefield WA (2021) Chronic toxicity of cobalt to marine organisms: application of a species sensitivity distribution approach to develop international water quality standards. Environ Toxicol Chem 40(5):1405–1418

Sánchez-Bayo F, van den Brink P, Mann R (eds) (2011) Ecological impacts of toxic chemicals. Bentham Books, p 288.

Stephan CE, Mount DI, Hansen DJ, Gentile JH, Chapman GA, Brungs WA (1985) Guidelines for deriving numerical national water quality criteria for the protection of aquatic organisms and their uses. US EPA Report No. PB-85–227049. Washington DC, US EPA, P 54. Available at: ► https://www.epa.gov/sites/default/files/2016-02/documents/guidelines-water-quality-criteria.pdf [Accessed 24 Feb 2022]

Suter II GW (2016) Ecological risk assessment. Boca Raton, CRC Press, p 680

Thorley J, Schwarz C (2018) ssdtools: species sensitivity distributions. R package version 0.0.1. Available at: ► https://github.com/bc-gov/ssdtools [Accessed 10 Oct 2021]

Underwood AJ (1994) Things environmental scientists (and statisticians) need to know to receive (and give) better statistical advice. In: Fletcher D, Manly B (eds) Statistics in ecology and environmental monitoring. University of Otago Press, Dunedin, pp 33–61

van Dam RA, Harford AJ, Houston MA, Hogan AC, Negri AP (2008) Tropical marine toxicity testing in Australia: a review and recommendations. Australasian J Ecotoxicol 14(2/3):55–88

van Dam JW, Trenfield MA, Harries SJ, Streten C, Harford AJ, Parry D, van Dam RA (2016) A novel bioassay using the barnacle *Amphibalanus amphitrite* to evaluate chronic effects of aluminium, gallium and molybdenum in tropical marine receiving environments. Mar Pollut Bull 112(1–2):427–435

van den Brink PJ, Boxall AB, Maltby L, Brooks BW, Rudd MA, Backhaus T, Spurgeon D, Verougstraete V, Ajao C, Ankley GT, Apitz SE (2018) Toward sustainable environmental quality: priority research questions for Europe. Environ Toxicol Chem 37(9):2281–2295

Van Dorst J, Wilkinson D, King CK, Spedding T, Hince G, Zhang E, Crane S, Ferrari B (2020) Applying microbial indicators of hydrocarbon toxicity to contaminated sites undergoing bioremediation on subantarctic Macquarie Island. Environ Pollut 259:113780

Wang Z, Kwok KWH, Lui GCS, Zhou GJ, Lee JS, Lam MHW, Leung KMY (2014) The difference between temperate and tropical saltwater species' acute sensitivity to chemicals is relatively small. Chemosphere 105:31–43

Warne MStJ (1998) Critical review of methods to derive water quality guidelines for toxicants and a proposal for a new framework. Supervising Scientist Report 135. Supervising Scientist, Canberra, Australia. ISBN 0 642 24338 7, p 92. Available at: ► https://www.environment.gov.au/system/files/resources/aed1c9d2-7115-44e1-a9e5-495e7da24eb5/files/ssr135.pdf [Accessed 15 Oct 2021]

Warne MStJ, Batley GE, van Dam RA, Chapman JC, Fox DR, Hickey CW, Stauber JL (2018) Revised method for deriving Australian and New Zealand water quality guideline values for toxicants—update of 2015 version. Prepared for the revision of the Australian and New Zealand Guidelines for Fresh and Marine Water Quality. Canberra: Australian and New Zealand Governments and Australian state and territory governments, p 48. Available at: ► https://www.waterquality.gov.au/sites/default/files/documents/warne-wqg-derivation2018.pdf [Accessed 24 Oct 2021]

Warne MStJ, Smith RA, Turner RDR (2020) Analysis of mixtures of pesticides discharged to the Great Barrier Reef, Australia. Environ Pollut 265 Part A:114088

Zhu B, Liu L, Li DL, Ling F, Wang GX (2014) Developmental toxicity in rare minnow (*Gobiocypris rarus*) embryos exposed to Cu, Zn and Cd. Ecotoxicol Environ Saf 104:269–277

Nutrients and Eutrophication

Michelle Devlin and Jon Brodie

Contents

4.1   **Introduction – 76**

4.2   **Nutrification and Eutrophication in Marine Waters – 77**
4.2.1   Definitions – 77
4.2.2   Nutrient Types – 78
4.2.3   Nutrient Limitation and Nutrient Ratios – 78
4.2.4   Sources and Causes – 79
4.2.5   Temperate Versus Tropical Waters – 82
4.2.6   Effects Related to Eutrophication – 82
4.2.7   Tropical Ecosystem Effects   84

4.3   **Case Studies – 85**
4.3.1   Baltic Sea – 85
4.3.2   Chesapeake Bay, USA – 87
4.3.3   Yellow Sea and Qingdao – 87
4.3.4   Caribbean Wide Algal Blooms and West Africa – 88
4.3.5   Brittany – 89
4.3.6   Tampa Bay, Florida, USA – 91
4.3.7   Kāne'ohe Bay, Oahu, Hawaii, USA – 91
4.3.8   Pago Pago Harbour, American Samoa – 92

4.4   **Time Lags and Non-linear Responses – 92**

4.5   **Management, Future Prospects and Conclusions – 92**

4.6   **Summary – 94**

4.7   **Study Questions and Activities – 95**

        References – 95

© The Author(s) 2023
A. Reichelt-Brushett (ed.), *Marine Pollution—Monitoring, Management and Mitigation*,
Springer Textbooks in Earth Sciences, Geography and Environment,
https://doi.org/10.1007/978-3-031-10127-4_4

Acronyms and Abbreviations

CoTS	Crown of Thorns Starfish
Chl-*a*	Chlorophyll-*a*
DIN	Dissolved inorganic nitrogen
DON	Dissolved organic nitrogen
DOP	Dissolved organic phosphorus
GBR	Great Barrier Reef
HABs	Harmful algal blooms
PIN	Particulate inorganic nitrogen
PON	Particulate organic nitrogen
POP	Particulate organic phosphorus
SAV	Submerged aquatic vegetation
STP	Sewage treatment plant
USA	United States of America
TM4-ECPL	Tracer Model 4 of the Environmental Chemical Processes Laboratory

4.1 Introduction

Excess nutrients from fertiliser application, pollution discharge and water regulations outflow through rivers from lands to oceans, seriously impact coastal ecosystems. Terrestrial runoff of waters polluted with nutrients (primarily **nitrogen [N]** and **phosphorus [P]** compounds) from point sources, such as sewage treatment plant (STP) discharges, and diffuse sources via river discharges, such as fertiliser losses, are having devastating adverse effects in coastal and marine ecosystems globally (Carpenter et al. 1998; Halpern et al. 2008; Crain et al. 2008; Smith and Schindler 2009). The nutrients can be dissolved such as dissolved nitrate and phosphate typically discharged from STPs or agricultural runoff or in a particulate form, often associated with soil erosion.

Biomass production of plant matter in coastal waters is often limited by the availability of nitrogen and/or phosphorus (light is a limiting factor in turbid zones). Conversely, the increased human-derived inputs of nutrients can lead to increased biomass production that can disturb the natural ecological balance in marine ecosystems. This disturbance, the process of **eutrophication**, is one of the biggest threats to marine ecosystem health. Eutrophication, like climate change, is a global issue with coastal regions throughout the world being impacted through the input of elevated nutrients (Galloway et al. 2014). Well-documented adverse ecological responses to increased nutrient discharge into coastal and marine waters include **harmful algal blooms (HABs)** (Hudnell 2008; Glibert and Burford 2017), changed preponderance and dominance of certain types of algae over other benthic plants (seagrass, coral, other algae) (Lapointe et al. 2018, 2019), hypoxia and subsequent **dead zones** (Diaz and Rosenberg 2008), habitat degradation and adverse changes in aquatic food webs (Carpenter et al. 1998; Gross and Hagy 2017).

Enrichment of both nitrogen and phosphorus is of concern, though the consensus that has evolved among much of the scientific community is that increased nitrogen is the primary driver of eutrophication in many coastal ecosystems (Howarth and Marino 2006). However, this has been challenged by recent scientific literature which acknowledges the need to reduce both nitrogen and phosphorus to control coastal eutrophication (Howarth and Marino 2006; Howarth and Paerl 2008; Riemann et al. 2016; Asmala et al. 2017). Successful reductions of phosphorus have occurred throughout freshwater systems through the banning of phosphorus in detergents, and a corresponding reduction in phosphorus is being measured in many coastal waters (Paerl 2006). While this is a hopeful trend, this has led to a global N:P imbalance in our coastal and marine ecosystems and an increasing **N: P ratio** which can impact the plankton community structure and phosphorus limitation of natural growth (Howarth and Paerl 2008; Paerl 2009). A comprehensive response needs to focus on consistent reductions in nitrogen to marine systems to alleviate this imbalance and will be the focus of this chapter.

The global **nitrogen cycle** is now greatly perturbed by human (anthropogenic) activity, particularly on land (Gruber and Galloway 2008; Rockström et al. 2009; Fowler et al. 2013). The increasing inputs of nitrogen from human activity, predominantly from land-based activities can modify oceanic, and even global, biogeochemical systems (Jickells et al. 2017). The estimated anthropogenic release of nitrogen into the global environment (160 Tg N/yr, Tg= Teragram = 10^{12} g) is now of similar magnitude to natural nitrogen fixation (250 Tg N/yr⁻) and is likely to increase in the future due to a growing global population (Gruber and Galloway 2008). Four of the primary sources of bioavailable (hence the term **reactive**) nitrogen to estuarine, coastal and marine waters are runoff and discharge from the land, upwelling on the continental shelf break; atmospheric deposition; and fixation by nitrogen-fixing microbes. The sources of the increased nutrient fluxes are associated with:

- fertiliser use and losses in agriculture such as grazing and cropping;
- human sewage discharges;
- farm animal wastes discharge; and
- fluxes to the atmosphere which are discharged to marine waters via rainfall and particulate matter deposition.

4.2 Nutrification and Eutrophication in Marine Waters

4.2.1 Definitions

Waters with low concentrations of nutrients and phytoplankton and hence low productivity are called **oligotrophic**, while those with high nutrient and/or phytoplankton (and benthic algae) concentrations and high productivity are **eutrophic**. Waters with an intermediate level of productivity are termed **mesotrophic**.

Nutrification is the action or process of nutrifying an environment with nutrients (generally nitrogen and/or phosphorus). Nutrification is of concern, however, enrichment alone does not necessarily confer an impact, and assessment of eutrophication typically needs to meet several other criteria before impact and disturbance can be measured (Tett et al. 2007; Ferreira et al. 2011; Brodie et al. 2011). The excessive input of anthropogenic nitrogen and phosphorus needs to cause additional impacts, for example, marine algal blooms above the consumption capacity of herbivores before we can conclude the system is eutrophic (Nixon 2009).

The term **eutrophication** refers to a process of increased production of biomass in an aquatic ecosystem, evolving over long timescales until the system is **full** of biomass (Bricker et al. 2008; Boyd 2020). However, the term is more commonly used now to refer to a process that has been accelerated by anthropogenic actions, resulting in the process occurring in short timeframes of years to decades (Nixon 1995). Eutrophication can be defined in different ways for different systems, but for marine and estuarine waterbodies, it is a process resulting from the input of excessive plant nutrients into an aquatic system. The excess nutrients lead to enhanced plant growth or changes in the composition and structure of communities and, as a consequence, the high plant growth reduces the penetration of light through the water. Light is essential for plant growth with light-limiting conditions resulting in plant death. This can cause ongoing adverse effects as the dead plant material is consumed by aerobic bacteria leading to high demands on the oxygen supply. Reductions in dissolved oxygen impacts all organisms and may result in a crash of the whole system. Criteria that are used to measure the impacts through this process include algal blooms and low-oxygen (hypoxic) waters that can kill fish, reduce essential fish habitats and result in epiphytic algae over-growth and death of marine plants, such as seagrass, through smothering and reducing its capacity to photosynthesise (■ Figure 4.1). Anthropogenic eutrophication thus can be defined as '*the overproduction of aquatic plant biomass/organic material induced by anthropogenic inputs of phosphorus and nitrogen*'.

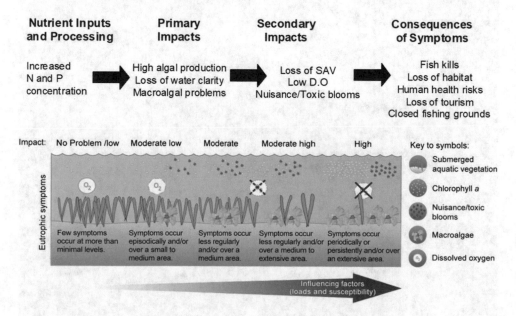

■ **Figure 4.1** The process of eutrophication and the resulting impacts on the marine ecosystems. Adapted from Devlin et al. 2011 by M. Devlin

This definition is used to overcome the difficulty of summarising in a few words the multitude of biogeochemical and biological responses (including direct and indirect effects) triggered by excessive nitrogen and phosphorus inputs (Devlin et al. 2011; Le Moal et al. 2019). Eutrophication can cause structural changes throughout the marine ecosystem and reduce ecosystem resilience (◘ Figure 4.2).

Eutrophication issues have often been divided into three descriptive terms:

- **Causative factors**: Factors which cause eutrophication such as nutrient inputs, elevated nutrient concentrations and imbalance in nutrient concentrations (see ▶ Section 4.2.3, where Redfield ratios are described).
- **Direct effects**: Effects which are caused directly by the increased nutrients such as impacts on primary producers (phytoplankton) and submerged aquatic vegetation.
- **Indirect effects**: Effects that are influenced by the direct effects and are known as secondary effects. These can be related to negative changes in zooplankton, fish and invertebrate benthic fauna (animals living on and in the seabed).

4.2.2 Nutrient Types

Nutrients enter the marine environment in many forms, including dissolved inorganic nitrogen (DIN), which includes ammonium (NH_4), nitrate NO_3^- and nitrite NO_2^-. Other sources of nitrogen include dissolved organic nitrogen (DON), particulate inorganic nitrogen (PIN, essentially ammonium ions attached to clay particles) and particulate organic nitrogen (PON). Phosphorus forms include phosphate PO_4^{3-}, which consist of orthophosphate or polyphosphates, dissolved organic phosphorus (DOP), particulate inorganic phosphorus (PIP, essentially phosphate ions attached to clay particles) and particulate organic phosphorus (POP). Rainfall and river flow will contain both ammonium and nitrate in solution, while dry deposition of dust can contain various forms of particulate nitrogen and phosphorus.

4.2.3 Nutrient Limitation and Nutrient Ratios

The nutrients essential for **primary production**, which are often present in a limiting amount, are nitrogen and phosphorus. The C:N:P (carbon to nitrogen to phosphorus) stoichiometric ratios of living organisms (especially plants, and in the ocean, phytoplankton) are fairly constant and are termed **Redfield ratios** (Geider and La Roche 2002). For both phytoplankton and zooplankton, the ratio of (C:N:P), known as the Redfield ratio is 106:16:1 with some variation between different organisms. In addition, silica (Si) is essential for diatom growth so the ratio of C to N to P to Si may also be

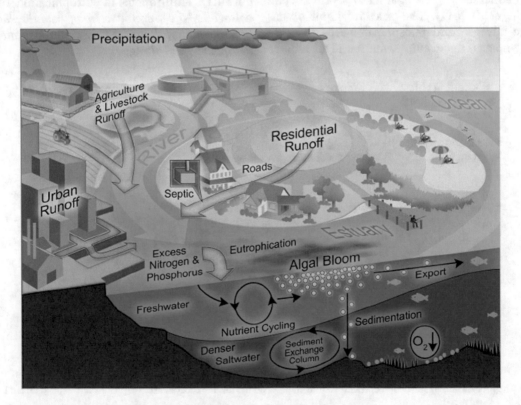

◘ **Figure 4.2** Schematic diagram of the different pathways of nutrient deposition into coastal waters and ensuing processes leading to eutrophication (algal blooms) and hypoxia. *Image*: Hans W. Paerl CC BY 2.0

important. Thus, instead of the traditional Redfield ratio of C:N:P as 106:16:1, a modified Redfield ratio to include silica becomes C:N:P:Si as 106:16:1:15, known as the **Redfield–Brzezinski ratio**, and is often used as a standard to understand nutrient limitation with respect to nitrogen, phosphorus or silicate for natural phytoplankton assemblages. Increased nutrient inputs generally entail a change in the ratio between dissolved nitrogen and phosphorus species in the water (i.e. the DIN:DIP ratio). A significantly lower ratio (than 16:1) can cause nitrogen limitation, whereas a higher ratio can lead to phosphorus limitation for phytoplankton primary production (Tett et al. 1985). Species that are less sensitive for their growth to require optimal DIN:DIP ratios can outcompete more sensitive species.

Nitrogen and phosphorus are the major limiting nutrients in most aquatic ecosystems (Conley et al. 2009). Primary production is frequently limited by nitrogen and phosphorus in freshwaters and by nitrogen in the ocean (Howarth and Marino 2006). The long-standing debate over nitrogen versus phosphorus limitations to ocean primary production had appeared to be settled in favor of nitrogen as a result of the substantial rates of denitrification recently reported in marine environments (Nixon 1995; Howarth and Marino 2006). Nevertheless, phosphorus appears to limit phytoplankton activity in some regions (Wu et al. 2000) and iron (Fe) and phosphorus appear to co-limit the growth of nitrogen-fixing *Trichodesmium* in the Atlantic Ocean (Mills et al. 2004). In addition, changing anthropogenic activities have caused imbalances in nitrogen and phosphorus loading, making it difficult to control eutrophication by reducing only one nutrient (Paerl 2006; Duarte et al. 2008; Howarth and Paerl 2008). The forms of nitrogen and the ratios of nitrogen and phosphorus in river discharge (from both agricultural and human waste sources) are also changing (Glibert 2017). A global increase in fertiliser nitrogen to phosphorus ratio has also occurred during 1961–2013, which may have global implications for the types and extent of marine eutrophication in the longer term (Lu and Tian 2017). For example, with the increasing use of urea as one of the cheapest and most readily available sources of nitrogen, losses of nitrogen are increasing from current applications of fertiliser in agriculture compared to older and less soluble forms of nitrogen fertiliser.

Changes in the ratio of nitrogen to phosphorus also have significant potential effects on phytoplankton and other algal growth and speciation in the marine environment. Upstream nutrient management actions (exclusively phosphorus controls) have exacerbated nitrogen-limited downstream eutrophication which can impact coastal plankton communities. These imbalances can lead to shorter trophic food webs with fewer predators, and potentially decreasing biodiversity and long-term management should consider controls on both

nutrients (Penuelas et al. 2013; Paerl et al. 2014; Burson et al. 2016).

4.2.4 Sources and Causes

Excess nitrogen and/or phosphorus is sourced from many anthropogenic processes including fertiliser run-off, human sewage effluent, animal waste discharge and atmospheric fallout in rain and precipitation. Iron, silica and other micronutrients may also be involved in nutrification, but case studies of adverse effects are less common (however, see silica to nitrogen ratio ▶ Section 4.2.3).

Increasing demands for nitrogenous **fertilisers** for use in agriculture (Lu and Tian 2017) and particularly urea in recent times, is largely responsible for the rapidly increasing discharge of nitrogen to the marine environment (Jickells and Weston 2011a, b) (◘ Figure 4.3). The share of total global anthropogenic nitrogen and use (187 Mt/yr) from agriculture has been estimated at 86% (Galloway et al. 2008). Many studies also reveal low nitrogen use efficiency in crops, with only approximately half of the nitrogen applied to croplands being incorporated into plant biomass, while the rest is lost through leaching (16%), soil erosion (15%) and gaseous emission (14%) (Liu et al. 2011; Liu et al. 2013a, b). Additional sources include nitrogen and phosphorus discharges to coastal seas from **domestic** wastewater **and groundwater** inputs driven by human population growth (Powley et al. 2016) with increased atmospheric deposition and rainfall inputs of phosphorus (Jickells et al. 2017). There is also increased **watershed erosion** (with particulate nitrogen and phosphorus content), especially in the tropics, associated with **deforestation** and **agricultural land development** (Bainbridge et al. 2018). Analysis of changes in the global freshwater nitrogen and phosphorus cycles in rivers and streams over the twentieth century suggests that, during this period, the global river nutrient transport to the ocean increased from 19 to 37 Tg N/yr and from 2 to 4 Tg P/yr (Seitzinger et al. 2005; Bouwman et al. 2009; Beusen et al. 2016).

From the 1940s to the 1980s, eutrophication was reported in the northern Adriatic Sea, the northwest continental shelf of the Black Sea (Mee 1992), the Kattegat betweenDenmark and Sweden (Rosenberg et al. 1996), Chesapeake Bay (Boesch et al. 2001) and many other areas in temperate northern hemisphere waters (Lotze et al. 2011a, b). Recent prominent and large-scale examples of eutrophication include the North China Sea (Qingdao) with massive algal blooms interfering with the aquatic events of the 2008 Beijing Olympic games (▶ Section 4.3.3); in the Caribbean and West Africa (Smetacek and Zingone 2013)

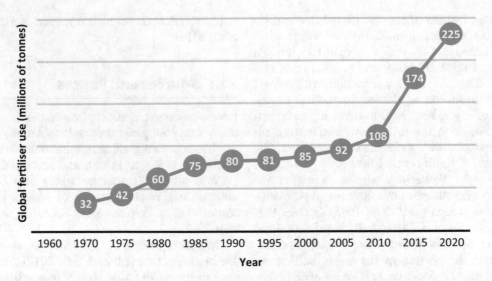

▣ Figure 4.3 World fertiliser consumption from 1970 to 2010 and projections to 2020. Based on available statistics (Fao 2012), and global estimates for 2013 and 2020 based on use projections *Data sources*: Heffer and Prud'homme 2012, 2013, 2014, 2015

(▶ Section 4.3.4); and further eutrophication across the Baltic Sea generally (Andersen et al. 2017). There are now numerous reports of macroalgal blooms with the most common algae involved being species of *Ulva* (green tides) and *Sargassum* (golden tides) worldwide in recent years. These blooms negatively impact tourism, particularly countries that have high economic dependence on tourism. The blooms may smother aquaculture operations (some of which are also a source of nutrients) or disrupt traditional artisanal fisheries (Smetacek and Zingone 2013).

Nutrient pollution is a leading global threat to coastal and marine ecosystems, including saltmarshes, mangroves, kelps, seagrasses and corals (Howarth and Paerl 2008). About half the global riverine nitrogen input (about 40 from the total 80 Tg of N yr^{-1}) is anthropogenic in origin (Beusen et al. 2016) and riverine fluxes of nitrogen have increased greatly (Bouwman et al. 2009; Beusen et al. 2016). Rivers in western Europe and eastern China have seen large increases in nitrogen fluxes (e.g. the Yangtze River had about four times more nitrogen load in 2010 than in 1991, while the amount of fertiliser used doubled, resulting in increased riverine DIN levels). The increased riverine DIN flux between 1991 and 2010 in the United States of America (USA) was affected primarily by nitrogen fertiliser use, while rivers in Europe and China have seen fertiliser use, human waste and atmospheric sources increase. These changes have also occurred in tropical waters with the total anthropogenic DIN exported to the Pacific Ocean increasing from 10 to 30% of the total, a higher rate than any other ocean (Liu et al. 2019).

Eutrophication from increased nutrient input is now recognised as **one of the most serious issues facing estuarine and coastal waters in many parts of the world**, with,

for example, 67% of the combined surface area of estuaries in the USA exhibiting moderate to high degrees of eutrophication (Potter et al. 2016), a trend also found elsewhere in coastal waters across the world (Duarte et al. 2008; Duarte 2009; Rabalais et al. 2009; Paerl et al. 2014). Breitburg et al. (2018) note a worrying trend in declining oxygen in the global ocean and coastal waters associated with watershed pollution as well as climate change.

Globally, nitrogen and phosphorus loadings to coastal and marine waters are expected to at least double by 2050 (Johnson and Harrison 2015; Kroeze and Seitzinger 1998) through the continued increase in the use of fertilisers (Heffer and Prud'homme 2012), increased coastal aquaculture, increased populations and associated sewage waste, animal wastes, further deposition of nitrogen associated with gaseous emissions from fossil fuel burning and other industrial discharges to the atmosphere (Johnson and Harrison 2015). Riverine nitrogen fluxes to the global ocean are estimated to be 23 Tg/N/yr for DIN and 11 Tg/N/yr for DON (Seitzinger et al. 2005, 2010). The total river input of nitrogen to coastal seas has approximately doubled over the last few hundred years (Seitzinger et al. 2005, 2010; Yan et al. 2010; Beusen et al. 2016). This input is also now dominated by nitrate, reflecting the influence of indirect land use inputs through fertiliser usage (Jickells and Weston 2011a, b; Jickells et al. 2017). Nitrogen use is now outside of the bounds of global planetary sustainability (Steffen et al. 2015) and poses a high risk to the Earth's systems (Johnson and Harrison 2015; Lu and Tian 2017).

Fertiliser Use and Losses from Agricultural Land
Rising agricultural demands for nitrogenous fertilisers (and particularly urea) in recent times is responsible

for the rise in reactive nitrogen (Galloway et al. 2008). Nitrogen fertiliser production increased from 15 million tonnes N/yr in 1860 to 187 million tonnes N/yr in 2005 (◘ Figure 4.3). Nitrogen and phosphorus fertiliser usage rates per unit of cropland area increased by approximately eight times and three times, respectively, since the year 1961 (Lu and Tian 2017). This increase in fertiliser nitrogen is compounded by the inefficient use of fertiliser on agricultural lands. More than half of the synthetic fertiliser applied to the world's fields has been applied in the past 30 years (Pearce 2018) but less than half of this fertiliser reach the intended crops, with the remainder running off into rivers and eventually into the ocean. Large increases in atmospheric nitrogen emissions have also occurred over the last 200 years associated with this human activity (◘ Table 4.1).

Fossil Fuel Combustion Emissions and Aerial Deposition
Combustion (especially of fossil fuels) is a major source of oxidised nitrogen which is transformed in the atmosphere to nitric acid and rained out as nitrate (◘ Table 4.1). Direct agricultural emissions from fertiliser use are a major source of ammonia (Duce et al. 2008). Rapid and efficient atmospheric transport allows these emissions to reach the open oceans within days, hence much faster and more effectively than fluvial inputs

(Fowler et al. 2013). Wang et al. (2015) estimate that combustion-related emissions (associated with fossil fuels) are 1.8 Tg/P/yr, which represent over 50% of global atmospheric sources of P. Using these estimates in models, they found that the total global emissions of atmospheric P (3.5 Tg/P/yr) were broken up into a deposited amount of 2.7 Tg/P/yr over land and 0.8 Tg/P/yr over the oceans.

Human Sewage Wstes
Global nitrogen and phosphorus emissions from human **sewage** for the period 1970–2050 have been estimated from the four Millennium Ecosystem Assessment scenarios. An increase in global sewage emissions is predicted, from 6.4 Tg of nitrogen and 1.3 Tg of phosphorus per year in 2000 to 12.0–15.5 Tg of nitrogen and 2.4–3.1 Tg of phosphorus per year in 2050. North America (strong increase), Oceania (moderate increase), Europe (decrease) and North Asia (decrease) show contrasting developments, and in the developing countries, sewage nitrogen and phosphorus discharge will likely increase by a factor of 2.5–3.5 between 2000 and 2050 (Bouwman et al. 2005; Van Drecht et al. 2009; Seitzinger et al. 2010). This is a combined effect of increasing population, urbanisation and development of sewage systems. Despite some optimistic scenarios for the development of wastewater treat-

◘ **Table 4.1** TM4-ECPL model estimated global atmospheric nitrogen emissions by source for 1850, 2005 and 2050, the latter based on the RCP6.0 scenario. Adapted from Jickells et al. 2017

Source	1850	2005	2050
NO_x			
Terrestrial anthropogenic NO_x	0.6	27	20.2
Shipping NO_x		5.3	3.1
Aircraft NO_x		0	0
Biomass burning NO_x	0.5	5.5	5.7
Natural NO_x soils and lightening	11.8	11.6	11.6
NH_x			
Terrestrial anthropogenic NH_x	5.4	32.9	43.7
Biomass burning NH_x	0.9	9.2	9.4
Natural NH_x soils	2.4	2.4	2.4
Natural NH_x ocean emissions	8.2	8.2	8.2
Total inorganic N	29.8	102.1	104.3
Organic N (ON)			
Anthropogenic and biomass burning	1.3	7.0	6.8
Natural biogenic particles and soil dust	9.3	9.3	9.3
ON insoluble on marine aerosol	1.1	1.1	1.1
ON soluble on marine aerosol and marine amines	5.8	5.8	5.8
Total ON	17.5	23.2	23
Total N emissions	47.2	125.2	127.3

ment systems, it is predicted the contributions of waste-water nutrients will contribute to high fluxes of global nitrogen and phosphorus fluxes for many years to come (Van Drecht et al. 2009).

Animal Wastes

The large quantity of manure produced by **intensive animal production** is generally applied to land as fertiliser, stacked in the feedlot, or stored in lagoons. Frequently, an oversupply of manure means that it is applied to crops more than is necessary, further exacerbating nutrient runoff and leaching (see WRI: ▶ https://www.wri.org/our-work/project/eutrophication-and-hypoxia/sources-eutrophication). In China, meat production rose by 127% between 1990 and 2002 (Fao 2012), but fewer than 10% of an estimated 14,000 intensive livestock operations have installed pollution controls (Ellis 2017).

Upwelling

The vertical distribution of nutrients in the sea shows, for both nitrates and phosphates, a surface minimum that sharply increases with depth during the first 100–500 m and is approximately steady in deeper waters. **Upwelling** occurs in the open ocean and along coastlines. Water that rises to the surface as a result of upwelling is typically colder and rich in nutrients (mainly nitrate and phosphates). These nutrients **fertilise** surface waters, meaning that these surface waters often have high biological productivity. Therefore, good fishing grounds typically are found where upwelling is common (see NOAA: ▶ https://oceanservice.noaa.gov/facts/upwelling.html). Upwelling of nutrients into shallow habitats is unlikely to have been increased by anthropogenic effects, although changed current regimes associated with climate change may affect this process in the future (Bakun et al. 2015).

4.2.5 Temperate Versus Tropical Waters

Differing Discharge Processes in the Tropics

Nitrogen pollution in aquatic systems is shaped by multiple sources and processes. Modelling of nitrogen budgets of basin–marine systems provides estimates that globally, land currently sequesters 11 (10–13)% of annual nitrogen input (Lee et al. 2019). **River basins can act as a buffer**, taking up greater than 50% of their nitrogen inputs, which can provide some protection to the coastal systems. However, activities such as deforestation, agricultural intensification and/or exports of land nitrogen storage in tropical systems can create large nitrogen pollution sources including erosion of nitrogen-rich soils. Particulate nitrogen (and phosphorus) discharges as a result of erosion are a major issue for the tropics and can contribute to the largest fraction of nitrogen and phosphorus river discharges (e.g. for the Great Barrier Reef (GBR) (Waterhouse et al. 2012)). The tropics produce $56 \pm 6\%$ of global land nitrogen pollution despite covering only 34% of global land area and receiving far lower amounts of fertilisers than the areas outside of the tropics. Tropical land use needs to be considered as a major mechanism in managing global nitrogen pollution (Lee et al. 2019).

Phytoplankton Speciation Differences

Phytoplankton species in the nutrient-depleted tropical waters are typically dominated by picocyanobacteria (often species of *Synechococcus* and/or *Prochlorococcus*) of very small cell size, while temperate waters have higher ratios of diatoms and dinoflagellates (influenced partially by temperature) (Odebrecht et al. 2018; Righetti et al. 2019). Polar seas can also be dominated by picocyanobacteria. In general, in tropical seas when large injections of nutrients occur from river discharge, sewage discharge or upwelling, phytoplankton speciation shifts from picocyanobacteria dominance to dominance by diatoms and dinoflagellates (Jacquet et al. 2006).

4.2.6 Effects Related to Eutrophication

Hypoxia, Dead Zones, Climate Change and Loss of Oceanic Oxygen

Human inputs of nutrients to coastal waters can lead to the excessive production of algae and an excess of **organic matter**, as part of the eutrophication process (see ▶ Section 4.2.1). Microbial consumption of this organic matter lowers oxygen levels in the water (Gilbert et al. 2010; Cai et al. 2011). The decomposing plant biomass causes an **oxygen deficit** and can produce toxic compounds such as hydrogen sulfide (H_2S) and ammonia (NH_3) in the anoxic sediments.

Oxygen concentrations in open ocean and coastal waters have been declining since at least the middle of the twentieth century. This change which is associated with the eutrophication process can be exacerbated by the increasing temperatures associated with increased CO_2 levels in the oceans and atmosphere. These changes are affecting the abundances and distributions of many marine species. **Low-oxygen zones, or dead zones**, in the ocean have expanded by several million square kilometres and hundreds of coastal sites now have oxygen concentrations low enough to limit the distribution and abundance of animal populations (Rabalais et al. 2009, 2014; Gilbert et al. 2010; Breitburg et al. 2018). There have been greater declines in marine oxygen levels in coastal seas compared to the open ocean (Gilbert et al. 2010); oxygen decline rates are more severe in a 30 km band near the coast than in the open ocean (>100 km from the coast) because of the influence of increased nutrient fluxes from rivers. In the 1990s, scientists reported coastal Hypoxia in northern Europe, North America and Japan. By the 2000s, there were more such reports in South America, southern Europe and Australia, as well as increasing dead zones in the Baltic Sea (Gilbert et al. 2010; Rabalais et al. 2014). Low-oxygen zones are now known as dead zones due to the detrimental impacts of low dissolved oxygen on benthic fauna which can culminate in mass mortality events (Diaz and Rosenberg 2008).

Coastal Hypoxia and the associated dead zones have been exacerbated by worldwide enhanced coastal primary production and eutrophication driven by increased riverine inputs of nitrogen and phosphorus, soil erosion of particulate nitrogen and phosphorus and the burning of fossil fuelsfossil fuels. These processes lead to an accumulation of particulate organic matter, which encourages microbial activity and the consumption of dissolved oxygen in bottom waters. Degradation of coastal water quality in the form of low dissolved oxygen levels (Hypoxia and anoxia) can harm biodiversity, ecosystem function and human well-being. Extreme hypoxic conditions along the coast, leading to dead zones, are known primarily in temperate and sub-tropical regions. Dead zones have now been reported from more than 400 ecosystems, affecting a total area of more than 245,000 km^2 (Diaz and Rosenberg 2008) with consequent impacts on marine ecosystems (Ekau et al. 2010; Altieri et al. 2017). However, less is known about the potential threat of Hypoxia in the tropics, even though

the known risk factors, including eutrophication and elevated temperatures, are common. Altieri et al. (2017) documented an unprecedented hypoxic event on the Caribbean coast of Panama and assessed the risk of dead zones to coral reefs worldwide. The event near Panama caused coral bleaching and massive mortality of corals and other reef-ahypoxiassociated organisms but observed shifts in community structure combined with laboratory experiments revealed that not all coral species are equally sensitive to . Analyses of global databases showed that coral reefs are associated with more than half of the known tropical dead zones worldwide, with >10% of all coral reefs at elevated risk of hypoxia based on local and global risk factors. Hypoxic events in the tropics and associated mortality events have likely been underreported, perhaps by an order of magnitude, because of the lack of local scientific capacity for their detection (Altieri et al. 2017).

Algal Proliferation and Subsequent Changes in Marine Plant communities

Algal blooms are a natural phenomenon, but their frequency, duration and geographical scope have been increasing since the 1950s, largely in response to fertiliser runoff and sewage discharge, and human-induced climate change. Increased competition from algal blooms can impact saltmarshes, mangroves, kelps, seagrasses and corals (Lefcheck et al. 2018) (■ Figure 4.4). For instance, partly as a result of increased nutrient inputs, the global cover of seagrasses has declined by over 29% in the last century (Waycott et al. 2009)

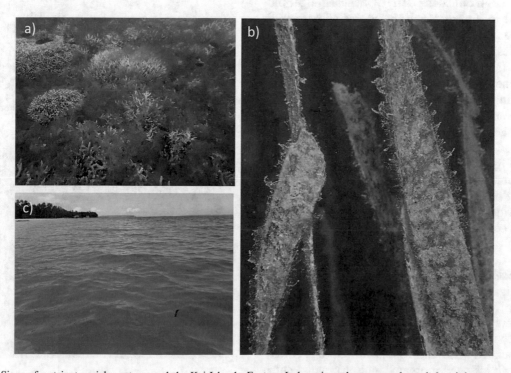

■ **Figure 4.4** Signs of nutrient enrichment around the Kei Islands, Eastern Indonesia **a** algae-covered corals **b** epiphyte growth on seagrass fonds inhibiting sunlight and photosynthesis of seagrass **c** green microalgae bloom visible in surface waters. *Photos*: A. Reichelt-Brushett

◘ **Figure 4.5**   Green algae blooms deposited on the beach at Byron Bay, NSW, Australia, inhibiting recreational activities. *Photo*: A. Reichelt-Brushett

(◘ Figure 4.4b). Localised issues of water quality, particularly sedimentation, can have negative impacts on seagrass cover (Petus et al. 2014; Brodie et al. 2020).

Green algal blooms (green tides), are formed by rapid growth and accumulation of unattached green macroalgae and are associated with nutrient-enhanced marine environments (◘ Figure 4.5). Over the last 50 years, green tides have been increasing in severity, frequency and geographic range, resulting in these events becoming a growing concern worldwide (Ye et al. 2011). High concentrations of beached algal biomass started to appear along the shores of industrialised countries through the 1970s. These became known as green tides and, over the next few decades, became a common sight along many beaches with increases in both frequency and magnitude of the green tides during the spring–summer growing season. Green algae blooming events harm shore-based activities and tourism as the sheer physical mass can cover the shoreline and the dense, drifting seaweeds prevent accessibility to the sea (◘ Figure 4.5). Over the growing season, if not manually removed, the algae can turn into a stinking morass, producing toxic hydrogen sulfide (H_2S) from its anoxic interior, and have major detrimental effects on the affected coastal ecosystems (Smetacek and Zingone 2013).

HABs and Red Tides

Harmful Algal Blooms (HABs), with the term often restricted to blooms of toxic algae, are increasing in coastal waters worldwide (Glibert and Bouwman 2012; Glibert 2017; Glibert and Burford 2017). These blooms can be associated with anthropogenic nutrient enrichment, through elevated inorganic and/or organic nutrient concentrations and modified nutrient ratios. Since

1950, their extent in coastal waters has increased (Anderson et al. 2002; Heisler et al. 2008) and the risks to coastal seas increased dramatically as was witnessed in the Gulf seas with mass mortality of coral reefs and fisheries associated with the proliferation of HABs and reduction of light climate (Richlen et al. 2010).

4.2.7   Tropical Ecosystem Effects

Crown of Thorns Starfish (CoTS)

CoTS are one of the major causes of coral mortality in the Great Barrier Reef (GBR) and generally on Indo-Pacific reefs (◘ Figure 4.6) (De'ath et al. 2012; Pratchett et al. 2017). **River nutrients can influence CoTS outbreak dynamics** (Schaffelke et al. 2017) as wet season nutrient inputs from the central GBR rivers, typically discharge when phytoplankton-feeding CoTS larvae are present in the water column (November to March) (Devlin et al. 2012, 2013). The increase in nutrients provides food for the phytoplankton blooms which allows a greater number of CoTS larvae to survive to a stage where they are able to settle out on a coral reef (Brodie et al. 2005; Fabricius et al. 2010; Brodie et al. 2017).

Waves of outbreaks are initiated when these phytoplankton food resource conditions are reinforced by favourable hydrodynamic conditions (Wooldridge and Brodie 2015) and sufficient coral cover to sustain

◘ **Figure 4.6**   Image of Crown of Thorns (CoTS) in process of consuming coral. *Photo*: A. Reichelt-Brushett

the outbreaks (Fabricius et al. 2010). Studies highlight that the number of outbreaks have increased through the period where the GBR inshore waters have experienced increases in nutrient loads from agriculture. This has resulted in the frequency of CoTS waves on the GBR moving from low frequencies of about every 50–80 years to about every 15 years (Brodie 1992; Fabricius et al. 2010; Brodie et al. 2017; Pratchett et al. 2017).

Macroalgae Versus Coral Diversity
Higher nutrient availability supports the **proliferation of macroalgae and can negatively affect coral physiology and ecosystem functioning** (D'Angelo and Wiedenmann 2014; Ulloa et al. 2017). High concentrations of Chlorophyll-*a* (Chl-*a*) (typically at concentrations greater than 0.45 µg/L) can indicate increased nutrient availability supporting the growth of macroalgae (De'ath and Fabricius 2010). High macroalgal biomass can have detrimental effects on corals which can include space competition (McCook et al. 2001), altering the microbial environment of corals which affects their metabolism (Hauri et al. 2010; Thurber et al. 2017) and larval survival (Morrow et al. 2017), reducing coral settlement (Birrell et al. 2008) and increasing the susceptibility of corals to disease (Vega Thurber et al. 2014).

Increased Coral Bleaching Susceptibility
DIN availability plays an important part in the coral–algae symbiosis, with elevated DIN concentrations disrupting the ability of the coral host to maintain an optimal population of algal symbionts (Wooldridge et al. 2015, 2017). Elevated DIN concentrations and changes in N:P ratios can increase the susceptibility of corals to bleaching from increased temperatures (Wooldridge 2009, 2017; Fabricius et al. 2013; Wiedenmann et al. 2013; D'Angelo and Wiedenmann 2014; Vega Thurber et al. 2014; Humanes et al. 2016; Rosset et al. 2017; Wooldridge et al. 2017).

Bioerosion
Coral, both living and dead, can be impacted by the process known as **bioerosion**. This can occur through a range of mechanisms involving many different organisms. Bioerosion can be caused by the very small, minute, primarily intra-skeletal organisms, the microborers (e.g. algae, fungi, bacteria) to larger and often externally visible macroboring invertebrates (e.g. sponges, polychaete worms, sipunculans, molluscs, crustaceans, echinoids) and fish (e.g. scarids, acanthurids) (Hutchings et al. 2005; Chazottes et al. 2017; Glynn et al. 2017). Nutrient enrichment can increase the growth of both types of borers. Increased DIN availability supports the growth of algal borers and the filter-feeding sponges, worms and bivalves are supported through the increased phytoplankton (and zooplankton) biomass

(Le Grand and Fabricius 2011). Eutrophication of reef waters by land-based sources of nutrient pollution can magnify the effects of ocean acidification through nutrient-driven bioerosion (Prouty et al. 2017). The combined impacts of increased bioerosion by the boring organisms and the reduced calcification due to ocean acidification can additively reduce reef net calcification (DeCarlo et al. 2015; Glynn et al. 2017).

Coral Diseases
Coral **diseases** are a considerable contributor to coral cover declines on coral reefs (Osborne et al. 2011) and are predicted to worsen with global pressures of increasing temperature and ocean acidification (Maynard et al. 2015; O'Brien et al. 2016). Coral disease manifests as a general response to multiple stressors (▶ Chapter 14) of corals and has been positively correlated to sedimentation, elevated concentrations of nutrients and organic matter and increased plastic pollution (Harvell et al. 2007; Haapkylä et al. 2011; D'Angelo and Wiedenmann 2014; Pollock et al. 2014; Thompson et al. 2014; Vega Thurber et al. 2014; Lamb et al. 2016, 2018; Zaneveld et al. 2016).

Light Reduction
Algal blooms can be associated with flood plumes (◘ Figure 4.7) due to inputs of river-derived nutrients (Devlin et al. 2001; Devlin and Schaffelke 2009; Brodie et al. 2013) and localised inputs of nutrients. Phytoplankton blooms, as well as non-algal, suspended particulate matter (e.g. detritus, clay particles) in flood plumes, reduce light availability for benthic plant communities including seagrass and coral (Bauman et al. 2010; Petus et al. 2014; Collier et al. 2016). In shallow waters, the reduction of in situ light penetration due to resuspended sediment is usually a more dominant effect, but in deeper waters (>15 m) where resuspension does not normally occur (except in cyclonic conditions), the light reduction due to phytoplankton (and zooplankton) may be an important factor for communities such as deep water seagrasses (Collier et al. 2016) and coral reefs (D'Angelo and Wiedenmann 2014).

4.3 Case Studies

4.3.1 Baltic Sea

Over the twentieth-century nutrient inputs to the Baltic Sea increased by factors of three and five for nitrogen and phosphorus, respectively, with consequent widespread eutrophication across the Baltic Sea (Gustafsson et al. 2012). Declining dissolved oxygen concentrations were noted in the Baltic Sea as early as the 1930s, with widespread reporting of this by the 1950s. This

4

▣ Figure 4.7   Riverine plume discharging into the Great Barrier Reef, Australia. This image was captured a few days after the torrential rain and shows the muddy waters flowing from the Burdekin River into the Coral Sea. *Image*: European Space Agency CC BY-SA 2.0 contains modified Copernicus Sentinel data (2019), processed by ESA, CC BY-SA 3.0 IGO

sustained increase in nutrients originates from farm fertiliser, industry, atmospheric deposition and waste water associated with population increases in the large catchment area of the Baltic countries.

Large amounts of nutrients in the water increase primary production and hence intensify phytoplankton growth. Dead algae sink to the bottom, where their decomposition consumes oxygen, leading to hypoxia. In hypoxic conditions, sediments can no longer retain previously stored nutrients which then start to **leak** from the sediments. This leakage increases the amount of available nutrients which, in turn, increases primary production. This so-called **vicious circle** is an important indirect effect of eutrophication (Andersen et al. 2017; Murray et al. 2019). Benthic animals cannot survive in these hypoxic (and eventually anoxic) conditions, and large areas on the sea floor become completely depleted of life.

In much of the Baltic Sea, the direct consequences of elevated nutrient concentrations are increased primary production and phytoplankton biomass, and often manifest as algal blooms (Murray et al. 2019). Subsequently, the increased deposition of dead algae has reduced oxygen concentrations. These dissolved oxygen sags have affected the benthic invertebrates, with high rates of mortality, and impacted the spawning success rate of cod, a commercially important fish species.

As the causes and consequences of eutrophication become better understood in the Baltic Sea, many policies have been implemented to reduce external nutrient inputs (Andersen et al. 2017). These policies include the Helsinki Commission (HELCOM) Baltic Sea Action Plan (BSAP), an ambitious program that established nutrient reduction targets to restore the ecological status of the Baltic marine environment by 2021. Additionally, a number of European Union (EU) policies legally require member states—eight of the nine coastal counties—to reduce nutrient inputs to surface waters in order to meet environmental goals (Borja 2005; Devlin et al. 2007; Borja et al. 2010a; Bermejo et al. 2012).

These policies and associated measures have seen nitrogen and phosphorus inputs to the Baltic Sea decrease by 9% and 14%, respectively, and human exposure to potential toxins has been reduced (Svendsen et al. 2018). The combined effects of nutrient and fisheries management have also resulted in top predator population recovery (including cod). Nutrient loads are decreasing, however, legacy pollution and different rates of load reductions have limited full ecosystem recovery with many serious problems still to be addressed for the Baltic Sea. Potentially toxic contaminants are still at levels of concern in wildlife and fish catches, and new contaminants continue to come into use, undesirable symptoms of eutrophication remain evident in many coastal areas; deep water oxygen deficiency is still recorded extensively through the Baltic Sea, and toxic blooms of cyanobacteria interfere frequently with tourism and recreation (Elmgren et al. 2015), and climate change impacts the fragile recovery (Elmgren et al. 2015; Cloern et al. 2016).

4.3.2 Chesapeake Bay, USA

Since USA1950, the population of the Chesapeake Bay watershed in the eastern USA has doubled to 18 million people, leading to expansion of agriculture and urbanised land use and adding to the substantial nutrient and sediment runoff from previously established urban and agricultural lands. From the 1950s through to the 1970s, tens of thousands of hectares of submerged aquatic vegetation (SAV) were lost in the largest decline documented in over 400 years with ongoing algal blooms ( Figure 4.8) (Harding 1994; Harding and Perry 1997; Boesch et al. 2001; Kemp et al. 2005). Concern over the loss of SAV and declines in the overall health and economy of the bay led to unparalleled cooperation among federal, state, local and scientific agencies, whose joint efforts identified nutrient pollution and subsequent loss of SAV as the two most critical issues facing Chesapeake Bay (Lefcheck et al. 2018). These agencies instituted measures to reduce nutrient inputs, as well as long-term monitoring programmes to gauge their effectiveness, thereby establishing the Chesapeake Bay as one of the few places on Earth where comprehensive long-term data exist to mechanistically link human impacts and ecological restoration at broad scales.

The sustained management actions that have evolved out of that cooperation have been successful in reducing N concentrations in the Chesapeake Bay by 23% with a recent study showing seagrass coverage in the Chesapeake Bay increased by 17,000 ha between 1984 and 2015, a 23% improvement (Lefcheck et al. 2018). This cooperative management demonstrates that nutrient reductions, improvements in water quality (Zhang et al. 2018) and biodiversity conservation are effective strategies to aid the successful recovery of degraded systems at regional scales, a finding which has been highly relevant to environmental management programs worldwide (Lefcheck et al. 2018).

4.3.3 Yellow Sea and Qingdao

Massive free-floating macroalgal blooms of *Ulva prolifera* occur in the Yellow Sea, covering thousands of square kilometres, with millions of tons of biomass and causing huge economic losses. These blooms have been identified as the world's largest green tide events, occurring annually from 2007 to 2017 along the coast of the Yellow Sea, China, seriously impacting the downstream marine environments and ecological services. One of the most prominent examples of this happened in 2008, when a large green tide covered Qingdao beaches, making it a prominent feature during the Beijing Olympics.

Figure 4.8 Organic inputs into rivers and coastal waters can increase turbidity and algal blooms. Turbid waters identified in the Potomac and Wilcomico Rivers section of Chesapeake Bay. *Image*: NASA Earth Observatory image by Joshua Stevens and Jesse Allen, using Landsat data from the U.S. Geological Survey

Masses of *Ulva* floated in from the open water of the Yellow Sea and beached a few weeks before the competition was due to start, ensuring prominent coverage by the international media (◘ Figure 4.9). Mitigation included the deployment of a 30-km-long boom to keep the masses of floating algae out of the bay, and the physical removal of more than a million tonnes of algae from the beaches involved 10,000 people at an estimated cost to the province of US$30 million. In addition, aquaculture operations along the shore suffered losses of US$100 million (Liu et al. 2013a).

The pelagic seaweed bloom, as well as those in subsequent years, could be traced in satellite images to the coastline some 200 km south of Qingdao, where aquaculture of the edible red alga *Porphyra yezoensis* (which is grown on rafts along the intertidal zone) has expanded rapidly since 2004. As the algae *Ulva prolifera* also grows profusely on the rafts, algal fragments dislodged and discarded in the sea during harvesting of *Porphyra* are the most likely seed source of the mid-summer green tide. It is estimated that 500 tonnes of *Ulva* algae, discarded from the *Porphyra* rafts, grow into one million tonnes in 6 weeks (Liu et al. 2013b). The floating algae are transported more than 200 km northward to the Shandong coast and proliferate sufficiently to generate this massive green tide.

Management of the Olympics bloom involved hand and mechanical clearance from the beaches but efforts to reduce the incidence of the blooms are also occurring (Yuan et al. 2017a, b).

4.3.4 Caribbean Wide Algal Blooms and West Africa

In recent years, *Sargassum* seaweed has been washing up in unprecedented quantities on beaches in the Caribbean, Florida and the Gulf of Mexico (Louime et al. 2017; Langin 2018; Gower and King 2019) (◘ Figure 4.10). NASA satellites recently observed the largest seaweed bloom in the world, stretching from West Africa to the Gulf of Mexico. A major cause of the algae bloom was likely to be nutrient discharge from deforestation and fertiliser use along the Amazon River (Wang et al. 2019). Fertiliser consumption in Brazil between 2011 and 2018 increased by about 67% compared to the rates in 2002, while the total forest loss along the Brazilian Amazon increased by 25%. In June 2018, Wang et al. (2019) documented that the 8850-km algal bloom contained >20 million metric tons of *Sargassum* biomass. The bloom of 2011 may be a result of the Amazon River discharge in previous years, but recent increases and interannual variability after 2011 appear to be driven by upwelling off west Africa during boreal winter, and by the Amazon River discharge during spring and summer, indicating a possible regime shift and raising the possibility that recurrent blooms in the tropical Atlantic and the Caribbean Sea may become the new norm.

During 2011, there was an ocean-scale build-up of *Sargassum* in the Caribbean that, at its peak, extended across the Atlantic Ocean and resulted in massive

◘ **Figure 4.9** Image of green tides in Qingdao beach, China, in 2010. *Photo*: Philip Roeland CC BY-NC-ND 2.0

Figure 4.10 Proliferation of *Sargassum* golden tide in a bay in the southern Caribbean. *Photo*: Mark Yokoyama CC BY-NC-ND 2.0

golden tides along the West African coast, from Sierra Leone to Ghana, and, on the other side of the Atlantic Ocean, from Trinidad to the Dominican Republic (Figure 4.12). It is believed *Sargassum* was unknown in north-west Africa before 2011, so the event came as a shock to the many afflicted fishing villages. A similar event occurred again in 2019 (Wang et al. 2019). Satellite images showed that the algal rafts had developed along the northern coast of Brazil, north of the mouth of the Amazon, from where they moved east and west, eventually stretching across the Atlantic Ocean (Figure 4.11). A notable event was the whole length of the western coastline in Ghana covered in *Sargassum* and extended offshore, clogging fishing nets and impacting small boat traffic and fishing. This resulted in food shortages for people living in villages dependent on artisanal fisheries for their livelihood (Smetacek and Zingone 2013). In the Caribbean, tourism has been negatively affected because of the closure of beaches and bays. These large-scale events seem to be unprecedented in this area (Louime et al. 2017; Langin 2018; Resiere et al. 2018).

4.3.5 Brittany

The increase in *Ulva* biomass on European and American beaches that began in the 1970s was linked to coastal eutrophication. These visible, rotting coastal blooms impacted tourism-based economies, smothered aquaculture operations and disrupted traditional artisanal fisheries (Smetacek and Zingone 2013). As the many harmful effects became evident, the affected countries took measures to understand the drivers and the extent of the problem (Newton et al. 2014; Perrot et al. 2014; Gaspar et al. 2017). In the popular tourist beaches of Brittany (Figure 4.12), the magnitude of green tides has been increasing since the 1970s (Charlier et al. 2008). Events have been managed through the collection of seaweed and use as fertiliser by local farmers, but this was untenable by the 1990s as the magnitude of the seaweed became unmanageable. There have been many incidents connected to the large volume of seaweed on the Brittany beaches including the death of a horse in 2009 from H_2S gas coming from rotting *Ulva*, and, in 2011, the death of around 30 wild boars. Both incidents were widely reported in the press admist rising public concerns about the toxicity of the algae. Tourism was severely impacted, with a loss of visits felt by the local economy, in addition to the costs of removing and disposing of 100,000 tonnes of beached algae (estimated up to US$150 per tonne).

The consensus among the scientific community is that eutrophication from the effluents of intensive stock rearing was one of the primary causes of the increase in the number and magnitude of green tides since the 1990s. Brittany is a wet region overloaded with nutrients released by the high density of animals—equivalent to those from 50 million people—and so eutrophication is inevitable because the manure is not being shipped back to the animal feed producers outside the province. The meat-producing and tourist industries are both mainstays of the provincial economy, and, following the animal deaths, con-

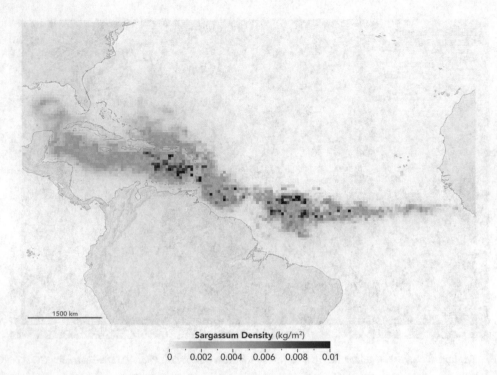

Figure 4.11   The Great Atlantic *Sargassum* Belt in July 2018. Scientists used NASA's Moderate Resolution Imaging Spectroradiometer (MODIS) on Terra and Aqua satellites to discover the Great Atlantic *Sargassum* Belt (GASB), which started in 2011. It has occurred every year, with the exception of 2013 and typically stretches from the west coast of Africa to the Gulf of Mexico. *Image*: NASA/Earth Observatory. Data provided by Mengqiu Wang and Chuanmin Hu, USF College of Marine Science

Figure 4.12   Image of green tides taken in South Coast, United Kingdom. *Photo*: Mike Best, Environment Agency

frontation between the two industries increased. In efforts to make the best out of a situation that is unlikely to change soon, *Ulva* biomass has been used as a raw material for biogas production, as an organic fertiliser and as an additive to animal and human food. However, the value barely meets the costs of current methods of algal collection and processing (Smetacek and Zingone 2013).

4.3.6 Tampa Bay, Florida, USA

In Tampa Bay, Florida, USA, large increases in population in the catchment area led to increased nutrient loads so that by the late 1970s the effects of eutrophic decline became obvious, including reduced water clarity, accumulations of macroalgae, noxious phytoplankton blooms, intermittent hypoxia and loss of about 50% of the seagrass meadows in Tampa Bay (Greening et al. 2014). The bay was already phosphorus-enriched due to catchment drainage from phosphorus ore mining operations. The ecosystem is strongly nitrogen-limited and thus management was focused on nitrogen removal from point source discharges of sewage and industrial wastes that, in the mid-1970s, comprising 60% of the total nitrogen load. Political responses at the state and local levels led the way, with the enactment of a 1978 Florida statute that required advanced treatment of water from all wastewater treatment plants discharging to Tampa Bay. Additional nutrient limits were required for stormwater discharges from 1985. This reduction in wastewater nitrogen loading of approximately 90% in the late 1970s lowered external total nitrogen loading by more than 50% within 3 years. Continuing nutrient management actions from public and private sectors were associated with a steadily declining total nitrogen load rate, despite an increase of more than 1 million people living within the Tampa Bay metropolitan area. Following recovery from an extreme weather event in 1997–1998, water clarity has increased significantly, and seagrass is expanding at a rate significantly different than before the event (Boesch 2019; Greening et al. 2018). Seagrass extent has increased by more than 65% since the 1980s, and in 2014 exceeded the recovery goal adopted in 1996 (■ Figure 4.13) (Greening et al. 2018).

Key elements supporting the nutrient management strategy and concomitant ecosystem recovery in Tampa Bay include:

- active community involvement, including agreement about quantifiable restoration goals;
- regulatory and voluntary reduction in nutrient loadings from point, atmospheric and nonpoint sources;
- long-term water quality and Seagrass extent monitoring; and
- a commitment from public and private sectors to work together to attain restoration goals.

4.3.7 Kāne'ohe Bay, Oahu, Hawaii, USA

Sewage discharges into Kāne'ohe Bay, Hawaii, increased from the end Second World War due to increasing population and urbanisation, and reached a peak of 20 ML/d in 1977. This chronic discharge into the lagoon introduced high levels of inorganic nitrogen and inorganic phosphorus, with the southern lagoon waters becoming increasingly rich in phytoplankton (■ Figure 4.14). Reefs closest to the outfall became overgrown by filter-feeding organisms, such as sponges, tube-worms and barnacles. Reefs in the centre of the bay further from the outfalls were overgrown by the indigenous green algae *Dictyosphaeria cavernosa*. After diversion of the outfalls into the deeper ocean in 1978, coastal nutrient levels were reduced with corresponding declines in the phytoplankton and zooplankton populations and *D. cavernosa* abundance. At the same time, increases in the abundance and distribution of coral species were reported, as the reefs slowly recovered (Bahr et al. 2015). A drastic decline in previously dominant *D. cavernosa* occurred in 2006, attributed to a gradual

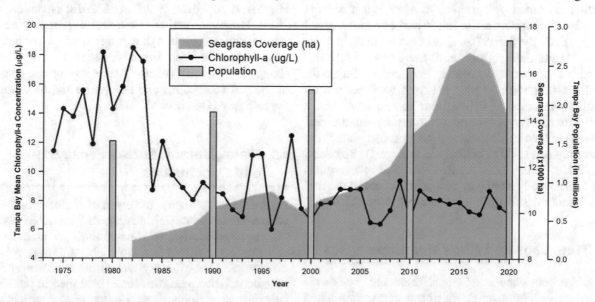

■ **Figure 4.13** Trends in mean annual chlorophyll-*a* concentrations and Secchi disk depth Seagrass extent and watershed population estimates for Tampa Bay. Produced by TBEP; data sources: Environmental Protection Commission of Hillsborough County (in public domain); Southwest Florida Water Management District (in public domain) and US Census Bureau (in public domain). *Image*: Greening et al. 2018, figure used with permission from Ed Sherwood, Executive Director. Tampa Bay Estuary Program

Figure 4.14   Kāne'ohe Bay with Moku O Loe island at right centre, Hawaii, USA. *Image*: NASA Earth Expeditions, National Aeronautics and Space Administration. NASA Official: Brian Dunbar

return to a coral-dominated state following relocation of the sewage outfall in 1978 that eliminated the sewage nutrient inputs that drove the initial phase shift to macroalgae in the 1970s. However, urban stormwater runoff continues to cause short-term eutrophication of the bay (Drupp et al. 2011) via spikes in nitrogen inputs and subsequent phytoplankton blooms (Stimson 2015).

4.3.8   Pago Pago Harbour, American Samoa

Diverse coral communities have been monitored at Aua village in Pago Pago Harbour, American Samoa (▣ Figure 4.15). Between the 1950s and 1980s, this area was seriously degraded by chronic pollution from two tuna canneries, fuel spills in the inner harbour and coastal development. By the 1970s, coral communities had declined substantially (Dahl and Lamberts 1977). Improved management of coastal development, fuel spills and the installation of a pipe to export wastewater from the tuna canneries to the harbour mouth have seen a significant recovery of coral communities on the reef crest and outer reef flat where there is consolidated reef substratum (up to 30 m behind the reef crest) (Birkeland et al. 2013). In contrast, it was found that recovery has been substantially slower or non-existent behind the reef crest, where the substratum is primarily loose rubble.

4.4   Time Lags and Non-linear Responses

In nutrient-enriched conditions, there are well-documented cases of eutrophic marine systems, dominated by algae, where reductions in nutrient loading have not returned the systems to their original ecological status

(Duarte et al. 2008; Lotze et al. 2011b; McCracken and Phillips 2017) or where only partial recovery was observed (Borja et al. 2010a; Elliott and Whitfield 2011). This can be partly attributed to the range of other factors in the system that have dramatically changed during the period of increased nutrient loading, such as human population increases, increased carbon dioxide in the atmosphere, changed catchment hydrology and discharge volumes, global temperature increases and fish stock losses where the functioning of the system is highly modified from the original pristine state.

In coral reef systems, the issues of reversibility, time lags and phase change have been the subject of much recent research (Bruno et al. 2009; Dudgeon et al. 2010; Hughes et al. 2011; Wolff et al. 2018; MacNeil et al. 2019). However, further research is required on ecosystem responses to changing water quality, particularly in combination with other stressors such as climate change, to quantify the likely time lags of the response of the reef ecosystems and the nature and trajectory of the response (Devlin et al. 2021).

4.5   Management, Future Prospects and Conclusions

Four decades following the onset of major efforts to reverse widespread eutrophication of coastal ecosystems via improved sewage treatment, fertiliser management and erosion controls (i.e. from about 1980), evidence of improvement of ecosystem status is growing. However, cumulative pressures have developed in parallel to eutrophication, including those associated with climate change, such as warming, deoxygenation, ocean acidification and increased runoff. These additional pres-

Figure 4.15   Pago Pago Harbor, American Samoa. *Image*: Tavita Togia, National Park Service of American Samoa, Wikimedia Commons

sures risk countering efforts to mitigate eutrophication and arrest coastal ecosystems in a state of eutrophication despite the efforts and significant resources already invested to revert coastal eutrophication (Duarte and Krause-Jensen 2018). With over 40% of the human population residing in coastal areas, ecosystem degradation in these areas can have disproportionate effects on society (Wright et al. 2006).

Given the seriousness of eutrophication, major efforts have been made to reduce nutrient inputs and hence **restore** ecosystems to their original state or at least to a better state (Conley et al. 2009) (▶ Chapter 15). However, there are concerns about the possibility of a full restoration or the time required for impacted systems improving to a more desirable state (Duarte 2009). Although reversing the effects of eutrophication and achieving some recovery of marine ecosystems requires actions beyond reducing nutrient loading (Duarte 2009), implementing coordinated and long-term management strategies has led to at least partial recovery in some systems, albeit over long time periods (Borja et al. 2008, 2010a, b; Jones and Schmitz 2009). Recent reviews, however, have shown that, in many cases, coastal ecosystems are failing to meet their recovery objectives (Jeppesen et al. 2005; Duarte 2009; Duarte et al. 2008; Kemp et al. 2009; Borja et al. 2010a; Verdonschot et al. 2013, Lefcheck et al. 2018). The recent review by McCrackin et al. (2017) of 89 case studies of nutrient reductions and recovery of lakes and coastal marine ecosystems from eutrophication showed that for coastal marine areas only 24% achieved baseline conditions after the cessation or partial reduction of nutrients with most taking decades to recover. In a similar study, Gross and Hagy (2017) identified 16 case studies where nutrient reductions had been achieved and found that improvements in 8 studies had fallen short of stated restoration goals. Five more were successful initially, but their conditions subsequently declined. Three of the case studies achieved their goals fully and are currently managing to maintain the restored condition. It is of noteworthy interest that of the marine examples identified in McCrackin et al. (2017) and Gross and Hagy (2017), only one is in the tropics (Kaneohe Bay, Hawaii) and one in the sub-tropics (Tampa Bay, Florida).

A study by Desmit et al. (2018) shows that a significant decrease in nitrogen fluxes from land to sea is possible by adapting human activities in the watersheds, which prevents at least part of the eutrophication symptoms in the adjacent coastal zones. The United Nations **Sustainable Development Goal (SDG) framework** recognises the importance of monitoring oceans with a dedicated goal on oceans (SDG 14). Sustainable Development Goal SDG 14 **Life below water** sets the aim to conserve and sustainably use the oceans, seas and marine resources for sustainable development. This includes targets dedicated to coastal eutrophication and marine debris, marine area management and conservation. SDG 14.1 states by 2025, countries should prevent and significantly reduce marine pollution of all kinds, in particular from land-based activities, including marine debris and nutrient pollution. To assist towards SDG 14.1, UNEP is implementing a global initiative to address excess nitrogen in the environment and its negative effects via a project titled '*Towards the Establishment of an International Nitrogen Management System*'. It aims to provide recommendations on strategies to reduce emissions of reactive nitrogen, including measures to make production systems, especially farms, more efficient in their use of fertiliser. However, recent analyses have concluded that new initiatives, not just relying on the reduction of nutrient loadings, will be required to solve coastal and marine eutrophication issues. Duarte and Krause-Jensen (2018) suggest (from the abstract) that

» *"the time has arrived for a broader, more comprehensive approach to intervening to control eutrophication. Options for interventions include multiple levers controlling major pathways of nutrient budgets of coastal ecosystems, i.e., nutrient inputs, which is the intervention mostcommonly deployed, nutrient export, sequestration in sediments, and emissions of nitrogen to the atmosphere as N² gas (denitrification). The levers involve local-scale hydrological engineering to increase flushing and nutrient export from (semi)enclosed coastal systems ecological engineering such as sustainable aquaculture of seaweeds and Mussels to enhance nutrient export and restoration of benthic habitats to increase sequestration in sediments as well as denitrification, and geo-engineering approaches including, with much precaution, aluminum injections in sediments."*

4.6 Summary

Eutrophication has been a key issue for coastal and marine waters for many years. The consequences of eutrophication are wide-ranging and can occur at both small and large scales, with multiple impacts on many parts of the marine environment. Negative impacts on the coastal and marine environment can result through the process of eutrophication as the marine environment becomes enriched with nutrients, increasing the amount of plant and algae growth to estuaries and coastal waters. Known consequences of nutrient enrichment in coastal and marine waters include increased primary production, increased biomass of primary producers such as phytoplankton and depletion of dissolved oxygen due to decomposition of accumulated biomass, resulting in local hypoxic or anoxic conditions. Other consequences can include shifts in species composition, blooms of nuisance and toxic algae and macroalgae, increased growth of epiphytic algae, red tides, water discolouration and foaming, loss of submerged vegetation due to shading and changes in benthic community structure due to oxygen deficiency or the presence of toxic phytoplankton species (Devlin et al. 2011). Our understanding of eutrophication has certainly improved over the last few decades, as long-term data sets provide a unique baseline to understand the changes and variability associated with long-term nutrient enrichment. Long-term studies have shown the impacts of eutrophication to be variable depending on the susceptibility of the coastal and marine system and require consideration of the many factors that influence that susceptibility and vulnerability (Cloern 2001; Cloern and Jassby 2009). Long-term data has also shown us that systems can recover, given enough time and ongoing management actions to reduce nutrients below acceptable thresholds.

Management of eutrophication has also improved over recent years, with programmes that focus across the catchment to the coast and look upstream to resolve the downstream eutrophication issues. Nutrient inputs to riverine and coastal systems come from a variety of diffuse sources (e.g. agricultural runoff and atmospheric deposition) and point sources (e.g. sewage treatment and industrial discharge). However, measures to reduce nitrogen and phosphorus inputs via targeted policies tend to focus on individual actions rather than addressing the wide range of activities that export nutrients into coastal waters. Future management should focus on parallel reductions in both nitrogen and phosphorus inputs to reduce coastal eutrophication and the impacts associated with an imbalanced nutrient system (Greenwood et al. 2019).

These long-term impacts on our coastal systems continue to degrade our coastal systems and impact coastal functioning. This is becoming increasingly more important as we recognise the importance of our coastal habitats in supporting biodiversity, carbon cycling, coastal protection and maintenance of a functioning food web. Management of eutrophication impacts must consider a changing baseline as climate change shifts coastal resilience with the cumulative and additive impacts of pollution and climate (Borja et al. 2010b). Management decisions must also reflect the recovery processes can be lengthy and require multiple facets of environmental management. Our coastal systems are integral to our environment, economy and community and urgently need long-term protection. These systems are facing an ever-increasing set of pressures, with climate change and extreme weather reducing the resilience of coastal waters. Eutrophication is an issue that can be solved, despite the complexity of the drivers and impacts, and the uncertainty and timing related to mitigation and recovery processes. There have, and continue to be, positive stories of systems recovering when nutrient inputs are reduced or eliminated.

Solutions are possible, though almost never simple, and rely on a combination of long-term strategies, sewage and groundwater infrastructure, best management practices around agriculture and aquaculture, detailed monitoring and assessment and close partnerships between all stakeholders, public users and government.

4.7 Study Questions and Activities

1. Research how sewage treatment plants work and create a diagram that shows the various steps in treatment processes.
2. Four common N-containing fertilisers are ammonia [NH_3], ammonium nitrate [NH_4NO_3], ammonium sulfate [$(NH_4)_2SO_4$] and urea [$(NH_2)_2CO$]. How much of each compound must be used to provide 1 kg of N?
3. Describe the process of eutrophication in your own words.
4. Using the various case studies described in ▶ Section 4.3 create a single table that summarises the causes, effects of nutrient enrichment and what solutions have been used.
5. Explore the recent media in your country and find an article about nutrient pollution. Critique the article and suggest some management options that will help mitigate the problem (see also ▶ Chapter 16).

References

Altieri AH, Harrison SB, Seemann J, Collin R, Diaz RJ, Knowlton N (2017) Tropical dead zones and mass mortalities on coral reefs'. Proc Natl Acad Sci 114(14):3660–3665

Andersen JH, Carstensen J, Conley DJ, Dromph K, Fleming-Lehtinen V, Gustafsson BG, Josefson AB, Norkko A, Villnäs A, Murray C (2017) Long-term temporal and spatial trends in eutrophication status of the Baltic Sea. Biol Rev 92(1):135–149

Anderson DM, Glibert PM, Burkholder JM (2002) Harmful algal blooms and eutrophication: nutrient sources, composition, and consequences. Estuaries 25(4):704–726

Asmala E, Carstensen J, Conley DJ, Slomp CP, Stadmark J, Voss M (2017) Efficiency of the coastal filter: nitrogen and phosphorus removal in the Baltic Sea. Limnol Oceanogr 62(S1):S222–S238

Bahr KD, Jokiel PL, Toonen RJ (2015) The unnatural history of Kāne'ohe Bay: coral reef resilience in the face of centuries of anthropogenic impacts. PeerJ 3:e950

Bainbridge Z, Lewis S, Bartley R, Fabricius K, Collier C, Waterhouse J, Garzon-Garcia A, Robson B, Burton J, Wenger A (2018) Fine sediment and particulate organic matter: a review and case study on ridge-to-reef transport, transformations, fates, and impacts on marine ecosystems. Mar Pollut Bull 135:1205–1220

Bakun A, Black BA, Bograd SJ, Garcia-Reyes M, Miller AJ, Rykaczewski RR, Sydeman WJ (2015) Anticipated effects of climate change on coastal upwelling ecosystems. Curr Clim Change Rep 1(2):85–93

Bauman AG, Burt JA, Feary DA, Marquis E, Usseglio P (2010) Tropical harmful algal blooms: an emerging threat to coral reef communities? Mar Pollut Bull 60(11):2117–2122

Bermejo R, Vergara JJ, Hernández I (2012) Application and reassessment of the reduced species list index for macroalgae to assess the ecological status under the water framework directive in the Atlantic coast of Southern Spain. Ecol Ind 12(1):46–57

Beusen AH, Bouwman AF, Van Beek LP, Mogollón JM, Middelburg JJ (2016) Global riverine N and P transport to ocean increased during the 20th century despite increased retention along the aquatic continuum. Biogeosciences 13(8):2441–2451

Birkeland C, Green A, Fenner D, Squair C, Dahl A (2013) Substratum stability and coral reef resilience: insights from 90 years of disturbances on a reef in American Samoa. Micronesica 6:1–16

Birrell CL, McCook LJ, Willis BL, Diaz-Pulido GA (2008) Effects of benthic algae on the replenishment of corals and the implications for the resilience of coral reefs. In: Oceanography and marine biology, pp 31–70. CRC Press

Boesch DF (2019) Barriers and bridges in abating coastal eutrophication. Front Mar Sci 6:123

Boesch DF, Brinsfield RB, Magnien RE (2001) Chesapeake Bay eutrophication. J Environ Qual 30(2):303–320

Borja Á (2005) The European water framework directive: a challenge for nearshore, coastal and continental shelf research. Cont Shelf Res 25(14):1768–1783

Borja A, Bricker SB, Dauer DM, Demetriades NT, Ferreira JG, Forbes AT, Hutchings P, Jia X, Kenchington R, Carlos Marques J, Zhu C (2008) Overview of integrative tools and methods in assessing ecological integrity in estuarine and coastal systems worldwide. Mar Pollut Bull 56(9):1519–1537

Borja Á, Dauer DM, Elliott M, Simenstad CA (2010a) Medium-and long-term recovery of estuarine and coastal ecosystems: patterns, rates and restoration effectiveness. Estuaries Coasts 33(6):1249–1260

Borja Á, Elliott M, Carstensen J, Heiskanen A-S, van de Bund W (2010b) Marine management—Towards an integrated implementation of the European Marine Strategy Framework and the water framework directives. Mar Pollut Bull 60(12):2175–2186

Bouwman A, Beusen AH, Billen G (2009) Human alteration of the global nitrogen and phosphorus soil balances for the period 1970–2050. Glob Biogeochem Cycles 23(4)

Bouwman A, Van Drecht G, Knoop J, Beusen A, Meinardi C (2005) Exploring changes in river nitrogen export to the worlds oceans. Glob Biogeochem Cycles 19(1)

Boyd CE (2020) Water quality—An introduciton, 3rd edn. Spinger, Cham, p 440

Breitburg D, Levin LA, Oschlies A, Grégoire M, Chavez FP, Conley DJ, Garçon V, Gilbert D, Gutiérrez D, Isensee K (2018) Declining oxygen in the global ocean and coastal waters. Science 359(6371):eaam7240

Bricker SB, Longstaff B, Dennison W, Jones A, Boicourt K, Wicks C, Woerner J (2008) Effects of nutrient enrichment in the nations estuaries: a decade of change. Harmful Algae 8(1):21–32

Brodie G, Brodie J, Maata M, Peter M, Otiawa T, Devlin M (2020) Seagrass habitat in Tarawa Lagoon, Kiribati: service benefits and links to national priority issues. Mar Pollut Bull 155:111099

Brodie J (1992) Enhancement of larval and juvenile survival and recruitment in Acanthaster planci: effects of terrestrial runoff—A review. Aust J Mar Freshw Res 43:539–554

Brodie J, Devlin M, Haynes D, Waterhouse J (2011) Assessment of the eutrophication status of the Great Barrier Reef lagoon (Australia). Biogeochemistry 106(2):281–302

Brodie J, Devlin M, Lewis S (2017) Potential enhanced survivorship of crown of thorns starfish larvae due to near-annual nutrient enrichment during secondary outbreaks on the central mid-shelf of the Great Barrier Reef, Australia. Diversity 9(1):17

Brodie J, Fabricius K, Death G, Okaji K (2005) Are increased nutrient inputs responsible for more outbreaks of crown-of-thorns starfish? An appraisal of the evidence. Mar Pollut Bull 51(1–4):266–278

Brodie J, Waterhouse J, Schaffelke B, Johnson J, Kroon F, Thorburn P, Rolfe J, Lewis S, Warne MStJ, Fabricius K (2013) Reef water quality scientific consensus statement 2013. Department of the Premier and Cabinet, Queensland Government, Brisbane

Bruno JF, Sweatman H, Precht WF, Selig ER, Schutte VG (2009) Assessing evidence of phase shifts from coral to macroalgal dominance on coral reefs. Ecology 90(6):1478–1484

Burson A, Stomp M, Akil L, Brussaard CP, Huisman J (2016) Unbalanced reduction of nutrient loads has created an offshore gradient from phosphorus to nitrogen limitation in the North Sea. Limnol Oceanogr 61(3):869–888

Cai W-J, Hu X, Huang W-J, Murrell MC, Lehrter JC, Lohrenz SE, Chou W-C, Zhai W, Hollibaugh JT, Wang Y (2011) Acidification of subsurface coastal waters enhanced by eutrophication. Nat Geosci 4(11):766

Carpenter SR, Caraco NF, Correll DL, Howarth RW, Sharpley AN, Smith VH (1998) Nonpoint pollution of surface waters with phosphorus and nitrogen. Ecol Appl 8(3):559–568

Charlier RH, Morand P, Finkl CW (2008) How Brittany and Florida coasts cope with green tides. Int J Environ Stud 65(2):191–208

Chazottes V, Hutchings P, Osorno A (2017) Impact of an experimental eutrophication on the processes of bioerosion on the reef. One Tree Island, Great Barrier Reef, Australia. Mar Pollut Bull 118(1–2):125–130

Cloern J (2001) Our evolving conceptual model of the coastal eutrophication problem. Mar Ecol Prog Ser 210:223–253

Cloern JE, Abreu PC, Carstensen J, Chauvaud L, Elmgren R, Grall J, Greening H, Johansson JOR, Kahru M, Sherwood ET (2016) Human activities and climate variability drive fast-paced change across the worlds estuarine-coastal ecosystems. Glob Change Biol 22(2):513–529

Cloern JE, Jassby AD (2009) Patterns and scales of phytoplankton variability in estuarine-coastal ecosystems. Estuaries Coasts 33(2):230–241

Collier C, Adams M, Langlois L, Waycott M, O'Brien K, Maxwell P, McKenzie L (2016) Thresholds for morphological response to light reduction for four tropical seagrass species. Ecol Ind 67:358–366

Conley DJ, Paerl HW, Howarth RW, Boesch DF, Seitzinger SP, Havens KE, Lancelot C, Likens GE (2009) Controlling eutrophication: nitrogen and phosphorus. Science 323(5917):1014–1015

Crain CM, Kroeker K, Halpern BS (2008) Interactive and cumulative effects of multiple human stressors in marine systems. Ecol Lett 11(12):1304–1315

D'Angelo C, Wiedenmann J (2014) Impacts of nutrient enrichment on coral reefs: new perspectives and implications for coastal management and reef survival. Curr Opin Environ Sustain 7:82–93

Dahl AL, Lamberts AE (1977) Environmental impact on a Samoan coral reef: a resurvey of Mayors 1917 transect. Pac Sci 31(3):309–319

De'ath D, Fabricius K (2010) Water quality as a regional driver of coral biodiversity and macroalgae on the Great Barrier Reef. Ecol Appl 20(3):840–850

De'ath G, Fabricius KE, Sweatman H, Puotinen M (2012) The 27-year decline of coral cover on the Great Barrier Reef and its causes. Proc Natl Acad Sci 109(44):17995–17999

DeCarlo TM, Cohen AL, Barkley HC, Cobban Q, Young C, Shamberger KE, Brainard RE, Golbuu Y (2015) Coral macrobioerosion is accelerated by ocean acidification and nutrients. Geology 43(1):7–10

Desmit X, Thieu V, Billen G, Campuzano F, Dulière V, Garnier J, Lassaletta L, Ménesguen A, Neves R, Pinto L, Silvestre M (2018) Reducing marine eutrophication may require a paradigmatic change. Sci Total Environ 635:1444–1466

Devlin M, Best M, Haynes D (2007) Implementation of the water framework directive in European marine waters. Mar Pollut Bull 55(1–6):1–2

Devlin M, Bricker S, Painting S (2011) Comparison of five methods for assessing impacts of nutrient enrichment using estuarine case studies. Biogeochemistry 106(2):177–205

Devlin M, Schaffelke B (2009) Spatial extent of riverine flood plumes and exposure of marine ecosystems in the Tully coastal region, Great Barrier Reef. Mar Freshw Res 60(11):1109–1122

Devlin M, Waterhouse J, Taylor J, Brodie J (2001) Flood plumes in the Great Barrier Reef: spatial and temporal patterns in composition and distribution. Great Barrier Reef Marine Park Authority, p 113. Available at: ► https://elibrary.gbrmpa.gov.au/jspui/handle/11017/354. Accessed 14 Jan 2022

Devlin MJ, da Silva E, Petus C, Wenger A, Zeh D, Tracey D, Álvarez-Romero JG, Brodie J (2013) Combining in-situ water quality and remotely sensed data across spatial and temporal scales to measure variability in wet season chlorophyll-a: Great Barrier Reef lagoon (Queensland, Australia). Ecol Processes 2(1):31

Devlin MJ, Lyons BP, Johnson JE, Hills JM (2021) The tropical Pacific Oceanscape: current issues, solutions and future possibilities. Mar Pollut Bull 166:112181

Devlin MJ, McKinna LW, Alvarez-Romero JG, Petus C, Abott B, Harkness P, Brodie J (2012) Mapping the pollutants in surface riverine flood plume waters in the Great Barrier Reef, Australia. Mar Pollut Bull 65(4–9):224–235

Diaz RJ, Rosenberg R (2008) Spreading dead zones and consequences for marine ecosystems. Science 321(5891):926–929

Drupp P, De Carlo EH, Mackenzie FT, Bienfang P, Sabine CL (2011) Nutrient inputs, phytoplankton response, and CO_2 variations in a semi-enclosed subtropical embayment, Kaneohe Bay, Hawaii. Aquat Geochem 17(4–5):473–498

Duarte CM (2009) Coastal eutrophication research: a new awareness. Hydrobiologia 629(1):263–269

Duarte CM, Conley DJ, Carstensen J, Sánchez-Camacho M (2008) Return to neverland: shifting baselines affect eutrophication restoration targets. Estuaries Coasts 32(1):29–36

Duarte CM, Krause-Jensen D (2018) Intervention options to accelerate ecosystem recovery from coastal eutrophication. Front Mar Sci 5:470

Duce RA, LaRoche J, Altieri K, Arrigo KR, Baker AR, Capone DG, Cornell S, Dentener F, Galloway J, Ganeshram RS, Geider RJ (2008) Impacts of atmospheric anthropogenic nitrogen on the open ocean. Science 320(5878):893–897

Dudgeon SR, Aronson RB, Bruno JF, Precht WF (2010) Phase shifts and stable states on coral reefs. Mar Ecol Prog Ser 413:201–216

Ekau W, Auel H, Pörtner H-O, Gilbert D (2010) Impacts of hypoxia on the structure and processes in pelagic communities (zooplankton, macro-invertebrates and fish). Biogeosciences 7(5):1669–1699

Elliott M, Whitfield AK (2011) Challenging paradigms in estuarine ecology and management. Estuar Coast Shelf Sci 94(4):306–314

Ellis JD (2017) J Appl Commun 101(1) Full Issue

Elmgren R, Blenckner T, Andersson A (2015) Baltic Sea management: successes and failures. Ambio 44(3):335–344

Fabricius KE, Cséke S, Humphrey C, De'ath G (2013) Does trophic status enhance or reduce the thermal tolerance of scleractinian corals? A review, experiment and conceptual framework. PloS One 8(1):e54399

Fabricius KE, Okaji K, De'ath G (2010) Three lines of evidence to link outbreaks of the crown-of-thorns seastar *Acanthaster planci* to the release of larval food limitation. Coral Reefs 29(3):593–605

FAO (Food and Agricutlure Organisation of the United Nations) (2012) FAO statistical yearbook. Available at: ► https://www.fao.org/3/i2490e/i2490e00.htm. Accessed 14 January 2022

Ferreira JG, Andersen JH, Borja A, Bricker SB, Camp J, Cardoso da Silva M, Garcés E, Heiskanen A-S, Humborg C, Ignatiades L, Lancelot C, Menesguen A, Tett P, Hoepffner N, Claussen U (2011) Overview of eutrophication indicators to assess environmental status within the European Marine Strategy Framework Directive. Estuar Coast Shelf Sci 93(2):117–131

Fowler D, Coyle M, Skiba U, Sutton MA, Cape JN, Reis S, Sheppard LJ, Jenkins A, Grizzetti B, Galloway JN (2013) The global nitrogen cycle in the twenty-first century. Philos Trans R Soc B: Biol Sci 368(1621):20130164

Galloway JN, Townsend AR, Erisman JW, Bekunda M, Cai Z, Freney JR, Martinelli LA, Seitzinger SP, Sutton MA (2008) Transformation of the nitrogen cycle: recent trends, questions, and potential solutions. Science 320(5878):889–892

Galloway JN, Winiwarter W, Leip A, Leach AM, Bleeker A, Erisman JW (2014) Nitrogen footprints: past, present and future. Environ Res Lett 9(11):115003

Gaspar R, Marques L, Pinto L, Baeta A, Pereira L, Martins I, Marques JC, Neto JM (2017) Origin here, impact there—The need of integrated management for river basins and coastal areas. Ecol Ind 72:794–802

Greening H, Janicki A, Sherwood ET, Pribble R, Johansson JOR (2014) Ecosystem responses to long-term nutrient management in an urban estuary: Tampa Bay, Florida, USA. Estuar Coast Shelf Sci 151:A1–A16

Geider R, La Roche J (2002) Redfield revisited: variability of C:N:P in marine microalgae and its biochemical basis. Eur J Phycol 37(1):1–17

Gilbert D, Rabalais N, Diaz R, Zhang J (2010) Evidence for greater oxygen decline rates in the coastal ocean than in the open ocean. Biogeosciences 7(7):2283

Glibert P, Bouwman L (2012) Land-based nutrient pollution and the relationship to harmful algal blooms in coastal marine systems. LOICZ Newslett Inprint 2:5–7

Glibert PM (2017) Eutrophication, harmful algae and biodiversity—Challenging paradigms in a world of complex nutrient changes. Mar Pollut Bull 124(2):591–606

Glibert PM, Burford MA (2017) Globally changing nutrient loads and harmful algal blooms: recent advances, new paradigms, and continuing challenges. Oceanography 30(1):58–69

Glynn PW, Manzello DP, Enochs IC (eds) (2017) Coral Reefs of the Eastern Tropical Pacific: persistence and loss in a dynamic environment. Springer, Dordrecht, p 548

Gower J, King S (2019) Seaweed, seaweed everywhere. Science 365(6448):27–27

Greening H, Janicki A, Sherwood ET (2018) Seagrass recovery in Tampa Bay, Florida (USA). In: The wetland book: II: distribution, description, and conservation, pp 495–506. Springer, Dordrecht

Greenwood N, Devlin MJ, Best M, Fronkova L, Graves CA, Milligan A, Barry J, Van Leeuwen SM (2019) Utilizing eutrophication assessment directives from transitional to marine systems in the Thames Estuary and Liverpool Bay, UK. Front Mar Sci 6:116

Gross C, Hagy JD (2017) Attributes of successful actions to restore lakes and estuaries degraded by nutrient pollution. J Environ Manage 187:122–136

Gruber N, Galloway JN (2008) An Earth-system perspective of the global nitrogen cycle. Nature 451(7176):293

Gustafsson BG, Schenk F, Blenckner T, Eilola K, Meier HM, Müller-Karulis B, Neumann T, Ruoho-Airola T, Savchuk OP, Zorita E (2012) Reconstructing the development of Baltic Sea eutrophication 1850–2006. Ambio 41(6):534–548

Haapkylä J, Unsworth RK, Flavell M, Bourne DG, Schaffelke B, Willis BL (2011) Seasonal rainfall and runoff promote coral disease on an inshore reef. PLoS ONE 6(2):e16893

Halpern BS, Walbridge S, Selkoe KA, Kappel CV, Micheli F, D'agrosa C, Bruno JF, Casey KS, Ebert C, Fox HE (2008) A global map of human impact on marine ecosystems. Science 319(5865):948–952

Harding L Jr, Perry E (1997) Long-term increase of phytoplankton biomass in Chesapeake Bay, 1950–1994. Mar Ecol Prog Ser 157:39–52

Harding LW (1994) Long-term trends in the distribution of phytoplankton in Chesapeake Bay: roles of light, nutrients and streamflow. Mar Ecol Prog Ser 104:267–267

Harvell D, Jordán-Dahlgren E, Merkel S, Rosenberg E, Raymundo L, Smith G, Weil E, Willis B (2007) Coral disease, environmental drivers, and the balance between coral and microbial associates. Oceanography 20:172–195

Hauri C, Fabricius KE, Schaffelke B, Humphrey C (2010) Chemical and physical environmental conditions underneath mat-and canopy-forming macroalgae, and their effects on understorey corals. PLoS ONE 5(9):e12685

Heffer P, Prud'homme M (2012) Fertilizer Outlook 2012–2016. International Fertilizer Industry Association (IFA), Paris, France. Available at ▶ http://www.anpifert.pt/Destaques/2012_doha_ifa_summary.pdf. Accessed 14 Jan 2022

Heffer P, Prud'homme M (2013) Fertiliser Outlook 2013–1017. In: Proceedings of the 81st international fertilizer industry association conference. Available at: ▶ https://www.fertilizer.org/Search?SearchTerms=81st+. Accessed 14 Jan 2022

Heffer P, Prud'homme M (2014) Fertilizer Outlook 2014–2018. International Fertilizer Industry Association (IFA). Paris, France. Avaiable at: ▶ https://www.fertilizer.org/images/Library_Downloads/2014_ifa_sydney_summary.pdf. Accessed 14 Jan 2022

Heffer P, Prud'homme M (2015) Fertilizer Outlook 2015–2019. In: 83rd IFA annual conference. International Fertilizer Industry Association (IFA). Istanbul, (Turkey). Available at: ▶ https://www.fertilizer.org/images/Library_Downloads/2015_ifa_istanbul_summary.pdf. Accessed 14 Jan 2022

Heisler J, Glibert PM, Burkholder JM, Anderson DM, Cochlan W, Dennison WC, Dortch Q, Gobler CJ, Heil CA, Humphries E (2008) Eutrophication and harmful algal blooms: a scientific consensus. Harmful Algae 8(1):3–13

Howarth R, Paerl HW (2008) Coastal marine eutrophication: Control of both nitrogen and phosphorus is necessary. Proc Natl Acad Sci 105(49):E103–E103

Howarth RW, Marino R (2006). Nitrogen as the limiting nutrient for eutrophication in coastal marine ecosystems: evolving views over three decades. Limnol Oceanogr 51(1 part2):364–376

Hudnell HK (2008) Cyanobacterial harmful algal blooms: state of the science and research needs. Springer, New York, p 950

Hughes T, Bellwood D, Baird A, Brodie J, Bruno J, Pandolfi J (2011) Shifting base-lines, declining coral cover, and the erosion of reef resilience: comment on Sweatman et al. (2011). Coral Reefs 30(3):653–660

Humanes A, Noonan SH, Willis BL, Fabricius KE, Negri AP (2016) Cumulative effects of nutrient enrichment and elevated temperature compromise the early life history stages of the coral *Acropora tenuis*. PLoS ONE 11(8):e0161616

Hutchings P, Peyrot-Clausade M, Osnorno A (2005). Influence of land runoff on rates and agents of bioerosion of coral substrates. Mar Pollu Bull 51(1–4):438–447

Jacquet S, Delesalle B, Torréton J-P, Blanchot J (2006) Response of phytoplankton communities to increased anthropogenic influences (southwestern lagoon, New Caledonia). Mar Ecol Prog Ser 320:65–78

Jeppesen E, Jensen JP, Soendergaard M, Lauridsen TL (2005) Response of fish and plankton to nutrient loading reduction in eight shallow Danish lakes with special emphasis on seasonal dynamics. Freshw Biol 50(10):1616–1627

Jickells T, Buitenhuis E, Altieri K, Baker A, Capone D, Duce R, Dentener F, Fennel K, Kanakidou M, LaRoche J (2017) A reevaluation of the magnitude and impacts of anthropogenic atmospheric nitrogen inputs on the ocean. Glob Biogeochem Cycles 31(2):289–305

Jickells T, Weston K (2011a) Nitrogen cycle–external cycling: losses and gains. Health Environ Res Online pp 261–278

Jickells TD, Weston K (2011b) 5.08 Nitrogen cycle—External cycling: losses and gains. In: Mclusky D, Wolanski E (eds) Treatise on Estuarine and Coastal Science. Academic Press, Waltham, pp 261–278

Johnson A, Harrison M (2015) The increasing problem of nutrient runoff on the coast: as development increases along coastlines worldwide, water quality—And everything that depends on it—Degrades. Am Sci 103(2):98–102

Jones HP, Schmitz OJ (2009) Rapid recovery of damaged ecosystems. PLoS ONE 4(5):e5653

Kemp W, Boynton W, Adolf J, Boesch D, Boicourt W, Brush G, Cornwell J, Fisher T, Glibert P, Hagy J (2005) Eutrophication of Chesapeake Bay: historical trends and ecological interactions. Mar Ecol Prog Ser 303(21):1–29

Kemp W, Testa J, Conley D, Gilbert D, Hagy J (2009) Temporal responses of coastal hypoxia to nutrient loading and physical controls. Biogeosciences 6(12):2985–3008

Kroeze C, Seitzinger SP (1998) Nitrogen inputs to rivers, estuaries and continental shelves and related nitrous oxide emissions in 1990 and 2050: a global model. Nutr Cycl Agroecosyst 52(2–3):195–212

Lamb JB, Wenger AS, Devlin MJ, Ceccarelli DM, Williamson DH, Willis BL (2016) Reserves as tools for alleviating impacts of marine disease. Philos Trans R Soc B: Biol Sci 371(1689):20150210

Lamb JB, Willis BL, Fiorenza EA, Couch CS, Howard R, Rader DN, True JD, Kelly LA, Ahmad A, Jompa J (2018) Plastic waste associated with disease on coral reefs. Science 359(6374):460–462

Langin K (2018) Seaweed masses assault Caribbean islands. Science 360(6394):1157–1158

Lapointe BE, Burkholder JM, Van Alstyne KL (2018) Harmful macroalgal blooms in a changing world: causes, impacts, and management. In: Shumway S, Burkholder J, Morton S (eds) Harmful algal blooms: a compendium desk reference. John Wiley & Sons, Hoboken, pp 515–560

Lapointe BE, Herren LW, Brewton RA, Alderman P (2019) Nutrient over-enrichment and light limitation of seagrass communities in the Indian River Lagoon, an urbanized subtropical estuary. Sci Total Environ 134068

Le Grand HM, Fabricius K (2011) Relationship of internal macrobioeroder densities in living massive Porites to turbidity and chlorophyll on the Australian Great Barrier Reef. Coral Reefs 30(1):97–107

Le Moal M, Gascuel-Odoux C, Ménesguen A, Souchon Y, Étrillard C, Levain A, Moatar F, Pannard A, Souchu P, Lefebvre A (2019) Eutrophication: a new wine in an old bottle? Sci Total Environ 651:1–11

Lee M, Shevliakova E, Stock CA, Malyshev S, Milly PC (2019) Prominence of the tropics in the recent rise of global nitrogen pollution. Nat Commun 10(1):1437

Lefcheck JS, Orth RJ, Dennison WC, Wilcox DJ, Murphy RR, Keisman J, Gurbisz C, Hannam M, Landry JB, Moore KA (2018) Long-term nutrient reductions lead to the unprecedented recovery of a temperate coastal region. Proc Natl Acad Sci 115(14):3658–3662

Liu X, Duan L, Mo J, Du E, Shen J, Lu X, Zhang Y, Zhou X, He C, Zhang F (2011) Nitrogen deposition and its ecological impact in China: an overview. Environ Pollut 159(10):2251–2264

Liu D, Keesing JK, He P, Wang Z, Shi Y, Wang Y (2013a) The worlds largest macroalgal bloom in the Yellow Sea, China: formation and implications. Estuar Coast Shelf Sci 129:2–10

Liu X, Zhang Y, Han W, Tang A, Shen J, Cui Z, Vitousek P, Erisman JW, Goulding K, Christie P (2013b) Enhanced nitrogen deposition over China. Nature 494(7438):459

Liu S, Xie Z, Zeng Y, Liu B, Li R, Wang Y, Wang L, Qin P, Jia B, Xie J (2019) Effects of anthropogenic nitrogen discharge on dissolved inorganic nitrogen transport in global rivers. Glob Change Biol 25(4):1493–1513

Lotze HK, Coll M, Dunne JA (2011a) Historical changes in marine resources, food-web structure and ecosystem functioning in the Adriatic Sea, Mediterranean. Ecosystems 14(2):198–222

Lotze HK, Coll M, Magera AM, Ward-Paige C, Airoldi L (2011b) Recovery of marine animal populations and ecosystems. Trends Ecol Evol 26(11):595–605

Louime C, Fortune J, Gervais G (2017) Sargassum invasion of coastal environments: a growing concern. Am J Environ Sci 13(1):58–64

Lu CC, Tian H (2017) Global nitrogen and phosphorus fertilizer use for agriculture production in the past half century: shifted hot spots and nutrient imbalance. Earth Syst Sci Data 9:181

MacNeil MA, Mellin C, Matthews S, Wolff NH, McClanahan TR, Devlin M, Drovandi C, Mengersen K, Graham NAJ (2019) Water quality mediates resilience on the Great Barrier Reef. Nat Ecol Evol 3:620–627

Maynard J, Van Hooidonk R, Eakin CM, Puotinen M, Garren M, Williams G, Heron SF, Lamb J, Weil E, Willis B (2015) Projections of climate conditions that increase coral disease susceptibility and pathogen abundance and virulence. Nat Clim Chang 5(7):688

McCook L, Jompa J, Diaz-Pulido G (2001) Competition between corals and algae on coral reefs: a review of evidence and mechanisms. Coral Reefs 19(4):400–417

McCracken K, Phillips DR (2017) Global health: an introduction to current and future trends, 2nd edn. Routledge: Taylor and Francis Group, p 466

McCrackin ML, Jones HP, Jones PC, Moreno-Mateos D (2017) Recovery of lakes and coastal marine ecosystems from eutrophication: a global meta-analysis. Limnol Oceanogr 62(2):507–518

Mee LD (1992) The Black Sea in crisis: a need for concerted international action. Ambio 21(4):278–286

Mills MM, Ridame C, Davey M, La Roche J, Geider RJ (2004) Iron and phosphorus co-limit nitrogen fixation in the eastern tropical North Atlantic. Nature 429(6989):292–294

Morrow KM, Bromhall K, Motti CA, Munn CB, Bourne DG (2017) Allelochemicals produced by brown macroalgae of the *Lobophora* genus are active against coral larvae and associated bacteria, supporting pathogenic shifts to *Vibrio* dominance. Appl Environ Microbiol 83(1):e02391-e2416

Murray CJ, Muller-Karulis B, Carstensen J, Conley DJ, Gustafsson B, Andersen JH (2019) Past, present and future eutrophication status of the Baltic Sea. Front Mar Sci 6:2

Newton A, Icely J, Cristina S, Brito A, Cardoso AC, Colijn F, Dalla Riva S, Gertz F, Hansen JW, Holmer M (2014) An overview of ecological status, vulnerability and future perspectives of European large shallow, semi-enclosed coastal systems, lagoons and transitional waters. Estuar Coastal Shelf Sci 140:95–122

Nixon SW (1995) Coastal marine eutrophication: a definition, social causes, and future concerns. Ophelia 41(1):199–219

Nixon SW (2009) Eutrophication and the macroscope. Hydrobiologia 629(1):5–19

O'Brien PA, Morrow KM, Willis BL, Bourne DG (2016) Implications of ocean acidification for marine microorganisms from the free-living to the host-associated. Front Marine Sci 3:47

Odebrecht C, Villac MC, Abreu PC, Haraguchi L, Gomes PD, Tenenbaum DR (2018) Flagellates versus diatoms: phytoplankton trends in tropical and subtropical estuarine-coastal ecosystems. In: Hoffmeyer M, Sabatini M, Brandini F, Calliari D, Santinelli N (eds) Plankton ecology of the Southwestern Atlantic. Springer, Cham, pp 249–267

Osborne K, Dolman AM, Burgess SC, Johns KA (2011) Disturbance and the dynamics of coral cover on the Great Barrier Reef (1995–2009). PLoS ONE 6(3):e17516

Paerl HW (2006) Assessing and managing nutrient-enhanced eutrophication in estuarine and coastal waters: interactive effects of human and climatic perturbations. Ecol Eng 26(1):40–54

Paerl HW (2009) Controlling Eutrophication along the freshwater–marine continuum: dual nutrient (N and P) reductions are essential. Estuaries Coasts 32(4):593–601

Paerl HW, Hall NS, Peierls BL, Rossignol KL (2014) Evolving paradigms and challenges in estuarine and coastal eutrophication dynamics in a culturally and climatically stressed world. Estuaries Coasts 37(2):243–258

Pearce F (2018) When the rivers run dry, fully revised and updated edition: water-the defining crisis of the twenty-first century. Beacon Press,, Boston, p 328

Penuelas J, Poulter B, Sardans J, Ciais P, Van Der Velde M, Bopp L, Boucher O, Godderis Y, Hinsinger P, Llusia J (2013) Human-induced nitrogen–phosphorus imbalances alter natural and managed ecosystems across the globe. Nat Commun 4(1):1–10

Perrot T, Rossi N, Ménesguen A, Dumas F (2014) Modelling green macroalgal blooms on the coasts of Brittany, France to enhance water quality management. J Mar Syst 132:38–53

Petus C, Collier C, Devlin M, Rasheed M, McKenna S (2014) Using MODIS data for understanding changes in seagrass meadow health: a case study in the Great Barrier Reef (Australia). Mar Environ Res 98:68–85

Pollock FJ, Lamb JB, Field SN, Heron SF, Schaffelke B, Shedrawi G, Bourne DG, Willis BL (2014) Sediment and turbidity associated with offshore dredging increase coral disease prevalence on nearby reefs. PLoS ONE 9(7):e102498

Potter IC, Veale L, Tweedley JR, Clarke KR (2016) Decadal changes in the ichthyofauna of a eutrophic estuary following a remedial engineering modification and subsequent environmental shifts. Estuar Coast Shelf Sci 181:345–363

Powley HR, Dürr HH, Lima AT, Krom MD, Van Cappellen P (2016) Direct discharges of domestic wastewater are a major source of phosphorus and nitrogen to the Mediterranean Sea. Environ Sci Technol 50(16):8722–8730

Pratchett M, Caballes C, Wilmes J, Matthews S, Mellin C, Sweatman H, Nadler L, Brodie J, Thompson C, Hoey J (2017) Thirty years of research on crown-of-thorns starfish (1986–2016): scientific advances and emerging opportunities. Diversity 9(4):41

Prouty NG, Cohen A, Yates KK, Storlazzi CD, Swarzenski PW, White D (2017) Vulnerability of coral reefs to bioerosion from land-based sources of pollution. J Geophys Res: Oceans 122(12):9319–9331

Rabalais NN, Cai W-J, Carstensen J, Conley DJ, Fry B, Hu X, Quinones-Rivera Z, Rosenberg R, Slomp CP, Turner RE (2014) Eutrophication-driven deoxygenation in the coastal ocean. Oceanography 27(1):172–183

Rabalais NN, Turner RE, Díaz RJ, Justić D (2009) Global change and eutrophication of coastal waters. ICES J Mar Sci 66(7):1528–1537

Resiere D, Valentino R, Nevière R, Banydeen R, Gueye P, Florentin J, Cabié A, Lebrun T, Mégarbane B, Guerrier G (2018) Sargassum seaweed on Caribbean islands: an international public health concern. The Lancet 392(10165):2691

Richlen ML, Morton SL, Jamali EA, Rajan A, Anderson DM (2010) The catastrophic 2008–2009 red tide in the Arabian gulf region, with observations on the identification and phylogeny of the fish-killing dinoflagellate *Cochlodinium polykrikoides*. Harmful Algae 9(2):163–172

Riemann B, Carstensen J, Dahl K, Fossing H, Hansen JW, Jakobsen HH, Josefson AB, Krause-Jensen D, Markager S, Stæhr PA (2016) Recovery of Danish coastal ecosystems after reductions in nutrient loading: a holistic ecosystem approach. Estuaries Coasts 39(1):82–97

Righetti D, Vogt M, Gruber N, Psomas A, Zimmermann NE (2019) Global pattern of phytoplankton diversity driven by temperature and environmental variability. Sci Adv 5(5):eaau6253

Rockström J, Steffen W, Noone K, Persson Å, Chapin FS III, Lambin EF, Lenton TM, Scheffer M, Folke C, Schellnhuber HJ (2009) A safe operating space for humanity. Nature 461(7263):472

Rosenberg R, Cato I, Förlin L, Grip K, Rodhe J (1996) Marine environment quality assessment of the Skagerrak-Kattegat. J Sea Res 35(1–3):1–8

Rosset S, Wiedenmann J, Reed AJ, D'Angelo C (2017) Phosphate deficiency promotes coral bleaching and is reflected by the ultra-structure of symbiotic dinoflagellates. Mar Pollut Bull 118(1–2):180–187

Schaffelke B, Collier C, Kroon F, Lough J, McKenzie L, Ronan M, Uthicke S, Brodie J (2017) Scientific consensus statement 2017. A synthesis of the science of land-based water quality impacts on the Great Barrier Reef, Chapter 1: The Condition of Coastal and Marine Ecosystems of the Great Barrier Reef and Their Responses to Water Quality and Disturbances. State of Queensland. Avaialbe at: ▶ https://www.reefplan.qld.gov.au/__data/assets/pdf_file/0030/45993/2017-scientific-consensus-statement-summary-chap01.pdf. Accessed 14 Jan 2022

Seitzinger S, Harrison J, Dumont E, Beusen AH, Bouwman A (2005) Sources and delivery of carbon, nitrogen, and phosphorus to the coastal zone: an overview of global Nutrient Export from Watersheds (NEWS) models and their application. Glob Biogeochem Cycles 19(4):GB4S01

Seitzinger S, Mayorga E, Bouwman A, Kroeze C, Beusen A, Billen G, Van Drecht G, Dumont E, Fekete B, Garnier J (2010) Global river nutrient export: a scenario analysis of past and future trends. Glob Biogeochem Cycles 24(4):GB0A08

Smetacek V, Zingone A (2013) Green and golden seaweed tides on the rise. Nature 504(7478):84–88

Smith VH, Schindler DW (2009) Eutrophication science: where do we go from here? Trends Ecol Evol 24(4):201–207

Steffen W, Richardson K, Rockström J, Cornell SE, Fetzer I, Bennett EM, Biggs R, Carpenter SR, De Vries W, De Wit CA (2015) Planetary boundaries: guiding human development on a changing planet. Science 347(6223):1259855

Stimson J (2015) Long-term record of nutrient concentrations in Kāne 'ohe Bay, O 'ahu, Hawai 'i, and its relevance to onset and end of a phase shift involving an indigenous alga, *Dictyosphaeria cavernosa1*. Pac Sci 69(3):319–340

Svendsen LM, Gustafsson B, Larsen SE, Sonesten L, Frank-Kamenetsky D (2018). Inputs of nutrients (nitrogen and phosphorus) to the Sub-basins of the Baltic Sea (2016). HELCROM core indicator report, p 30

Tett P, Droop M, Heaney S (1985) The Redfield ratio and phytoplankton growth rate. J Mar Biol Assoc UK 65(2):487–504

Tett P, Gowen R, Mills D, Fernandes T, Gilpin L, Huxham M, Kennington K, Read P, Service M, Wilkinson M (2007) Defining and detecting undesirable disturbance in the context of marine eutrophication. Mar Pollu Bull 55(1):282–297

Thompson A, Schroeder T, Brando VE, Schaffelke B (2014) Coral community responses to declining water quality: Whitsunday Islands, Great Barrier Reef, Australia. Coral Reefs 33(4):923–938

Thurber RV, Payet JP, Thurber AR, Correa AM (2017) Virus–host interactions and their roles in coral reef health and disease. Nat Rev Microbiol 15(4):205

Ulloa MJ, Álvarez-Torres P, Horak-Romo KP, Ortega-Izaguirre R (2017) Harmful algal blooms and eutrophication along the Mexican coast of the Gulf of Mexico large marine ecosystem. Environ Dev 22:120–128

Van Drecht G, Bouwman A, Harrison J, Knoop J (2009) Global nitrogen and phosphate in urban wastewater for the period 1970–2050. Glob Biogeochem Cycles 23(4)

Vega Thurber RL, Burkepile DE, Fuchs C, Shantz AA, McMinds R, Zaneveld JR (2014) Chronic nutrient enrichment increases prevalence and severity of coral disease and bleaching. Glob Change Biol 20(2):544–554

Verdonschot P, Spears B, Feld C, Brucet S, Keizer-Vlek H, Borja A, Elliott M, Kernan M, Johnson R (2013) A comparative review of recovery processes in rivers, lakes, estuarine and coastal waters. Hydrobiologia 704(1):453–474

Wang M, Hu C, Barnes BB, Mitchum G, Lapointe B, Montoya JP (2019) The great Atlantic Sargassum belt. Science 365(6448):83–87

Wang R, Balkanski Y, Boucher O, Ciais P, Peñuelas J, Tao S (2015) Significant contribution of combustion-related emissions to the atmospheric phosphorus budget. Nat Geosci 8(1):48

Waterhouse J, Brodie J, Lewis S, Mitchell A (2012) Quantifying the sources of pollutants in the Great Barrier Reef catchments and the relative risk to reef ecosystems. Mar Pollut Bull 65(4–9):394–406

Waycott M, Duarte CM, Carruthers TJ, Orth RJ, Dennison WC, Olyarnik S, Calladine A, Fourqurean JW, Heck KL, Hughes AR (2009) Accelerating loss of seagrasses across the globe threatens coastal ecosystems. Proc Natl Acad Sci 106(30):12377–12381

Wiedenmann J, D'Angelo C, Smith EG, Hunt AN, Legiret F-E, Postle AD, Achterberg EP (2013) Nutrient enrichment can increase the susceptibility of reef corals to bleaching. Nat Clim Chang 3(2):160

Wolff NH, Mumby PJ, Devlin M, Anthony KR (2018) Vulnerability of the Great Barrier Reef to climate change and local pressures. Glob Change Biol 24(5):1978–1991

Wooldridge SA (2009) Water quality and coral bleaching thresholds: formalising the linkage for the inshore reefs of the Great Barrier Reef, Australia. Mar Pollut Bull 58(5):745–751

Wooldridge SA (2017) Instability and breakdown of the coral–algae symbiosis upon exceedence of the interglacial pCO$_2$ threshold (>260 ppmv): the "missing" Earth-System feedback mechanism. Coral Reefs 36(4):1025–1037

Wooldridge SA, Brodie JE (2015) Environmental triggers for primary outbreaks of crown-of-thorns starfish on the Great Barrier Reef, Australia. Mar Pollu Bull 101(2):805–815

Wooldridge SA, Brodie JE, Kroon FJ, Turner RD (2015) Ecologically based targets for bioavailable (reactive) nitrogen discharge from the drainage basins of the Wet Tropics region, Great Barrier Reef. Mar Pollu Bull 97(1–2):262–272

Wooldridge SA, Heron SF, Brodie JE, Done TJ, Masiri I, Hinrichs S (2017) Excess seawater nutrients, enlarged algal symbiont densities and bleaching sensitive reef locations: 2. A regional-scale predictive model for the Great Barrier Reef, Australia. Mar Pollu Bull 114(1):343–354

Wright A, Stacey N, Holland P (2006) The cooperative framework for ocean and coastal management in the Pacific Islands: effectiveness, constraints and future direction. Ocean Coast Manag 49(9–10):739–763

Wu J, Sunda W, Boyle EA, Karl DM (2000) Phosphate depletion in the western North Atlantic Ocean. Science 289(5480):759–762

Yan W, Mayorga E, Li X, Seitzinger SP, Bouwman A (2010) Increasing anthropogenic nitrogen inputs and riverine DIN exports from the Changjiang River basin under changing human pressures. Glob Biogeochem Cycles 24(4):GB0A06

Ye NH, Zhang XW, Mao YZ, Liang CW, Xu D, Zou J, Zhuang ZM, Wang QY (2011) 'Green tides' are overwhelming the coastline of our blue planet: taking the world's largest example. Ecol Res 26(3):477–485

Yuan D, Li Y, Wang B, He L, Hirose N (2017a) Coastal circulation in the southwestern Yellow Sea in the summers of 2008 and 2009. Cont Shelf Res 143:101–117

Yuan Y, Yu Z, Song X, Cao X (2017b) Temporal and spatial characteristics of harmful algal blooms in Qingdao Waters, China. Chin J Oceanol Limnol 35(2):400–414

Zaneveld JR, Burkepile DE, Shantz AA, Pritchard CE, McMinds R, Payet JP, Welsh R, Correa AM, Lemoine NP, Rosales S (2016) Overfishing and nutrient pollution interact with temperature to disrupt coral reefs down to microbial scales. Nat Commun 7:11833

Zhang Q, Murphy RR, Tian R, Forsyth MK, Trentacoste EM, Keisman J, Tango PJ (2018) Chesapeake Bay's water quality condition has been recovering: insights from a multimetric indicator assessment of thirty years of tidal monitoring data. Sci Total Environ 637:1617–1625

Metals and Metalloids

Amanda Reichelt-Brushett and Graeme Batley

Contents

5.1   Introduction – 103

5.2   Sources of Trace Metals – 104
5.2.1   Natural Sources – 104
5.2.2   Anthropogenic Atmospheric Inputs – 105
5.2.3   Mining Operations – 106
5.2.4   Mineral Processing – 110
5.2.5   Urban and Industrial Discharges – 110
5.2.6   Other Sources – 111

5.3   Metal Behaviour in Marine Waters – 112
5.3.1   Metal Speciation – 112
5.3.2   Evaluating Metal Speciation and Bioavailability
         in Marine Waters – 114

5.4   Metal Behaviour in Marine Sediments – 114
5.4.1   Metal Forms in Sediments – 114
5.4.2   Metal Bioavailability in Sediments – 115

5.5   Metal Uptake by Marine Organisms – 116
5.5.1   Transport Across Biological Membranes – 117
5.5.2   Other Uptake Routes – 117
5.5.3   Metal Detoxification – 117
5.5.4   Metal Depuration – 117

5.6   Metal Toxicity to Marine Organisms – 118
5.6.1   Mercury Toxicity to Marine Biota – 120
5.6.2   Copper Toxicity to Marine Biota – 120

© The Author(s) 2023
A. Reichelt-Brushett (ed.), *Marine Pollution—Monitoring, Management and Mitigation*,
Springer Textbooks in Earth Sciences, Geography and Environment,
https://doi.org/10.1007/978-3-031-10127-4_5

5.7   **Managing Metal Pollution – 121**
5.7.1   What Is 'Pollution' – 121
5.7.2   Guideline Values – 121

5.8   **Summary – 121**

5.9   **Study Questions and Activities – 122**

           References – 122

Acronyms and Abbreviations

AChE	Acetylcholinesterase activity
ASM	Artisanal and small-scale mining
ASGM	Artisanal and small-scale gold mining
ASS	Acid sulfate soils
AVS	Acid volatile sulfide
BLM	Biotic ligand model
CEC	Cation exchange capacity
DGT	Diffusive gradients in thin films
DGV	Default guideline value
DSTP	Deep-sea tailings placement
EC10	Concentration of a toxicant that causes a measured negative effect to 10% of a test population
GST	Glutathione S-transferase
ISA	International Seabed Authority
NOEC	No observed effect concentration
PNG	Papua New Guinea
POM	Particulate organic matter
TBT	Tributyltin
STD	Submarine tailings disposal (also known as DSTP)
USA	United States of America
USEPA	United States Environmental Protection Agency
USGS	United States Geological Survey

5.1 Introduction

This chapter introduces you to **metals** and **metalloids** that are a concern to the health of marine ecosystems. It provides a general chemical understanding of important metals and metalloids, their sources, behaviour, impacts and management. Metals, metalloids and non-metals all make up the periodic table (Appendix II) and are classified into these categories according to their properties. Metals are good conductors of heat and electricity and are malleable and ductile, making them very useful to humans and therefore economically valuable. Metalloids sit on the periodic table in a jagged line at the division between metals and non-metals and have intermediate properties.

You will come across various terms when studying metal pollution. **Trace metals** are generally referred to as those metals that are found in trace quantities in the environment although the term may also refer to those metals that are required in trace quantities in biological systems. The term **heavy metal** generally refers to density and excludes lighter metals (such as sodium and potassium) but is imprecise and has been questioned as useful (Chapman 2007, 2012; Batley 2012). Both terms are used to describe **metals of environmental concern** including aluminium (Al), vanadium (V), chromium (Cr), manganese (Mn), iron (Fe), cobalt (Co), nickel (Ni), copper (Cu), zinc (Zn), molybdenum (Mo), silver (Ag), cadmium (Cd), gold (Au), mercury (Hg), tin (Sn) and lead (Pb). Metalloids of environmental concern include boron (B), arsenic (As) and antimony (Sb). Selenium (Se) is also sometimes referred to as a metalloid. As the electronics industry advances, rare earth elements are becoming more useful, and in the future, these elements may also be of environmental concern due to poor management of e-waste and other waste sources (e.g. Herrmann et al. 2016; Trapasso et al. 2021; Brewer et al. 2022). To assist with the flow of this chapter, metals and metalloids will generally be referred to as **metals** except when specific distinctions are necessary.

Most metals occur naturally in the environment, and at some places, they are found naturally in very high concentrations (e.g. in geological formations such as ancient volcanoes and deep ocean hydrothermal vents). They are found naturally in ocean waters (albeit at extremely low concentrations), sediments and rocks, and are transported to the ocean from terrestrial sources. The abundance and distribution of metals in the ocean are a function of their solubility in seawater and their degree of involvement in abiotic and biotic processes and oceanic circulation (Allen 1993). Some metals are essential to life and are required in small quantities, others have no known biological function (◘ Table 5.1).

◘ Table 5.1 Essential and non-essential biotic requirements for metals of environmental concern

Typical concentration	No known biological function in marine species	Probably essential (for some species)	Proven essential (for many species)
Trace (µg/g)			Fe, Zn, Cu
Ultra-trace (ng/g)	Au, Pb, Hg,	Ni, V, Sb, Cd[b], As[a]	Mn, Co, Se, Mo, Cr

Adapted from Brady et al. (2015), other sources [a] Saunders et al. (2019), [b] Lane and Morel (2000)

5.2 Sources of Trace Metals

Metals are naturally found in the marine environment, but through anthropogenic activities, they have increased in concentration in waters, sediments and biota. There are many thousands of research publications, from all areas of the world's rivers and oceans, that demonstrate the wide range of metal sources and their impacts on marine biota. An updated assessment of global emissions of metals to the environment (◘ Table 5.2) has confirmed that anthropogenic sources far exceed natural sources with releases to soils being greater than those to water and the atmosphere (e.g. Salam, 2021).

5.2.1 Natural Sources

Most of the Earth's crust is composed of silicate minerals and rarer minerals that have been concentrated by one of the rock-forming processes (crystallisation, metamorphosis and erosion and sedimentation). The formation of mineral deposits such as sulfide minerals of lead and zinc, and bauxite deposits of aluminium from the weathering of igneous rock, result in metal-enriched soils and sediments as they break down by natural weathering processes over geologic time. For this reason, natural **background concentrations** of metals in waters and sediments **vary depending on the geological features** in the related environments. For example, bauxite forms in rainy tropical climates and are associated with laterites and aluminous rock, depending upon climatic conditions in which chemical weathering and leaching are pronounced (Tarbuck and Lutgens 1987); deposits of nickel and cobalt are also found in laterites that develop from igneous rocks with high ferromanganese mineral contents (Tarbuck and Lutgens 1987). Maus et al. (2020) provided a timely update on mining activities at a global level which highlights where naturally rich mineral deposits exist. Globally, there are over 5650 mines associated with metals of en-

◘ Table 5.2 Global emissions of metals to the environment

Metal	Natural	Anthropogenic			Total
		Atmosphere	Water	Soil	
		10³ tonnes/y			
Mercury	2.5	2.0	4.6	8.3	14.9
Lead	12.0	119	138	796	1050
Arsenic	12.0	5.0	41.0	82.0	128
Cadmium	1.3	3.0	9.4	22.0	34.4
Zinc	45.8	57.0	226	1372	1655
Copper	28.0	25.9	112	955	1030
Selenium	9.3	4.6	41.0	41.0	86.6
Antimony	2.4	1.6	18.0	26.0	45.6
Tin	44.0	14.7	142	896	1940
Chromium	317	11.0	262	1670	194
Manganese	30.0	95.3	113	325	533
Nickel	26.0	240	12.0	132	384
Vanadium	3.0	2.6	11.0	88.6	102

Adapted from Pacyna et al. (2016)

□ **Table 5.3** Dissolved metal concentrations in ocean waters

Metal	Coastal ocean water (ng/L)	References
Aluminium	46	Rijkenberg et al. (2014)[a]
Manganese	5	Angel et al. (2010)
Iron	50	Rijkenberg et al. (2014)[a]
Cobalt	3	Shelley et al. (2012)[b]
Nickel	110	Angel et al. (2010)
Copper	30	Apte et al. (1998)
Zinc	22	Apte et al. (1998)
Arsenic	1.5	Apte et al. (1998)
Selenium	<73	Apte et al. (1998)
Silver	<0.5	Apte et al. (1998)
Cadmium	2	Apte et al. (1998), Angel et al. (2010)
Mercury	<1	Apte et al. (1998)
Lead	9	Apte et al. (1998)

[a] Data for western Atlantic Ocean
[b] Data for Sargasso Sea. All other data for NSW and Queensland coastal waters

vironmental concern in over 100 countries (Maus et al. 2020). The United States Geological Survey (USGS) provides an interesting interactive map of global mineral resources: ▶ https://mrdata.usgs.gov/general/map-global.html. In addition, there are rich mineral deposits of economic interest (but not yet mined) on the ocean floor (e.g. Heffernan 2019; Milinovic et al. 2021).

Metals from terrestrial sources can be transported to marine environments by dust, in catchment runoff and through the atmosphere. Volcanic eruptions contribute to the release of metals into the atmosphere and subsequent deposition into the marine environment (e.g. Gaffney and Marley 2014). In 1989, it was estimated that biogenic sources (natural sources) of metals contributed 30–50% of total metal emissions to the atmosphere (Nriagu 1989). As with most contaminants in the environment, the ocean is the ultimate sink for the vast majority of trace metals. Concentrations of met-

als of environmental concern in ocean waters are generally in the nanogram/litre range (□ Table 5.3) which are very difficult to quantify. The analytical process requires ultra-trace metal sampling and analysis procedures to ensure reliable estimations of baseline concentration against which to assess riverine and estuarine inputs.

5.2.2 Anthropogenic Atmospheric Inputs

Inputs of metals via the atmosphere from anthropogenic activities to the marine environment is an important pathway and, for mercury, it is the foremost transport pathway (Marx and McGowan 2011; Driscoll et al. 2013) (▶ Box 5.1). Interestingly, atmospheric lead was reportedly deposited in ice layers in Greenland between 500 BC and 300 AD and was expected to be a result of emissions from Roman mines and smelters (Nriagu 1989, 1996). Atmospheric distribution processes result in the deposition of metals throughout marine environments even where the human populations are very small (e.g. polar environments, Barrie et al. 1992; Rudnicka-Kępa and Zaborska 2021).

The **combustion of coal** liberates traces of Hg, Pb, Cr, Cd, Sn, Sb, Se, As, Mn and Ti, and exhaust emissions from the combustion of oil is a major source of nickel and vanadium (Pacyna and Pacyna 2001; Munawer 2018). The **combustion of leaded gasoline** was for many years determined to be the major source of atmospheric lead emissions (Pacyna and Pacyna 2001), although the phasing out of leaded fuels in most countries has seen these emissions decline. **Non-ferrous metal production** also contributes to atmospheric As, Cd, Cu, Sn and Zn (e.g. Pacyna and Pacyna 2001).

The largest atmospheric emissions of anthropogenic metals were estimated to come from Asia as a result of growing demands for energy in the region and increasing industrial production and limited regulatory controls (Pacyna and Pacyna 2001). Atmospheric deposition has been attributed to long-distance transport of Hg, Pb, Cd, As and Fe with enhanced loadings measured in polar regions and likely sources from Russia, China and Europe (Barrie et al. 1992; Driscoll et al. 2013; De Vera et al. 2021; Thorne et al. 2018).

Box 5.1: The Mercury Cycle

Mercury (Hg) is transported through the atmosphere from coal burning, oil refining, natural gas combustion, artisanal and small-scale gold mining, the chlor-alkali industry that produces chlorine and caustic soda, and waste incineration. Consumer goods such as batteries, electric switches, fluorescent lamps etc. all contain mercury) (Gaffney and Marley 2014) (□ Figure 5.1). Land and ocean processes play an important role in the redistribution of Hg through the environment, including to marine ecosystems. Toxic effects and biomagnification potential result from the net conversion of Hg(II) to monomethylmercury (CH_3Hg^+) and dimethylmercury ($(CH_3)_2Hg$). This conversion mostly occurs near the sediment:water

5

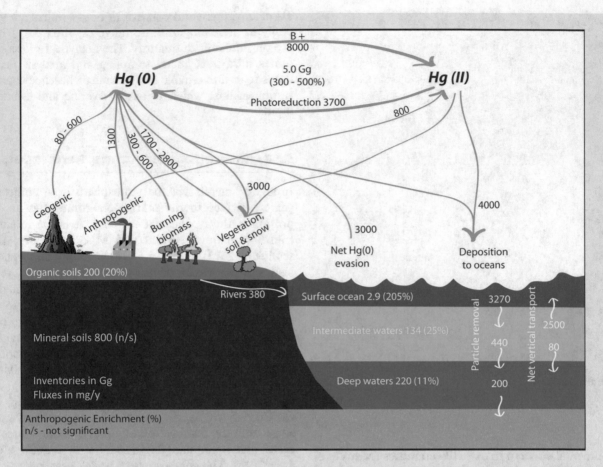

□ Figure 5.1 ► Box 5.1: Current estimates of the fluxes (mg/y), pools and enrichment (%) of mercury at the Earth's surface. adapted from Driscoll et al. (2013) and citations therein by A. Reichelt-Brushett. This is an unofficial adaptation of an article that appeared in an ACS publication. ACS has not endorsed the content of this adaptation or the context of its use

interface and primarily in anoxic environments with sulfate-reducing bacteria (Scwartzendruber and Jaffe 2012). Such conditions are commonly found in wetlands, in river sediments, in the coastal zones and the upper ocean (Driscoll et al. 2013; Gerlach 1981). The production of methylmercury drives the major human exposure route via the consumption of fish, particularly higher order fish with the greatest potential for biomagnification (Driscoll et al. 2013). Initiatives such as the United Nations Global Mercury Partnership, set up in 2005, are helping global efforts to protect human health and the environment from mercury emission to the atmosphere, water and land (► https://www.unep.org/globalmercurypartnership/).

5.2.3 Mining Operations

Metal ore deposits are a vital resource for mineral processing facilities that recover purified metals for human use. The extraction and processing of ores enhances the mobilisation and distribution of metals throughout the environment. Mining operations on land areas adjacent or close to marine waters are potential sources of marine pollution. The **major contamination source arises from waste rock and mine tailings**. Depending on the local geology, these can be impounded in tailings dams, disposed of on nearby land in erodible dumps or transported to the ocean for **deep-sea tailings placement (DSTP)**.

Many ore deposits contain a combination of several metals and all of these can be contaminants at a single mining or processing site. In Thailand, for example, elevated concentrations of Pb, Zn, Cu and Fe were all found near tin mining and processing operations (Brown and Holley 1982). Other examples of mining, ore processing and/or tailings disposal that impinge on marine environments include copper in Chile, Indonesia and Papua New Guinea (PNG); manganese on Groote Island, Australia, and North Maluku Province, Indonesia; gold on Lihir Island, PNG, and Buyat Bay, Indonesia; aluminium in Gladstone, Australia; nickel in New Caledonia and PNG. Yanchinski (1981) noted that there were 56 large-scale mining operations in the Caribbean region alone. Ultimately, the marine environment is a major sink for terrestrial runoff and river and ocean discharges from mining activities.

Adequate waste management in mining operations is important for the protection of surrounding ecosystems and, in tropical regions, the restrictions on mining waste disposal are often related to the seasonal variations in rainfall (e.g. Holdway 1992). In many cases, there are **agreed acceptable levels of discharge** of overburden into the environment. The mining of copper and gold at Ok Tedi in PNG is an example of the difficulties associated with managing mine waste. Gold and copper mining on the Ok Tedi River (a tributary of the Fly River) was estimated to contribute 750,000 tonnes per day of copper-rich mine tailings and 90,000 tonnes of sediment per day to the river (Apte and Day 1998). High sediment loads containing significant concentrations of copper could be detected some 600 km downstream and beyond the mouth of the Fly River into the ocean (Apte et al. 1995). Elevated concentrations of certain metals reported in seafood commonly eaten by Torres Strait Islanders prompted ongoing monitoring of Torres Strait metal concentrations (e.g. Gladstone 1996). Further study showed Ni, Cr, and As were elevated in sediments from the Gulf of Papua but less so in the Torres Strait (Haynes and Kwan 2002).

Significant **unintentional impacts from landslides and erosion** have occurred in mining operations in mountainous terrain with high rainfall (▶ Box 5.2). Such accidents highlight a need to develop sustainable approaches to **mine tailings management** and a range of alternatives such as tailings thickening and paste or cement production may be viable for some types of tailings (e.g. Adianyah et al. 2015; Saedi et al. 2021). Furthermore, such innovative technologies have the potential to address environmental problems for both the cement industry and tailings management (Saedi et al. 2021).

Box 5.2: Mariana Dam Disaster (Samarco Mine Tailing Disaster), Brazil

Dr. Pelli Howe, Environmental Scientist.

The collapse of an iron ore tailings dam in Mariana, Brazil, on the 5th of November 2015, has been described as Brazil's worst environmental disaster. Nineteen people were killed and the village of Bento Rodrigues was destroyed. 60 million m^3 of iron-rich waste was released and contaminated 620 km of freshwater ecosystems before arriving at the Atlantic Ocean (via the Doce River mouth) 17 days after the collapse. The United Nations reported the immediate death of 11 million tonnes of fish, and that the flow of mud had destroyed 1469 ha of riparian forest. The plume spread over 2580 km^2 in surface waters, two times the natural plume observed two months before the incident and high concentrations of dissolved metals (Pb, Mn, and Se) were also detected in the plume (Frainer et al. 2016) and further studies indicate future metal bioavailability and contamination risk in estuarine soils (Queiroz et al. 2018).

The Doce River mouth is recognised in the Ramsar Convention (2016) due to its extremely high biodiversity. Serious concerns were raised for local populations of thousands of marine flora and fauna, including the two most endangered cetaceans of the Southwestern Atlantic Ocean: the Guiana dolphin (*Sotalia guianensis*) and the Franciscana dolphin (*Pontoporia blainvillei*) (Frainer et al. 2016; Miranda and Marques 2016).

Manslaughter charges were laid due to the evidence of negligence. However, on the 25th January 2019, another tailings dam in Brazil, Brumadinho Dam, owned by the same company, collapsed, releasing 11 million tonnes of tailings and killing an estimated 270 people (Cionek et al. 2019) (◘ Figure 5.2).

◘ **Figure 5.2** ▶ Box 5.2: Tailings smother the land Mariana, Brazil. *Photo*: Senado Federal—Bento Rodrigues, Mariana, Minas Gerais, CC BY 2.0

5

Deep-Sea Tailings Placement

Continental margins or slopes are the boundary zones between the shallow shelf regions that surround most continents and the deeper **abyssal plains** of the sea floor. These areas have a steep profile, deep canyons and rugged topography (Ramirez-Llodra et al. 2010), which are the very features that make them attractive for DSTP, also known as submarine tailings disposal (STD). At the site of disposal (the end of a pipeline), which is **usually between 50 and 150 m in depth**, tailings spread over benthic communities (in the impact zone) (■ Figure 5.3). The pipeline is preferably near a submarine canyon, and once discharged, tailings are expected to travel downslope to the deep-sea floor and settle. Tailings density, local upwelling, currents and other conditions will influence the likelihood of tailings redistribution and settlement (Reichelt-Brushett 2012).

DSTP operations currently occur in Chile, France, Turkey, Indonesia, PNG and Norway. Most are unconfined discharges into the deep ocean, but many, such as in Norway, use confined disposal into deep fjords at 30–300 m depth. In the coral triangle, a hot spot of global marine biodiversity, 19 past, current and proposed DSTP sites exist (e.g. Reichelt-Brushett 2012).

The load of tailings to the ocean from a single STD operation is in the order of 10–100 s of thousands of tonnes a day, with the actual amount being site-spe-

cific. Once tailings are disposed of at continental margins and into deep-sea environments, the metal availability and toxicity to organisms will depend on the physicochemical conditions specific to the location.

There are various other scientific considerations that should be considered in the **risk assessment of DSTP** (see Vare et al. 2018; Stauber et al. 2022). For example, the **continental margins** in general are characterised by many **species-rich** deep-sea communities, mostly dependent on food produced in the upper layers of the ocean (Glover and Earle 2004; Ramirez-Llodra et al. 2010). Coral and sponge communities flourish in these areas where currents carry food to them. The heads of canyons are often productive nursery areas for fish (Yoklavich et al. 2000; Howard et al. 2020). Most publications on deep-sea biodiversity highlight a limited understanding and the need for further studies, (e.g. Etter et al. 1999; Brandt et al. 2007; Baker et al. 2010; German et al. 2011; Ramirez-Llodra 2020). Furthermore, canyon topography influences current patterns and local upwelling, pumping nutrients into the euphotic zone which stimulates primary productivity (Fernandez-Arcaya et al. 2016 and references therein). Events such as large storm waves and underwater earthquakes along with dense water cascades and hyperpycnal waters may trigger mass failures of unstable deposits in canyon heads and shelf edges (Fernandez-Arcaya et al. 2016 and references therein).

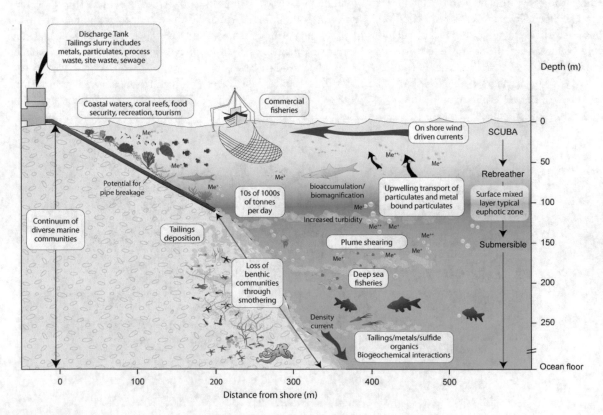

■ **Figure 5.3**   Conceptual diagram of submarine tailings disposal. *Image*: Reichelt-Brushett 2012, ■ Figure 5.3, CC BY 4.0: ► https://creativecommons.org/licenses/by/4.0

There is an important need to develop standardised risk assessment protocols that consider, environment, communities and cost–benefit analysis of alternatives. Precautionary principles should also be applied where knowledge is lacking, such as impacts of smothering, changes in water quality and contamination loads on ecosystem structure and function and diversity (many species are currently unknown to science).

Artisanal and Small-Scale Mining (ASM)

Between 10 and 15 million people in virtually all developing countries are involved in extracting over 30 different minerals using rudimentary techniques (Veiga and Baker 2004). Gold is the predominant metal extracted in **artisanal and small-scale mining (ASM)** (more specifically known as artisanal and small-scale gold mining (ASGM)) due to its high value and **easy extraction from ore using mercury** (◘ Figure 5.4). Koekkoek (2013) projected the annual amount of mercury released by ASGM in 70 countries to be 1608 tonnes. Such mining operations are often deemed illegal but provide pathways from poverty for rural communities. This extraction process requires large volumes of water for flushing and results in the deposition of fine sediments and mercury in river systems and eventually the ocean, along with many other environmental and social problems (Velasquez-Lopez et al. 2010; Male et al 2013) (◘ Figure 5.4b, c). Furthermore, the processing of the mercury–gold amalgam results in the volatilisa-

tion of mercury to the atmosphere. Unmapped legacy sites (◘ Figure 5.4d), are commonly close to rivers and provide a source of mercury to the marine environment via catchment runoff. Mercury can then get into the food chain including commercial and small-scale fisheries (► Box 5.1). In some countries like Indonesia, with its many islands, large population and limited farmland, communities rely heavily on the ocean for protein resources and have high consumption rates, and in some communities, seafood is part of every meal (◘ Figure 5.4e, f).

On Buru Island, Indonesia, gold was discovered in 2011 and ASGM commenced soon after. Sediment samples collected from the Wae Apu River and offshore from the river mouth just one year after the commencement of mining contained elevated mercury concentrations (Male et al. 2013). Several years later mercury concentrations in sediments had increased dramatically at some sites and some seafood sourced from the local fish markets also showed mercury concentrations of concern to human health (Reichelt-Brushett et al. 2017a).

Deep Seabed Mining

A new threat to marine ecosystems is the actual mining of the deep seabed. **Deep seabed mining** was raised as a possibility in the 1970s in the context of mining manganese nodules, but, at the time, technology and metal prices did not make the operations viable. Today, we

◘ **Figure 5.4** Artisanal gold mining and food resources on Buru Island, Eastern Indonesia: **a** one of the mine sites (Gogrea) in operation; **b** trommel operations to crush ore and extract the with mercury. Water is used to flush the spent ore to the tailings ponds, **c** tailings ponds are designed with small trenches to overflow to the river, **d** abandoned trommel operations on the Wae Apu River bank, **e** up to 90% of protein comes from the marine environment in many areas of Eastern Indonesia, Buru Island fish markets, **f** wild harvest of mangrove molluscs. *Photos*: A. Reichelt-Brushett

have reached a point where such initiatives are economically viable and technological developments have aided in accessibility to the deep sea. Geologic exploration of the deep sea has identified many sites rich in a wide range of mineral resources. In PNG alone, there are 60–100 exploration leases in deep waters around the island archipelagos. In 2018 one mine was in the verge of commercial operation in the sea near New Britain, PNG (Nautilus Minerals was developing the Solwara 1 copper and gold project, which is located at 1600 m depth). More recently, the mineral rich Clarion-Clipperton Zone in the Pacific, controlled by Nauru, has considerable commercial interest to extract cobalt and other metals. As with DSTP operations, deep seabed mining is another risk to the health of marine ecosystems that we do not fully understand. The **deep sea represents the largest and the least explored environment on Earth** (e.g. Ramirez-Llodra et al. 2010). Along with the limited biological assessment mentioned earlier, less than 20% of the deep ocean floor has been mapped (seabed2030.org) and only a small fraction of it has been studied to assess its environmental, economic and social values. Studies are ongoing and new benthic and pelagic species and habitats are continuously being discovered.

Impacts of seabed mining may include the removal and compaction of the substrate and the generation of large sediment plumes, possibly containing toxic metals released from the sediments (Hauton et al. 2017; Washburn et al. 2019). The ecotoxicological effects on mid-water and benthic communities exposed to environmental changes such as these are generally not well understood (Drazen et al. 2020; Mestre et al. 2017; Washburn et al. 2019). Some information exists on the specialised biological communities and functioning of deep seabed ecosystems, but it is insufficient to properly assess the impacts of these pressures on them or on the services they may provide for the well-being of humans (van den Hove and Moreau 2007). There are knowledge gaps and transdisciplinary challenges associated with deep seabed mining which need to be addressed to ensure unexpected and unacceptable negative effects do not result (e.g. Reichelt-Brushett et al. 2022).

The International Seabed Authority (ISA) was established in 1994 (see also ▶ Chapter 16). It is comprised of 167 Member States, and the European Union is mandated under the UN Convention on the Law of the Sea to organise, regulate and control all mineral-related activities in the international seabed area for the benefit of mankind as a whole. In so doing, ISA has the duty to ensure the effective protection of the marine environment from harmful effects that may arise from deep seabed-related activities.

Drill Cuttings
The exploration and production of oil and gas reservoirs have resulted in large quantities of **drill cuttings** (drilling mud, speciality chemicals and fragments of reservoir rock) being deposited onto the seafloor. Elevated concentrations of Cr, Cu, Ni, Pb, Zn and Ba relative to the natural (background) concentrations in sediment have been measured in North Sea drill cutting accumulations (Breuer et al. 2004) and some drilling muds have been shown to be toxic to biota (Tsventnenko et al. 2000).

5.2.4 Mineral Processing

It is usual to transport ore concentrates from what are usually remote mine locations to more accessible mainland facilities where the ore is refined to produce pure metals. These facilities are typically located at coastal sites for shipping access and are a major source of trace metal contamination from ore spillage, site runoff and other discharges.

Largely due to the presence of one of the world's largest zinc smelters, the Derwent estuary was for many years the most polluted water body in Australia and arguably the world, resulting in some of the highest reported metal concentrations in sediments and shellfish (Macleod and Coughanowr 2019). Contaminants included Zn, Hg, Cd, Pb, Cu and As were contributed to also by discharges from Australia's largest paper mill.

Lake Macquarie in New South Wales, Australia, suffered extreme lead, zinc, cadmium and selenium contamination from the 100-year operation of a lead–zinc smelter in the north of the lake (Batley 1987), again with residual high concentrations in sediments affecting shellfish. The lead smelter at Port Pirie in South Australia (Lent et al. 1992) is a further example of historical impacts that remain a concern today. Internationally, there are many such examples of legacy contamination. Contamination sources are hopefully now being better managed, but the costs of remediating many years of sediment contamination are generally prohibitive.

5.2.5 Urban and Industrial Discharges

Urban harbours and waterways have long been the recipient of metal contaminants from a variety of sources including shipping, licensed industrial discharges, sewer overflows and sewage treatment plant discharges and stormwater. There are activities worldwide that are attempting to better manage these sources (Steinberg et al. 2016).

Elevated metal concentrations including (but not limited to) Zn, Ni, Pb, Hg, Cu and Cr in sediments and organisms have been related to discharges from **sewage outfalls** (e.g. Kress et al. 2004; Echavarri-Erasun et al. 2007) and the less well-developed the sewage treatment facilities the more likely for adverse effects. In China alone, the amount of **industrial sewage** discharged into the aquatic environment was estimated to be 21.7 billion tonnes in 2008 (NBSC 2009 in Pan and Wang 2012). The Yangtze River, the Pearl River and the Minjiang River are the main rivers that carry metals into coastal areas, all of which contributed over 78% of the total discharge of metals in 2008 resulting in alarmingly high metal concentrations in sediment, water and biota at some coastal locations in China (Pan and Wang 2012).

Power Stations

Coal-fired power stations represent a significant industrial source of metal contaminants to estuarine waterways. The direct discharges of cooling waters frequently contribute copper and zinc from brass fittings, while arsenic and selenium as leachable components of coal ash are present in overflows or releases from ash dams (Schneider et al. 2014).

Stormwater

Stormwater is a significant contributor to metal contaminants. Increased urbanisation has meant that stormwater that would have been absorbed on land is now being directed via gutters and drains to the nearest waterways. Sediment traps and artificial wetlands offer partial solutions in selected areas, but within major urbanised catchments, stormwaters remain the major source of metal contaminants to sediments (e.g. Lau et al. 2009; Birch et al. 2015; Becouze-Lareure et al. 2019).

5.2.6 Other Sources

Shipping

Most large ships (cruise ships, cargo ships, container ships, tankers and ore carriers) are today equipped with exhaust gas scrubbers that discharge contaminants to the sea that might otherwise be emitted to the atmosphere. Washwater discharges from these scrubbers contain vanadium and nickel (derived from fuel oil combustion) together with copper and zinc as the major metal contaminants (Turner et al. 2017).

All ships use antifouling paints to prevent marine growth on their hulls. For a long time, tributyltin (TBT) was the major biocide used until its banning on small ships in the late 1990s with a slower decline in its use on bigger vessels. With a leaching rate near 5 μg/cm^2/day, it is a significant source of dissolved copper to the marine environment from both small ships in marinas and large vessels (Turner et al. 2017) (see ▶ Chapter 7 for detail on metal biocides and ▶ Chapter 8 for additional detail on TBT). Today, most antifouling paints are copper-based usually together with an organic biocide.

To protect steel hulls of large ships from corrosion, it is usual to fit sacrificial anodes, typically made of zinc or aluminium. As the anode supplies electrons to the cathode, it gradually dissolves, with the result that the steel cathode becomes negatively charged and protected against corrosion (Netherlands National Water Board 2008). For zinc anodes, release rates are typically 50–80 μg/cm^2/day. Zinc is a ubiquitous environmental contaminant, so it is difficult to estimate the contribution of this source to sediments in ports and harbours.

Dredging

Dredging is an activity that has the potential to release metals into the marine environment both from the dredging sites in ports and harbours (e.g. Reichelt and Jones 1994; Montero et al. 2013), and from the dredge spoil disposal that typically occurs in relatively deep (<100 m) offshore waters. Such activities are controlled by the London Dumping Convention (NAGD 2009) (see also ▶ Chapter 16) and the dredged sediment is contained within an agreed spoil ground. Consideration of the metals and their concentrations must be done prior to dredging activity in ports and harbours and dredge spoil dumping. Dredging physically disturbs and redistributes sediments, mobilising associated metals.

Shipwrecks and Dumping Sites

Shipwrecks are another source of metals. For example, the Gulf of Gdańsk, Poland, was an important place in Baltic trade routes and military activity, and numerous **shipwrecks** have been identified on its sea bed. Data published by the National Maritime Museum and the Maritime Office in Gdynia describe 25 wrecks in the Gulf of Gdańsk (NMM 2018 in Zaborska et al. 2019). Scientists observed that oil derivatives and metals from the *SS Stuttgart* wreck located near the entrance to the Port of Gdynia have contaminated a large part of the nearby sea bed (Rogowska et al. 2010, 2015).

Many other solid metal wastes have been dumped into the ocean, for example, the famous wreck dive site called Million Dollar Point in Vanuatu was created when the USA army dumped bulldozers, jeeps, trucks, semi-trailers, fork lifts and tractors off the point when they failed to come to a deal with the local community to buy the equipment and it was deemed cheaper to dump in the ocean rather than transport it back to the USA.

Agricultural Runoff

There are several sources of metals in **agricultural runoff**. For example, copper-based fungicides such as copper oxychloride are used in the agricultural industry and these may contribute to the contaminants in agricultural runoff. In addition, phosphate fertilisers naturally contain elevated concentrations of cadmium (Roberts 2014), and the cadmium concentration is directly correlated with the amount of total phosphorus in the fertiliser (e.g. Roberts 2014; Rayment 2011). Based on a nutrient budget for the tropical Port Moresby catchment, Eyre (1995) suggested that agricultural practices have caused a 2–fivefold increase in the phosphorus flux. This provides a potential source of cadmium to marine waters from land runoff, particularly during the wet season. The regulation of cadmium in commercial fertilisers has helped reduce the quantities of it entering cropping systems (Rayment 2011). See ► Chapter 7 for more detail on metal-based pesticides and biocides.

Acid Sulfate Soils (ASS)

Acid sulfate soils (ASS) are soils or sediments that contain highly acidic soil horizons or layers affected by the oxidation of iron sulfides (actual ASS), and/or soils or sediments containing iron sulfides or other sulphidic materials that have not been exposed to air and oxidised (potential ASS). The term acid sulfate soil generally refers to both actual and potential ASS. The acidic leachates and dynamic porewater chemistry influence metal cycling and behaviour (e.g. Gröger et al. 2011).

Acid sulfate soils are found in North America, South America, Asia, Africa, Oceania and Europe. They are expansive through the east coast of the USA, the east and west coasts of Mexico and Africa, the northern and eastern countries of South America, Vietnam, India, Bangladesh, China, Indonesia, PNG and much of Australia (Proske et al. 2014).

In eastern Australia, most ASS layers were deposited in the Holocene Epoch (10,000 years ago to the present) as a consequence of post-glacial sea level rise and the subsequent stillstand (a period of stable sea level), during which there was an infilling of estuarine embayment by marine and fluviatile sediments (Powell and Martens 2005). An estimated 666,000 ha of ASS occur within the Great Barrier Reef (GBR) catchments of Queensland, Australia. Extensive areas have been drained causing acidification, metal contamination, deoxygenation and iron precipitation in reef receiving waters (Powell and Martens 2005).

Landfills

Historical **coastal landfills** are potential sources of diffuse pollution due to leaching of contaminants through groundwater. For example, the United Kingdom alone has approximately 20,000 historical landfill sites without engineered waste management and leachate control (e.g. Cooper et al. 2012). Many of the historical landfill sites around the Thames River, London are in low-lying, flood-prone areas and recent sampling of sediments showed Cu, Pb and Zn contamination from anthropogenic sources (O'Shea et al. 2018). These legacy sites are problematic but enhanced environmental regulations have halted uncontained landfill sites in many countries. There are risks of increased **landfill leachates** impacting marine ecosystems in the future due to limited environmental regulatory controls or limited enforcement of them in some countries, although some mitigation reuse prospects for leachates are developing (Wijekoon et al. 2022). Cash-poor, low- and middle- income countries also accept (for a price) a large portion of the world's difficult-to-manage waste such as e-waste and known toxicants (e.g. Makam et al. 2018).

Desalination Plants

Desalination plants treat seawater to extract freshwater from the ocean. Metals are introduced to marine waters from desalination plants in waste brine with corrosion of metallic surfaces of the desalination system (e.g. Sadiq 2002) resulting in changes to community structure (Roberts et al. 2010). Desalination has become a reliable solution to water stress by supplying potable water in regions where freshwater supply is restricted. Some work is being done on brine management and pre-treatment to minimise the impacts of desalination from both brine and metal toxicity (Khan and Al-Ghouti 2021).

5.3 Metal Behaviour in Marine Waters

5.3.1 Metal Speciation

Metals enter aquatic systems in both dissolved and particulate forms. Of concern are the chemical species that make up these forms, their stability and possible transformations and transport that can occur over time. The chemical (and physical) speciation can be approached in several ways as will be discussed, but ultimately the concern is for their potential to cause biological effects to aquatic biota, i.e. their **bioavailability**, or potential to be taken up by aquatic organisms with the likelihood of toxic effects.

The speciation of dissolved metals in its simplest form involves the free metal ion, e.g. Cu^{2+}, and metals that are complexed or bound to complexes, both inorganic (e.g. sulfate, carbonate) or organic (e.g. natural humic and fulvic acids or other anthropogenic organic contaminants) (e.g. Rashid 1985; Florence and Batley 1988; Allen 1993; Batley et al. 2004). Hydrous iron and manganese oxides form binding sites for many metals, particularly in estuarine waters where these exist as colloidal species, often in heterogeneous mixtures with organic complexes. In some instances, these forms aggregate and are transported to bottom sediments.

The greatest bioavailability has been shown to involve the free metal ion, whereas **complexes** with dissolved organics are considerably less bioavailable. It is typical to use the term **lability** to describe the ability of metal–organic complexes to dissociate at a biological membrane and exert toxic effects. Strong metal complexes are usually non-labile, whereas weak complexes are typically labile. Lability is, however, operationally defined, so measurements of the labile fraction determined using a particular technique need to be assessed for their link to toxicity to sensitive biota. Organometallic complexes such as methylmercury where the metal is covalently bound to a carbon atom, are usually lipid-soluble (unless charged) and are directly transported across biological membranes and so have greater toxicity than other complexed forms.

The bioavailability of metals in estuarine and marine waters will be controlled by pH, salinity and redox potential, together with the presence of dissolved organic matter and its metal-binding constant (Luoma 1996; Batley et al. 2004). Many metals can exist in solution in different oxidation states, in particular Fe, Mn, Cr, As and Se, and these have different bioavailabilities and toxicities. Often both oxidation states can co-exist with transformations between forms highly dependent on redox potential. Manganese is a typical example, where in oxic waters, it exists as colloidal MnO_2, whereas in anoxic waters, Mn^{2+} prevails. Since MnO_2, as with hydrous iron (III) oxides, is able to adsorb metals, redox potential changes can significantly affect this association.

Metal speciation and toxicity (particularly of Cu, Pb, Ni, Zn and Cd) in natural waters depend on the pH, and the type and concentration of potential complexing **ligands**. The tendency for metals to form certain complexes is largely pH-dependent. Because the pH is easily changed in freshwater a large range of complexes are possible (Turner et al. 1981). In contrast, seawater is well buffered at a pH of 8.1–8.2 (Sadiq 1992), and the range of complexes that can form is more limited compared to freshwater. Variations in pH occur in coastal and estuarine environments due to freshwater mixing (e.g. Riba et al. 2003), groundwater inputs (e.g. Santos et al. 2011), and interactions with floodplain soils (see acid sulfate soils in this chapter).

There are several extensive reviews of metal chemistry in marine and aquatic waters which discuss metal behaviour in detail (e.g. Batley 1989; Sadiq 1992; Tessier and Turner 1995). This section provides a basis to build your knowledge upon and the literature cited are good places to seek more detailed information.

Metal Complexation
Complexes in freshwater are formed predominantly by oxygen-containing ligands (nitrates, phosphates, sulfates and organic acids), whereas most metal complexes in seawater are chloro- and carbonate or bicarbonate complexes (Kester 1986). Cadmium and copper exhibit the most notable differences in their toxicities between fresh and salt water. Cadmium is considered to be an extremely toxic metal (Hawker 1990; Sadiq 1992; Baird and Cann 2012) and this is true in freshwater environments, where the chloride concentration is low, and cadmium forms complexes with oxygen-containing ligands. These cadmium oxo-complexes are more labile and more bioavailable. With the abundance of chloride in seawater, more thermodynamically stable cadmium complexes are formed, which may be less bioavailable. Conversely, copper in seawater forms more labile chloro- and carbonate complexes (Steemann Nielsen and Wium-Andersen 1970).

A reduction in salinity, due to freshwater influxes from rainfall can be extreme during major weather events. Reduced salinity may extend far offshore and remain for several weeks, interfering with the dominance of metal–chloride complexes and subsequently altering trace metal availability.

Many metal ions including Fe, Co, Ni, Cu, Zn and Cd, are complexed by organic ligands in seawater which influences their speciation. For iron, these include siderophores (low-molecular-weight ligands produced by marine bacteria), humic and fulvic substances and microbial exopolymeric substances (porphyrins, saccharides and humic-like substances), while for copper, protein-based phytoplankton exudates, thiols and humic substances appear to dominate (Sato et al. 2021).

The dissolved organic matter content of marine waters is very low except in areas close to river discharges where it is more abundant and the nutrient availability affects the abundance of planktonic masses. Planktonic and other biotic interactions have been reported to affect copper speciation due to the complexing capacity of the associated organic molecules (e.g. Jones and Thomas 1988; Florence and Batley 1988). Hence, large temporal and spatial variations in the copper complexing capacity of seawater are expected and may cause large variations in the speciation of copper in seawater (Coale and Bruland 1990; Sadiq 1992).

Metal Interactions with Suspended Particles
Adsorption to the surfaces of **suspended particles** plays an important role in the removal of metals from seawater. The capacity for metals to bind to these surfaces depends upon the size, composition and abundance of the particles, concentration of other ions in solution, the charge of the metal ion and pH of the solution. Metal adsorption onto suspended particles is a significant mechanism controlling their solubility and dispersion (Batley and Gardner 1978; Florence 1986; Sadiq 1992; Reichelt-Brushett et al. 2017b). **Flooding events** can transport suspended sediment and freshwater loads far offshore (e.g. Devlin and Schaffelke 2009).

Positive and negative charges can be present simultaneously on solid surfaces of colloidal particles. It is commonly supposed that the **adsorption of ionic species** occurs in response to attraction by solids of opposite electrical charge. However, this oversimplification does not take into account of adsorption of non-electrolytes, selectivity between ions of like charge, adsorption of ionic species on solids of like charge or the reversal of charge that occurs when an excess of certain ionic species is adsorbed (Parks 1975). The binding capacity of colloidal material to trace metals depends on the net charge density of the particle.

Clays carry both a positive charge and a negative charge, and the magnitude of the charge depends on the type of clay. Positive charges are a result of the isomorphous replacement of structural oxygen by the hydroxyl groups: this leaves a negative charge deficiency. Negative charges are largely due to the isomorphous replacement of the structural silicon by aluminium or ferric iron, or the replacement of structural aluminium by magnesium or ferrous iron (Yariv and Cross 1979). Negative charges on clays are usually more common than positive charges. Positively charged metallic exchangeable cations are adsorbed in the inter-layer spaces (Yariv and Cross 1979). The capacity of clay minerals to adsorb ions is primarily governed by the degree of electrostatic attraction or **cation exchange capacity** (**CEC**) (e.g. Gambrell et al. 1976; Davranche and Bollinger 2001), which shows a linear relationship with particle size (Ormsby et al. 1962). Hydroxides and hydrous oxides of polyvalent cations such as aluminium, iron and manganese often cover clay minerals and some are potentially able to attract positively charged metal ions or species from seawater (e.g. Drever 1982).

Despite humic compounds and clays both being negatively charged, they do not necessarily repel one another: organo-clays can form as a result of intricate and varied forms of bonding involving physical and chemical forces. The reaction involved depends on the nature of the humic material, the type of clay minerals, the ionic composition of seawater and pH conditions. The chemical bonds associated with the organo-clays influence trace metal adsorption and desorption from particles. Some of the most prominent bonds are ionic bonds, coordinate bonds or ligand exchange and hydrogen bonds.

5.3.2 Evaluating Metal Speciation and Bioavailability in Marine Waters

Geochemical Modelling
There are a range of geochemical models that have been used to estimate the **equilibrium speciation of dissolved metals** (Batley et al. 2004) and the findings are not necessarily consistent. A challenge remains with the accommodation of binding to colloids and to natural organic ligands. Modelled complexation of Cu^{2+} in seawater varies widely with different major species predicted (e.g. Kester (1986) suggested that 90% of copper in seawater forms carbonate complexes; Hawker (1990)

90% of copper in seawater is in the form of copper hydroxide; and Sunda and Hanson (1987) suggested that organic complexation plays a major role). While the outputs of such models are of interest, they provide little information about metal bioavailability.

The Biotic Ligand Model (BLM)
A major advance in identifying the bioavailable concentration of metals in natural waters was offered by the **biotic ligand model** (**BLM**) as an extension of the free ion activity model (Pagenkopf 1983). The BLM is based on the assumption that metal bioavailability and toxicity are controlled by the binding of metals to a fish gill or cell membrane surface via a biotic ligand (BL). There is competition for this ligand between the free metal ion, protons, other metal ions, and organically and inorganically bound metals. **Application of the BLM requires a chemical speciation model and derived equilibrium constants for the metal–BL complexes.** The BLM has been applied extensively to metals in freshwaters, but there have been limited applications to marine waters apart from that for copper (Arnold et al. 2005). Limitations to current approaches to marine waters have been discussed by de Polo and Scrimshaw (2012). BLM models usually only predict metal toxicity to within a factor of 2.

Speciation Measurement
Measurement techniques offer a dynamic approach to the estimation of metal bioavailability, compared to the equilibrium approaches offered by modelling. In essence, these involve the measurement of an **operationally defined labile metal fraction** that is able to be related to the bioavailable or toxic form. Measurement techniques, as described by Batley et al. (2004), include separations using a chelating resin, electroanalytical techniques such as anodic stripping voltammetry and the use of diffusive gradients in thin films (DGT) to sample a metal fraction that diffuses via a gel membrane to a chelating resin-binding phase.

Toxicity Testing
The ultimate test of whether the chemical species are in forms that are potentially toxic requires the use of a sensitive bioassay (► Chapter 3).

5.4 Metal Behaviour in Marine Sediments

5.4.1 Metal Forms in Sediments

Metals in sediments are distributed among a range of chemical forms. In particular, these include metals adsorbed to iron and manganese oxyhydroxides often in association with organic matter in stabilised colloids in

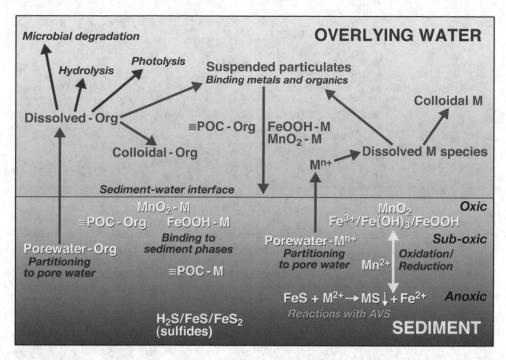

Figure 5.5 Conceptual model of major metal contaminant processes in sediments (where M indicates 'metal', POC is particulate organic carbon, and Org refers to organic compounds, so POC—Org is organics associated with POC). *Image*: Simpson, Stuart; Batley, Graeme, editors. Sediment quality assessment: A practical guide. CSIRO; 2016

surface waters, that ultimately aggregate and precipitate, particularly as the salinity increases to that of seawater, ultimately settling to bottom sediments. In anoxic waters, sulfides metals such as Cu, Cd, Ni, Pb and Zn form sulfides with low solubility products that will precipitate, thereby becoming enriched in marine sediments (Chester 1990). For this reason, sediments are referred to as a **sink** for metals with metals being most often found in higher concentrations in sediments than in marine waters at any particular site (Förstner 1987) (◘ Figure 5.5).

Typically, a zone of oxygenation extends from the sediment:water interface to about 1–5 cm below the sediment surface. This is known as the oxic zone. Below this is an intermediate sub-oxic zone of reduction overlying an anoxic zone, where dissolved oxygen is minimal and sulfate-reducing bacteria are active (◘ Figure 5.6a). If reduced sediments, high in metals are mobilised, the sulfide is oxidised to sulfate and the associated metals can be released from the sediments into the water column. Bioturbation (◘ Figure 5.6b–d) results in a mixing of the oxic and anoxic zones and benthic organisms can be in close contact with sediments, pore water (water that sits between sediment particles) and associated metals.

A number of selective extraction schemes have been devised to quantify the metal phases in sediments. These typically consider an exchangeable fraction, separate fractions for carbonates, organics (and sulfides) and metal oxyhydroxides, and a residual fraction comprising inert mineralised forms (Hass and Fine 2010). While these are useful for comparing sediments, the an-

alytical techniques are operationally defined and not truly selective and, more importantly, they do not relate to metal bioavailability. Analysis of metals in sediments typically uses a total acid digestion, however, a cold, dilute acid extraction has been shown to best relate to the bioavailable fraction and discriminate from the mineralised forms. This will dissolve iron and manganese oxyhydroxides and metal sulfides (Simpson and Batley 2016).

5.4.2 Metal Bioavailability in Sediments

Sediment Grain Size

In the metric scale sediment grain size range from clays ($<2\ \mu m$ diameter) to silts (2–$<63\ \mu m$) and sand ($<63\ \mu m$–2 mm). Gravel, rocks and other coarse material exceed 2 mm. Metal concentrations are highest in the finer clay and silt particles which have a greater surface area and hence more binding sites for metals. It is therefore important when reporting metal contamination to indicate the grain size. Most sediment quality guideline values apply to clay/silt sediments. The same metal concentration in a sandy sediment would potentially have greater bioavailability than that in a clay/silt sediment.

Pore Waters

Pore waters (or interstitial waters) are the waters occupying the spaces between sediment particles, typically comprising 30–80% of the sediment volume, depending

5

◘ Figure 5.6 Sediment redox interactions with organisms; **a** black reducing sediment just below the surface in a mangrove area. The aerial roots of mangroves are called pneumatophores and take up oxygen in these reducing sedimentary environments; **b** some fish such as a number of goby species excavate burrows to live in, sometimes they also share these burrows with shrimp who help in the excavation; **c** large sediments mounds (~25 cm diameter) processed by benthic organisms; **d** high-density benthic burrowers. *Photos* A. Reichelt-Brushett

on the grain size. Because they are in close association with sediments, porewater contaminants are in chemical equilibrium with those in sediments. Pore waters represent a diffusive pathway for metals to overlying waters. The speciation and bioavailability of porewater metals will be largely controlled by redox potential and pH. Burrowing organisms (◘ Figure 5.6) can introduce oxygenated waters into anoxic sediments, oxidising iron and manganese and other metal sulfides and releasing metals that can diffuse to overlying waters. The changing physicochemical conditions associated with bioturbation influence metal bioavailability and toxicity.

Acid Volatile Sulfides (AVS)
In sub-oxic sediments, amorphous iron and manganese monosulfides, so-called **acid-volatile sulfides (AVS)** (because they dissolve in dilute acids) can react readily with dissolved metals (e.g. Cd, Cu, Ni, Pb and Zn) forming insoluble metal sulfides. This means that if there are metals in the sediments or pore waters that can exchange with AVS, then there should be no bioavailable metals, and hence no toxicity, provided AVS is in excess of the available exchangeable metals. The exchangeable metals (so-called simultaneously extractable metals (SEM)) and AVS are both measured after dilute acid extraction of the sediments to determine if AVS > SEM (Simpson and Batley 2016 and citations therein).

5.5 Metal Uptake by Marine Organisms

The topic of bioaccumulation of metals in marine biota is very broad. The general principles of bioaccumulation are provided in ▶ Chapter 3 and further details on measuring rates of accumulation are provided in ▶ Chapter 6. These same principles apply to metals but the ways in which they interact with biota need to be specifically considered given that some metals are essential in small quantities for life. Furthermore, metals generally do not biomagnify as they are not lipophilic (there are a few exceptions, such as mercury when it is methylated).

Metal uptake by organisms not only depends on metal chemistry in the different environmental compartments the organism utilises (water, sediment and biota) but also on the metal interactions within an organism (e.g. an organism's ability to take up, regulate and detoxify accumulated metals). Such abilities vary between taxonomic groups and the different life stages of a species. Some filter-feeding marine organisms such as bivalve molluscs have been utilised in biomonitoring studies of pollution because they readily bioconcentrate and bioaccumulate contaminants (see ▶ Chapter 2, Box 2.1).

Importantly, metal ion **assimilation** (the processes of uptake) is essential for organisms and the pathways of uptake and methods of regulation help to satisfy their dietary requirements of essential metals while avoiding toxicity, known as **homeostasis**. It is possible for some non-essential metals to be taken up via these pathways and also regulated. When metal concentrations exceed an organism's ability to store and regulate them, then the organism exhibits toxic responses (Morrison et al. 1989). Here, we discuss processes of metal uptake and methods of regulation.

5.5.1 Transport Across Biological Membranes

There are three main metal uptake pathways by which metals enter organisms. The simplest route is via **passive diffusion** where metals diffuse through aqueous pores in cell membranes. The rate of diffusion is a function of the size of the molecule with larger colloidal species excluded. **Active transport** is driven by potential ionic gradients across the membrane, known as membrane-bound ion channels and higher metal concentrations can overwhelm their function (Morrison et al. 1989). Metal uptake termed **carrier-mediated transport** is facilitated by carrier molecules that involve interaction with the cell membrane (Morrison et al. 1989; Rainbow et al. 1990; Riba et al. 2003). Additionally, siderophores, organic chemicals excreted by organisms such as phytoplankton and bacteria, complex metals in seawater which can then be taken across the membrane (Vraspir and Butler 2009). Once inside an organism, diffusible metal species are able to bind to non-diffusible, intracellular ligands, and may be transferred to blood proteins and transported away from the uptake site (Rainbow et al. 1990).

5.5.2 Other Uptake Routes

Examples of other ways that organisms may accumulate metals include **ingested from food sources** when metals are bound to ingested sediment particles, or directly in the food they consume.

Once metals are ingested by organisms, the internal body conditions may then play a role in changing the metal speciation, as seen in the pearl oysters *Pinctada carchariarium* in Shark Bay, Australia (McConchie and Lawrence 1991). Cadmium concentrations in these oysters exceeded health guidelines, but there was no apparent anthropogenic or geologic contamination of the environment. It was discovered that cadmium in the water had adsorbed onto fine particles of negatively charged colloidal hematite (Fe_2O_3). During normal filter-feeding, oysters ingested these metal-loaded particles, and the lower pH conditions in the gut of the oyster induced a reversal of the hematite charge which caused cadmium to be released from the particles (i.e. become bioavailable) and subsequently absorbed by the oysters.

5.5.3 Metal Detoxification

Some organisms are able to regulate metal uptake through detoxification processes such as **sequestration in granules**, or by temporary storage in granules that are later excreted or made available for use (Rainbow et al., 1990). Similarly, **lysosomes** are used by many invertebrates such as crustaceans to sequester metals (e.g. Sterling et al. 2007). Lysosomes are organelles that regulate cellular waste. Other organisms can regulate and detoxify metals through the production of **metallothionein proteins** which can be enhanced by increased metal loads (Roesijadi and Robinson 1994; Roseijadi 1996). Metallothionein proteins not only play a role in the homeostasis of essential metals such as copper and zinc but can also be induced by non-essential metals such as cadmium (Stillman et al. 1999). Another detoxification system used by some algae is the production of a layer of **metal hydroxides** such as $Fe(OH)_3$ on the outside of the cell which adsorbs metals and thus renders them less toxic.

Some elements can provide protection from toxicity of other metals. A rather well-known example of this is the protective effects that selenium (an essential element) seemingly plays with mercury for some marine mammals (e.g. Kchrig et al. 2016) and seabirds (e.g. Ikemoto et al. 2004). The presence of selenium reduces the availability of some metal ions by forming insoluble compounds (Feroci et al. 2005).

Processes of detoxification require energy that is diverted from other needs or organisms such as sourcing food, growth and reproduction.

5.5.4 Metal Depuration

Depuration is the process that removes metals from the organism's body and is helpful in understanding the longer term ability of organisms to regulate metal loads and recover from toxicity after exposure. Many studies on the uptake and toxicity of metals now incorporate a recovery phase where organism health is monitored for a period after the exposure to the toxicant has ended. Depuration can occur as a reverse of passive and active diffusions (▶ Section 5.4.1), in organism waste, shedding of exoskeletons, reproductive outputs (e.g. eggs, sperm and offspring) and suckling of young in marine mammals. Some pathways of depuration need to be considered in biomonitoring studies (▶ Box 5.3).

Box 5.3: Cautious Considerations for Using Some Species as Biomonitors

Caution needs to be taken for some species intended for use in biomonitoring studies and consideration of depuration pathways is important. For an interesting example, corals have often been considered useful as biomonitors because they are sessile, easy to collect and the same genetic colonies can be subsampled over time. However, corals and some other marine species such as anemones, jelly fish and giant clams, contain symbiotic dinoflagellates (Symbiodiniaceae). Some thoughts about using corals as biomonitors:

- when corals are stressed they may bleach resulting in a loss of the Symbiodiniaceae;
- gametes take place 5–9 months to develop and can amount to about 80% of the tissue weight of a coral; the time of year sampling takes places in relation to annual spawning will influence the contribution of the gametes to the overall sample mass;
- clear differences exist for different metals in terms of the uptake and partitioning between the coral tissue, symbiotic dinoflagellates, gametes and skeleton as summarised by Reichelt-Brushett and McOrist (2003) and further investigated in Hardefeldt and Reichelt-Brushett (2015); and
- the density of the dinoflagellates can naturally vary widely within and between colonies depending on factors such as exposure of the coral surface to sunlight, therefore repeated sampling of the same colony is unlikely to have consistent ratios of host tissue and dinoflagellates.

For the reasons above, each type of biological material should be assessed separately or at least their mass contribution to the sample taken into the consideration in the assessment (◼ Figure 5.7).

◼ **Figure 5.7**   ▶ Box 5.3: Metals in corals can be lost from the colony though bleaching, coral may recover from a bleaching event and will slowly regain Symbiodiniaceae; **a** coral bleaching; **b** clear linear assemblages of Symbiodiniaceae in *Acropora muricata*. *Photos*: A. Reichelt-Brushett

5.6 Metal Toxicity to Marine Organisms

Metals can affect many factors associated with the health of marine organisms and the mode of action (▶ Chapter 3) will vary between taxonomic groups. ◼ Table 5.4 provides a summary of the types of effects measured in organisms after exposure to metals. These responses have been measured in a combination of field and laboratory studies and specific responses have only been measured in some species (e.g. moulting is a typical feature of crustaceans but is not common in other taxa). ◼ Table 5.4 provides some insight into what might be useful organism responses to be meas-

ured in future studies. Metal toxicity in marine waters and sediments can be considered in relation to the concentrations that cause detrimental effects and can be generally categorised in order of toxicity (◼ Table 5.5), although the order may vary depending on the environmental conditions as explained above. Values are based on the 95% species protection values except for the two metals that are known to biomagnify, mercury and cadmium, for which the 99% species protection value is recommended as default guideline values. The two metals, mercury and copper, that are among the greatest concern in marine waters in relation to toxicity and current sources will be discussed further.

■ **Table 5.4** Example of types of responses exhibited by marine biota when exposed to elevated metal concentrations in marine waters (content sourced from Weis 2014 and citations there in)

Response mechanisms	Examples of sublethal responses of organisms after metal exposure
Osmoregulation	Loss of osmoregulatory capacity through organ damage, inhibition of Na^+, K^+ and ATPase inhibition
Excretion	Reduced ammonia excretion due to decreased food intake Increased ammonia excretion due to increased protein catabolism (digestion) Decreased excretion rates and faeces production
Respiration and metabolism	Enhanced or reduced respiration rates due to changes in enzyme activity Enhanced mucous production Reduced metabolism is a strategy to minimise metal uptake or a result of damage to organs such as gills Affects metabolic biomarkers (e.g. ATP, histidine)
Feeding	Growth can be inhibited due to reduced feeding which may be a manifestation of changed behaviour (e.g. bivalves may remain closed to avoid poor water quality)
Digestion	Inhibition of digestive enzymes
Reproduction and development	Endocrine disruption (impacting the nervous system and reproductive system) Reduced sperm motility and reduced fertilisation success Delayed moulting Delayed sexual maturity Reduced larval motility Interference with metamorphosis
Embryonic development	Reduced fecundity Inhibition of embryo development Failure to hatch Deformities Genotoxicity
Growth	Reduced growth Impaired limb regeneration Reduced pigment production Abnormal growth Weakened bones and reduced calcification Tumours
Behaviour	Disruption of mating activity Erratic swimming Increased swimming speed Decreased motility Lack of interest in optimal habitat Reduced valve closing speed in molluscs Active avoidance of contamination (e.g. reduced burrowing and burying in contaminated sediments) Decreased prey capture Reduced escape ability Reduced olfactory (food odour) responses Reduced response to water-borne alarm substances resulting in increased vulnerability to predators Retardation of schooling behaviour and migration Neurotoxicity

■ **Table 5.5** General order of metal toxicity in marine environments (based on Australian and New Zealand default guideline values (DGVs) for 95% species protection for marine waters (99% for Hg and Cd) and the related DGVs for sediments, expressed as molar concentrations. Available at: ► https://www.waterquality.gov.au/anz-guidelines/guideline-values/default/water-quality-toxicants)

	General order of metal toxicity (most toxic first)
Marine water[a]	TBT[b] > Hg[c] > Cd > Cu, Ag > Pb, Se > Cr(VI), Co > Ni, As > Zn > V > Cr(III)
Marine sediments[a]	TBT[b] > Hg[c] > Ag, Cd > As, Pb > Ni > Cu > Cr > Zn

[a] where metals are not shown, insufficient data exist, [b] as nM Sn/kg, [c] inorganic Hg

5

5.6.1 Mercury Toxicity to Marine Biota

Data on the acute toxicity of mercury (II) chloride ($HgCl_2$) in marine water to biota was summarised by the US EPA (1985) and values ranged from 3.5 to 1700 µg/L, depending on the species. Hg (II) concentrations ranging from 10 to 160 µg/L inhibited growth and photosynthetic activity of marine plants (ANZECC/ARMCANZ 2000). Wu and Wang (2011) showed that an inorganic mercury concentration between 15 and 36 µg/L inhibited the growth of three marine algae species, and effects from organometallic forms of mercury were similar but interspecies variations were evident. Marine molluscs are relatively resistant to the effects of mercury exposure, but some life stages are sensitive. Fertilisation success of the European clam (*Ruditapes decussatus*) was significantly reduced compared to controls at 32 µg/L, the EC50 (EC50 is defined in ▶ Chapter 3) for embryonic development was 21 µg/L, and larval survival was affected at 4 µg/L after 11 days exposure (Fathallah et al. 2010). Responses of crustaceans to mercury exposure can be variable. The proteasome systems (a protein complex which degrades unneeded or damaged proteins) of the lobster *Homarus gammarus* and crab *Cancer pagurus* were severely inhibited by mercury at concentrations of 2 and 5 mg/L respectively (Götze et al. 2014) but these concentrations are unlikely to be reached in the environment.

Dietary pathways of exposure to mercury are also an important consideration for toxicity. Mercury (II) exposure via the diet of post-larvae *Penaus monodon* after 96 h resulted in changed swimming behaviour and this endpoint was more sensitive than biochemical biomarker endpoints including glutathione S-transferase (GST) and acetylcholinesterase activity (AChE) (Harayashiki et al. 2016). Dietary exposure of inorganic mercury concentrations below 2.5 µg/g to juvenile *P. monodon* for up to 12 days did not increase the body burden or impact AChE activity but resulted in a suppression of CAT activity at 2.5 µg/g (Harayashiki et al. 2018).

Mercury is generally less toxic to fish than some other metals, such as Cu, Pb, Cd or Zn. The main danger is diet-derived methylmercury, which accumulates in internal organs and exerts its effects by disruption of the central nervous system. Harayashiki et al. (2019) studied the effects of dietary exposure to inorganic mercury on fish activity and brain biomarkers on yellowfin bream (*Acanthopagrus australis*) and found that swimming activity increased for the test population after dietary exposure to food containing 2.4 and 6 µg/g although there was some variably between concentrations. Additionally, GST activity was also higher in mercury-exposed fish relative to controls, but differences were not found for other biomarkers.

Bioaccumulation of mercury from water may also be an issue. Bioconcentration factors of 5000 have been reported for mercury (II); factors for methylmercury ranged from 4000 to 85,000 (US EPA 1986). Further studies could focus on reproductive success resulting from maternally derived mercury to embryonic and larval stages.

5.6.2 Copper Toxicity to Marine Biota

EC10 values and no observed effect concentrations (NOECs) for the chronic effects of copper on marine algae range from 0.2–10 µg/L (ANZECC/ARMCANZ 2000) (examples provided in ☐ Table 5.6). The acute toxicity of copper to marine animals is also wide-rang-

☐ **Table 5.6** Some examples of marine species' sensitivity to copper (µg/L)

Species	Endpoint	Duration	EC10	Source
Algae				
Nitzschia closterium	Exponential growth	72 h	8	Johnson et al. (2007)
Phaeodactylum tricornutum	Exponential growth	72 h	1.5	Osborn and Hook (2013)
Echinoderms				
Evechinus chloroticus	Larval development	96 h	2.1	Rouchon (2015)
Corals				
Acropora aspera	Fertilisation success	5 h	5.8	Gissi et al. (2017)
Mollusc				
Mytilus galloprovincialis	Embryo development	48 h	5	Zitoun et al. (2019)
Haliotis iris	Larval development	96 h	0.7	Rouchon (2015)
Fish				
Sparus aurata	Juvenile growth	30 d	290 (NOEC)	Minghetti et al. (2008)
Atherinops affinis	Embryo development	12 d	62 (NOEC)	Anderson et al. (1991)

ing from 5.8 µg/L for blue mullet to 600 µg/L for green crab (US EPA 1986). Invertebrates, particularly crustaceans, corals and sea anemones are sensitive to copper. Fertilisation success is a sensitive endpoint for copper across a wide range of marine invertebrates. A good summary was provided by Hudspith et al. (2017), who reported that EC50 estimates ranged between 1.9 and 10,030 µg/L, with most species of corals, echinoderms, polychaetes, molluscs and crustaceans tested exhibiting EC50 estimates of <70 µg/L.

Gastropods seem to be more tolerant to copper and can accumulate quite high concentrations without toxic effects and typical 96-h LC50 values for snails are 0.8–1.2 mg Cu/L (ANZG 2018). Marine bivalves, including the mussel *Mytilus edulis* are more sensitive to copper, with a 96-h LC50 of 480 µg/L (Amiard-Triquet et al. 1986). Reduced growth and larval development were found at copper concentrations as low as 3 µg Cu/L in bivalves (ANZG 2018 and references therein).

Marine fish appear to be relatively tolerant of copper (ANZG 2018). In general, embryos of marine fish are more sensitive than their larvae, whereas larvae of freshwater fish are more sensitive than embryos.

5.7 Managing Metal Pollution

5.7.1 What Is 'Pollution'

Pollution is the introduction of harmful materials into the environment. These harmful materials are called pollutants. Many environmental scientists prefer the term **contaminants** as all pollutants are contaminants but not all contaminants are pollutants (see ▶ Chapter 1 for further details). The challenge is in defining what level of contamination constitutes pollution. Metal concentrations that are close to guideline values are deemed contaminated, but we don't have an accepted metric that defines polluted or heavily contaminated. Nevertheless, it is common among the general public to refer to water pollution and air pollution as representing something bad that needs management. We need to keep that in mind when we are talking about mildly contaminated waters and say they are polluted, as it over-exaggerates the problem.

5.7.2 Guideline Values

Many countries have guideline values (or similar) to protect marine ecosystems from contaminants including metals in marine waters (◘ Table 5.7) and separate guidelines for sediments (see also ▶ Chapter 3). Some countries have also developed protocols to determine

state/province or site-specific guidelines. The guideline values for waters are usually derived from rigorous toxicological testing, usually laboratory-based, using multiple aquatic species (e.g. Warne et al. 2018; ANZG 2018; Gissi et al. 2020). Most long-term guideline values are based on chronic toxicity testing, whereas short-term effects use acute toxicity data.

Chronic toxicity is defined as a lethal or adverse sub-lethal effect that occurs after exposure to a chemical for a period of time that is a substantial portion of the organism's life span (>10%) or an adverse effect on a sensitive early life stage. Acute toxicity is a lethal or adverse sub-lethal effect that occurs after exposure to a chemical for a short period relative to the organism's life span (Warne et al. 2018).

When chronic toxicity data are used in **species sensitivity distributions (SSDs)** to derive guideline values, it is usual to apply the 95% species protection value to most waters, defined as slightly to moderately contaminated, while the more conservative 99% species protection value is reserved for high conservation value waters, e.g. in a national park (see ▶ Chapter 6 for further details in SSDs). Species protection levels of 90 and 80% are both reserved for highly disturbed ecosystems and it is these values that likely constitute pollution as the stated goal with such waters is a continual improvement (ANZG 2018; ANZECC/ARMCANZ 2000).

For sediments, guideline values are commonly based on the 10th percentile of a ranking of effects data (Simpson and Batley 2007, 2016). Limited approaches to the chronic toxicity testing of whole sediments have been undertaken, e.g. for copper (Simpson et al. 2011), hampered until recently by the availability of a sufficient number of whole sediment test species (Simpson and Batley 2016).

5.8 Summary

Most metals and metalloids are found naturally in the marine environment in very low concentrations and many are essential to life. Anthropogenic inputs from atmospheric emissions, mining, mineral processing and urban and industrial discharges increase the concentrations in marine environments. Coastal waters are at greater risk of predominantly terrestrially derived metal sources.

Each metal behaves differently and organo-metallic metal forms are generally the most toxic to marine organisms. Understanding the sediment and water interactions, chemical behaviour and pathways of metal uptake in marine organisms are important for understanding toxic effects. Toxic effects vary between different metals and are different for different taxonomic groups.

◻ Table 5.7 Examples of water quality guidelines and protocols used in different countries and how to access them

Country	Guideline name and access location
Australia and New Zealand	Australian and New Zealand Guidelines for Fresh and Marine Waters ▶ www.waterquality.gov.au/anz-guidelines/guideline-values/default
Canada	Canadian Water Quality Guidelines for the Protection of Aquatic Life ▶ www.ccme.ca/en/summary-table
European Union	UK Technical Advisory Group, Water Framework Directive, Proposals for environmental quality standards for Annex VIII substances 2008
USA	National Recommended Water Quality Criteria -Aquatic life ▶ https://www.epa.gov/wqc/national-recommended-water-quality-criteria-aquatic-life-criteria-table

5.9 Study Questions and Activities

1. Select one metal and describe the sources, fate and consequences in the marine environment. This may be a metal that is explored in the chapter, or you may select a different metal of interest to you. Use diagrams if you wish.
2. Investigate the cycle and fluxes of a metal of environmental concern and explain it in your own words or create your own conceptual model (for an example see ▶ Box 5.1).
3. Investigate the guideline values for metals in marine waters in your region or country. What are three key points you notice about them in the context of this chapter?
4. What metal do you think is of most concern in the marine environment? Justify your answer.

References

Adiansyah JS, Rosano M, Vink S, Keir G (2015) A framework for a sustainable approach to mine tailings management: disposal strategies. J Clean Prod 108:1050–1062

Albarano L, Costantini M, Zupo V, Lofrano G, Guida M, Libralato G (2020) Marine sediment toxicity: a focus on micro- and mesocosms towards remediation. Sci Total Environ 708:134837

Allen HE (1993) The significance of trace metal speciation for water, sediment and soil quality criteria and standards. Sci Total Environ 134:23–45

Amiard-Triquet C, Berthet B, Metayer C, Amiard C (1989) Contribution to the ecotoxicological study of cadmium, copper and zinc in the mussel Mytilus edulis. Mar Biol 92:7–13

Anderson BS, Middaugh DP, Hunt JW, Turpen SL (1991) Copper toxicity to sperm, embryos and larvae of topsmelt Atherinops affinis, with notes on induced spawning. Mar Environ Res 31:17–35

Angel BM, Hales LT, Simpson SL, Apte SC, Chariton AA, Shearer DA, Jolley DF (2010) Spatial variability of cadmium, copper, manganese, nickel and zinc in the Port Curtis Estuary, Queensland, Australia. Mar Freshw Res 61:170–183

ANZECC/ARMCANZ (Australian and New Zealand Environment Conservation Council/Agriculture and Resource Management Council of Australia and New Zealand). (2000) Australian and New Zealand guidelines for fresh and marine water quality. Canberra, Australia. Available at: ▶ https://www.waterquality.gov.au/anz-guidelines/resources/previous-guidelines/anzecc-armcanz-2000. Accessed 12 Jan 2022

ANZG (Australian and New Zealand Governments) (2018) Australian and New Zealand guidelines for fresh and marine water quality. Australian and New Zealand Governments and Australian state and territory governments, Canberra ACT, Australia. Available at: ▶ www.waterquality.gov.au/anz-guidelines. Accessed 14 Nov 2021

Apte SC, Benko WI, Day GM (1995) Partitioning and complexation of copper in the Fly River, Papua New Guinea. J Geochem Explor 52:67–79

Apte SC, Batley GE, Szymczak R, Rendell PS, Lee R, Waite TD (1998) Baseline trace metal concentrations in New South Wales coastal waters. Mar Freshw Res 49:203–214

Apte SC, Day GM (1998) Dissolved metal concentrations in the Torres Strait and Gulf of Papua. Mar Pollut Bull 36:298–304

Arnold WR, Santore RC, Cotsifas JS (2005) Predicting copper toxicity in estuarine and marine waters using the Biotic Ligand Model. Mar Pollut Bull 50:1634–1640

Baird C, Cann M (2012) Environmental chemistry, 5th edn. W.H. Freeman and Company, New York, p 776

Baker MC, Ramires-Llodra EZ, Tyler PA, German CR, Booetius A, Cordes EE, Dubilier N, Fisher CR, Levin LA, Metaxas A, Rowden AA, Santos RS, Shank T, Van Dover CL, Young CM, Watén A (2010) Biogeography, ecology, and vulnerability of chemosynthetic ecosystems in the deep sea. In: McIntyre A (ed) Life in the world's oceans: diversity, distribution, and abundance. Wiley-Blackwell, Chichester, p 384

Barrie LA, Gregor D, Hargrave B, Lake R, Muir D, Shearer R, Tracey B, Bidleman T (1992) Arctic contaminants: sources, occurrence and pathways. Sci Total Environ 122:1–74

Batley GE, Gardner D (1978) Copper, lead and cadmium speciation in some estuarine and coastal marine waters. Estuar Coast Mar Sci 7:59–70

Batley GE (1989) Trace element speciation, analytical methods and problems. CRC Press Inc., Florida, p 360

Batley GE (1987) Heavy metal speciation in waters, sediments and biota from Lake Macquarie, NSW. Aust J Mar Freshw Res 38:591–606

Batley GE (2012) "Heavy metal"—A useful term. Integr Environ Assess Manag 8:215–215

Batley GE, Apte SC, Stauber JL (2004) Speciation and bioavailability of trace metals in water: progress since 1982. Aust J Chem 57:903–939

Batley GE, Simpson SL (2016) Introduction. In: Simpson S, Batley G (eds) Sediment quality assessment—A practical guide, 2nd edn. CSIRO Press, Clayton South, pp 1–14

Becouze-Lareure C, Dembélé A, Coquery M, Cren-Olivé C, Bertrand-Krajewski J-L (2019) Assessment of 34 dissolved and particulate organic and metallic micropollutants discharged at the outlet of two contrasted urban catchments. Sci Total Environ 651:1810–1818

Birch GF, Lean J, Gunns T (2015) Historic change in catchment land use and metal loading to Sydney estuary, Australia (1788–2010). Environ Monit Assess 187:594

Brady JP, Ayoko GA, Martens WM, Goonetilleke A (2015) Development of a hybrid pollution index for heavy metals in marine and estuarine sediments. Environ Monit Assess 187:306

Brandt A, De Broyer C, De Mesel I, Ellingsen KE, Gooday AJ, Hilbig B, Linse K, Thomas MRA, Tyler PA (2007) The biodiversity of the deep Southern Ocean benthos. Philos Trans R Soci Lond, Ser B, Biol Sci 362:39–66

Breuer E, Stevenson AG, Howe JA, Carroll J, Shimmield GB (2004) Drill cutting accumulations in the Northern and Central North Sea: a review of environmental interactions and chemical fate. Mar Pollut Bull 48:12–25

Brewer A, Dror I, Berkowitz B (2022) Electronic waste as a source of rare earth element pollution: leaching, transport in porous media, and the effects of nanoparticles. Chemosphere 287:132217

Brown BE, Holley MC (1982) Metal levels associated with tin dredging and smelting and their effect on intertidal reef flats at Ko Phuket, Thailand. Coral Reefs 1:131–137

Chapman PM (2007) Heavy metal—music, not science. Environ Sci Technol 41:6C

Chapman PM (2012) "Heavy metal"—cacophony, not symphony. Integr Environ Assess Manag 8:216–216

Chester R (1990) Marine geochemistry. Allen Unwin, Australia, p 698

Cionek VM, Alves GHZ, Tófoli RM, Rodrigues-Filho JL, Dias RM (2019) Brazil in the mud again: lessons not learned from Mariana dam collapse. Biodivers Conserv 28:1935–1938

Coale KH, Bruland KW (1990) Spatial and temporal variability in copper complexation in the North Pacific. Deep Sea Res Part a, Oceanogr Res Pap 37:317–333

Cooper N, Bower G, Tyson R, Flikweert J, Rayner S, Hallas A (2012) Guidance on the anagement of landfill sites and land contamination on eroding or low-lying coastlines (C718). CIRIA. Classic House, London, p 174

Davranche M, Bollinger J-C (2001) A desorption–dissolution model for metal release from polluted soil under reductive conditions. J Environ Qual 30:1581–1586

de Polo A, Scrimshaw MD (2012) Challenges for the development of a biotic ligand model predicting copper toxicity in estuaries and seas. Environ Toxicol Chem 31:230–238

De Vera J, Chandan P, Landing WM, Stupple GW, Steffen A (2021) Amount, sources, and dissolution of acrosol trace elements in the Canadian Arctic. ACS Earth and Space Chemistry 5(10):2686–2699

Devlin M, Schaffelke B (2009) Spatial extent of riverine flood plumes and exposure of marine ecosystems in the Tully coastal region, Great Barrier Reef. Mar Freshw Res 60:1109–1122

Drazen J, Smith C, Gjerde K, Haddock S, Carter G, Choy C, Clark M, Dutrieux P, Goetze E, Hauton C, Hatta M, Koslow A, Leitner A, Pacini A, Perelman J, Peacock T, Sutton T, Watling L, Yamamoto H (2020) Opinion: Midwater ecosystems must be considered when evaluating environmental risks of deep-sea mining. Proc Natl Acad Sci USA 117:17455–17460

Drever JI (1982) The geochemistry of natural waters. Prentice-Hall, New Jersey, p 239

Driscoll CT, Mason RP, Chan H-M, Jacob DJ, Pirrone N (2013) Mercury as a global pollutant: sources, pathways, and effects. Environ Sci Technol 47:4967–4983

Echavarri-Erasun B, Juames JA, García-Castrillo G, Revilla JA (2007) Medium-term responses of rocky bottoms to sewage discharges from a deepwater outfall in the NE Atlantic. Mar Pollut Bull 54(7):941–954

Etter RJ, Rex M, Chase MC, Quattro JM (1999) A genetic dimension to deep-sea biodiversity. Deep Sea Res I 46:1095–1099

Fathallah S, Medhioub MN, Medhioub A, Kraiem MM (2010) Toxicity of Hg, Cu and Zn on early developmental stages of the European clam (*Ruditapes decussatus*) with potential application in marine water quality assessment. Environ Monit Assess 171:661–669

Ferndez-Arcaya U, Drazen JC, Murua H, Ramierz-Llodra E, Bahamon N, Recasens L, Rotlant G, Company JB (2016) Bathymetric gradients of fecundity and egg size in fishes: a Mediterranean case study. Deep-sea Res 116:106–117

Feroci G, Badiello R, Fini A (2005) Interactions between different selenium compounds and zinc, cadmium and mercury. J Trace Elem Med Biol 18:227–234

Florence TM (1986) Electrochemical approaches to trace element speciation in waters: a review. Analyst 111:489–505

Florence M, Batley G (1988) Chemical speciation and trace element toxicity. Chem Aust, pp 363–366

Förstner U (1987) Sediment-associated contaminants -an overview of scientific bases for developing remedial option. Hydrobiologica 149:221–246

Frainer G, Siciliano S, Tavares DC (2016) Franciscana calls for help: the short and long-term effects of Mariana's disaster on small cetaceans of South-eastern Brazil. In: Conference: International Whaling Commission. Bled, Slovenia Volume: SC/66b/SM/04

Gaffney JS, Marley NA (2014) In-depth review of atmospheric mercury: sources, transformations, and potential sinks. Energy Emission Control Technol 2:1–21

Gambrell RP, Kahlid RA Patrick WH Jr (1976) Physicochemical parameters that regulate mobilisation and immobilisation of toxic heavy metals. In: Proceedings of the speciality conference on dredging and its environmental effects. American Society of Civil Engineers, New York, pp 418–434

Gerlach SA (1981) Marine pollution, diagnosis and therapy. Springer, Berlin, p 228

German CR, Ramirez-Llodra E, Baker MC, Tyler PA (2011) Deep realm research beyond the census of marine life: a trans-Pacific road map. In: Oceans'11 MTS/IEEE, Kona, Waikoloa, HI, USA. Available at: ► https://ieeexplore.ieee.org/document/6106995. Accessed 13 Dec 2021

Gissi F, Stauber J, Reichelt-Brushett A, Harrison P, Jolley DF (2017) Inhibition in fertilisation of coral gametes following exposure to nickel and copper. Ecotoxicol Environ Saf 145:32–41

Gissi F, Wang Z, Batley GE, Leung KMY, Schlekat CE, Garman ER, Stauber JL (2020) Deriving a chronic guideline value for nickel in tropical and temperate marine waters. Environ Toxicol Chem 39:2540–2551

Gladstone W (1996) Trace metals in sediments, indicator organisms and traditional seafoods of the Torres Strait. Final report of the Torres Strait baseline study. Great Barrier Reef Marine Park Authority, Townsville. Available at: ► https://elibrary.gbrmpa.gov.au/jspui/handle/11017/262. Accessed 15 Nov 2021

Glover L, Earle S (eds) (2004) Defying ocean's end—An agenda for action. Island Press, Washington, p 250

Götze S, Bose A, Sokolova IM, Abele D, Saborowski R (2014) The proteasomes of two marine decapod crustaceans, European lobster (*Homarus gammarus*) and edible crab (*Cancer pagurus*), are differently impaired by heavy metals. Comp Biochem Physiol Part C: Toxicol Pharmacol 162:62–69

Gröger J, Proske U, Hanebuth TJJ, Hamer K (2011) Cycling of trace metals and rare earth elements (REE) in acid sulfate soils in the Plain of Reeds, Vietnam. Chem Geol 288:162 177

Harayashiki CY, Reichelt-Brushett AJ, Butcher P, Lui B (2016) Behavioural and biochemical alterations in *Penaeus monodon* post-larvae diet-exposed to inorganic mercury. Chemosphere 164:241–247

Harayashiki CAY, Reichelt-Brushett A, Butcher P, Benkendorff K (2018) Ingestion of inorganic mercury by juvenile black tiger prawn (*Penaeus monodon*) diet alters biochemical biomarkers. Ecotoxicology 27:1225–1236

Harayashiki CAY, Reichelt-Brushett A, Benkendorff K (2019) Behavioural and brain biomarker responses in yellowfin bream (*Acanthopagrus australis*) after inorganic mercury ingestion. Mar Environ Res 144:62–71

Hardefeldt J, Reichelt-Brushett A (2015) Unravelling the role of zooxanthellae in the uptake and depuration of an essential metal in *Exaiptasia pallida*; an experiment using a model cnidarian. Mar Pollut Bull 96:294–303

Hauton C, Brown A, Thatje S, Mestre NC, Bebianno MJ, Martins I, Bettencourt R, Canals M, Sanchez-Vidal A, Shillito B, Ravaux J, Zbinden M, Duperron S, Mevenkamp L, Vanreusel A, Gambi C, Dell'Anno A, Danovaro R, Gunn V, Weaver P (2017) Identifying toxic impacts of metals potentially released during deep-sea mining—A synthesis of the challenges to quantifying risk. Front Mar Sci 4:368

Haynes D, Kwan D (2002) Trace metals in sediments from Torres Strait and the Gulf of Papua: concentrations, distribution and water circulation patterns. Mar Pollut Bull 44:1296–1313

Hawker DW (1990) Bioaccumulation of metallic substances and organometallic compounds. In: Connell D (ed) Bioaccumulation of xenobiotic compounds. CRC Press, Boca Raton, pp 187–207

Heffernan O (2019) Deep-sea dilemma. Nature 571:465–468

Herrmann H, Nolde J, Berger S, Heise S (2016) Aquatic ecotoxicity of lanthanum—A review and an attempt to derive water and sediment quality criteria. Ecotoxicol Environ Saf 124:213–238

Hass A, Fine P (2010) Sequential selective extraction procedures for the study of heavy metals in soils, sediments, and waste materials -a critical review. Crit Rev Environ Sci Technol 40:365–399

Holdway DA (1992) Control of metal pollution in tropical rivers in Australia. In: Connell D, Hawker D (eds) Pollution in tropical aquatic systems. CRC Press, London, pp 231–246

Howard IL, Vahedifard F, Williams JM, Timpson C (2018) Geotextile tubes and beneficial reuse of dredged soil: applications near ports and harbours. Proc Inst Civ Eng Ground Improv 171(4):244–257

Howard P, Parker G, Jenner N, Holland T (2020) An assessment of the risks and impacts of seabed mining on marine ecosystems. Fauna and Flora International: Cambridge, United Kingdom, p 336. Available at: ▶ https://cms.fauna-flora.org/wp-content/uploads/2020/03/FFI_2020_The-risks-impacts-deep-seabed-mining_Report.pdf. Accessed 14 Nov 2021

Hudspith M, Reichelt-Brushett AJ, Harrison PL (2017) Factors affecting the sensitivity of fertilization to trace metals in tropical marine broadcast spawners: a review. Aquat Toxicol 184:1–13

Ikemoto T, Kunito T, Tanaka H, Baba N, Miyazaki N, Tanabe S (2004) Detoxification mechanism of heavy metals in marine mammals and seabirds: interaction of selenium with mercury, silver, copper, zinc, and cadmium in liver. Arch Environ Contam Toxicol 47:402–413

Johnson HL, Stauber JL, Adams MS, Jolley DF (2007) Copper and zinc tolerance of two tropical microalgae after copper acclimation. Environ Toxicol 22:234–244

Jones GB, Thomas FG (1988) Effects of terrestrial and marine humics on copper speciation in an estuary on the Great Barrier Reef lagoon. Aust J Mar Freshw Res 39:19–31

Khan M, Al-Ghouti MA (2021) DPSIR framework and sustainable approaches of brine management from seawater desalination plants in Qatar. J Clean Prod 319:128485

Kehrig HA, Hauser-Davis RA, Seixas TG, Pinheiro AB, Di Beneditto APM (2016) Mercury species, selenium, metallothioneins and glutathione in two dolphins from the southeastern Brazilian coast: mercury detoxification and physiological differences in diving capacity. Sci Total Environ 186:95–104

Kester DR (1986) Equilibrium models in seawater: applications and limitation. In: Bernhard M, Brinckman F, Sadler P (eds) The importance of chemical speciation in environmental processes. Spinger-Verlag, Berlin, pp 337–363

Koekkoek B (2013) Measuring global progress towards a transition away from mercury use in artisanal and small-scale gold mining. M.A. Royal Roads University, Canada, p 74. ▶ https://viurrspace.ca/bitstream/handle/10170/567/Koekkoek_brenda.pdf?sequence=1&isAllowed=y. Accessed 8 Nov 2021

Kress N, Herut B, Galil BS (2004) Sewage sludge impact on sediment quality and benthic assemblages off the Mediterranean coast of Israel -a long-term study. Mar Environ Res 57(3):213–233

Lane TW, Morel MM (2000) A biological function for cadmium in marine diatoms. Proc Natl Acad Sci USA 97:4627–4631

Lau S-L, Han Y, Kang J-H, Kayhanian M, Stenstrom MK (2009) Characteristics of highway stormwater runoff in Los Angeles: metals and polycyclic aromatic hydrocarbons. Water Environ Res 81(3):308–318

Lent RM, Herczeg AL, Welch S, Lyons WB (1992) The history of metal pollution near a lead smelter in Spencer gulf, South Australia, Toxicol Environ Chem 36:139–153

Luoma SN (1996) The developing framework of marine ecotoxicology: pollutants as a variable in marine ecosystems. J Exp Mar Biol Ecol 200:29–55

Macleod C, Coughanowr C (2019) Heavy metal pollution in the Derwent estuary: history, science and management. Reg Stud Mar Sci 32:100866

Makam AN, Puneeth MK, Varalakshmi, Jayarekha P (2018) E-waste management methods in Bangalore. In: Proceedings of 2nd international conference on green computing and Internet of Things, ICGCIoT, 6–10, 8572976

Male Y, Reichelt-Brushett AJ, Pocock M, Nanlohy A (2013) Recent mercury contamination from artisanal gold mining on Buru Island, Indonesia—Potential future risks to environmental health and food safety. Mar Pollut Bull 77:428–433

Marx AK, McGowan HA (2011) Long-distance transport of urban and industrial metals and their incorporation into the environment: sources, transport pathways and historical trends. In: Zereini F, Wisemand C (eds) Urban airborne particulate matter. Springer, Berlin, pp 103–126

Maus V, Giljum S, Gutschlhofer J, da Silva DM, Probst M, Gass SLB, Luckeneder S, Leiber M, McCallum I (2020) A global-scale data set of mining areas. Sci Data 7:289

McConchie DM, Lawrence IM (1991) The origin of high cadmium loads to some bivalve molluscs from Shark Bay, Western Australia: a new mechanism for cadmium uptake by filter feeding organisms. Arch Environ Contam Toxicol 21:1—8

Mestre NC, Rocha T, Canals M, Cardoso C, Danovaro R, Dell'Anno A, Gambi C, Regoli F, Sanchez-Vidal A, Bebianno MJ (2017) Environmental hazard assessment of a marine mine tailings deposit site and potential implications for deep-sea mining. Environ Pollut 228:169–178

Milinovic J, Rodrigues FJL, Barriga FJAS, Murton BJ (2021) Ocean-floor sediments as a resource of rare earth elements: an overview of recently studied sites. Minerals 11:142

Minghetti M, Leaver MJ, Carpene E, George SG (2008) Copper transporter 1, metallothionein and glutathione reductase genes are differentially expressed in tissues of sea bream (*Sparus aurata*) after exposure to dietary or waterborne copper. Comp Biochem Physiol, c: Comp Pharmacol Toxicol 147:450–459

Miranda LS, Marques AC (2016) Hidden impacts of the Samarco mining waste dam collapse to Brazilian marine fauna-an example from the staurozoans (Cnidaria). Biota Neotrop 16:e20160169

NMM (National Maritime Museum) (2018) ▶ http://www.en.nmm.pl/underwaterarchaeology/wrecks-researched-by-nmm

Montero N, Belzunce-Segarra MJ, Gonzalez J-L, Nieto O, Franco J (2013) Application of toxicity identification evaluation (TIE) procedures for the characterization and management of dredged harbor sediments. Mar Pollut Bull 71:259–268

Morrison GMP, Batley GE, Florence TM (1989) Metal speciation and toxicity. Chem Br, pp 791–796

Munawer ME (2018) Human health and environmental impacts of coal combustion and post-combustion wastes. J Sustain Min 17:87–96

NAGD (2009) National assessment guidelines for dredging. Department of the Environment, Water, Heritage and the Arts, Commonwealth of Australia, Canberra, ACT

Netherlands National Water Board (2008) Sacrificial anodes, merchant shipping and fisheries. Emission estimates for diffuse sources. Netherlands Emission Inventory. Available from ► http://www.emissieregistratie.nl/erpubliek/documenten/Water/Factsheets/English/Sacrificial%20anodes,%20merchant%20shipping.pdf. Accessed 19 Nov 2021

Nriagu JO (1989) A global assessment of natural sources of atmospheric trace metals. Nature 338:47–49

Nriagu JO (1996) A history of global metal pollution. Sci, New Ser 272:223–224

Ormsby WC, Shartsis JM, Woodside KH (1962) Exchange behaviour of kaolins of varying degrees of crystallinity. J Am Ceram Soc 45:361–366

Osborn HL, Hook SE (2013) Using transcriptomic profiles in the diatom *Phaeodactylum tricornutum* to identify and prioritize stressors. Aquat Toxicol 138:12–25

O'Shea FT, Cundy AB, Spenser KL (2018) The contaminant legacy from historic coastal landfills and their potential as sources of diffuse pollution. Mar Pollut Bull 128:446–455

Pacyna JM, Pacyna EG (2001) An assessment of global and regional emissions of trace metals to the atmosphere from anthropogenic sources worldwide. Environ Rev 9:269–298

Pacyna JM, Sundseth K, Pacyna EG (2016) Sources and fluxes of harmful metals. In: Pacyna J (ed) Environmental determinants of human health. Springer, New York, pp 1–26

Pagenkopf GK (1983) Gill surface interaction model for trace-metal toxicity to fishes: role of complexation, pH, and water hardness. Environ Sci Technol 7:342–347

Pan K, Wang W-X (2012) Trace metal contamination in estuarine and coastal environments in China. Sci Total Environ 421–422:3–16

Parks GA (1975) Adsorption in the marine environment. In: Riley J, Skirrow G (eds) Chemical oceanography, vol 1, 2nd edn. Academic Press, London, pp 341–308

Powell B, Martens M (2005) A review of acid sulfate soil impacts, actions and policies that impact on water quality in Great Barrier Reef catchments, including a case study on remediation at East Trinity. Mar Pollut Bull 51:149–164

Proske U, Heijnis H, Gadd P (2014) Using x-ray fluorescence core scanning to assess acid sulfate soils. Soil Res 52:760–768

Queiroz HM, Nóbrega GN, Ferreira TO, Almeida LS, Romero TB, Santaella ST, Bernardino AF, Oterob XL (2018) The Samarco mine tailing disaster: a possible time-bomb for heavy metals contamination? Sci Total Environ 637–638:498–506

Rainbow PS, Phillips DJH, Depledge MH (1990) The significance of trace metal concentrations in marine invertebrates—The need for laboratory studies in accumulation strategies. Mar Pollu Bull 21(7):321–324

Ramirez-Llodra E, Brandt A, Danovaro R, De Mol B, Escobar E, German CR, Levin LA, Martinez Arbizu P, Menot L, Buhl-Mortensen P, Narayanaswamy BE, Smith CR, Tittensor DP, Tyler PA, Vanreusel A, Vecchione M (2010) Deep, diverse and definitely different: unique attributes of the world's largest ecosystem. Biogeosciences 7:2851–2899

Ramirez-Llodra E (2020) Deep-sea ecosystems: biodiversity and anthropogenic impacts. In: Banet C (ed) The law of the Seabed—Access, uses and protection of seabed resources. 90: Publications on Ocean Development, Brill, Nijhoff, pp 924–1922

Rashid MA (1985) Geochemistry of marine humic compounds. Springer-Verlag, New York, p 380

Rayment GE (2011) Cadmium in sugar cane and vegetable systems of northeast Australia. Commun Soil Sci Plant Anal 36:597–608

Reichelt AJ, Jones GB (1994) Trace metals as tracers of dredging activities in Cleveland Bay—Field and laboratory study. Aust J Mar Freshw Res 45:1237–1257

Reichelt-Brushett AJ, McOrist G (2003) Trace metals in the living and nonliving components of scleractinian corals. Mar Pollut Bull 46(12):1573–1582

Reichelt-Brushett AJ (2012) Risk assessment and ecotoxicology—Limitations and recommendations in the case of ocean disposal of mine waste. Oceanography 25(4):40–51. ► https://doi.org/10.5670/oceanog.2012.66

Reichelt-Brushett AJ, Stone J, Thomas B, Howe P, Clark M, Male Y, Nanlohy A, Butcher P (2017a) Geochemistry and mercury contamination in receiving environments of artisanal mining waste and identified concerns for food safety. Environ Res 152:407–418

Reichelt-Brushett AJ, Clark MW, Birch G (2017b) Physical and chemical factors to consider when studying historical contamination and pollution in estuaries. In: Weckström K, Saunders K, Gell P, Skillbeck G (eds) Applications of paleoenvironmental techniques in estuarine studies. Springer, Dordrecht, pp 239–276

Reichelt-Brushett A, Hewitt J, Kaiser S, Kim RE, Wood R (2022) Deep seabed mining and communities: A transdisciplinary approach to ecological risk assessment in the South Pacific. Integr Environ Assess Manag 18(3):664–673

Riba I, García-Luquea RE, Blasco J, DelValls TA (2003) Bioavailability of heavy metals bound to estuarine sediments as a function of pH and salinity values. Chem Speciat Bioavailab 15:101–114

Rijkenberg MJA, Middag R, Laan P, Gerringa LJA, van Aken HM, Schoemann V, de Jong JTM, de Baar HJW (2014) The distribution of dissolved iron in the west Atlantic Ocean. PLoS ONE 9:e101323

Roberts DA, Johnston EL, Knott NA (2010) Impacts of desalination plant discharges on the marine environment: a critical review of published studies. Water Res 44:5117–5128

Roberts TL (2014) Cadmium and phosphorous fertilizers: the issues and the science. Procedia Eng 83:52–59

Rogowska J, Wolska L, Namieśnik J (2010) Impacts of pollution derived from ship wrecks on the marine environment on the basis of s/s "Stuttgart" (Polish coast, Europe). Sci Total Environ 408:5775–5783

Rogowska J, Kudłak B, Tsakovski S, Gałuszka A, Bajger-Nowak G, Simeonov V, Konieczka P, Wolska L, Namieśnik J (2015) Surface sediments pollution due to shipwreck s/s "Stuttgart": a multidisciplinary approach. Stoch Env Res Risk Assess 29:1797–1807

Roesijadi G (1996) Metallothionein and its role in toxic metal regulation. Comp Biochem Physiol, Part C 113:117–123

Roesijadi G, Robinsonson WE (1994) Metal regulation in aquatic animals: mechanisms of uptake, accumulation and release. In: Malins D, Ostrander K (eds) Aquatic toxicology: molecular, biochemical and cellular perspectives. CRC/Lewis Publishers, Boca Raton, pp 387–420

Rouchon A (2015) Effects of metal toxicity on the early life stages of the sea urchin Evechinus chloroticus. Ph.D. thesis, Victoria University of Wellington, Wellington: New Zealand, p 211

Rudnicka-Kępa P, Zaborska A (2021) Sources, fate and distribution of inorganic contaminants in the Svalbard area, representative of a typical Arctic critical environment–a review. Environ Monit Assess 193:724

Sadiq M (1992) Toxic metal chemistry in the marine environment. Marcel Dekker, New York, p 390

Sadiq M (2002) Metal contamination in sediments from a desalination plant effluent outfall area. Sci Total Environ 287:37–44

Saedi A, Jamshidi-Zanjani A, Khodadadi Daran A (2021) A review of additives used in the cemented paste tailings: environmental aspects and application. J Environ Manage 289:112501

Salam MA, Dayal SR, Siddiqua SA, Muhib MI, Bhowmik S, Kabir MM, Rak AALE, Srednicki G (2021) Risk assessment of heavy metals in marine fish and seafood from Kedah and Selangor coastal regions of Malaysia: a high-risk health concern for consumers. Environ Sci Pollut Res 28:55166–55175

Santos IR, Glud RN, Maher D, Erler D, Eyre B (2011) Diel coral reef acidification driven by porewater advection in permeable carbonate sands, Heron Island, Great Barrier Reef. Geophys Res Lett 38:L03604

Sato M, Ogata N, Wong KH, Obata H, Takeda S (2021) Photodecomposition of natural organic metal-binding ligands from deep seawater. Mar Chem 230:103939

Saunders JK, Fuchsman CA, KcKay C, Rocap G (2019) Complete arsenic-based respiratory cycle in the marine microbial communities of pelagic oxygen-deficient zones. Proc Natl Acad Sci USA 116:9925–9930

Schneider L, Maher W, Potts J, Gruber B, Batley G, Taylor A, Chariton A, Krikowa F, Zawadzki A, Heijnis H (2014) Recent history of sediment metal contamination in Lake Macquarie, Australia, and an assessment of ash handling procedure effectiveness in mitigating metal contamination from coal-fired power stations. Sci Total Environ 490:659–670

Scwartzendruber P, Jaffe D (2012) Sources and transport—A global issue. In: Banks M (ed) Mercury in the environment—pattern and processes. University of California Press, London, pp 3–18

Shelley RU, Sedwick PN, Bibby TS, Cabedo-Sanz P, Church TM, Johnson RJ, Macey AI, Marsay CM, Sholkovitz ER, Ussher SJ, Worsfold PJ, Lohan MC (2012) Controls on dissolved cobalt in surface waters of the Sargasso Sea: comparisons with iron and aluminum. Glob Biogeochem Cycles 26:GB2020

Simpson S, Batley G (2007) Predicting metal toxicity in sediments: a critique of current approaches. Integr Environ Assess Manag 3:18–31

Simpson SL, Batley GE, Hamilton IL, Spadaro DA (2011) Guidelines for copper in sediments with varying properties. Chemosphere 75:1487–1495

Simpson S, Batley G (2016) Sediment quality assessment—A practical guide, 2nd edn. Clayton South, CSIRO Press, p 346

Stauber JL, Adams MS, Batley GE, Golding L, Hargreaves I, Peeters L, Reichelt-Brushett A, Simpson S (2022) A generic environmental risk assessment framework for deep-sea tailings placement using causal networks. Sci Total Environ 845:157311

Steelman Neilsen E, Wium-Andersen S (1970) Copper ion as poison in the sea and freshwater. Mar Biol 6:93–97

Steinberg PD, Airoldi L, Banks J, Leung KMY (2016) Special issue on the World Harbour Project—Global harbours and ports: different locations, similar problems? Reg Stud Mar Sci 8:217–370

Sterling KM, Mandal PK, Roggenbeck BA, Ahearn SE, Gerencser GA, Ahearn GA (2007) Heavy metal detoxification in crustacean epithelial lysosomes: role of anions in the compartmentalization process. J Exp Biol 210:3484–3493

Stillman MJ, Green AR, Gui Z, Fowle D, Presta PA (1999) Circular dichroism, emission and EXAFS studies of Ag(I), Cd(II), Cu(I), and Hg(II) binding to metallothioneins and modeling the metal binding site. In: Klaassen C (ed) Metallothionein IV. Birkhauser Verlag, Basel, pp 23–35

Sunda WG, Hanson AK (1987) Measurement of free cupric ion concentration in seawater by ligand competition technique involv-

ing copper sorption onto C_{18} SEP-PAK cartridges. Limnol Oceanogr 32:537–551

Tarbuck EJ, Lutgens FK (1987) The Earth—An introduction to physical geology. Merrill Publishing Company, Ohio, p 590

Tessier A, Turner D (eds) (1995) Metal speciation and bioavailability in aquatic systems. Wiley, Chichester, p 679

Thorne RJ, Pacyna JM, Sunddetth K, Paycyna EG (2018) Fluxes of trace metals on a global scale. In: DellaSala D, Goldstein M (eds) Encyclopedia of the Anthropocene. Elsevier, Waltham, pp 93–103

Trapasso G, Chiesa S, Freitas R, Pereira E (2021) What do we know about the ecotoxicological implications of the rare earth element gadolinium in aquatic ecosystems? Sci Total Environ 781:146273

Turner DR, Whitefield M, Dickson AG (1981) The equilibrium speciation of dissolved components if freshwater and seawater at 25 °C and at one atmosphere pressure. Geochim Cosmochim 45:855

Turner DR, Hassellov I-M, Ytreberg E, Rutgersson A (2017) Shipping and the environment: smokestack emissions, scrubbers and unregulated oceanic consequences. Elementa: Sci Anthropocene 5:45

Tsventnenko YB, Black AJ, Evans LH (2000) Development of marine sediment reworker tests with Western Australian species for toxicity assessment of drilling mud. Environ Toxicity 15:540–548

US EPA (United States Environmental Protection Agency) (1985) Ambient water quality criteria for mercury—1984. Criteria and Standards Division, US Environmental Protection Agency, Washington, Report No. EPA-440/5-85-026. Available at: ► https://www.epa.gov/sites/default/files/2019-03/documents/ambient-wqc-mercury-1984.pdf. Accessed 23 Nov 2021

US EPA (United States Environmental Protection Agency) (1986) Quality criteria for water. US Department of Commerce, National Technical Information Service, Springfield, PB87-226759, EPA 440/5 86-001. Available at: ► https://www.epa.gov/wqc/national-recommended-water-quality-criteria-aquatic-life-criteria-table. Accessed 23 Nov 2021

Vare L, Baker M, Howe J, Levin L, Neira C, Ramirez-Llodra E, Reichelt-Brushett A, Rowden A, Shimmield T, Simpson S, Soto E (2018) Scientific considerations for the assessment and management of mine tailing disposal in the deep sea. Deep-sea tailing deposition: what is at risk? Front Mar Sci 5:17

Veiga MM, Baker RF (2004) Protocols for environmental and health assessment of mercury released by artisanal and small-scale gold miners. GEF/UNDP/UNIDO Global Mercury Project, Vienna, p 289. Available at: ► https://delvedatabase.org/resources/protocols-for-environmental-and-health-assessment-of-mercury-released-by-artisanal-and-small-scale-gold-miners. Accessed 8 Nov 2021

Velasquez-Lopez PC, Veiga MM, Hall K (2010) Mercury balance in amalgamation in artisanal and small-scale gold mining: identifying strategies for reducing environmental pollution in Portovelo-Zaruma, Ecuador. J Clean Prod 18(3):226–232

Vraspir JM, Butler A (2009) Chemistry of marine ligands and siderophores. Ann Rev Mar Sci 1:43–63

Warne MStJ, Batley GE, van Dam RA, Chapman JC, Fox DR, Hickey CW, Stauber JL (2018) Revised method for deriving Australian and New Zealand water quality guideline values for toxicants—Update of 2015 Version. Prepared for the revision of the Australian and New Zealand Guidelines for Fresh and Marine Water Quality. Australian and New Zealand Governments and Australian state and territory governments, Canberra, p 48. Available at: ► https://www.waterquality.gov.au/anz-guidelines/guideline-values/derive/laboratory-data. Accessed 24 Nov 2021

Washburn TW, Turner PJ, Durden JM, Jones DO, Weaver P, Van Dover CL (2019) Ecological risk assessment for deep-sea mining. Ocean Coast Manag 176:24–39

Weis JS (2014) Physiological, developmental and behavioural effects of marine pollution. Springer, Dordrecht, p 452

Wijekoon P, Arundathi Koliyabandara P, Cooray AT, Lam SS, Athapattu BCL, Vithanage M (2022) Progress and prospects in mitigation of landfill leachate pollution: risk, pollution potential, treatment and challenges. J Hazard Mater 421:126627

Wu Y, Wang W-X (2011) Accumulation, subcellular distribution and toxicity of inorganic mercury and methylmercury in marine phytoplankton. Environ Pollut 159:3097–3105

Yanchinski S (1981) Who will save the Caribbean? New Sci 92:388–389

Yariv S, Cross H (1979) Geochemistry of colloidal systems, for Earth scientists. Springer-Verlag, New York, p 449

Yoklavich MM, Greene HG, Cailliet GM, Sullivan DE, Lea RN, Love MS (2000) Habitat associations and deep-water rockfishes a submarine canyon: an example of a natural refuges. Fish Bull 98:625–641

Zaborska A, Siedlewicz G, Szymczycha B, Dzierzbicka-Głowacka L, Pazdro K (2019) Legacy and emerging pollutants in the Gulf of Gdańsk (southern Baltic Sea)—Loads and distribution revisited. Mar Pollut Bull 139:238–255

Zitoun R, Clearwater SJ, Hassler C, Thompson K, Albert A, Sander S (2019) Copper toxicity to blue mussel embryos (*Mytilus galloprovincialis*): the effect of natural dissolved organic matter on copper toxicity in estuarine waters. Sci Total Environ 653:300–314

Oil and Gas

Angela Carpenter and Amanda Reichelt-Brushett

Contents

6.1   Introduction – 130

6.2   Sources of Oil in the Marine Environment – 131
6.2.1   Naturally Seeped Oil – 131
6.2.2   Oil from Land-Based Sources – 131
6.2.3   Oil from Shipping Activities – 132
6.2.4   Oil from Exploration and Exploitation Activities – 134
6.2.5   Oil from Atmospheric Sources – 137
6.2.6   Natural Gas – 138

6.3   Fate of Oil in the Marine Environment – 138
6.3.1   Physical Factors Influencing Oil Degradation – 138
6.3.2   Oil Clean-Up and Recovery Activities – 140
6.3.3   Oil Spill Monitoring Activities – 141

6.4   Consequences of Oil Pollution – 141
6.4.1   Impact of Oil on Marine Ecosystems – 143
6.4.2   Impact of Oil on Marine Taxa – 143
6.4.3   Economic Damage from Oil Pollution – 146

6.5   Planning for, and Responding to, Oil Pollution
         Incidents – 147
6.5.1   Context – 147
6.5.2   Oil Pollution Preparedness and Response Co-operation (OPRC) – 147
6.5.3   Contingency Planning, Risk Assessment, and Emergency
         Response – 147

6.6   Summary – 148

6.7   Study Questions and Activities – 150

         References – 150

© The Author(s) 2023
A. Reichelt-Brushett (ed.), *Marine Pollution—Monitoring, Management and Mitigation*,
Springer Textbooks in Earth Sciences, Geography and Environment,
https://doi.org/10.1007/978-3-031-10127-4_6

Acronyms and Abbreviations

EEA	European Environment Agency
EMSA	European Maritime Safety Agency
EPC	Environmental Pollution Centres
EU	European Union
FEPA	Food and Environment Protection Act
GBRMPA	Great Barrier Reef Marine Park Authority
GESAMP	Joint Group of Experts on the Scientific Aspects of Marine Environmental Protection
GNOME	General NOAA Operational Modelling Environment
IMO	International Maritime Organisation
IRC	Incident Response Contract
IOPC	International Oil Pollution Compensation Funds
IPIECA	International Petroleum Industry Environmental Conservation Association
ITOPF	International Tanker Owners Pollution Federation
LNG	Liquified natural gas
MARPOL	International Convention for the Prevention of Pollution from Ships
nmVOCs	Non-methane Volatile Organic Compounds
NOAA	National Oceanic and Atmospheric Administration
NRC	National Research Council
OCHA	United Nations Office for the Coordination of Humanitarian Affairs
OPEP	Oil pollution emergency plan
OPRC	International Convention on Oil Pollution Preparedness, Response and Co-operation
OSPAR	Oil Spill Prevention, Administration and Response
PAHs	Polycyclic aromatic hydrocarbons
PSSA	Particularly Sensitive Sea Area
PTTER	PTT Exploration and Production Public Company Limited
REMPEC	Regional Marine Pollution Emergency Response Centre for the Mediterranean Region
UNEP	United Nations Environment Programme
USA	United States of America
US EPA	United States Environmental Protection Agency
VOCs	Volatile organic compounds

6.1 Introduction

Oil is a generic term that can cover a very wide range of natural hydrocarbon-based substances and also refined petrochemical products. **Crude oil** and **petroleum products** can have a range of physical properties on the basis of which their behaviour in the marine environment can differ widely. These properties range from viscosity (the rate at which liquid flows), density, and specific gravity (density relative to water) (e.g. Hollebone 2017). The chemistry of crude oil can also differ widely, and includes *saturates* (aliphatics including alkanes) which are composed of carbon and hydrogen only, **aromatics** such as benzene, tolune, and xylenes, and polycyclic aromatic hydrocarbons (PAHs) such as naphthalene; and **asphaltenes** which are a mix of very large organic compounds (Hollebone 2017). The latter can have a tar-like consistency at low temperature, cannot be distilled, and can form **tar balls** (also known as surface residual oil balls; see ◘ Figure 6.1) that wash ashore and are persistent in the environment (e.g. Lorenson et al. 2009).

Oil can enter the marine environment from a range of sources, both natural and anthropogenic (human activities) and, depending on type, can persist for a long period of time (persistent) or disperse fairly rapidly (non-persistent). In a definition adopted by the International Oil Pollution Compensation Funds (IOPC), non-persistent oil (volatile in nature and tending to dissipate rapidly by Evaporation) consists of hydrocarbon fractions at least 50% of which, by volume, distils at a temperature of 350 °C and at least 95% of which distils at a temperature of 370 °C (Anderson 2001). Spilled oil can have a range of consequences on the environment, often depending on its level of persistence. This chapter considers sources, fate, and consequences of oil pollution, together with a brief discussion of gas pollution. It then considers some of the mitigation strategies available to minimise impacts from shipping and oil ex-

Figure 6.1 Tar balls are seen washed ashore on Okaloosa Island in Fort Walton Beach, Florida on June 16, 2010 (*Image* Gatrfan ► www.drewbuchanan.com—Own work, Public Domain, ► https://commons.wikimedia.org/w/index.php?curid=10668105)

ploration and exploitation activities, and to prepare for and respond to oil pollution in the event of an emergency (see also ► Chapter 16).

6.2 Sources of Oil in the Marine Environment

Oil is a naturally occurring substance that can enter the marine environment from a range of sources. These include seeps from the seabed; industrial and urban runoff into coastal waters and into rivers; and anthropogenic sources (human activities) including shipping and oil exploration and exploitation activities, and atmospheric sources through incomplete combustion of petroleum products from cars and aircraft. The total volume of oil entering the marine environment annually is unclear (GESAMP 2007). A study by the National Research Council (NRC 2003) estimated that more than 1.3 million tonnes (76 million gallons) of oil enter the marine environment each year, while a study by the European Environment Agency (EEA 2007) estimated global oil inputs of between 1 and 3 million tonnes (264–793 million gallons). However, a 2001 study estimated that around 4.63 million tonnes (1223 million gallons) a year entered the marine environment from transport and oil production activities (Clark 2001). Despite the lack of any accurate estimate of inputs, and the wide variation

between estimates, these figures highlight that oil pollution poses a threat to the marine environment. This section will review the various sources of oil—both crude oil and refined oil (petroleum and other hydrocarbon products) entering the marine environment.

6.2.1 Naturally Seeped Oil

It is estimated that around 45% of crude oil entering the global marine environment comes from natural seepages through faults and cracks leading to geological formations under the seabed associated with oil reservoirs; that figure rises to around 60% in North American waters (NRC 2003). Globally, these natural seeps, which may have lasted for hundreds of thousands of years, are estimated to release around 600,000 tonnes (159 million gallons) of crude oil annually (NRC 2003). A study of natural offshore seepages and the accumulation of tar balls along the California coastline (Lorenson et al. 2009) identified that there are "*prolific, frequently chronic, onshore, and offshore shallow oil seeps*" in the area where tar balls wash ashore, together with oil seeping from rocky outcrops and cliff faces. This is in addition to anthropogenic sources such as shipping and offshore drilling rigs in the area. High concentrations of methane gas in the water column around natural seeps were also identified in that study. ☐ Figure 6.2 illustrates the fate and distribution of such naturally seeped oil showing that lighter petroleum hydrocarbons migrate to the ocean surface, together with methane, and enter the atmosphere. Heavier petroleum hydrocarbons either form a slick on the ocean surface, from which they can wash ashore, or fall back to the seabed.

6.2.2 Oil from Land-Based Sources

Land-based sources of oil include municipal wastewaters, urban runoff, and river discharges, together with industrial discharges, including non-refinery discharges and refinery discharges. River discharges can, according to the NRC (2003), come from both inland basins draining via major rivers to the sea, and from coastal basins which discharge directly to the sea. Urban runoff, such as untreated or insufficiently treated municipal sewage and stormwater, comes from a range of sources including cars, machinery, fuel spills, and waterborne or airborne pollutants that fall onto hard surfaces and are washed into rivers flowing into the sea (US EPA 2017). Pollution from land-based oil refineries can also occur as a result of accidents or operational discharges from coastal refineries or power plants, as in the case of the Jiyeh Power-Plant spill in Lebanon in 2006 (► Box 6.1) and intentionally

6

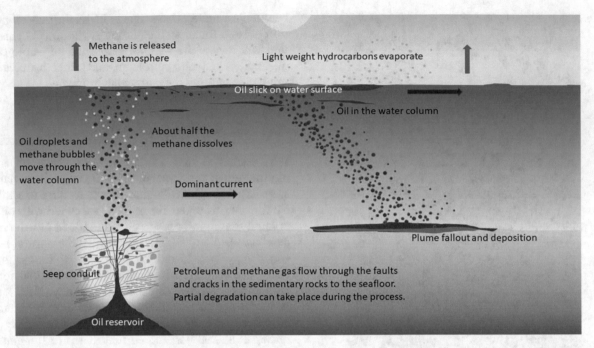

■ Figure 6.2 Fate and distribution of naturally seeped oil in the marine environment. Adapted from Cook, Woods Hole Oceanographic Institution by A. Reichelt-Brushett

---

Box 6.1: Jiyeh Power Plant Spill, July 2006

A large release of oil came from a land-based source occurred from the Jiyeh power-plant, a coastal plant located 28 km south of Beirut in Lebanon in July 2006. After a missile attack on fuel tanks at the power plant, an estimated 12,000–15,000 tonnes of heavy fuel oil entered the marine environment of the eastern Mediterranean Sea (Greenpeace 2007). However, efforts by the Lebanese Army, Civil Defence and other agencies, prevented approximately 20,000 tonnes of heavy fuel oil from leaking into the sea (UNEP/OCHA 2006). The ongoing conflict between Israel and Lebanon, including a naval blockade, meant that clean-up operations to deal with the spill were delayed, and more than 150 km of Lebanese coastline and 10 km of Syrian coastline were contaminated by oil as the spill was carried out to sea, and also dispersed along the coast of Lebanon (Greenpeace 2007). As a result, some sandy beaches and rocky shorelines were extremely contaminated by oil, while others were only moderately or lightly contaminated (Greenpeace 2007).

(e.g. with the destruction of over 730 oil wells by retreating Iraqi forces in February 1991 where huge volumes of hydrocarbons were released directly into the marine environment while additional volumes entered the marine environment indirectly as fall-out from numerous oil fires (Saenger 1994). The Gulf War oil spill was over 10 times greater than the Torrey Canyon spill (see also ▶ Chapter 1) in Britain in 1961 (Saenger 1994).

Due to the nature of land-based inputs, there is little data to estimate loads from these sources (NRC 2003), although recommendations were made in the NRC report to improve that situation. Those recommendations included a requirement for sampling of both petroleum hydrocarbons and PAHs, establishment of regular monitoring sites on major rivers, collection of stormwater

samples from urban coastal cities, and determining how much of the inputs are from petroleum-derived PAHs rather than total petroleum hydrocarbon inputs that included non-petroleum-derived PAHs (Saito et al. 2010).

6.2.3 Oil from Shipping Activities

There are three main sources of oil pollution entering the marine environment from shipping activities, operational, accidental, and illegal (intentional) sources.

Operational pollution takes place during normal ship operations. In the case of oil tankers, this includes discharging bilge water containing oil (up to 15 mg/L in water) through an oily water separator, and discharging oily waters from cargo tanks. These are innovations introduced under the MARPOL Convention, Annex

Case 1: *Atlantic Empress*, West Indies, 1979—Collision

On 19 July 1979 two fully loaded very large crude carriers, the *Atlantic Empress* and the *Aegean Captain*, collided around 10 miles off the island of Tobago during a tropical storm. Both vessels started to leak oil and both caught fire, resulting in loss of life. While the fire on the *Aegean Captain* was brought under control, and the vessel was towed to Curacao where its oil cargo was removed, the *Atlantic Empress* continued to burn and was towed 300 nautical miles from land between 21 and 22 July. Following a large explosion on board it sank on 2 August. An estimated 287,000 tonnes of oil spilled from the *Atlantic Empress*, making it the largest ship source spill of all time (ITOPF 2018a).

Case 2: *Torrey Canyon*, United Kingdom, 1967—Grounding

On 18 March 1967 the oil tanker *Torrey Canyon* ran aground on Pollard Rock on the Seven Stones Reef near Lands' End, Cornwall. The entire cargo of 119,000 tonnes of Kuwait crude oil spilled from ruptured tanks over a 12 day period. Although various measures were attended to mitigate the slick (including aerial bombardment to try and burn the oil), a large oil slick reached the south west coast of England and the beaches and harbours of the Channel Islands and Brittany. The *Torrey Canyon* was the first oil spill to receive major coverage in the media, ultimately leading to International Conventions covering compensation for damage caused by tanker spills (ITOPF 2018b).

1, and have led to a decrease in operational pollution from shipping (IMO 2018a). Specific limits on inputs in various sea regions are also in place under the MARPOL Convention, with operational discharges permitted only outside those areas. Those limits are discussed in ▶ Section 6.4 of this chapter.

Accidental pollution can include oil spills from shipping accidents, including from collisions at sea (▶ Box 6.2, Case 1), or from a vessel sinking (e.g. sinking in severe weather conditions). It can also come from derelict vessels (unseaworthy vessels that are tied up and abandoned), vessels that have run aground (▶ Box 6.2, Case 2), and from historic wrecks [where residual fuel seeps out of a vessel that has sunk; commonly wrecks from World War II (NRC 2003)]. Poorly coordinated responses and challenges with international boundaries exasperate the impacts of accidental spills (▶ Box 6.3). Accidental pollution can also result from mechanical failure during loading and offloading operations in ports (although this may also be categorised as operational pollution).

The Torrey Canyon disaster discussed in ▶ Box 6.2 (Case 2) ultimately led to the introduction of the MARPOL Convention (International Convention for the Prevention of Pollution from Ships, 1973, as modified by the Protocol of 1978 relating thereto (also known as MARPOL 73.78). This Convention, operated under the aegis of the IMO, covers accidental and operational oil pollution under Annex I. Its other Annexes cover pollution from chemicals (Annex II); goods in packaged form (Annex III): sewage (Annex IV); garbage (Annex V); and from air pollution (Annex VI)[1] (See also ▶ Chapter 16).

Dr. Alexis Vizcaino, Marine Geologist and Geophotograher.
On November 13th, 2003, oil tanker MV *Prestige* burst offshore Galicia in Spain. Six days later, at 133 nautical miles off the Spanish coast in international waters, the *Prestige* broke in two and sank. Between the alarm call and the sink, the oil tanker was erratically sailing. The Spanish, Portuguese, and French authorities denied docking the boat somewhere protected from the heavy sea. The *Prestige* sailed 243 nautical miles (473 km) in those six days. Because of the semi-erratic path, the *Prestige* spilled around 70,000 tons (63,503 tonnes) of heavy oil in the open ocean (◻ Figure 6.3). The extensive oil slick spread throughout the Galician Coast and beyond, becoming the largest oil-spill catastrophe in Spain. The oil spill, helped by currents and wind, had an acute effect on seabird mortality (e.g. shags population dropped 11%; Martinez-Abraín et al. 2006) because of the extent of the pollution and the time of the year (many seabirds get trapped because of lives in contact with the sea surface). It also affected other species (e.g. sea turtles), although limited information is known.
Both stern and bow now sit on the southwestern edge of the Galicia Bank, at 3565 and 3830 m water depths, respectively (Ercilla et al. 2006). Numerous studies have been carried out to understand the associated risks. In 2004, the Spanish government hired Repsol (Spanish Oil Company) to extract the 13,700 tons (12,428 tonnes) left in the boat. The total clean-up cost USD 12 billion, being the third most expensive in the world's history.

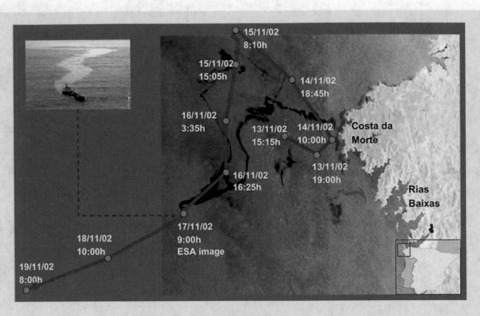

⊡ Figure. 6.3 ▶ Box 6.3: Composite image published in Marine Pollution Bulletin by Albaigés et al. (2006). The main picture corresponds to a satellite image of the spill taken on November 17th, 2002, two days before the sinking location with the conspicuous heavy oil slick. In red, track of oil tanker SM Prestige between 13th and19th November 2002. On the upper left corner, there is an image of SM Prestige and its oil slick on November, 17th. *Image*: Envisat ASAR instrument on 17 November 2002 and processed by the Earth Watching team on 20 November at ESRIN in Italy and finally published by ESA (▶ https://earth.esa.int/web/earth-watching/natural-disasters/oil-slicks/content/-/asset_publisher/71yyBC1MdfOT/content/galicia-spain-november-2002-april-2003/ (European Space Agency (ESA), licensed under CC BY-SA 3.0 IGO)

The *Prestige* was produced by Hitachi Shipbuilding Engineering in Japan, registered in Greece with Bahamas flag owned by the Liberian company Mare Shipping, run by the Greek company Universe Maritime and the load owned by the Russian Crown Resources. The boat was in poor ship structure conditions (▶ http://www.shipstructure.org/case_stud ies/prestige/Prestige.pdf). It sank after a complete hull failure in international waters off Galicia because no action was taken to avoid the accident or mitigate its impact.

Illegal oil pollution can occur during the normal operations of a ship (e.g. where the ship needs to clean out its bilges or cargo tanks and intentional discharges of oily waters and other noxious substances occur in restricted areas or in facilities provided by ports). In 2006, it was estimated that around 3000 major illegal hydrocarbon dumping incidents occurred annually in European waters (UNEP 2003). Many of these illegal discharges take place during the hours of darkness, where the likelihood of detection is low. However, the introduction of satellite monitoring for oil spills (e.g. the CleanSeaNet service operated by the European Maritime Safety Agency [EMSA 2018]), has increased the chance of a vessel being caught, or a spill being associated with a specific vessel. This has led to reduced rates of illegal discharges (see the discussion on oil spill monitoring in the North Sea in ▶ Box 6.4).

6.2.4 Oil from Exploration and Exploitation Activities

According to Devold (2013), oil production platforms in the sea include shallow-water complexes where several independent platforms are linked by bridges, large concrete fixed structures placed on the sea bottom (with oil storage cells resting on the sea bottom), and floating production platforms where crude oil is pumped from sub-sea wells and stored on board until it can be offloaded on to shuttle tankers. ⊡ Figure 6.4 illustrates a range of typical oil and gas production facilities.

Oil pollution from exploration and exploitation activities can be both operational (including from drilling activities or pipelines transporting oil to the shore) and accidental.

Operational pollution can come from (1) drilled cuttings (solid material removed from drilled rock, to-

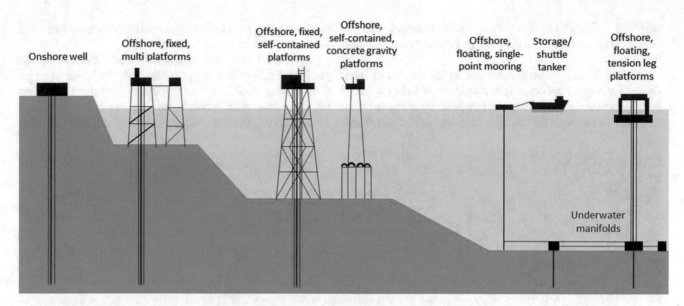

Figure 6.4 Basic schematic of oil and gas production facilities and infrastructure 2013. Adapted from Devold by A. Reichelt-Brushett

gether with muds and chemicals) which can contain oil-based muds, (2) produced water (water that comes from the reservoir as a by-product of oil and gas extraction), and (3) displacement water (seawater used for ballasting the storage tanks of offshore installations which is discharged into the sea when oil is loaded into those tanks). Piles of drilled cuttings at the base of a platform can have persistent, chronic, and long-lasting impacts (Henry et al. 2017). Cuttings were the major source of oil entering the marine environment of the North Sea from oil production activities between 1984 and 1999 (OSPAR Commission 2001). Stricter standards in that region have, however, resulted in virtually no oil entering the sea from that source since 2012 (OSPAR Com-

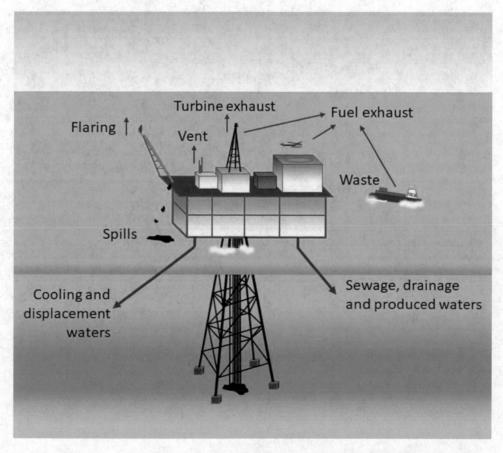

Figure 6.5 Sources of oil inputs from oil and gas exploration activities 2010. Adapted from OSPAR by A. Reichelt-Brushett

mission 2010, 2014), illustrating that measures can be taken to reduce or even halt oil pollution from specific sources.

Operational pollution can also come from the drains, sewage, and cooling water outflows on oil platforms, while oil can also be released by atmospheric deposition through flaring (the burning off of gas from the reservoir), and exhaust gases from the platform and the vessels serving it (◘ Figure 6.5).

Accidental pollution can come from spills from the platform itself, or occur during the transfer of oil to ships, from storage tanks, or from pipelines. The most devastating oil spills occur, however, when there is a major accident on an oil platform, examples are set out in ► Box 6.4.

Box 6.4: Examples of Accidental Oil Spills from Oil Platforms

Case 1: *Deepwater Horizon*, Manaconda, Gulf of Mexico, 2010

The drilling rig exploded and sank with the loss of 11 lives on 20 April, 2010. Over the next 87 days, and until the well was capped on 15 July, a spill of approximately 4.9 million barrels (1 barrel=159L) of oil occurred, despite the use of a range of technologies including skimmer ships, floating booms, in situ burning, and the use of dispersants. Oil from the accident dispersed widely (see ◘ Figure 6.5) and ultimately came ashore on the northern Gulf coasts of Louisiana, Mississippi, Alabama and Florida, with heavy oiling occurring along much of that coast and the most severe oiling observed in November 2010 (National Commission on the BP Deepwater Horizon Oil Spill and Offshore Drilling 2011). The *Deepwater Horizon* oil spill had an acute impact on marine ecosystems, marine biota, and commercial fisheries in the Gulf of Mexico. For example, it had a detrimental effect on the abundance and composition of bacterial communities in beach sands in the Gulf of Mexico (Kostka et al. 2011), as well as on marsh vegetation in coastal salt-marshes in the Barataria Bay of Louisiana in the northern Gulf (Lin and Mendelssohn 2012). It also had a major impact on the wetlands of the Mississippi River Delta system in Louisiana, an area responsible for approximately one third of US commercial fish production (Mendelssohn et al. 2012).

◘ **Figure 6.6**   ► Box 6.4: With its ability to penetrate clouds and haze, this radar image taken by the ASAR instrument on the Envisat satellite illustrates the usefulness of radar imagery for oil pollution detection and mapping. Oil slicks and sheen from the ongoing BP/Deepwater Horizon spill—patchy in places—are spread across an area of 67,476 km² in the northeast Gulf of Mexico. The western edge of the area of slicks and sheen extends beyond the left side of the radar image. Radar image was taken at 03:48 UTC (10:48 pm June 21 local time). *Image*: "Deepwater Horizon Oil Spill—Envisat ASAR Image, June 21, 2010" by SkyTruth is licensed under CC BY-NC-SA 2.0

Case 2: *Montara* Wellhead Platformoil rig spill, Australian Continental Shelf, 2009 (by marine pollution student Dean Senti, 2019, Southern Cross University).

Located 250 km northwest of the Australian mainland (12° 41′S 124° 32′E), and close to Ashmore Reef and Cartier Island, the H1 Well of the *Montara* Wellhead Platform blew on 21 August 2009 ( Figure 6.7). Over the next 74 days, it released approximately 23.5 million litres of light crude oil into the Timor Sea (Spies et al. 2017). Oil flowed into Australia's Commonwealth waters and, after several days, into Indonesian waters (Spies et al. 2017). Estimates of volumes of oil discharge range from 1500 to 2000 barrels (238,500 to 318,000 L) per day. The oil surface slick was visible from space, even remaining visible for weeks after the well was sealed on January 13, 2010.

Impacts of the spill included: oiled and dead seabirds; dolphins, seabirds, turtles, and sea snakes all interacting within the oil slick; reduction of seagrass and fisheries in Indonesia; and copious amounts of oil nearshore and at sea were also recorded (Spies et al. 2017). PTT Exploration and Production Public Company Limited (PTTEP) Australasia, owner of the *Montara* Wellhead Platform commissioned an environmental monitoring program, establishing a world-class body of independent scientific research (PTTEP Australasia 2013) in response to the incident. This research has studied marine life and ecosystems of the Timor Sea, making it the most comprehensive database ever generated for this region (PTTEP Australasia 2013). See also: ▶ https://www.awe.gov.au/environment/marine/marine-pollution/montara-oil-spill/scientific-monitoring-studies

◘ **Figure 6.7**   ▶ Box 6.4: Aerial photo of the *Montara* offshore oil platform and West Atlas mobile drilling rig. On August 21, 2009, a well on the platform blew out as a new well was being drilled, and both the rig and the platform were immediately evacuated. Oil and condensate are spewing uncontrolled into the Timor Sea off Western Australia, and will continue to do so for at least 7–8 weeks until a new rig can be brought into the vicinity to drill a relief well. *Photo*: Chris Twomey, courtesy of WA Today "Montara Oil Spill—August 25, 2009" by SkyTruth is licensed under CC BY-NC-SA 2.0

6.2.5   Oil from Atmospheric Sources

As mentioned in previous sections, oil can also enter the marine environment from atmospheric sources, including the deposition of oil from incomplete combustion from car engines or from aircraft, and from flaring and exhaust gases from oil production platforms and their service vessels. The NRC (2003) identifies that volatile compounds escape to the atmosphere during production, transport and refining of hydrocarbons, with heavier compounds (Volatile Organic Compounds (VOCs)) being deposited to the sea surface.

VOCs are also emitted by tankers at all stages including loading, tank cleaning and during a voyage, and the amounts discharged depend on properties of the cargo, the degree of mixing and temperature variations during a voyage, and whether vapour recovery systems are used (NRC 2003). The International Petroleum Industry Environmental Conservation Association (IPIECA 2018) identifies that there are two types of recovery systems that can reduce VOC emissions by up to 90% from oil storage ships: active recovery units use compression, condensation, absorption, and/or adsorption to recover VOCs, while passive recovery units use vapour-balanced loading/unloading with non-methane VOCs (nmVOCs) as a blanket gas for storage vessels.[1]

6.2.6 Natural Gas

Natural gas is a fossil fuel that has been mined from the sea floor for many years. Its formation, along with oil, depends on the ambient conditions in the reservoirs where the remains of animals and plants sank to the ocean floor, were compressed under deep layers of sediment, and then were converted by bacteria (aided by pressure and temperature) into precursor substances, and ultimately into hydrocarbons (WOR 2010a). The main offshore natural gas deposits are located in the Middle East, with the South Pars/North Dome located on the Iran/Qatar border considered to hold an estimated 38 trillion cubic metres making it the largest natural gas reserve in the world. Other offshore natural gas fields are located in the North Sea (currently the world's most important gas producing area), the Gulf of Mexico, Australia, Africa, and in the Commonwealth of Independent States (CIS; made up of Russia, Belarus, Ukraine, Armenia, Azerbaijan, Kazakhstan, Kyrgyzstan, Moldova, Turkmenistan, Tajikistan, and Uzbekistan) and also off India, Bangladesh, Indonesia and Malaysia (WOR 2010a). Between 2001 and 2007, 25% of natural gas came from the North Sea, 25% from Australasia, and 15% from each of the Gulf of Mexico and the Middle East (WOR 2010a).

Natural gas is transported by sea in its cooled form as liquified natural gas (LNG), where it has been liquified at about minus 160 °C to make it more easily transportable. While the process of liquification consumes energy and adds to transportation costs, it is cheaper to ship LNG by sea in tankers rather than through pipelines (WOR 2010a). However, by doing so, emissions to air from ships, and at LNG facilities, will contribute to atmospheric pollution and, ultimately, marine-pollution (WOR 2010b). Notably, fire tests are mostly conducted on land and of the few tests on water the results differ markedly due to the turbulent mixing of the LNG and water resulting in greater heat transfer (Hissong 2007). In a response to these differences it has been recommended that modelling of evaporation and burning considers; the use of time-varying release rates, the use of physical properties of LNG, not methane, to use a time-step analysis that captures the time-varying release rates and the changes in properties resulting from composition changes as the LNG vaporises or burns, and to use parameters that reflect actual spill conditions, including turbulence between the water and LNG (Hissong 2007).

In a study by Malačič and co-authors (2008) on a proposed LNG terminal in the Gulf of Trieste in the northern Adriatic Sea 4 potential areas of concern were highlighted including products of chlorinated water, inputs of toxic mercury and other metals by sediment resuspension, toxicity of aluminium compounds in seawater due to galvanic protection of metal constructions, cooled and chlorinated seawater released by terminals spreading around the Gulf. It was suggested that the proposed technology could be improved to reduce the environmental impact, for example, construction of diffusers at the end of outfall pipes could lead to lower the resuspension and the use of alternative methods to battle the fouling (e.g. ultrasound as an antifouling approach) (Malačič et al. 2008). Furthermore, technology is available that allows for underwater platforms and pipes, that if used, would avoid the need for the offshore terminal, however, these would have their own impacts. Furthermore, air could be used as a heating medium instead of seawater (Malačič et al. 2008).

6.3 Fate of Oil in the Marine Environment

Contamination by oil fractions may persist in the marine environment for many years after an oil spill, depending on characteristics of oil such as type, spill size, and location (Tansel 2014). Kingston (2002) identifies that while salt marsh and mangrove swamp areas may recover within 2–10 years of a spill, in other areas where the oil is not physically removed, it can persist for more than 25 years. While physical factors can influence the speed at which oil disperses and oiled areas recover, it is also necessary to take action to clean up and recover oil in some areas. These aspects of oil pollution, together with activities to monitor oil pollution so that it can be dealt with rapidly, are discussed in this section.

6.3.1 Physical Factors Influencing Oil Degradation

According to ITOPF, (2018c) the principal factors influencing oil degradation, both on and in the sea, by weathering are spreading, evaporation, dispersion, emulsification, dissolution, oxidation, sedimentation/sinking, and biodegradation (◘ Figure 6.8).

ITOPF (2018c) describes the main processes involved in weathering as follows:

- **Spreading**—this takes place as soon as oil is spilled and the rate of spread depends on viscosity of the oil, its composition, and the ambient temperature. Low viscosity oils spread faster than high viscosity,

---

1   For further information on VOC recovery systems see: ▶ http://www.ipieca.org/resources/energy-efficiency-solutions/units-and-plants-practices/voc-recovery-systems/.

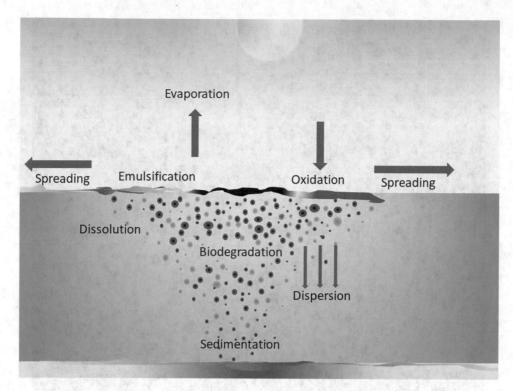

◻ Figure 6.8   Weathering processes acting on oil at sea. Adapted from ITOPF (undated a) by A. Reichelt-Brushett

and oil becomes more viscous at low temperatures. After a few hours a slick may start to break up due to the action of wind, waves, and water turbulence, and this occurs more rapidly in strong currents and at high temperatures.

– **Evaporation**—light and volatile compounds (e.g. kerosene, gasoline, and diesel) evaporate more rapidly than heavier compounds (e.g. heavy fuel oil). As oil spreads, it evaporates faster due to the larger surface area, while surface wave conditions, wind speed, and temperature can also influence the rate of evaporation. The heavier compounds tend to form a thicker layer that is less likely to dissolve naturally.

– **Dispersion**—where waves and turbulence cause a slick to break up into droplets. Depending on their size these droplets will either remain suspended in seawater or rise to the surface and reform as a thin film on the sea surface known as a sheen. Dispersion rates depend on the nature of the oil and the sea state so that light, low viscosity oil will disperse rapidly in rough seas. The use of chemical dispersants, discussed in the next section, can speed up this process.

– **Emulsification**—where seawater and oil combine and seawater droplets are suspended as a water-in-oil suspension. Such an emulsion is very viscous and more persistent than the original oil, and resemble chocolate mousse in appearance, and have a light foamy texture. They weather more slowly than the original oil; mousses with 70% volume of seawater are thixotropic and may solidify when pumped into a salvage vessel or storage tank (Bridié et al. 1980).

– **Dissolution**—where water-soluble compounds such as light aromatic hydrocarbons (benzene, toluene) dissolve in seawater. However, the majority of light compounds are normally weathered more rapidly by evaporation and so dissolution is a less significant weathering process.

– **Oxidation**—where oils react chemically with oxygen to form either soluble products or form persistent tar compounds. These tars can form tar balls (such as those that occur along the California coastline from natural seeps (see Lorenson et al. 2009) where oxidation of thick layers of high viscosity oils or emulsions forms a protective outer coating of heavy compounds, and a softer, less weathered centre. Tar balls are generally small and can last for a long time after a spill.

– **Sedimentation/Sinking**—this generally occurs when oil is approaching the shore, once lighter compounds have evaporated, and the slick has been weathered. If the oil has a similar density to that of seawater, then floating, semi-submerged, or dispersed oil can come into contact with sediments and bind to them. In addition, oil that has washed ashore can mix with sand and sediments and then be washed back out to sea and sink. Where large amounts of sediments mix with spilled oil dense **tar mats** can form on the seabed.

– **Biodegradation**—this is where micro-organisms in seawater that use hydrocarbons as an energy source can partially or completely degrade oil to water-soluble compounds. Biodegradation can only take place at the sea surface since the process requires oxygen and takes place at a later stage than other

6

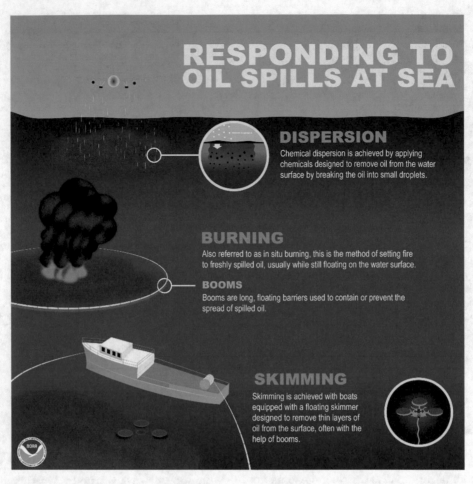

RESPONDING TO
OIL SPILLS AT SEA

DISPERSION
Chemical dispersion is achieved by applying chemicals designed to remove oil from the water surface by breaking the oil into small droplets.

BURNING
Also referred to as in situ burning, this is the method of setting fire to freshly spilled oil, usually while still floating on the water surface.

BOOMS
Booms are long, floating barriers used to contain or prevent the spread of spilled oil.

SKIMMING
Skimming is achieved with boats equipped with a floating skimmer designed to remove thin layers of oil from the surface, often with the help of booms.

◘ **Figure 6.9** Various methods for dealing with oil spills at sea. *Image*: NOAA/ORR, undated, ▶ https://response.restoration.noaa.gov/about/media/how-do-oil-spills-out-sea-typically-get-cleaned.html

processes as it requires a slick to disperse and oil droplets to be created, allowing the micro-organisms to attach themselves to the oil.

6.3.2 Oil Clean-Up and Recovery Activities

A number of technologies are available to **clean-up** following an oil spill, including for the recovery of oil. These technologies include mechanical on-water containment and recovery systems such as booms and skimmers, the use of chemical dispersants, and in situ (i.e. on site) burning (see ◘ Figure 6.9).

The type of response method used will depend on factors such as the type of oil spilled, and the environmental condition where the spill is located (close to shore, in a harbour, near a protected area, or out at sea for example).[2] ITOPF (undated b) identifies a range of

techniques available to contain and recover floating or beached oil including:

- protective booming in calm water or low currents where floating oil poses a threat to sensitive areas since booms can restrict oil from reaching those areas;
- using pumps and skimmers to remove floating oil that has not yet dispersed, and has not been mixed with debris;
- mechanical collection of high viscosity slicks, and those close to shore or stranded on the shoreline, can be done using excavators, bulldozers, and vessel-based cranes, for example; and
- manual collection by hand, with personnel wearing protective equipment and using hand tools and buckets.

The use of chemical dispersants can rapidly remove large quantities of oil from the sea surface in weather conditions that are too rough for containment and recovery, as dispersants sprayed from aircraft or ships will break up slicks and produce smaller oil droplets that biodegrade more rapidly than large droplets. However, dispersants work more effectively on low viscosity oils

2   For further reading on dealing with clean up and recovery activities for marine oil spills, see the range of ITOPF Technical Information Papers available at: ▶ http://www.itopf.org/knowledge-resources/documents-guides/technical-information-papers/.

Box 6.5: Oil Spill Monitoring in the North Sea

Since the late 1980s **aerial surveillance** has been conducted by the Bonn Agreement to monitor the North Sea for oil pollution (and pollution from other hazardous substances) (Bonn Agreement 2001). Over time a number of developments have occurred so that aerial surveillance data has become more accurate (Carpenter 2019). These developments include:

- from 1992 onwards the data includes daylight and night time surveillance activities;
- from 1997 the source of a spill has been attributed to either a ship, oil platform, or unknown sources (the latter generally being considered as illegal discharges, often taking place at night time); and
- from 2003 onwards, the number of observed spills makes a distinction between detections and confirmed mineral oil spills, the latter being spills where visual verification from an aircraft has taken place.

More recently, the Bonn Agreement has also made use of **satellite surveillance** imagery, provided by the European Maritime Safety Agency (EMSA) under its *CleanSeaNet* programme (EMSA 2018). This programme provides near real-time radar images to contracting parties of potential spills using synthetic aperture radar (SAR) satellites. Potential spills are reported to coastal states within approximately 30 min of detection (EMSA 2018). Spills detected using Side-Looking Airborne Radar (SLAR; imaging radar that point perpendicular to the direction of flight, mounted on an aircraft or satellite) are verified visually by the aircrew conducting Bonn Agreement surveillance flights (Carpenter 2019).

Monitoring of spills from oil production activities takes place through **direct monitoring** using sampling equipment on board manned and unmanned platforms, with samples taken to determine the average concentration of hydrocarbons discharges in produced water, displacement water, ballast water and drainage water (OSPAR Commission 2011). Samples have, since 2007, been assessed using gas chromatography, and before that infrared detection was used to measure oil in water concentrations. As a result of increasingly stringent emissions standards imposed by the OSPAR Commission, volumes of oil entering the sea from platforms has decreased significantly: almost 14 million tonnes of oil were discharged to sea in produced water in 2001, down to just under 4 million tonnes in 2012 (less than 30% of the 2001 volumes); and 262.2 tonnes of oil entered the sea in discharge water in 2001, down to 61.4 tonnes in 2012 (less than 20% of 2001 levels) (Carpenter 2019).

and are largely ineffective at higher viscosities. There are limitations on the use of dispersants close to shore or near coral reefs and mariculture (ITOPF, undated b).

In situ burning can be used on floating slicks where oil is freshly spilled, and can rapidly remove large amounts of oil from the water surface; however, a minimum thickness of oil is required to achieve such a burn and it will produce large quantities of smoke (ITOPF, undated b). As was described in ▶ Box 6.2 (Case 2), in the case of the *Torrey Canyon* an attempt was made to burn off the oil spill but this was unsuccessful as the attempt only took place several days after the spill, and after dispersants had been used on that spill (BBC 2008).

6.3.3 Oil Spill Monitoring Activities

Oil spill monitoring can be conducted in a number of ways including by the use of aerial surveillance, where a trained observer on an aircraft can spot a slick and determine whether it is oil or not, through the use of satellite imagery, and through direct monitoring on

board oil platforms. The North Sea provides an example of a region where all three types of activity are undertaken, and there has been a significant decline in oil spills in the region since the mid-1980s (▶ Box 6.5).

◼ Figure 6.10 illustrates trends in flight hours, observed slicks, and the ratio between the two for oil spills identified in the North Sea by Bonn Agreement and EMSA activities. It is apparent that the number of oil slicks has declined significantly in the region since surveillance activities commenced, and the number of slicks observed for every flight hour has also significantly fallen.

6.4 Consequences of Oil Pollution

The impacts of oil entering the marine environment can be acute, where there is an immediate short-term effect from a single exposure in relation to the life-span of an organism (GESAMP 1993). They can also be chronic, where sub-lethal effects of exposure are long term (10% or more of the life-span of the organism in question), and

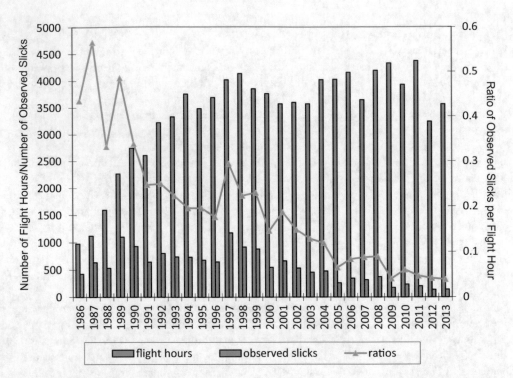

Figure 6.10   Bonn Agreement Aerial Surveillance Data for all North Sea countries, 1986–2013.[3] *Image*: A. Carpenter

Figure 6.11   Oil washes ashore at Grand Isle State Park, Grand Isle, La. *Photo*: US EPA—by Eric Vance. ► https://commons.wikimedia. org/wiki/File:June_4,_Oil_washing_ashore_at_Grand_Isle_State_Park,_La_(4683067430).jpg

---

3   NOTE: 2008–2013 Annual Reports were, at the time of writing, available from the Bonn Agreement Secretariat at: ► http://www. bonnagreement.org/publications Reports for earlier years are available by writing to the Bonn Agreement Secretariata.

where it takes a significant length of time for the toxic effect to be observable (Fingas 2012). Oil can impact on marine ecosystems such as shorelines (see ◘ Figure 6.11), marshes and mangrove swamps (Kingston 2002; Duke 2016), and on a range of marine biota including mammals (Fair et al. 2000), seabirds (Schultz et al. 2017) invertebrates, and plankton (Brussaard et al. 2016). It can also cause commercial damage to fisheries (Mendelssohn et al. 2012) and aquaculture such as mussel beds, together with wild mussels (Soriano et al. 2006). In addition, the use of dispersants can have additional impacts on marine organisms such as copepods (Cohen et al. 2014).

6.4.1 Impact of Oil on Marine Ecosystems

WOR (2010b) presents an overview of how oil damages different habitats, ranging from exposed rocky and sandy shores (where regeneration of the shoreline can take anywhere from a few months to years), to salt marsh areas (where regeneration can take anywhere from 2 to more than 20 years). For protected rocky shores and coral reefs, WOR (2010b) indicates that regeneration can take anywhere from 2 to more than 10 years. The more sheltered a shore, the longer oil will remain in the environment.

Certain areas are at much higher risk of damage than others. For example, the Wadden Sea is an area between Denmark, Germany, and the Netherlands which contains the world's largest tidal flats system, large areas of coastal salt-marshes, accommodates over 5000 species of flora and fauna, and attracts over 10 million migratory seabirds annually (► https://www. waddensea-worldheritage.org/our-world-heritage). An oil spill in that region would potentially cause severe and long-lasting damage. The Wadden Sea, therefore has, since 2002, held Particularly Sensitive Sea Area (PSSA) status from the IMO under the MARPOL Convention. It fulfils ecological criteria such as ecosystem diversity and vulnerability to degradation by natural and human activities (IMO 2018b). This protection goes further than that for Special Areas under MARPOL (IMO 2018c) under which, for technical reasons relating to their oceanographical and ecological condition and to their sea traffic, the adoption of special mandatory methods for the prevention of sea pollution is required. A range of areas have Special Status under MARPOL Annex I: Oil, these include the Mediterranean, Baltic, Black and Red Seas, the "Gulfs" area, the Antarctic area, North West European Waters, and Southern South African waters. PSSA Status has been granted to a wide range of areas including the Great Barrier Reef, Australia, the Wadden Sea, the Galapagos Archipelago, and the Tubbataha Reefs Natural Park in the Sulu Sea, Philippines (the area most recently awarded PSSA status).

An example of long-term impacts on marine ecosystems the heavy oiling from the *Deepwater Horizon* oil spill (see ► Box 6.3) has had a direct, significant, and long-lasting impact on marsh vegetation in coastal salt-marshes in the Barataria Bay area of Louisiana (Lin and Mendelssohn 2012). It also impacted the wetlands of the Mississippi River Delta system in Louisiana, an area responsible for approximately one-third of commercial fish production in the USA (Mendelssohn et al. 2012).

Oil spills can have different impacts on freshwater versus marine environments, for example, on free flowing streams and rivers compared to standing water in areas such as wetlands and salt-marshes. An overview of impacts in freshwater environments provides an understanding oil spills in these environments (US EPA 1999).

6.4.2 Impact of Oil on Marine Taxa

The most visible impact of an oil spill is oiled birds which have been directly coated by oil washing ashore on beaches (◘ Figure 6.11). However, there are many less obvious effects on marine taxa such as plankton, corals, and marine copepods (small or microscopic aquatic crustaceans that are food organisms for small fish, whales, turtles, and a range of crustaceans). There are also impacts on benthic invertebrates (bivalves such as clams and mobile crustaceans such as crabs, shrimp, and lobster) that live on the seafloor, and on intertidal/sub-tidal species that live in the zone between high and low tides (► Box 6.6). Fingas (2012) identifies a number of impacts on sub-tidal species:

- immobile species such as barnacles and mussels are most vulnerable to oil spills as they will become smothered with oil on each high tide, as also are shoreline plants and algae growing on rocks and sediments;
- some sub-tidal plants such as *Fucus* in North America can survive initial oiling, unless it is by heavy oil, but may be impacted on by long-term sub-lethal effects;
- kelp species, which live in deeper waters, are rarely covered with oil as they live in deeper waters, but they may also be impacted by sub-lethal levels of oil resulting in changes in leaf colour, reproduction and growth rates; and
- Seagrasses, for example, Eelgrass, are rarely directly oiled as they live in low intertidal areas; they take up hydrocarbons from the water column, however, which can result in death within a few hours at moderate levels or at low concentrations over a number of days. Seagrass beds may take several years to recover following an oil spill.

6

Box 6.6: The Effects of a Small Oil Spill at Macquarie Island, Subantarctic
Adjunct Professor Stephen Smith, Marine Biologist, Southern Cross University, Australia.

World Heritage Listed Macquarie Island is located in the subantarctic region, halfway between Tasmania and Antarctica, and has been an important Australian research station since 1948. During a re-supply voyage in the austral summer of 1987, adverse weather led to the grounding of the re-supply ship *Nella Dan* resulting in the spillage of 270,000 L of light marine diesel oil into the sea and adjacent rocky shores. While no mortalities of megafauna (mammals and birds) were recorded, thousands of dead marine invertebrates washed up on oiled shores within days of the spill. Subsequent assessments of the impact focused on assemblages on open rocky shores (Pople et al. 1990; Simpson et al. 1995), and the diverse fauna inhabiting the holdfasts of bull kelp (*Durvillaea antarctica*) which dominate the lower shore. There were significant differences in the biotic assemblages between oiled and control sites in all habitats. In particular, patterns of assemblage structure in holdfasts were markedly different, with opportunistic species of worm (polychaetes and oligochaetes) dominating samples from oiled sites, and peracarid crustaceans (amphipods and isopods) dominating at control sites (Smith and Simpson 1995). The impact of oil on the population of the dominant isopod *Limnoria stephenseni* was a primary driver for these differences. This species feeds on the holdfast tissue, excavating tunnels and chambers that provide habitat for other species. In the absence of this keystone species, the internal spaces of holdfasts became filled with oil-contaminated sediment. Follow-up studies 7 years after the oil spill indicated that, while assemblages on the open shore had recovered, differences in holdfast assemblages persisted between control sites and some of the oiled sites (Smith and Simpson 1998). Traces of oil were still detectable in holdfast sediments and worms continued to dominate these samples. This series of studies demonstrates that even a relatively small oil spill can have long-lasting consequences in some marine settings.

◼ **Figure 6.12** ▶ Box 6.6: **a** The bull kelp *Durvillaea antarctica* dominates the lower shore at Macquarie Island, **b** Sections of holdfasts showing the differences between oiled and unoiled sites: *Limnoria stephenseni* in situ in freshly excavated tunnels at unoiled sites, **c** sediment-filled spaces in holdfasts from an oiled site. *Photos*: S. Smith

Acute and chronic toxicity of petroleum hydrocarbons on marine organisms is dependent on a number of factors. These include concentration and length of exposure; persistence and bioavailability of specific hydrocarbons; how organisms accumulate and metabolise those hydrocarbons; the fate of those metabolised products; and how hydrocarbons or metabolised products interfere with normal metabolic processes such as growth, reproduction, and ability to survive (NRC 2003; Fingas 2012). For example, the *Deepwater Horizon* spill had a detrimental effect on the abundance and composition of bacterial communities in beach sands in the Gulf of Mexico (Kostka et al. 2011). Offshore drilling activities, and the accumulation of large amounts of drilling cuttings, can also have chronic impacts including a significant reduction in the number of taxa, abundance, biomass, and diversity around oil platforms (Trannum et al. 2010).

Fish, birds, and some species such as seals and dolphins, are often able to avoid surface slicks and move to other areas, although some birds mistake slicks for calm water and are oiled as a result (Fingas 2012) and air breathing organisms can be impacted due to their need to break the water surface. The immediate impacts on birds and surface breathing animals are highly visible (◼ Figure 6.13), Peterson et al. (2003) also suggest that almost a decade after the *Exxon Valdez* tanker struck Prince William Sound's Bligh Reef in Alaska, in March 1989, chronic impacts were still being seen in a number of marine birds such as harlequin ducks (For further details of the impacts of the *Exxon Valdez* oil spill, see ▶ Box 6.7).[4]

---

4   For an illustrated timeline of recovery from the *Exxon Valdez* oil spill, 25 years after the event, see ▶ https://aamboceanservice.blob.core.windows.net/oceanservice-prod/podcast/mar14/exxon-valdez-timeline-large.jpg.

▣ **Figure 6.13** **a** "Gulf-Spill-2010-Washing-Oiled-Pelican-22" by IBRRC, licensed under CC BY 2.0 (*Photo*: Brian Epstein). **b** *Photo* "Rescuing a pelican" by lagohsep is licensed under CC BY-SA 2.0. *Photos*: courtesy of Louisiana Department of Wildlife and Fisheries. June 4 2010 Biologists from the Louisiana Department of Wildlife and Fisheries responded to 60 calls reporting oiled birds in and around Grand Isle Thursday June 3, resulting in the successful location and capture of 35 brown pelicans and 15 gulls. All of the birds were collected from areas in the Deepwater Horizon oil spill impact zone. **c** Gulf-Spill-2010-Washing-Spoonbill-28. *Photo*: Gulf Oil Spill Bird Treatment in Louisiana provided by IBRRC, Brian Epstein by IBRRC, licensed under CC BY 2.0. **d** "Oiled Turtled Rescued May 21" by lagohsep, licensed under CC BY-SA 2.0. *Photo*: courtesy of Louisiana Department of Wildlife and Fisheries

Dispersants used on oil spills can also have an impact on marine species such as copepods and can have acute effects including increased mortality. One example comes from the *Deepwater Horizon* spill where a dispersant used to break up the spill led to increased mortality rates amongst the common coastal copepod *Labidocera aestival* (Cohen et al. 2014).

Box 6.7: Short Term and Long-Term Impacts of Oil Spills

Exxon Valdez, Alaska, 1989—Grounding

On 24 March 1989 the oil tanker *Exxon Valdez* grounded on Bligh Reef in Prince William Sound, Alaska, and around 37,000 tonnes of Alaska North Slope crude oil escaped into the sound and spread widely. Limited dispersant spraying took place, as well as in situ burning. The at sea response concentrated on containment and recovery, but despite massive efforts less than 10% of the spill was recovered from the sea surface. The spill came ashore across 1000 km in Prince William Sound, along the south coast of Alaska, and as far west as Kodiak Island. It affected a range of shore types including rock and cobble (ITOPF 2018d).

Exxon Valdez—Impacts.

According to Peterson et al. (2003), the acute mortality (short term) phase of the spill had a number of severe impacts on marine taxa:

- mass mortality of between 1000 and 2800 sea otters initially, together with up to 250,000 birds within days of the spill. These mammals and birds came into contact with floating oil leading to loss of insulation which can result in death from hypothermia, smothering, drowning, and ingestion of toxic hydrocarbons;
- around 300 harbour seals were killed, most likely as a result of inhaling toxic fumes causing brain lesions, stress and disorientation; and
- mass morality among macroalgae and benthic invertebrates on oiled shores from a combination of chemical toxicity, smothering and displacement from habitat by after-spill pressure washing of rocky beaches.

Long-term population impacts of the spill included:
- chronic exposure over many years in sediment-affiliated species such as fish, sea otters, and sea ducks (in the latter exposure was related to sediments used for egg laying and foraging);
- chronic exposure to partially weathered oil identified in fish embryos and larvae; elevated mortality of incubated pink salmon eggs in oiled streams at least 4 years post-spill;
- limited to no recovery of sea otter populations in various areas, plus higher mortality rates in animals born after the spill; and
- higher mortality rates in harlequin ducks overwintering in the region identified in 1998; in 1999, elevated rates of an enzyme CYP1A found in the livers of adult pigeon guillemots feeding on shallow-water benthic invertebrates when compared to chicks fed only on fish.

6.4.3 Economic Damage from Oil Pollution

Oil gives fish and other animals an unpleasant smell and taste and, as noted previously, can remain in the environment for long periods of time with continued detrimental effects. Commercial fisheries are at particular risk of harm from oil pollution, particularly where a slick occurs near to farmed fish or shellfish operations, or close to breeding grounds where fish eggs and larvae are vulnerable to oil pollution (e.g. Whitehead et al. 2012). In such cases it may be impossible to sell the fish or shellfish produced in an area impacted by a spill. An example of this was following the *Sea Empress* oil spill off Milford Haven, Wales in 1996 (▶ Box 6.8).

The Environmental Pollution Centres (EPC) note that other economic impacts include loss of tourism as people stay away from visibly oiled areas, or areas where there has recently been a spill (EPC 2017). This can have a negative impact on local jobs, commercial enterprises, and accommodation and food providers. Fishermen and associated onshore support (fish handling, transport) can also lose their jobs while fishing bans are in place. Property values can decline as properties in an area close to a very large spill may also be at risk of being polluted.

Box 6.8: Economic Impacts of Oil Pollution on Fishing

Sea Empress, Wales, UK, 1996—Grounding
On 15 February 1996 the oil tanker *Sea Empress* ran aground in the entrance to Milford Haven, South Wales. While the tanker was quickly refloated, serious damage was caused to its centre and starboard tanks and around 72,000 of its 130,000 tonnes cargo (Forties Blend North Sea crude) and 370 tonnes of heavy fuel oil was released between initial grounding and final refloating. Around 200 km of coastline was contaminated, and required major shoreline clean-up efforts. Much of the coastline was within the Pembrokeshire Coast National Park, while main tourist beaches were also impacted, two months before the Easter holidays in the UK (Source: ITOPF 2018e).
Commercial impacts of the *Sea Empress* spill included a ban on both commercial and recreational fishing. Under the UK Food and Environmental Protection Act, 1985 (FEPA) monitoring of a voluntary ban on mussel harvesting was undertaken, as FEPA officials determined that mussels in the Milford Haven/Pembrokeshire Coast area had accumulated dangerous levels of oil. 200 km^2 were unfishable and mussel harvesting was discontinued until 12 September 1997 when all bans were finally lifted (Environment and Society Portal, undated).

6.5 Planning for, and Responding to, Oil Pollution Incidents

6.5.1 Context

No matter what safety measures are in place to prevent marine oil pollution, there is always a risk of an incident occurring from ships or offshore installations. As a result, a range of measures and strategies are in place to plan for such incidents, and to respond to them when they occur. Contingency planning, emergency management and response planning, and oil pollution monitoring are all necessary components in being ready to deal with marine-pollution by oil.

6.5.2 Oil Pollution Preparedness and Response Co-operation (OPRC)

The International Convention on Oil Pollution Preparedness, Response and Co-operation (OPRC) was adopted in November 1990, following a conference in Paris in July 1989 at which the IMO was asked to develop additional measures to prevent pollution from ships. OPRC entered into force in May 1995. The Convention aimed to provide *"a global framework for international cooperation in combating major incidents or threats of marine-pollution"*.[5] All nation states that are signatories to OPRC are required to put in place measures to deal with pollution incidents from oil (or from hazardous and noxious substances—a separate Protocol known as OPRC-HNS). These measures may involve a response at national level or in cooperation with other parties to the Convention.

Each State party is required to (1) establish a national system for responding to oil (or HNS) pollution incidents; (2) have a designated national authority, national contact point, and national contingency plan; and (3) have a minimum level of response equipment, communications plans, regular training, and exercises. They are also encouraged to develop bilateral or mul-

tilateral agreements to augment their own national capacity to respond to incidents.

OPRC has specific requirements for ships including that they carry on board an oil pollution emergency plan and that they report any incident of pollution to coastal authorities. For offshore installations in State waters, those installations must have Oil Pollution Emergency Plan (OPEP) response to oil pollution incidents. These should be co-ordinated with national agencies so that they are dealt with promptly and effectively. Seaports, oil terminals, pipelines, and other oil handing facilities are also required to have OPEPs and to deal with national authorities in the event of a spill.

6.5.3 Contingency Planning, Risk Assessment, and Emergency Response

Contingency planning for oil (or chemical) spills can help deal with such a spill in an efficient and effective way and help minimise the impact on the environment. ITOPF (2018f) identifies a number of factors that need to be considered in a plan. The factors fall under four main headings: risk assessment, strategic policy, operational procedures, and information directory. Factors that need to be considered include: determining the risks of spills and expected consequences; defining roles and responsibilities; establishing procedures when a spill occurs; and collecting supplementary information (contact details of relevant agencies, equipment inventory, sensitive area maps, restrictions on dispersant use, guidelines on preferred response techniques, sources of funding, for example). Maps showing areas which are most in need of protection are particularly important for areas where there may be a high ecological risk, a risk to commercial fisheries and aquaculture activities, or there are industrial plants such as power stations that use seawater for cooling. Examples of contingency plans and a risk assessment are set out in ▶ Box 6.9 and see also US EPA (1999).

---

5 For more information, please see the IMO website at: ▶ http://www.imo.org/en/About/Conventions/Pages/International-Convention-on-Oil-Pollution-Preparedness,-Response-and-Co-operation-(OPRC).aspx (accessed on 18 October 2021).

Box 6.9: Contingency Plans and Risk Assessment

Contingency Planning—Mediterranean Sea Region
Contingency plans have been developed for the Mediterranean Sea under Article 4 of the Barcelona Convention (Convention for the Protection of the Mediterranean Sea Against Pollution 1976) (see UNEP/MAP, undated). This includes various protocols such as the Dumping Protocol (dealing with pollution dumped from ships and aircraft; UNEP (1972)) and the Offshore Protocol (dealing with pollution from exploration and exploitation activities (UNEP, undated). Under the auspices of REMPEC (the Regional Marine Pollution Emergency Response Centre for the Mediterranean Region) a range of plans for national preparedness and response plans, including contingency planning aspects, have been developed by the seventeen Mediterranean states that are contracting parties to the Convention (REMPEC, undated; Carpenter et al. 2017).

Rsk Assessment—Australia
An oil spill risk assessment for coastal waters of Queensland and the Great Barrier Reef Marine Park was undertaken in around 2000 (Queensland Transport and GBRMPA 2000). This resulted in the identification of several high risk areas including Torres Strait, Port of Cape Flattery, Moreton Bay, and the Whitsunday Islands. A range of maps covering shipping incident data, port and coastal traffic data, navigational hazards, and oil spill risk profile maps, were also developed at that time.

As an example of emergency response to oil pollution in European waters (with assistance provided to both European Union (EU) and non-EU states), the EMSA helps provide technical and scientific assistance in the area of ship-source pollution and in responding to pollution incidents. EMSA has in place a network of stand-by oil spill response vessels located in ports around Europe (see ◻ Figure 6.14).

The ships forming the network keep trading in the vicinity of the area where they are based but once mobilised they should stop their commercial operations and be ready for pollution response activity within 24 h. Before entering the network, they are adapted to undertake oil spill response activities. They offer a large heated storage capacity to stay longer on operations and for easy discharge the recovered oil. They make use of a range of oil recovery systems such as rigid sweeping arms (a mechanical oil spill containment system consisting of a floating pontoon and an oil collection chamber), booms (temporary floating containment barrier to prevent oil from spreading), and skimmers (equipment which can recover oil from the water surface).

The choice of equipment used in a spill depends on factors such as weather conditions, type of oil, and the coverage area. The ships in the EMSA network are equipped with local radar-based oil slick detection systems and are ready to sail within 24 h of an Incident Response Contract being signed (e.g. EMSA 2019). This contract, between the ship operator and the affected State, includes details of the actual oil recovery operation and the cost of hiring vessels. Since oil pollution at sea is transboundary in nature, EMSA ships can also be mobilised by non-EU countries sharing a regional sea with the EU. For other examples of transboundary cooperation in dealing with oil pollution see Kelly (2016) for cooperation between the US and

Mexico, the Pacific States/British Columbia Oil Spill Task Force (2011) for cooperation between the US and Canada, and IMO (2017) for cooperation between west, central and southern Africa.

6.6 Summary

Oil is a generic term that can cover a very wide range of natural hydrocarbon-based substances and also refined petrochemical products while natural gas is a fossil fuel that has been mined from the seafloor for many years. Both were formed when the remains of animals and plants sank to the ocean floor and were compressed under deep layers of sediment. Processes including the actions of bacteria, together with pressure and temperature, over a long period of time, converted those remains into hydrocarbon. Oil and gas reservoirs are found widely around the globe, both on land and beneath the seabed. These hydrocarbon deposits are extracted and become the fuel for cars, aircraft and ships, or are used to heat homes, or are converted into a wide range of chemicals for industrial processes.

Oil can enter the marine environment from a range of sources, both natural such as seeps from the seabed, or via human activities including shipping and oil exploration and exploitation activities, from the incomplete combustion of petroleum products from cars and aircraft, or via urban runoff via sewage and stormwater where pollutants fall on hard surfaces and are washed into rivers which eventually flow into the sea. Oil pollution incidents can range from very large, highly visible spills from an incident such as a tanker accident (*Torrey Canyon, Sea Empress, Exxon Valdez* for example) or from a major oil rig disaster such as the *Deepwater Horizon*, Manaconda accident. They

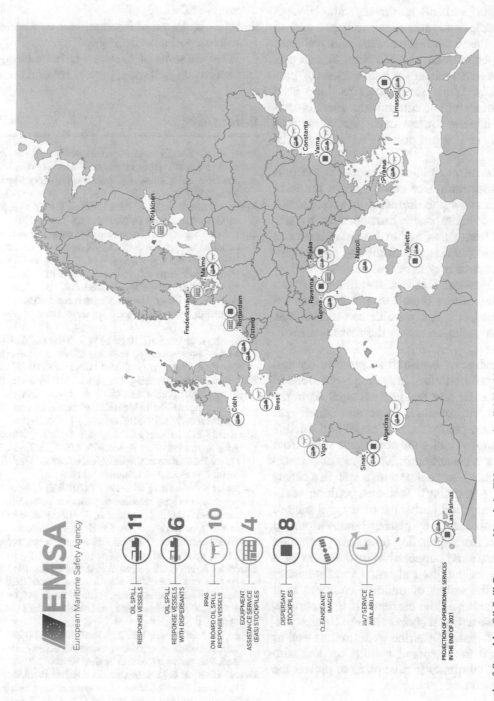

■ **Figure 6.14** Network of Stand-by Oil Spill Response Vessels in EU waters, April 2021 2021. *Source:* EMSA—regularly updated, see the website for the most recent version: ▶ http://emsa.europa.eu/we-do/sustainability/pollution-response-services/oil-recovery-vessels.html

can also be much smaller such as when a ship illegally dumps oil out to sea while sailing at night or outside territorial waters.

Depending on the type of oil, ocean/sea surface and weather conditions, and the location of a spill (at sea or close to the shore), oil can have a rapid impact on marine life including seabirds, mammals, immobile species such as barnacles and mussels, for example, and on sub-tidal plants. Based on factors such as of oil type, sea and weather conditions, and location, oil can persist in the marine environment for long periods of time or disperse rapidly. While some estimates have been made about volumes of oil entering the marine environment from different sources each year, these estimates vary widely so it is not possible to determine the actual amount.

The main natural gas pollution source is the production and transportation of cooled gas in the form LNG, where natural gas is liquified at about minus 160 °C to make it more easily transportable by sea. Emissions to air from ships, and at LNG facilities, can contribute to atmospheric and, ultimately, marine-pollution as LNG vaporises or is burned during the operation of those ships and facilities. Even more so than with oil pollution, where there are at least some estimates of volumes, it is not possible to estimate how much natural gas pollution enters the marine environment each year.

There are a range of measures in place to try and reduce oil (and gas) pollution, including regulations to make oil tankers safer and less likely to spill oil in the event of an accident, measures to monitor discharges from oil production platforms, and aerial and satellite surveillance to spot an oil spill or prevent ships from illegally dumping oil overboard. There are also measures in place to plan how to deal with a spill in a timely way (contingency planning), and also making available specialist equipment and ships to try and remove as much oil as possible, or by placing booms around a spill to prevent it from spreading (emergency response planning). International cooperation in the event of a spill is necessary as pollution at sea is transboundary and can impact the waters of multiple countries as it is moved on ocean and tidal currents. It is important, therefore, that international cooperation to combat major incidents or threats of marine pollution, as well as well-planned local and regional actions to deal with smaller incidents, continues to take place to protect the marine environment from pollution.

6.7 Study Questions and Activities

1. What are the main anthropogenic (human) sources of oil entering the marine environment?

2. What types of equipment can be used to clean up or minimise the spread of an at sea oil spill?

3. Along with oil seeping from the seabed as a natural seep which greenhouse gas is often released from such a seep?

4. How does biodegradation work as a weathering process of spilled oil?

5. What is the main international convention dealing with oil pollution from ships?

6. What are some of the short-term and long-term impacts of oil pollution, both at sea and on land?

References

Anderson C (2001) Persistent versus non-persistent oils: what you need to know. Article in Beacon (Skuld Newsletter), July 2001. Available at ▶ https://fliphtml5.com/vepv/prno/basic. Accessed 20 Jan 2022

Albaigés J, Morales-Nin B, Vilas F (2006) The Prestige oil spill: a scientific response. Mar Pollut Bull 53(5–7):205–207

Bonn Agreement (2001) Agreement for cooperation in dealing with pollution of the North Sea by oil and other harmful substances, 1983 as amended by a decision of 21 September 2001 by the contracting parties to enable the accession of Ireland to the agreement. In: Text of the Bonn agreement, vol 3, Chapter 29. Available at ▶ http://www.bonnagreement.org/site/assets/files/3831/chapter29_text_of_the_bonn_agreement.pdf. Accessed 18 Oct 2021

Bridié AL, Wanders ThH, Zegveld W, van der Heijde HB (1980) Formation, prevention and breaking of sea water in crude oil emulsions 'chocolate mousses.' Mar Pollut Bull 11(12):343–348

BBC (British Broadcasting Corporation) (2008) On this day—29 March 1967: bombs rain down on Torrey Canyon. Available at ▶ http://news.bbc.co.uk/onthisday/hi/dates/stories/march/29/newsid_2819000/2819369.stm. Accessed 18 Oct 2021

Brussaard CPD, Peperzak L, Beggah S, Wick LY, Wuerz B, Weber J, Arey S, van der Burg B, Jonas A, Huisman J, van der Meer JR (2016) Immediate ecotoxicological effects of short-lived oil spills on marine biota. Nat Commun 7:11206

Carpenter A, Donner P, Johansson T (2017) The role of REMPEC in prevention of and response to pollution from ships in the Mediterranean Sea. In: Carpenter A, Kostianoy A (eds) Oil pollution in the Mediterranean Sea: Part I—The international context. Handbook of environmental chemistry series, vol 83. Springer, Cham, pp 167–190

Carpenter A (2019) Oil pollution in the north sea: the impact of governance measures on oil pollution over several decades. SI North Sea open science conference. Hydrobiologia 845(1):109–117

Clark RB (2001) Marine pollution, 5th edn. Oxford University Press, Oxford, p 248

Cohen JH, McCormick LR, Burkhardt SM (2014) Effects of dispersant and oil on survival and swimming activity in a marine copepod. Bull Environ Contam Toxicol 92:381–387

Devold H (2013) Oil and gas production handbook: an introduction to oil and gas production, transport, refining and petrochemical industry. 3rd edn. ABB Oil and Gas, Oslo, August 2013. Available at ▶ https://library.e.abb.com/public/34d5b70e18f7d6c-8c1257be500438ac3/Oil%20and%20gas%20production%20handbook%20ed3x0_web.pdf. Accessed 18 Oct 2021

Duke NC (2016) Oil spill impacts on mangroves: recommendations for operational planning and action based on a global review. Mar Pollut Bull 109(2):700–715

Environment and Society Portal (undated) The sea empress oil spill in Milford Haven. Contributed by Denger C. Available at: ▶ http://www.environmentandsociety.org/tools/keywords/sea-empress-oil-spill-milford-haven Accessed on 18 October 2021

EPC (Environmental Pollution Centres) (2017) Oil spills' effects on human life. Available at ▶ https://www.environmentalpollutioncenters.org/oil-spill/humans/. Accessed 18 Oct 2021

Ercilla G, Córdoba D, Gallart J, Gràcia E, Muñoz JA, Somoza L, Vázquez JT, Vilas R (2006) Geological characterization of the Prestige sinking area. Mar Pollut Bull 53(5–7):208–219

EEA (European Environment Agency) (2007) Europe's environment—The fourth assessment. European Environment Agency, Copenhagen. Available at ▶ https://www.eea.europa.eu/publications/state_of_environment_report_2007_1. Accessed 18 Oct 2021

EMSA (European Maritime Safety Agency) (2018) CleanSeaNet service. Available at ▶ http://emsa.europa.eu/csn-menu.html. Accessed 18 Oct 2021

EMSA (European Maritime Safety Agency) (2019) Pollution response services: supporting pollution response for cleaner European seas. Available at ▶ www.emsa.europa.eu/newsroom/latest-news/item/3175-pollution-response-services-supporting-pollution-response-for-cleaner-european-seas-2.html. Accessed 18 Oct 2021

EMSA (European Maritime Safety Agency) (2021) EMSA network of Sandby oil spill response vessels, EAS and dispersants stockpiles. Available at ▶ http://www.emsa.europa.eu/newsroom/infographics/item/3537-emsa-network-of-standby-oil-spill-response-vessels-eas-dispersants-stockpiles-2.html. Accessed 30 Jan 2022

Fair PA, Becker PR (2000) Review of stress in marine mammals. J Aquat Ecosyst Stress Recover 7:335–354

Fingas M (2012) The basics of oil spill cleanup, 3rd edn. CRC Press, Boca Raton, p 286

GESAMP (Joint Group of Experts on the Scientific Aspects of Marine Environmental Protection) (2007) Estimates of oil entering the marine environment from sea-based activities. Reports and Studies GESAMP (75). International Maritime Organization, London. Available at ▶ www.gesamp.org/publications/estimates-of-oil-entering-the-marine-environment-from-sea-based-activities. Accessed 18 Oct 2021.

GESAMP (Joint Group of Experts on the Scientific Aspects of Marine Environmental Protection) (1993) Impact of oil and related chemicals and wastes on the marine environment. Reports and studies GESAMP (50). International Maritime Organization, London. Available at ▶ http://www.gesamp.org/publications/impact-of-oil-based-substances-on-the-marine-environment. Accessed 18 Oct 2021.

Greenpeace (2007) The Mediterranean: from crimes to conservation—A call for protection. Greenpeace, July 2007. No longer available online

Henry L-A, Harries D, Kingston P, Roberts JM (2017) Historic scale and persistence of drill cuttings impacts on North Sea benthos. Mar Environ Res 129:219–228

Hissong DW (2007) Keys to modelling LNG spills on water. J Hazard Mater 140:465–477

Hollebone B (2017) Oil physical properties: measurement and correlation. In: Fingas M (ed) Oil spill science and technology, 2nd edn. Elsevier, pp 185–207

IMO (International Maritime Organization) (2017) Cooperation for oil spill preparedness in West, Central and Southern Africa. Briefing 31 of 17 November 2017. Available at ▶ http://www.imo.org/en/MediaCentre/PressBriefings/Pages/31-GIWCAFconference.aspx. Accessed 18 Oct 2021

IMO (International Maritime Organization) (2018a) MARPOL Annex I—Prevention of pollution by oil. Available at ▶ http://www.imo.org/en/OurWork/Environment/Pages/Oil-Pollution-Default.aspx. Accessed 18 Oct 2021

IMO (International Maritime Organization) (2018b) Particularly sensitive sea areas. Available at ▶ http://www.imo.org/en/OurWork/Environment/Pages/PSSAs.aspx. Accessed 18 Oct 2021

IMO (International Maritime Organization) (2018c) Special areas under MARPOL. Available at ▶ http://www.imo.org/en/OurWork/Environment/Pages/Special-Areas-MARPOL.aspx. Accessed 18 Oct 2021

IPIECA (International Petroleum Industry Environmental Conservation Association) (2018) VOC recovery systems. Webpage. Topic last Reviewed 10 April 2013. Available at ▶ http://www.ipieca.org/resources/energy-efficiency-solutions/units-and-plants-practices/voc-recovery-systems/. Accessed 18 Oct 2021

ITOPF (International Tanker Owners Pollution Federation Ltd) (2018a) Atlantic Empress, West Indies, 1979. Available at ▶ http://www.itopf.org/in-action/case-studies/atlantic-empress-west-indies-1979/. Accessed 18 Oct 2021

ITOPF (International Tanker Owners Pollution Federation Ltd) (2018b) Torrey Canyon, United Kingdom, 1967. Available at ▶ http://www.itopf.org/in-action/case-studies/torrey-canyon-united-kingdom-1967/. Accessed 18 Oct 2021

ITOPF (International Tanker Owners Pollution Federation Ltd) (2018c) Fate of oil spills—Weathering. Available at ▶ http://www.itopf.org/knowledge-resources/documents-guides/fate-of-oil-spills/weathering/. Accessed 18 Oct 2021

ITOPF (International Tanker Owners Pollution Federation Ltd) (2018d) Exxon Valdez, Alaska, United States, 1989. Available at ▶ http://www.itopf.org/in-action/case-studies/exxon-valdez-alaska-united-stated-1989/. Accessed 18 Oct 2021

ITOPF (International Tanker Owners Pollution Federation Ltd) (2018e) Sea Empress, Wales, UK, 1996. Available at ▶ http://www.itopf.org/in-action/case-studies/sea-empress-wales-uk-1996/. Accessed 18 Oct 2021

ITOPF (International Tanker Owners Pollution Federation Ltd) (2018f) Contingency planning for marine oil spills. In: Technical information paper 16. Available at ▶ https://www.itopf.org/knowledge-resources/documents-guides/tip-16-contingency-planning-for-marine-oil-spills/. Accessed 18 Oct 2021

ITOPF (International Tanker Owners Pollution Federation Ltd) (undated a) Fate of oil spills—Technical information paper 2. Available at ▶ https://www.itopf.org/knowledge-resources/documents-guides/tip-02-fate-of-marine-oil-spills/. Accessed 18 Oct 2021

ITOPF (International Tanker Owners Pollution Federation Ltd) (undated b) Leadership, command and management of marine oil spills. Technical information paper 10. Available at ▶ http://www.itopf.org/knowledge-resources/documents-guides/tip-10-leadership-command-management-of-oil-spills/. Accessed 18 Oct 2021

Kelly W (2016) Potential impacts of a US/Mexico trans-boundary oil spill. SPE-179708-MS. Society of Petroleum Engineers. SPE Mexico Health, Safety, Environment, and Sustainability Symposium, 30–31 Ma 2016, Mexico City, Mexico. Available at ▶ https://www.onepetro.org/conference-paper/SPE-179708-MS. Accessed 18 Oct 2021

Kingston PF (2002) Long-term environmental impact of oil spills. Spill Sci Technol Bull 7:53–61

Kostka JE, Prakash O, Overholt WA, Green SJ, Greyer G, Canion A, Delgardio J, Norton N, Hazen TC, Huettel M (2011) Hydrocarbon-degrading bacteria and the bacterial community response in Gulf of Mexico beach sands impacted by the Deepwater Horizon oil spill. Appl Environ Microbiol 77:7962–7974

Lin Q, Mendelssohn IA (2012) Impacts and recovery of the Deepwater Horizon oil spill on vegetation structure and function of coastal salt marshes in the northern Gulf of Mexico. Environ Sci Technol 46(7):3737–3743

6

Lorenson TD, Hostettler FD, Rosenbauer RJ, Peters KE, Kvenvolden KA, Dougherty JA, Gutmacher CE, Wong FL, Normark WR (2009) Natural offshore seepage and related Tarball accumulation on the California Coastline; Santa Barbara Channel and the Southern Santa Maria Basin; source identification and inventory. U.S. Geological Survey Open-File 2009–1225 and MMS report 2009–030, p 116. Available at ► https://pubs.usgs.gov/of/2009/1225/of2009-1225_text.pdf. Accessed 30 Jan 2022

Malačič V, Faganeli J, Malej A (2008) Environmental impact of LNG terminals in the Gulf of Trieste (Northern Adriatic). NATO security through science series C: environmental security. Springer, Berlin, pp 375–395

Martínez-Abraín A, Velando A, Genovart M, Gerique C, Bartolomé MA, Villuendas E, Sarzo B, Oro D (2006) Sex-specific mortality of European shags during an oil spill: demographic implications for the recovery of colonies. Mar Ecol Prog Ser 318:271–276

Mendelssohn IA, Andersen GL, Baltz DM, Caffey RH, Carman KR, Fleeger JW, Joye SD, Lin Q, Maltby E, Overton EB, Rozas LP (2012) Oil impacts on coastal wetlands: implications for the Mississippi River Delta ecosystem after the Deepwater Horizon oil spill. Bioscience 62:562–574

National Commission on the BP Deepwater Horizon Oil Spill and Offshore Drilling (2011) Deep water: The Gulf oil disaster and the future of offshore drilling –Report to the President. United States Government Printing Office. Available at ► https://www.gpo.gov/fdsys/pkg/GPO-OILCOMMISSION/pdf/GPO-OIL-COMMISSION.pdf. Accessed 18 Oct 2021

NRC (National Research Council) (2003) Oil in the sea III: inputs, fates, and effects. NRC Committee on oil in the sea. National Academies Press, Washington, p 277

NOAA/ORR (National Oceanic and Atmospheric Administration and Office of Response and Restoration (undated) How do oil spills out at sea typically get cleaned up? Available at ► https://response.restoration.noaa.gov/about/media/how-do-oil-spills-out-sea-typically-get-cleaned.html. Accessed 18 Oct 2021

OSPAR Commission (Oil Spill Prevention, Administration and Response Commission) (2001) Discharges, waste handling and air emissions from offshore installations for 1998–1999. For 1984–1999 data follow the link to the excel file "98_99 Offshore Report Tables and Figures.xls". Available at ► https://www.ospar.org/about/publications?q=%22Discharges,%20Waste%20Handling%20and%20Air%20Emissions%20from%20Offshore%20Installations%20for%201998-1999%22. Accessed 18 Oct 2021

OSPAR Commission (Oil Spill Prevention, Administration and Response Commission) (2010) The quality status report 2010. London. Available at ► https://qsr2010.ospar.org/en/ch07_01.html. Accessed 11 Mar 2022

OSPAR Commission (Oil Spill Prevention, Administration and Response Commission) (2011) OSPAR reference method of analysis for the determination of dispersed oil content in produced water. Agreement 2005–15 (amended in 2011). OSPAR Commission, London

OSPAR Commission (Oil Spill Prevention, Administration and Response Commission) (2014) OSPAR report on discharges, spills and emissions from offshore oil and gas installations in 2012. Available at ► https://odims.ospar.org/en/submissions/ospar_discharges_offshore_2012_01/. Accessed 18 Oct 2021

Pacific States/British Columbia Oil Spill Task Force (2011) The stakeholder workgroup review of planning and response capabilities for a marine oil spill on the U.S./Canadian transboundary areas of the Pacific Coast—Project report, April 2011. Available at ► http://oilspilltaskforce.org/docs/Final_US_Canada_Transboundary_Project_Report.pdf. Accessed 18 Oct 2021

Peterson CH, Rice SD, Short JW, Esler D, Bodkin JL, Ballachey BE, Irons DB (2003) Long-term ecosystem response to the Exxon Valdez Oil spill. Science 303:2082–2086

PTTEP Australasia (PTT Exploration and Production Australasia) (2013) Montara environmental monitoring program: report of research, p 108. Available at ► https://www.au.pttep.com/wp-content/uploads/2013/10/2013-Report-of-Research-Book-vii.pdf. Accessed 25 Feb 2022

Pople A, Simpson RD, Cairns SC (1990) An incident of Southern Ocean oil pollution: effects of a spillage of diesel fuel on the rocky shore of Macquarie Island (sub-Antarctic). Aust J Mar Freshw Res 41:603–620

Queensland Transport and GBRMPA (Great Barrier Reef Marine Park Authority) (2000) Oil spill risk assessment for the coastal waters of Queensland and the Great Barrier Reef Marine Park, August 2000. Available at ► https://www.msq.qld.gov.au/Marine-pollution/Oil-spill-risk-assessment. Accessed 18 Oct 2021

REMPEC (Regional Marine Pollution Emergency Response Centre for the Mediterranean Sea) (undated) Mandate. Available at ► https://www.rempec.org/en/about-us/mandate. Accessed 18 Oct 2021

Samuels WB, Amstutz DE, Bahadur R, Ziemniak C (2013) Development of a global oil spill modelling system. Earth Sci Res 2(2):52–61

Saenger P (1994) Cleaning up the Arabian Gulf: aftermath of an oil spill. Search 25:19–22

Saito L, Rosen MR, Roesner L, Howard N (2010) Improving estimates of oil pollution to the sea from land-based sources. Mar Pollut Bull 60:990–997

Schulz M, Fleet DM, Camphuysen KCJ, Schulze-Dieckhoff M, Laursen K (2017) Oil pollution and seabirds. In: Kloepper S, Baptist M, Bostelmann A, Busch J, Buschbaum C, Gutow L, Janssen G, Jenson K, Jørgensen H, de Jong F, Lüerßen G, Schwazer K, Strempel R, Thieltges D (eds) Wadden Sea quality status report 2017. Common Wadden Sea Secretariat, Wilhelmshaven, Germany. Last updated 01 Mar 2018. Available at ► https://qsr.waddensea-worldheritage.org/reports/oil-pollution-and-seabirds. Accessed 18 Oct 2021

Soriano JA, Viñas L, Franco MA, González JJ, Lortiz L, Bayona JM, Albaigés J (2006) Spatial and temporal trends of hydrocarbons in wild mussels from the Galician coast (NW Spain) affected by the Prestige oil spill. Sci Total Environ 370(1):80–90

Simpson RD, Smith SDA, Pople AR (1995) The effects of a spillage of diesel fuel on a rocky shore in the sub-Antarctic region (Macquarie Island). Mar Pollut Bull 31:367–371

Smith SDA, Simpson RD (1995) Effects of the Nella Dan oil-spill on the fauna of Durvillaea antarctica holdfasts. Mar Ecol Prog Ser 121:73–89

Smith SDA, Simpson RD (1998) Recovery of benthic communities at Macquarie Island (sub-Antarctic) following a small oil spill. Mar Biol 131:567–581

Spies RB, Mukhtasor M, Burns KA (2017) The Montara oil spill: a 2009 well blowout in the Timor Sea. Arch Environ Contam Toxicol 73:55–62

Tansel B (2014) Propagation of impacts after oil spills at sea: categorization and quantification of local vs regional and immediate versus delayed impacts. Int J Disaster Risk Reduction 7:1–8

Trannum HC, Nilsson HC, Schaanning MT, Øxnevad S (2010) Effects of sedimentation from water-based drill cuttings and natural sediment on benthic macrofaunal community structures and ecosystem processes. J Exp Mar Biol Ecol 383:111–121

UNEP (United Nations Environment Programme) (1972) Protocol for the prevention of pollution of the Mediterranean sea by dumping from ships and aircraft. Available at ► https://www.unep.org/unepmap/who-we-are/contracting-parties/dumping-protocol-and-amendments. Accessed 18 Oct 2021

UNEP (United Nations Environment Programme) (undated) Protocol for the Protection of the Mediterranean Sea against pollution resulting from exploration and exploitation of the continental shelf and the seabed and its subsoils. Available at ► https://www.

unep.org/unepmap/who-we-are/contracting-parties/offshore-protocol. Accessed 18 October 2021.

UNEP (United Nations Environment Programme) (2003) Environment alert bulletin 7: illegal oil discharge in European seas. Available at ► http://2012-2018.unepgrid.ch/products/3_Reports/ew_oildischarge.en.pdf. Accessed 18 Oct 2021

UNEP/MAP (United Nations Environment Programme/Mediterranean Action Plan) (undated) Who we are: the barcelona convention and its protocols. Available at ► http://www.unep.org/unepmap/index.php/who-we-are. Accessed 18 Oct 2021

UNEP/OCHA (United Nations Environment Programme/United Nations Office for the Coordination of Humanitarian Affairs) (2006) Environmental emergency response to the lebanon crisis: consolidated report on activities undertaken through the joint UNEP/OCHA environment unit. Available at ► https://www.unocha.org/sites/dms/Documents/Report_on_response_to_the_Lebanon_Crisis.pdf. Accessed 18 Oct 2021

US EPA (United States Environmental Protection Agency) (1999) Understanding oil spills and oil spill response. Brochure. Available at ► http://nepis.epa.gov/Adobe/PDF/10001XNZ.PDF. Accessed 18 Oct 2021

US EPA (United States Environmental Protection Agency) (2017) Polluted runoff: nonpoint source (NPS) pollution. Available at ► https://www.epa.gov/nps/nonpoint-source-urban-areas. Accessed 18 Oct 2021

Whitehead A, Dubansky B, Bodinier C, Garcia TI, Miles S, Piley C, Raghunathan V, Roach JL, Walker N, Walter RB, Rise CD, Galvez F (2012) Genomic and physiological footprint of the *Deepwater Horizon* oil spill on resident marsh fishes. Proc Natl Acad Sci 109(50):20298–20302

WOR (World Ocean Review) (2010a) Living with the oceans. A report on the state of the world's oceans. Chapter 7—Energy. Available at ► https://worldoceanreview.com/en/wor-1/energy/. Accessed 18 Oct 2021

WOR (World Ocean Review) (2010b) Living with the oceans. A report on the state of the world's oceans. Chapter 4—Pollution. Available at ► https://worldoceanreview.com/en/wor-1/pollution/oil/2/. Accessed 18 Oct 2021

Pesticides and Biocides

Michael St. J. Warne and Amanda Reichelt-Brushett

Contents

7.1 Introduction – 156

7.2 A Brief History of Pesticide Use – 157

7.3 Types of Pesticides – 158
7.3.1 Classification by Target Organism – 158
7.3.2 Classification by Chemical Structure – 158
7.3.3 Classification by Mode of Action (MoA) – 159

7.4 Quantities of Pesticides Used – 162

7.5 Environmentally Relevant Properties – 162
7.5.1 Molecular Weight – 163
7.5.2 Aqueous Solubility and Hydrophobicity – 163
7.5.3 Partition Coefficients – 163
7.5.4 Volatility – 166
7.5.5 Degradation and Persistence – 166

7.6 Pesticide Distribution in the Marine Environment – 167
7.6.1 Transport to Marine Environments via River Waters and Sediments – 167
7.6.2 Transport of Pesticides to Marine Waters via the Atmosphere – 172
7.6.3 Potential Impacts of Climate Change on Transport of Pesticides to and Within Marine Waters – 173

7.7 Marine Biocides – 174
7.7.1 Impacts of TBT Use and Regulation – 175
7.7.2 Advancing Technologies – 176

7.8 Effects of Pesticides in Marine Environments – 177

7.9 Summary – 177

7.10 Study Questions and Activities – 180

References – 180

© The Author(s) 2023
A. Reichelt-Brushett (ed.), *Marine Pollution—Monitoring, Management and Mitigation*,
Springer Textbooks in Earth Sciences, Geography and Environment,
https://doi.org/10.1007/978-3-031-10127-4_7

Acronyms and Abbreviations

2,4-D	2,4-Dichlorophenoxyacetic acid
2,4,5-T	2,4,5-Trichlorophenoxyacetic acid
AChR	Acetyl cholinesterase receptor
ATPase	A group of enzymes that catalyse the hydrolysis of a phosphate bond in ATP to form adenosine diphosphate (ADP)
ATP	Adenosine triphosphate
BAF	Bioaccumulation factor
BCF	Bioconcentration factor
BMF	Biomagnification factor
BSAF	Biota-sediment accumulation factor
CAT	Catalase activity
CCA	Copper, chrome and arsenic wood treatment
DDD	1,1-Dichloro-2,2-bis(p-chlorophenyl)ethane
DDE	1,1-Dichloro-2,2-bis(p-chlorophenyl)ethylene
DDT	1-Chloro-4-[2,2,2-trichloro-1-(4-chlorophenyl)ethyl]benzene
EC50	Effective concentration to cause a sublethal effect to 50% of the test population
EQY	Effective quantum yield
FIFRA	Federal Insecticide, Fungicide, and Rodenticide Act
FOA	Food and Agriculture Organisation of the United Nations
GST	Glutathion-S-transferase
HCH	Hexachlorocyclohexane
IMI	Imidacloprid
IMO	International maritime organisation
Koa	Octanol-water partition coefficient
Koc	Organic carbon–water partition coefficient
Kow	Octanol-water partition coefficient
LC50	Lethal concentration to cause 50% mortality to a test population
LPO	Lipid oxidation levels
MCPA	2-Methyl-4-chlorophenoxyacetic acid
MoA	Modes of action
nAChR	Nicotinic acetylcholine receptor
OC	Organochlorine
OP	Organophosphate
OTCs	Organotin compounds
PBT	Persistent, bioaccumulative, toxic chemical
POPs	Persistent organic pollutants
PSII	Photosystem II inhibiting
QB	Plastoquinone B protein binding site on the D1 protein in Photosystem II
REACH	Regulation concerning the Registration, Evaluation, Authorisation and Restriction of Chemicals (a European Union regulation)
ROS	Reactive oxygen species
TBT	Tributyltin
TSCA	Toxic Substances Control Act
vBvP	Very bioacculumative very persistent chemical
USA	United States of America
US EPA	United States Environmental Protection Agency

7.1 Introduction

Pesticides are chemicals that have been specifically synthesised to "*kill pests, including insects, rodents, fungi and unwanted plants*" (WHO 2020). They are generally used with the aim of protecting plants or plant products and this is why they are referred to in European Union legislation as Plant Protection Products. **Biocides** are chemicals that are also designed to have the same properties as pesticides but they are not used to protect plants or plant products. Examples of biocides include: wood preservatives, repellents, antifouling paint for boats and chemicals used to prevent biofouling on underwater structures such as discharge pipes. While pesticides and biocides are technically different types of chemicals they will be discussed collectively in this chapter as they are both designed to kill or inhibit organisms.

Pesticides are predominantly **organic chemicals**, the vast majority of which are synthetic. They are produced by chemists in large-scale industrial plants and almost always have a carbon basis derived from petroleum hydrocarbons. There are no natural background concentrations of synthetic pesticides; however, many pesticides are now found virtually in all biological and environmental samples that have been analysed. For example, organochlorine pesticides including DDT and its breakdown products DDD and DDE have been found globally in water, soil, sediment, animal tissue (e.g. Mansouri et al. 2017) and human tissue (Jaga and Dharmani 2003). These, usually low, concentrations of pesticides are termed ambient concentrations.

In this chapter, we introduce the history of pesticide use; discuss types of pesticides and those of most concern to the marine environment in terms of exposure, environmental fate, behaviour and toxicity.

◘ **Figure 7.1**   Rachel Carson, the author of *Silent Spring* (1962), the book that highlighted the toll of pesticides on the environment. She began her career with the US Fish and Wildlife Service (CC BY 2.0)

7.2   A Brief History of Pesticide Use

We often think that pesticides are recent inventions, but humans have been using pesticides based on arsenic, copper, mercury and sulphur as well as plant derivatives for over 3000 years. For example, sulphur was burned to fumigate homes in Greece around 1000 B.C. (Baird and Cann 2012) as well as to purify temples. Matthews (2018) refers to some early methods used around 1600–1800s to manage pests on plants using vinegar, salt, cow urine, boiled herbs, and tobacco that were first reported by Lodeman in 1896.

In 1885, copper in the form of the Bordeaux mixture (a copper sulfate pentahydrate and lime mixture) became the first large-scale fungicide applied to plants. In fact, pesticides based on copper and mercury are still used in agriculture today. Bordeaux mixture and copper hydroxide are two fungicide/s permitted to be applied to organic farms in many regions while mercury (as methoxy ethyl mercury chloride) is also still used in some countries as a fungicide.

Plant-derived insecticides such as pyrethrins are reported to have been used around 400 B.C. in Persia (modern day Iran) becoming popular in Europe in the early 1800s, and today there are over 2000 globally registered products containing pyrethrin (Mathews 2018).

One of the first (produced as early as 1874) and most well-known synthetically produced insecticides is DDT. Even though the compound had been synthesised earlier, its insecticidal activities were only discovered in 1939, by chemist Paul Hermann Müller, for which he was awarded the Nobel Prize for Medicine in 1948. It has been reported that DDT has saved millions of lives when used to control malaria, a dis-

ease transmitted by mosquitoes in many tropical regions. It became a common agricultural pesticide along with other organochlorine pesticides such as dieldrin, lindane, endrin and chlordane. DDT is infamous for its ability to biomagnify (concentrations are passed from prey to predator and increase as food chains are ascended) and to cause egg-shell thinning in birds of prey as described in the 1962 *Silent Spring* by Rachel Carson (◘ Figure 7.1). Because of its dramatic adverse effects on non-target organisms, DDT was included in the Stockholm Convention that aimed to reduce and ultimately ban Persistent Organic Pollutants (POPs). However, it can still be used for disease vector control (e.g. malarial mosquitoes).

Organophosphate pesticides such as malathion and trichlorfon were developed alongside organochlorine pesticides in the 1930s and 1940 s but ultimately replaced organochlorines as they were less toxic to mammals and were less persistent. Another class of pesticides known as carbamates were introduced around 1958 (Matthews 2018).

As the chemical control of weeds was also becoming a necessity, due to the industrial revolution, the growth of the **herbicide** industry began in the 1940s. This included the development of synthetic herbicides such as MCAP, 2,4-D and 2,4,5-T (a component of agent orange the powerful defoliant used extensively in the Vietnam War), atrazine, amitrole, and diuron. Glyphosate was first registered in the USA in 1974 and is one of the most commonly used herbicides today. The first synthetic fungicide was thiram that was used as a seed treatment and its success led to the development of other fungicide such as fentin and ferbam.

Within six years of the patenting and use of DDT concerns about its impacts on non-target species were being raised (e.g. Coburn and Treichler 1946; Mitchell 1946). Rachel Carson's book *Silent Spring* highlighted the detrimental environmental effects of some pesticides and created a groundswell of public awareness associated with the impacts of pesticides on the environment and to humans. Prior to 1962, the US government mainly regulated pesticides to ensure the efficacy of chemical preparations (e.g. the Insecticide Act of 1910 and the Federal Insecticide, Fungicide, and Rodenticide Act (FIFRA) of 1947). In 1952, an amendment to the Food, Drug, and Cosmetic Act established an approach for setting tolerances for chemical residues in food, feed, and fibre. The Toxic Substances Control Act (TSCA) of 1976 required the United States Environmental Protection Agency (US EPA) to prevent *"unreasonable risk of injury to health or the environment"*. Because of this, the US EPA banned or severely restricted aldrin, chlordane, DDT, dieldrin, endrin and heptachlor and assumed responsibility for assessing the risk posed by new chemicals. Similar regulatory changes occurred in many countries and the pesticides mentioned above and additional persistent organic pollutants (POPs) have now been banned or are to be phased out by signatories of the Stockholm Convention 2001 (for further detail see ► Chapter 8).

When *Silent Spring* was published in 1962 over 500 new pesticides were entering the market annually. In more recent times, the process of registering a pesticide has required rigorous environmental and human health assessment. This has dramatically increased the cost and time for bringing a pesticide to market. The estimated cost of discovery, development and registration for bringing a new pesticide to market in 2006 was US$180 million requiring a timeframe of 8–10 years (Whitford et al. 2006). In addition, Bayer CropScience (► https://cropscience.bayer.co.uk/tools-and-services/stewardship-food-and-environment/bringing-products-to-market) estimate that only 1 in every 139,000 potential active ingredients makes it to commercial markets. Associated with this has been a decrease in the number of active ingredients introduced into the commercial market from approximately 19 per year in 1997 to approximately 8 per year in 2018 (Phillips 2020). It is likely that the stringent and costly, but necessary, processes for registration of new pesticides have limited more new products being approved for use.

Neonicotinoids (e.g. acetamiprid, clothianidin, imidacloprid and thiamethoxam) are a relatively new group of insecticides, which were commercialised in the 1990s. The structure of neonicotinoids is based on nicotine, which is a natural chemical synthesised by a range of plants that has powerful insecticidal properties. Neonicotinoids are the most extensively used group of insecticides globally (Jeschke et al. 2011; Simon-Delso et al. 2015). Originally thought to act specifically on target organisms (i.e. insects), there is growing evidence

that negative impacts are occurring on non-target estuarine and marine species including crustaceans (prawns/shrimps) and molluscs (oysters) (e.g. Hook et al. 2018; Butcherine et al. 2019; Ewere et al. 2019).

7.3 Types of Pesticides

Pesticides are classed in different ways for different reasons but all are designed to target specific groups of pest organisms. The chemical structure is a driver of the **mode of action** (MoA—the means by which pesticides exert their toxic effects) and these characteristics influence what the target organisms are. Most commercial pesticides are synthetic but as noted earlier some are derived from natural products. In the constant battle for survival many plants synthesise a range of chemicals that kill, repel or inhibit bacteria, fungi and insects (i.e. they produce natural pesticides). For example, pyrethrins are secondary metabolite terpenoids produced by sunflowers and other plants, azadiractin is a compound found in the neem tree and Sero-X is based on cyclotides (a group of circular mini-proteins) that are naturally produced by the butterfly pea plant (*Clitoria ternatea*). Sometimes, the structure of natural pesticides provides precursors or inspiration to develop similar but synthetic pesticides (e.g. pyrethroids were inspired by the molecular structure of naturally found pyrethrins).

7.3.1 Classification by Target Organism

The wide-ranging biology and life-cycle characteristics of different taxonomic groups influence the chemical characteristics of an effective pesticide. Yet the very similarities within taxonomic groups enable a pesticide to be effective on multiple species of that or similar types of organisms. For these reasons, the broad pesticide grouping of target organisms is influenced by the effectiveness of the mode of action on similar traits and characteristics within taxonomic groups. ◨ Table 7.1 shows the range of target organisms by which pesticides are classified into different types. Of course, not all these are directly relevant to the marine environment so some relevance has also been provided so you can focus your attention on those pesticides which are of most concern.

7.3.2 Classification by Chemical Structure

Another way of classifying pesticides is based on their **chemical structure**. Pesticides with the same key chemical structures are grouped together. Usually, pesticides with the same chemical structure will have the same

Table 7.1 Pesticides, their target organism and relevance to the marine environment

Pesticide type	Target organisms	Relevance to the marine environment
Acaricides	Mites/spiders	Usually applied very locally in terrestrial environments
Algicides	Algae	Used to eradicate nuisance algae (most commonly in freshwater) and in marine aquariums, they are taxonomically relevant
Avicides	Birds	Seagull and other nuisance marine bird control. They are taxonomically relevant
Bactericides	Bacteria	Used in mariculture and are taxonomically relevant
Disinfectant	Microorganisms	Usually applied locally in terrestrial environments. Pandemic situations such as COVID-19 dramatically increase usage which may enhance risk via sewage treatment plants
Fungicides	Fungi	Catchment discharges, they are taxonomically relevant
Herbicides	Plants	Catchment discharges, they are taxonomically relevant
Insecticides	Insects	Catchment discharges, they are taxonomically relevant
Larvicides	Insect larvae	Catchment discharges, they are taxonomically relevant
Molluscicides	Molluscs	Discharge pipe management, they are taxonomically relevant
Piscicides	Fish	Applied to eradicate invasive fish (most commonly in freshwater), they are taxonomically relevant
Rodenticides	Rodents	Isolated irradiation programs on islands, transport accidents
Termiticides	Termites	Usually very localised terrestrial application
Marine applications		
Antifoulants	Broad spectrum	Fouling communities
Parasiticides	Sea lice	Mariculture
Nematicides	Nematodes	Mariculture

mode of action and are therefore likely to affect the same type of organisms. There are seven main groups of pesticides including organochlorines, organophosphates, carbamates, pyrethrin and pyrethroids, neonicotinoids, phenylpyrazoles and triazines, along with some elemental-based inorganic pesticides (■ Table 7.2).

7.3.3 Classification by Mode of Action (MoA)

MoA is operationally defined and a number of different MoA schemes have been developed. They generally describe the key biological changes that occur as part of the toxic response of an organism. However, for pesticides, the MoAs are generally defined in terms of the biological receptor that they interact with to cause toxicity (HRAC 2020; FRAC 2020; IRAC 2019). Some of the major groupings of pesticides based on their mode of action are: acetylcholinesterase inhibitors; photosystem II inhibiting herbicides; nicotinic acetylcholine receptor (nAChR) competitive modulators; and synthetic auxins. The mode of action of a pesticide is controlled by its chemical structure, specifically its three-dimensional shape and volume. This is because the toxic effect is caused by the pesticide having a shape similar to a natural protein or chemical that binds to a binding receptor and triggers a biochemical reaction. A pesticide can bind either reversibly to a receptor (e.g. it can be bound and then released from the receptor) or irreversibly (once bound it is not released from the receptor). Despite binding to a receptor, pesticides do not trigger the normal biochemical reaction, rather they typically prevent normal reactions from occurring. In such cases, the pesticides are competing with the correct protein to bind to the receptor.

If a pesticide inhibits a particular biochemical pathway then it is described as having a specific MoA. Examples of the classification of pesticides by target organism; chemical structure and mode of action are presented in ■ Table 7.2. Whereas, some pesticides react with many different biological molecules and these are termed as having a non-specific MoA. It is possible for a pesticide to have both specific and non-specific MoAs. For example, pesticides such as ametryn, atrazine, diuron, hexazinone, simazine and tebuthiuron have a specific MoA in plants of binding to the plastoquinone B (QB) protein binding site on the D1 protein in Photosystem II) and hence are called Photosystem II inhibiting (PSII) herbicides. However, they also

■ **Table 7.2** The main pesticide groupings, structure, behaviour, target organisms and mode of action

Chemical group	Characteristic structure or example	Common examples	Characteristics/Behaviour	Target purpose	Mode of action
Organic					
Organochlorines (organic compounds ofoften with ring structures (e.g. a benzene ring) and at least one but up to six covalently bonded chlorine atoms)		DDT 2,4,5-T dieldrin lindane chlordane heptachlor	Chlorinated hydrocarbon compounds and known persistent organic pollutants (POPs) Likely to biomagnify Toxic to marine life, humans and other animals	Insecticide	Most are neurotoxins, affecting the central nervous system DDT disrupts the transfer of nerve impulses by inhibiting the K^+ and Ca^{2+} ATPase which controls the active transfer of ions through membranes
Organophosphates (the central structure is $O=P(OR)_3$, with the P atom being central and having alkyl or aromatic groups attached)		Chlorpyrifos valathion diazion	Readily biodegrade Widely toxic to insects Identified human toxicity	Insecticide	Acetyl cholinesterase receptor (AChR) inhibitors. These pesticides bind to acetyl cholinesterase that leads to the accumulation of acetyl choline in nerve cell synapses and uncontrolled muscle contraction, depletion of cellular energy and death
Carbamates (structures are similar to carbamic acid)		Carbaryl aldicarb carbofuran oxamyl methomyl	Functional group: carbamate esters More degradable than organophosphates	Insecticide Herbicide	Reversible inactivation of the enzyme AChR Similar to organophosphates but with different reactive properties
Pyrethroids and pyrethrins	(1R,3R)- or (+)-trans-chrysanthemic acid.	Allethrin resmethrin permethrin cyfluthrin esfenvalerate	Properties derived from ketoalcoholic esters of chrysanthemic and pyrethroic acids isolated from flowers of pyrethrums	Insecticide	Affect the sodium channels and lead to paralysis of the organism
Neonicotinoids (structures are similar to nicotine)	nicotine	Imidacloprid thiacloprid clothianidin	Many are water soluble, slower breakdown in soil Absorbed by plants and provide protection during growth Photodegrades, half-life minimum around 3 months	Insecticide	Chemically related to nicotine and like nicotine these compounds act on receptors in the nerve synapse Nicotinic acetylcholine receptor (nAChR) competitive modulators
Phenylpyrazoles	fipronil	Fipronil pyriprole aceoprole	Characterised by a pyrazole ring and attached phenyl group	Insecticide	Block glutamate-activated chloride channels in insects

(continued)

■ Table 7.2 (continued)

Chemical group	Characteristic structure or example	Common examples	Characteristics/Behaviour	Target purpose	Mode of action
Triazines (a benzene ring with three carbon atoms replaced by nitrogen atoms)		Atrazine cyanazine propazine simazine ametryn	Nitrogen containing hetrocycles. Soluble in water. Considered as POPs resisting biological and chemical degradation. High affinity for soil organic matter and may be transported through soil	Herbicide	Photosystem II inhibiting herbicide preventing electron transfer between photosystem I and II. Photosynthetic energy deviators compete for electron flow at the reducing end of photosystem I. They accept electrons that would usually have been passed on to iron-sulphur proteins that mediate electron transfer
Phenoxy (carboxylic acid)		2,4-D MCPA fenoprop	High volatility Slightly soluble in water	Herbicide	They are also inhibitors of chloroplast development and block carotenoid synthesis. Acts by mimicking the plant hormones such as auxins, cytokinins and abscisic acid, interfering with natural plant growth regulators
Inorganic					
Mercury		Methoxy ethyl mercury chloride	Lipophilic and biomagnifies (organo-metallic forms) Does not readily biodegrade Species and forms vary in toxicity	Fungicide	Mercury has a strong affinity for sulfhydryl groups (SH) in proteins, enzymes, haemoglobin and serum albumin. The central nervous system is affected by damage to the blood–brain barrier; transfer of metabolites such as amino acids are affected
Copper	Cu²⁺	Copper sulphate, copper hydroxide, copper oxychloride	Persistent Does not readily biodegrade	Algicide Fungicide Bactericide Biocide	Interference of complex homeostasis, for example stimulating free radical production in cells which induces lipid peroxidation and disturbs antioxidant capacity
Arsenic		Copper arsenate, CCA (copper, chromium, arsenic) wood treatment herbicide	A metalloid Does not readily biodegrade As³⁺ and As⁶⁺ vary in toxicity	Insecticide	Coagulates proteins, forms complexes with coenzymes, inhibits the production of ATP. Like cadmium and mercury, it can substitute for phosphorous in some biochemical processes
Tin		Organotin, TBT	Slow degradation in reducing environments	Biocide	Multiple depending on species e.g endocrine disruption -inhibiting cytochrome P450, inhibit oxidative phosphorylation and alter mitochondrial structure and function. Causes imposex and shell thickening in molluscs

have a non-specific MoA in plants of indirectly increasing the concentration of reactive oxygen species (ROS) that can cause irreversible cell damage and ultimately lead to cell death (apoptosis) (Vass 2011). The MoA for a pesticide can also change depending on the organism being exposed. For example, PSII herbicides have a different MoA in amphibians where they cause endocrine-disrupting effects (Mnif et al. 2011; DEPA 2015) (See ▶ Chapter 13 for more detail on endocrine disruption). Thus, determining the MoA of pesticides is not straightforward.

7.4 Quantities of Pesticides Used

A recent estimate of annual global pesticide usage by the Food and Agriculture Organisation of the United Nations (FAO) is 4.2 million tonnes for 2019 (FAO 2021). Pesticide usage has risen steadily since 1990 and nearly doubled in that time (FAO 2021) (◼ Figure 7.2). Despite a plateau being reached in recent years, total pesticides use increased in the 2010s by more than 50% compared to the 1990s, with pesticides use per area of cropland increasing from 1.8 to 2.7 kg/ha (FAO 2021). Asia is the largest user of pesticides accounting for on average 52.8% of global usage, followed by the Americas with 30.2% and Europe with 13.8%. The largest single country user of pesticides is China that uses an annual average of 1.77 million tonnes followed by 408,000 tonnes by the USA and 377,000 tonnes by Brazil (FAO 2021). Some countries such as Italy, Portugal, Austria, Czechoslovakia and Denmark have decreased their use of pesticides in recent times (Worldatlas 2018; Sharma et al. 2019).

The trends in pesticide usage over time are highly variable between countries (◼ Table 7.3). There are numerous factors that influence the amounts of pesticides used by any one country, some examples include amount of land dedicated to agriculture, type of crops grown, population, trade regulations, climate, developments in integrated pest management, the extent of non-pesticide reliant forms of agriculture (e.g. organic farming and permaculture).

Global pesticide use is reported to have significantly increased with time during 1990 and 2007 (Zhang 2018). However, the trend changed in 2007 showing two phases, 1990–2007, and 2007–2014. Specifically, total global insecticide use has significantly declined since 2007 with the use of chlorinated hydrocarbons decreasing since 1990, carbamate use decreased since 2007 and organophosphates have generally decreased over time (Zhang 2018). Furthermore, Zang (2018) reports that plant growth regulators and other more novel pesticides have increased continuously since 1990, herbicide use has mostly increased since 1990 and fungicide and bactericide use have stabilised.

7.5 Environmentally Relevant Properties

As with all chemicals, the **environmental fate** and effects of pesticides are determined by their chemical structure and physicochemical properties and the physicochemical properties of the media where the pesticide is located. Environmentally relevant physicochemical properties include: molecular weight; aqueous solubility and hydrophobicity; partition coefficients; bioaccumulation and bioconcentration factors; volatility; and degradability (or persistence). A brief description of each of these properties and how they affect environmental fate follows (See also ▶ Box 7.1).

◼ **Table 7.3** Changes in the usage of all pesticides since 1990 in selected countries

Country	Trend in total pesticide usage between 1990 and 2018 (compared to 1990 usage, tonnes)
Denmark and Italy	Decreased to ≈ 50%
Australia	Increased to 350%
Austria	Decreased to 75% in 2014, Increased to 122% since 2014
India	Decreased to 20% in 2008, Increased to 75% since 2009
Germany	Increased to 166%
China	Increased to 200%
Brazil	Increased to 800%
Argentina	Increased to 1000% till 2010, Decreased to 665% since 2011

Data source: FAO ▶ http://www.fao.org/faostat/en/#data/RP

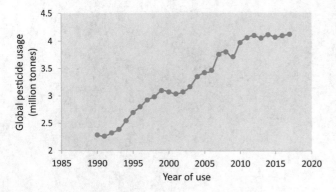

◼ **Figure 7.2** Global use of pesticides from 1990 to 2017. *Data source*: FAO ▶ http://www.fao.org/faostat/en/#data/RP

7.5.1 Molecular Weight

Typically, as the molecular weight of a chemical increases a range of other physicochemical properties change. For example, **as the molecular weight increases the chemical becomes less water soluble** (because it requires more energy to dissolve the chemical), becomes more soluble in plant and animal tissue, becomes less volatile (as it takes more energy to volatilise the chemical), and is more likely to bind to organic matter and particles.

7.5.2 Aqueous Solubility and Hydrophobicity

The solubility of a pesticide is the maximum mass of the pesticide that can be dissolved in a solvent at a specified temperature and pressure. An important factor in determining the solubility of a pesticide is the polarity (charge) of the pesticide and the solvent. A general rule of thumb is that like dissolves like. So **polar pesticides will dissolve in polar solvents and conversely non-pola chemicals will dissolve in non-pola solvents**. Water is polar and thus the more polar the pesticide the greater it's aqueous solubility and the lower the pesticide solubility in a non-polar solvent such as animal tissue. In order for a pesticide to enter an animal it must pass a cell membrane that is composed of a non-pola lipid bilayer. The more non-pola a pesticide the greater its solubility in the cell membrane and its ability to enter the organism.

Chemicals that have high aqueous solubility are termed **hydrophilic** (water-loving) or conversely **lipophobic** (fat-hating). Chemicals that are highly soluble in tissue are termed **lipophilic** (fat-loving) or conversely **hydrophobic** (water-hating). Lipophilic pesticides, once ingested by organisms, are more likely to become stored in the fatty tissues. Here they will bioaccumulate, and if this organism, is prey for another, then the higher order predator receives an already accumulated dose, and so on up the food chain, this is known as biomagnification

(see ▶ Chapter 2). The potential for biomagnification is of great concern to regulatory agencies, and there are tests that are used to assess the potential of a chemical to biomagnify.

How hydrophobic or hydrophilic a pesticide is will affect its distribution in animals and plants. The more hydrophobic a pesticide the greater the proportion that will be found in the non-aqueous parts of animals and plants (e.g. cell walls, fat cells). Conversely, the more hydrophilic a pesticide the greater the proportion that will be found in aqueous parts of animals and plants (e.g. blood plasma, cytoplasm). It is important to note that irrespective of the aqueous solubility or hydrophobicity of a pesticide some will always be found in the less favoured phase (water or tissue). The aqueous solubility and hydrophobicity of pesticides also play a role in how pesticides are excreted from the animal or plant. Hydrophilic pesticides will mainly be excreted by urine and to a lesser degree sweat or any other process that leads to the loss of water. Conversely, hydrophobic pesticides will mainly be excreted by processes that remove solid material (e.g. such as faeces and loss of dead cells, eggs, leaves or branches). An exception to this is mammalian mothers milk which is high in fat and therefore likely to have related lipophilic contaminants.

7.5.3 Partition Coefficients

Octanol–water partition coefficient

A partition coefficient is the ratio of the concentration of a chemical in two different media once equilibrium has been reached. The magnitude of a partition coefficient depends on the solubility of a chemical in the two different media. Because partition coefficient values can be very small and very large they are usually expressed as a logarithm. A widely used partition coefficient in environmental science is the octanol–water partition coefficient (Kow or its logarithm, Log Kow). This partition coefficient is calculated using Eq. 7.1.

$$\text{LogKow} = \log\left(\frac{\text{concentration in octanol}}{\text{concentration in water}}\right) \quad (7.1)$$

Box 7.1: Important Physicochemical Properties of Organic Pesticides That Control Their Environmental Behaviour

🔲 Table 7.4 ▶ Box 7.1: shows important physical and chemical properties of some common pesticides.

🔲 **Table 7.4**   ▶ Box 7.1: Properties of common pesticides that influence their environmental behaviour

Pesticide	Pesticide type	Molecular weight (amu)	Aqueous solubility (mg/L)	Log octanol–water partition coefficient	Vapour pressure (mPa)	Log bioconcentration factor
Atrazine	PSII herbicide	215.7	35	1.59–2.34	0.038	0.63
Imidacloprid	Neonicotinoid insecticide	255.7	0.61	0.57	4×10	-0.21
Diuron	PSII herbicide	233.1	~35	2.87	1.1×10^{-3}	0.98
Chlorothalonil	Fungicide	266	0.81	2.92–2.94	0.076	2.00
DDT	Organochlorine insecticide	354.5	0.006	6.91	0.025	3.50
Chlorpyrifos	Organophosphate insecticide	350.6	~1.4	4.7	2.7	3.14
Cybutryne	PSII herbicidal biocide	253.4	7.0	3.95	0.09	2.20
Deltamethrin	Pyrethroid insecticide	505.2	0.0002	4.6	1.2×10^{-5}	3.15
Dieldrin	Cyclodiene insecticide	380.9	0.14	3.7	0.024	4.54
Fipronil	Phenylpyrazole insecticide	437.2	1.9–2.4	4.0	2×10^{-3}	2.50

[1]amu = atomic mass units. *Data sources*: MacBean (2012) and the Pesticide Property Database (University of Hertfordshire 2013)

An example calculation of the octanol–water partition coefficient (Kow) for a hypothetical pesticide (Q) is presented below:

Solubility (s) of Q in water = 0.1 mg/L

Solubility of Q in octanol = 158,489 mg/L

$$
\begin{aligned}
Kow &= \text{solubility in octanol/aqueous solubility} \\
&= \frac{158,489\,\text{mg/L}}{0.1\,\text{mg/L}} \\
&= 1,584,890
\end{aligned}
$$

Log Kow = 6.19

At its simplest, Kow is determined by mixing the chemical of interest in a flask containing water and octanol then letting the system stabilise until equilibrium is reached (i.e. the concentrations of the chemical in the octanol and water are stable over time). While this sounds quite easy there are challenges in obtaining accurate estimates as the Kow increases. With increasing Kow, the aqueous solubility becomes very small and thus small differences in the measured aqueous concentration can result in large differences in Kow. For example, if the aqueous solubility of pesticide Q was 10% smaller or larger than 0.1 mg/L (▶ Box 7.1) then the Kow values would be 1,760,000 or 1,440,000, respectively. Other difficulties in measuring Kow and other partition coefficients have been identified during their 40–50 years of application (Hermens et al. 2013).

The Kow is the main physicochemical property used to express the lipophilicity of a compound and it is widely used as a surrogate of the likelihood of a chemical to accumulate in fatty tissue. The relationship between Log Kow and accumulation into animal tissue is shaped like an upside-down U—a positive linear re-

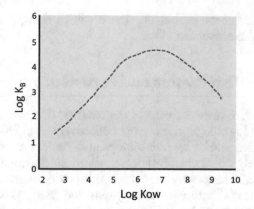

Figure 7.3 A typical relationship between the logarithm of the octanol–water partition coefficient (log Kow) and the logarithm of by uptake by tissue (log K_B). Adapted from Connell and Hawker 1988 by M St. J. Warne

lationship up to Log Kow values of about 6 at which point it plateaus off and then above log Kow values of 7 accumulation decreases (Figure 7.3). This could be due to decreased solubility in animal tissue because of the increased size of the pesticide molecule and there being so little dissolved in the water to accumulate into the tissue. Some argue that the decrease in accumulation is due to the fact that as log Kow increases the rate of accumulating decreases and that if the duration of the accumulation experiments was continued until equilibrium was reached that the relationship would remain linear beyond log Kow values of 6.

Organic Carbon–Water Partition Coefficient

The organic carbon–water partition coefficient (Koc or its logarithm, Log Koc) is a measure of how much of a chemical will bind to the organic carbon in soil or in sediment as opposed to being dissolved in soil or sediment pore water at equilibrium. The larger the Koc value the greater the proportion of the pesticide that will bind to the organic carbon and the less that is likely to be transported off-site dissolved in surface or groundwater. Conversely, the greater the Koc the more of the pesticide that can be transported bound to eroded soil particles or bound to suspended solids in water. This partition coefficient is calculated using Eq. 7.2.

$$LogKoc = log\left(\frac{concentration\ in\ organic\ carbon}{concentration\ in\ water}\right) \quad (7.2)$$

Octanol–Air Partition Coefficient

The octanol–air partition coefficient (Koa or its logarithm, Log Koa) is a measure of how much of an organic chemical will bind to air particles as opposed to the gaseous phase of air at equilibrium. The octanol is used as a surrogate for the organic component of air particles—which is what the organic chemical binds to. The larger the Koa value the greater the proportion of the pesticide that will bind to air particles and the less

that will be in a gaseous form. This partition coefficient is calculated using Eq. 7.3.

$$LogKoa = log\left(\frac{concentration\ in\ octanol}{concentration\ in\ air}\right) \quad (7.3)$$

Bioaccumulation, Bioconcentration and Biomagnification Factors

These terms are sometimes used interchangeably by people. However, they are not the same. Bioconcentration is the process of a chemical moving from the surrounding ambient media into plants or animals. Thus, for aquatic organisms, bioconcentration is the movement of a chemical from water into the organism. For terrestrial organisms, it is the movement of a chemical from the air into the organism. The bioconcentration factor (BCF) is thus the ratio of the concentration of the chemical in the organism to that in the ambient Eq. 7.4 media.

$$BCF = \left(\frac{concentration\ in\ the\ organism}{concentration\ in\ the\ ambient\ media}\right) \quad (7.4)$$

The size of the BCF depends on such properties as the aqueous solubility and Kow of the chemical being considered.

Biomagnification is the process of a chemical moving into an organism solely from the food that it eats. With biomagnification, there is an increase in the organism concentration in those organisms higher in food chains. A classic example of biomagnification is DDT where concentrations in algae were low but increased in zooplankton, herbivorous, carnivorous fish and finally in fish-eating birds of prey. The biomagnification factor (BMF) is the ratio of the chemical in the organism to that in its food Eq. 7.5.

$$BMF = \left(\frac{concentration\ in\ the\ organism}{concentration\ in\ the\ eaten\ food}\right) \quad (7.5)$$

Thus, the BMF will vary for different species in a single food chain.

Bioaccumulation is the process of a chemical moving from the ambient environment and/or from food into an organism. Bioaccumulation does not differentiate the source of the chemical, unlike bioconcentration and biomagnification. Bioconcentration and Biomagnification factors can be determined experimentally in laboratories using well-established protocols (e.g. OECD 2012). In contrast, bioaccumulation factors are usually measured in the field where it is not possible to determine the source of the chemical in the organism. Despite considering uptake from both the ambient environment and food, the bioaccumulation factor (BAF) is calculated using the same formula as the BCF (► https://www.epa.gov/pesticide-science-and-assessing-pesticide-risks/kabam-version-10-users-guide-and-technical-3#:~:-text=Bioaccumulation%20factors%20(BAF)%20are%20calculated,which%20the%20pesticide%20was%20taken).

Another relevant factor for pesticides in marine ecosystems is the biota-sediment accumulation factor (BSAF). The BSAF is the ratio of the concentration of the chemical in the organism to that in the sediment Eq. 7.6.

$$BSAF = \left(\frac{\text{concentration in the organism}}{\text{concentration in the sediment}} \right) \quad (7.6)$$

Different species and types of organisms have different lipid contents and as lipid is where bioaccumulating chemicals are stored this will affect the magnitude of the BCF, BMF, BSAF and BAF values. Normalising the organism concentration data to the lipid content of the organism removes the effect of lipids and assists with inter-species comparisons. The lipid correction is done by expressing the organism concentration as mass per kg lipid. For example, a lipid corrected BCF is calculated using Eq. 7.7:

$$BCF(lipid) = \left(\frac{\text{concetration in the organism's lipid}}{\text{concentration in the ambient media}} \right) \quad (7.7)$$

The same logic applies to BCF, BAF and BMF values for plants, except that they are corrected for the organic carbon content of the plant. For the BSAF, there are two normalisation steps—the correction for the lipid content of the organism and a correction for the organic carbon content of the sediment.

If the BCF, BAF, BMF and BSAF values are greater than one it means the chemical preferentially partitions into the organism. These factors can be very large and therefore are often expressed as log10 values (e.g. a BCF of 1,000,000 would have a log BCF of 6). At what BAF, BCF or BMF value a chemical is considered to accumulate varies amongst different organisations. However, in order for a chemical to be classed as a persistent, bioaccumulative, toxic chemical (PBT) or a very bioacculumative, very persistent chemical (vBvP), their BCF must be at least 2000 and 5000, respectively (▶ https://reachonline.eu/reach/en/annex-xiii-1.html).

7.5.4 Volatility

Volatility is a measure of the ease with which a chemical changes state from a solid or liquid phase into the gaseous phase. One measure of volatility is vapour pressure. It is the pressure exerted by a vapour when it is in equilibrium with its solid and/or liquid forms. The vapour pressure increases with increasing temperature and decreasing atmospheric pressure. The more volatile a pesticide the greater the amount that will be in the gaseous compartment of the environment under any set of environmental conditions (particularly temperature and pressure) as opposed to staying in the aqueous, soil, sediment, animal or plant compartments. The greater a pesticide's volatility the greater the probability that it will be transported by the mass movement of air.

7.5.5 Degradation and Persistence

There are various forms of degradation that are caused by biological (biotic) factors collectively called biodegradation and/or by non-biological (abiotic) factors collectively called chemical degradation. Biodegradation is generally caused by the metabolic activity of micro-organisms including bacteria, yeasts and fungi. However, most plants and animals also have various mechanisms for degrading or metabolising contaminants. Chemical degradation includes breakdown by water (hydrolysis) and by sunlight (photolysis). Typically, biological and chemical degradation break pesticides down into smaller chemicals with reduced toxicity and increased aqueous solubility compared to the original (parent) compound. The faster a pesticide is degraded to non-toxic chemicals the shorter the period that organisms are exposed to it and the lower its ability to cause toxic effects. The chemicals produced through degradation are called **degradation products** or **degradates**.

The word persistence (the opposite of degradability) was introduced into the scientific literature on pesticides to describe their continuing existence in the environment, and it was only after this that the term was applied to any organic chemical that is biologically active (Greenhalgh 1980). In a practical sense, a pesticide needs to have some persistence to ensure that it remains biologically active for sufficient time to act on target pests. However, the longer the persistence the greater the chance that the pesticide will be biologically active on or in the food grown for human consumption and the greater the chance of being transported from the site of application and exerting harmful effects to non-target organisms. This is why persistence is viewed as a key characteristic of pesticides in assessing their potential to cause environmental harm.

Persistence is a measurable property and represents a chemical resistance to change of its chemical structure and is a variable, which is a function of many interactions such as sunlight, heat and microbiological decay that result in oxidation, reduction, hydrolysis, photolysis and substitution (Grenhalgh 1980). The physical properties of a chemical inducing vapour pressure, solubility in water, dissociation constant, partition coefficient, sorption to soil and volatility will influence its persistence. Understanding these properties is essential for product registration today. Further to this, characteristics of the receiving environment including soil/sediment particle size, soil moisture content, organic matter content, pH, microbial biomass and temperature play a role in environmental persistence. Hence, persistence is site and condition-specific. Persistence

Table 7.5 Classification of the rate of degradation of pesticides in water

Half-life (days)	Persistence
<1	Fast degradation
1–14	Moderately fast degradation
14–30	Slow degradation
>30	Stable in water

Data source: University of Hertfordshire (2013)

Table 7.6 Pesticides must have one or more of the listed properties to be classified as either a persistent, bioaccumulative and toxic (PBT) chemical or as a very persistent and very bioaccumulative (vPvB) chemical 2021

PBT	vPvB
Half-life in marine water > 60 days	Half-life in marine water, fresh or estuarine water > 60 days
Half-life in fresh or estuarine water > 40 days	
Half-life in marine sediment > 180 days	Half-life in marine, fresh or estuarine water sediment > 180 days
Half-life in fresh or estuarine water sediment > 120 days	
Half-life in soil > 120 days	Half-life in soil > 180 days
Examples include: aldrin, chlordane, heptachlor, methoxychlor and toxaphene (all of which are organochlorine pesticides)	Examples: TBT

Adapted from REACH (2021) and US EPA (undated)

is usually measured in the laboratory under standardised experimental conditions that are optimal for degradation. They should therefore be used cautiously as they may not reflect environmental degradation rates. For example, laboratory-based shake flask degradation studies of PSII herbicides in seawater have typically reported half-lives of one to three months (see references in Mercurio et al. 2015). Yet Mercurio et al. (2015) using field-based mesocosms in tropical marine waters found that the half-lives were all greater than 1 year.

Persistence is usually expressed in terms of the chemical's half-life ($t_{1/2}$)—the period of time required for the concentration of the chemical to be halved. The $t_{1/2}$ of a chemical can be measured in any abiotic environmental compartment (e.g. soil, water, sediment, air). A commonly used scheme for classifying the rate of degradation of pesticides is that of the Pesticide Property DataBase (PPDB) (University of Hertfordshire 2013) (■ Table 7.5).

The European Union (REACH 2021) and the United States of America (USA) (US EPA, undated) have classifications for persistent chemicals called persistent, bioaccumulative and toxic (PBT) chemicals and very persistent, very bioaccumulative (vPvB) chemicals. The minimum persistence required to be classified as a PBT or vPvB chemical is presented in ■ Table 7.6. Some PBT and vPvB chemicals may also meet the requirements to be classed as a POP as their requirements partially overlap.

7.6 Pesticide Distribution in the Marine Environment

Most pesticides measured in marine environments originate from terrestrial sources (■ Table 7.1), with several notable exceptions such as antifoulants. Pesticides are transported from their source via water and sediment (■ Figure 7.4 and ► Section 7.6.1) and via the atmosphere (► Section 7.6.2). Pesticides have been measured in waters, sediment and biota in marine environments throughout the world for many decades (■ Tables 7.7, 7.8 and 7.9). Once pesticides are taken up by organisms (■ Table 7.9), they too act as a transport pathway in a spatial sense and also through food chains.

7.6.1 Transport to Marine Environments via River Waters and Sediments

Most pesticides are applied to land, particularly in agriculture. For example, over 70% of pesticide use in Australia is applied to agricultural land (IbisWorld 2016). However, pesticides are quite extensively used in urban areas for the control of termites, insects and weeds in and around houses, for fleas and parasitic worms in domestic pets and to control weeds in parks and water-

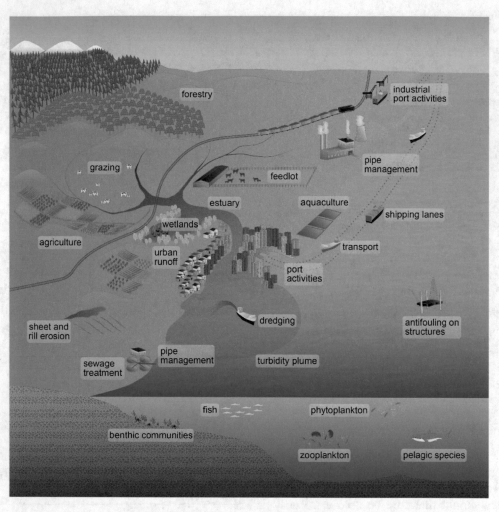

◨ **Figure 7.4**   Sources of pesticides to the marine environment. Adapted from ◨ Figure 1.1>

◨ **Table 7.7**   Examples of the global distribution of pesticides in marine waters

Location	Examples of pesticides measured	Source
Great Barrier Reef Lagoon, Australia	ametryn, atrazine, diuron, hexazinone, tebuthiuron, bromacil, fluometuron, metribuzin, prometryn, propazine, simazine, terbuthylazine, terbutryn, desethyl atrazine, metolachlor, 2,4-D, MCPA, fluroxypyr, imazapic, imidacloprid, metsulfuron-methyl, tebuconazole, propiconazole, pendimethalin, chlorpyrifos, endosulfan, malathion	Lewis et al. (2009) Thai et al. (2020)
Mediterranean Sea	atrazine	Nödler et al. (2013)
Baltic Sea	atrazine	Nödler et al. (2013)
Caribbean and Pacific Surface water slicks	DDTs[1], cyclodiene pesticides, chlordane-related compounds, hexachloro cyclo hexanes, chlorinated benzene	Menzies et al. (2013)
Jiaozhou Bay, China	atrazine and acetochlor	Ouyang et al. (2019)
Japan Sea, North Pacific and Arctic Oceans	α-HCH, β-HCH, ϒ-HCH, δ-HCH, heptachlor, aldrin, heptachlor epoxide, α-endosulfan, dieldrin, endrin, β-endosulfan, p,p'- DDD, p,p'-DDE, endrin aldehyde, p,p'-DDT, methoxychlor and endosulfan sulfate	Minggang et al. (2010)
North Pacific to Arctic oceans	chloroneb, simazine, atrazine, alachlor, dacthal, chlorobenzilate, methoxychlor, and permethrin	Gao et al. (2019)
Cape Town, South Africa	atrazine, alachlor, simazine, metolachlor, and butachlor	Ojemaye et al. (2020)

[1]DDTs = DDT, DDD, DDE

◻ Table 7.8 Examples of the global distribution of pesticides in marine sediments

Location	Types of pesticides measured	Source
Singapore coastline	DDTs[1], HCH, chlordane, Hheptachlor, aldrin, dieldrin, endrin, endosulfan, mirex, methoxychlor	Wurl and Obbard (2005)
Jiaozhou Bay, China	Atrazine and acetochlor	Ouyang et al. (2019)
Osaka Bay, Japan	DDT, HCH, chlordane	Iwata et al. (1994)
Aegean coast	DDT, DDE, lindane, heptachlor, aldrin, dieldrin, endrin	Muzyed et al. (2017)
Australia	Organochlorines and PSII herbicides	Haynes et al. (2000)
Hong Kong waters	DDT, HCH, chlordane	Richardson and Zheng (1999)
Tyrrhenian Sea, Italy	HCH, DDTs, endosulfan, aldrin, dieldrin, endrin	Qu, et al. (2018)
Mar Menor lagoon, eastern Spain	Organophosphorus and triazine pesticides	Moreno-González and León (2017)
West coast of Tanzania	DDTs, HCH, endrin, dieldrin, naphthalene, acenaphthylene, anthracene, heptachlor, pyrene	Mwevura et al. (2020)
San Blas Bay Multiple Use Nature Reserve, Argentina	Endosulfan, heptachlors, DDT, chlordane, HCH	Commendatore et al. (2018)
North coast of Vietnam	DDTs, HCH	Nhan et al. (1999)
Antarctica	DDTs, HCB, HCH	Zhang et al. (2015)
Bearing Sea, Chukchi Sea and adjacent Arctic areas	DDTs, HCH	Jin et al. (2017)
Gulf of Mexico	Organophosphate pesticides (e.g. chlorpyrifos, diazinon, dimethoate, ethion, malathion, parathion, terbufos)	Ponce-Vélez and de la Lanza-Espino (2019)

[1]DDTs = DDT, DDD, DDE

ways. These urban pesticides are predominantly discharged to waterways via surface runoff, storm water drains and wastewater treatment plants.

The main means of off-site transport is surface water runoff and subsequent delivery to estuaries and marine waters. Groundwater also transports pesticides offsite but their contribution is generally much less than surface water, although in dry periods, the contribution may be significant. Pesticides are registered for use on certain crops or types of agriculture at a maximum application rate. Therefore, the types of crops or agriculture and the percentage of the catchment on which they occur will determine the type, mass and concentration of pesticides that are transported from agricultural land to marine waters. Recent work has also shown that land use controls the number of pesticides present (Warne et al. 2020a), the toxicity of pesticide mixtures and the risk that pesticides pose (Warne et al. 2020b) at the point where rivers discharge to marine waters. The distance from the point of pesticide application to the ocean influences how much and what types of pesticides actually end up in marine waters. Rainfall volume and intensity combined with environmental properties such as the soil moisture content, the amount of vegetative ground cover, slope of the ground and the extent of riparian vegetation are key factors that control the amount and the rapidity with which surface runoff and eroded soil enters waterways. The same transport pathways occur in urban areas but in addition pesticides are transported to waterways via wastewater treatment plants (e.g. Bailey et al. 2000; Zhang et al. 2020) and stormwater drains (e.g. Chen et al. 2019).

Pesticides in runoff will either be dissolved in water (including being bound to dissolved organic carbon/matter) or bound to suspended sediment particles. The proportion of dissolved and bound forms will vary depending on the physicochemical properties of the pesticides and of the water. For example, Davis et al. (2012) and Packett (2014) found that between 10% and approximately 33% of a range of pesticides were bound to suspended sediment in freshwaters. Bound pesticides are generally not available to water column dwelling organisms (e.g. fish and algae) but are available to benthic organisms or filter feeders. Not all the pesticides that enter a waterway will reach the ocean. There are numerous processes that remove pesticides. Dissolved pesticides can be biologically or chemically degraded, they may bind to suspended or bottom sediment, be absorbed by biota, or be volatilised. The distance that bound pesticides are transported depends on the size (diameter) of the particle they are bound to and the ve-

Table 7.9 Examples of the global distribution of pesticides in marine organisms

Location and organisms	Types of pesticides measured	Source
India; commercial marine fishes, seaweeds	HCH, DDTs[1], heptachlor, endosulfan, dieldrin	Muralidharan et al. (2009); Sundhar et al. (2020)
Northeastern Brazil; King mackerel	Organochlorines	Miranda and Yogui (2016)
Northern Florida Reef Tract, USA; coral, fish, sponge	DDTs, endrin, mirex, deildrin, aldrin, chlordane, nonachlor, heptachlor, BHC,	Glynn et al. (1995)
Cambodia, China, India, Indonesia, Japan, South Korea, Philippines, Malaysia, Russia, Singapore and Vietnam; mussels	Organochlorines	Monirith et al. (2003)
Mediterranean Sea; bluefin tuna	HCB, DDTs	Klinčić et al. (2020)
Australia; coral, dolphins, fish, invertebrates, molluscs, seagrass, turtles,	PSII herbicides, OCs, lindane, aldrin, atrazine, chlorpyrifos, endosulfan (α, β and endosulfan sulfate), dieldrin, DDTs, diuron, hexachlorobenzene, heptachlor, heptachlorepoxide, lindane	Klumpp and Von Westernhagen (1995); Heffernan et al. (2017); Vijayasarathy et al. (2019); Haynes et al. (2000); Cagnazzi et al. (2020)
Aegean Coast; fish	DDTs, lindane, heptachlor, aldrin, dieldrin, endrin	Muzyed et al. (2017)
Irish Sea (Liverpool Bay); molluscs, crustaceans, fish, sea stars	DDTs, dieldrin	Riley and Wahby (1977)
North Sea off Netherlands Mediterranean Sea, North Pacific and Bering Sea, Eastern Arabian Sea, North West Pacific, Gulf of Mexico, Atlantic Ocean, Caribbean; zooplankton	Depending on location: DDTs, HCH, chlordane, dieldrin, endrin	Day (1990) and citations therein
Incheon North Harbour and Kyeonggi, Republic of Korea; zooplankton, oyster, crab, goby	Organochlorines: HCHs, DDTs, chlordane	Kim (2020)
North Atlantic; killer whales	Organochlorine pesticides	Pedro et al. (2017)
Argentina; dolphins	Organochlorine pesticides (DDE, endosulfan, endosulfan sulfate, heptachlor) Organophosphate and pyrethroid pesticides	Romero et al. (2018)
Northwest Pacific; birds and marine mammals	Organochlorine pesticides	Tsygankov et al. (2018)
Antarctica; humpback whales	Organochlorine pesticides	Das et al. (2017)
Oregon/Washington coast; shrimp, molluscs, fish	DDT, DDE, dieldrin	Claeys, et al. (1975)
Arctic; beluga whale, fish, seal	Endosulfan	Weber et al. (2010)

[1]DDTs=DDT, DDD, DDE

locity of the water—with larger particles settling out of suspension at higher flow velocities while fine particles will remain in suspension until the flow velocities are low. This particle size and velocity related transport of pesticides can lead to pesticide bound sediments moving down a waterway in a number of steps over a number of flood events. Similarly, in flood plumes pesticides bound to larger particles will not be transported as far from the river mouth as those bound to finer particles or dissolved in the water. In addition, smaller particles have a much larger surface area to mass ratio and thus the same mass of smaller particles will have higher mass of pesticides bound than larger particles.

Rainfall and related climatic conditions (such as monsoon seasons) influence the distance freshwater discharges (flood plumes) are transported in the marine waters. For example, flood plumes from rivers on the eastern coast of Queensland and adjacent to the Great Barrier Reef usually extend up to several kilometres off shore into the lagoon but under extreme conditions have been measured some 75 km offshore (Prekker 1992; see also Devlin et al. 2001). While the Amazon River, which discharges more freshwater than other river, affects the Atlantic Ocean's density and optical properties for more than 3500 km from the river mouth (Hellweger and Gordon 2002).

Pesticides bound to particles are in equilibrium with the concentration of pesticides in water—pesticides are continually binding to and being released from the particles. Therefore, if bound pesticides are deposited in areas with lower aqueous concentrations of pesticides, the particles can act as a source of pesticides to the water due to the release of pesticides from the particles. Similarly, if sediment contaminated with pesticides are resuspended either by natural causes (e.g. storms, strong tidal movement or wind) or by human causes (e.g. dredging), pesticides will desorb from the particles and increase the aqueous concentration. Such increased pesticide concentrations typically persist for a fairly short period of time as the pesticides rapidly resorb onto suspended solids and sediment.

The pattern of pesticide transported to marine waters varies depending on the size and geographical characteristics of the catchment and the rainfall pattern. Typically, waterways draining small catchments deliver pesticides in relatively short duration pulses, while those draining large catchments tend to have considerably longer periods of elevated pesticide concentrations. The proportion of a pesticide that is transported off agricultural land is largely controlled by the length of the period between pesticide application and the next rain event. The longer the period between application and rain the smaller the percentage of the applied pesticide that will be transported off-site. Another key feature of pesticides being transported to marine waters is that they almost always occur in mixtures (▸ Box 7.2).

Box 7.2: Aqueous Transport of Pesticides to the Great Barrier Reef, Australia

The Great Barrier Reef (GBR), located on the north-east coast of Australia, is the world's largest coral reef ecosystem and is a National Marine Park and a World Heritage Listed site. The reef is exposed to multiple stressors—one of which is the quality of water entering the reef from adjacent land (Waterhouse et al. 2017). This water has three main contaminants: suspended solids, nutrients (nitrogen and phosphorus) and pesticides (Waterhouse et al. 2017). In order to improve the health and resilience of the reef, the Australian and Queensland governments have developed a series of Reef Water Quality Improvement Plans that have set land use management and pollutant reduction targets. The pesticide reduction target is to protect at least 99% of aquatic species throughout the wet season (November to April) from the harmful effects of pesticide mixtures at the mouth of waterways that discharge to the GBR (Australian and Queensland governments 2018). Progress to achieving the pesticide target is determined by monitoring pesticide concentrations in rivers that discharge to the reef, wetlands, in flood plumes and in-shore waters of the reef itself. The number of sampling sites and the number of pesticides analysed for have varied over time. Currently, over 80 pesticides and degradates are monitored in 20 waterways from grab samples collected by automated samplers. In-shore sampling at sites is conducted at 11 sites, up to 50 km from the nearest river mouth, using passive samplers.

Flood plumes are monitored on an ad-hoc basis. Both in-shore and flood plume samples are analysed for 45 pesticides and degradates. Up to 50 pesticides have been detected in rivers. Of the approximately 2600 river samples collected between 2011 and 2015 over 99.8% contained more than one pesticide. The maximum, mean and median number of pesticides in each sample was 20, 5.1 and 4 respectively (Warne et al. 2020a). Similarly, 59 pesticides have been detected in 22 coastal wetlands and each wetland contained 12 to 30 pesticides with a mean of 21 (Vandergragt et al. 2020). At least 24 pesticides were detected in flood plumes while 27 pesticides were detected in in-shore samples in 2017/2018 and all flood plume and in-shore samples contained pesticide mixtures (◖ Figure 7.5). Pesticides are detected year-round in both the rivers and in-shore with higher concentrations occurring in the wet season.

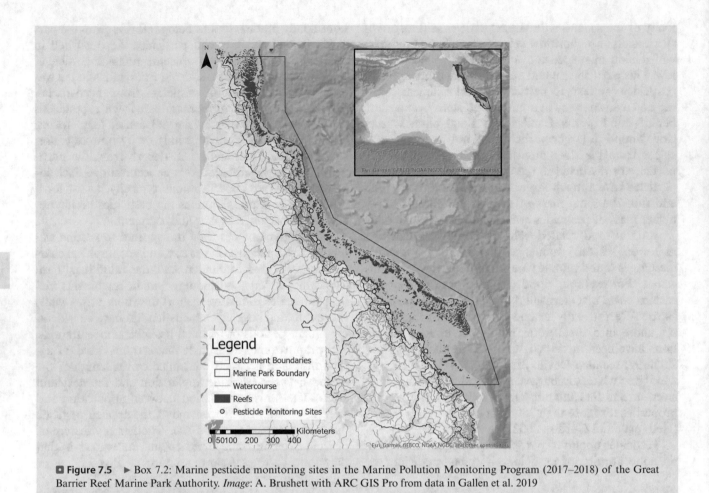

◘ Figure 7.5 ▶ Box 7.2: Marine pesticide monitoring sites in the Marine Pollution Monitoring Program (2017–2018) of the Great Barrier Reef Marine Park Authority. *Image*: A. Brushett with ARC GIS Pro from data in Gallen et al. 2019

7.6.2 Transport of Pesticides to Marine Waters via the Atmosphere

As long ago as 1975 Goldberg cited earlier work by Lloyd Jones who determined the evaporation rate of DDT to show that if the annual amount of DDT used was applied evenly over the Earth's land that all the DDT could evaporate (Goldberg 1975). He also coined the phrase **global distillation** to describe the volatilisation of pesticides, their mass movement by wind systems throughout the world and subsequent transfer to marine waters by precipitation. The term was later expanded to include the transport of pesticides from warm climates (particularly the tropics) to colder polar regions (◘ Figure 7.6).

In hot climates, a percentage of applied pesticides will evaporate from plants, soil and water. They are then moved by large-scale air movements north or south towards the colder polar regions. Once the air temperature at which condensation occurs is reached the pesticides will condense onto any surfaces including biota, ice, snow, soil and water. Thus, there is a spatial separation of pesticides with those with the lower condensation point being transported further towards the poles than pesticides with higher condensation points. This is called **global fractionation**. Air temperatures vary annually with the season. In winter, the condensation temperature for a pesticide may be reached leading to its condensation. In summer, the higher air temperature could again lead to the pesticide evaporating and being transported further towards the poles until its temperature of condensation is again reached. This annual evaporation, transport and condensation cycle is called the **grasshopper effect** (◘ Figure 7.5). The rate of degradation in the air also affects the distance that pesticides will be transported—with less persistent chemicals having a reduced likelihood of being transported long distances. Global distillation accounts for the exposure of Arctic wildlife to many POPs, including pesticides, and subsequent toxic effects (Sonne et al. 2017).

Wania (2003) divides pollutants into four groups based on their physicochemical properties (see earlier text) and likelihood of participating in global distillation:

- **Fliers** are so volatile they remain in the air and are not deposited on land or water even at the poles;
- **Single hoppers** are sufficiently volatile to be transported to and deposited at the poles in a single event;

Figure 7.6 Global atmospheric transport of Persistent pesticides by global distillation with more volatile (lower condensation temperature) pesticides being transported further towards the poles (A to C) and annual movement of pesticides by the grasshopper effect. Adapted from Semeena 2005 by M St. J. Warne

— **Multiple hoppers** will be transported toward the poles by a series of evaporation and condensation cycles; and

— **Swimmers** have very low volatility but relatively higher aqueous solubility and are therefore transported by oceans currents.

This classification is helpful as it indicates the main mechanism(s) that are responsible for the transport of pesticides.

Interestingly, while there is considerable evidence for global distillation occurring in the northern hemisphere, the evidence is limited for the southern hemisphere (e.g. Iwata et al. 1994; Corsolini et al. 2002, 2003). Sadler and Connell (2012) argue that this is probably due to three reasons. First, the southern hemisphere contains much less land and more ocean than the northern hemisphere. Second, the two large land masses that could provide a means of transfer from tropical to near polar conditions (i.e. Africa and South America) are triangular in shape with less land closer to the poles. Third, the pesticides that are deposited in water are likely to bind to particulate matter and settle at depth onto the ocean floor.

7.6.3 Potential Impacts of Climate Change on Transport of Pesticides to and Within Marine Waters

The effects of climate change on the transport of pesticides have received relatively little attention and are far from conclusive. Nonetheless, some potential ef-

fects have been identified (e.g. Wang et al. 2016) and are discussed below (see also ▶ Chapter 14). Increased air temperatures will increase the volatility of pesticides so if the same amounts of pesticides are used a greater proportion will be subject to atmospheric transport. Increased air temperature will (and this has been shown to be occurring) decrease the amount of ice which is likely to lead to the release of pesticides present in melting ice increasing aqueous pesticide concentrations and the amount available to enter polar ecosystems. Also, as ice is a major site for the deposition of pesticides in polar regions, warmer air temperature could lead to less pesticides being deposited on ice. Increased air temperatures may result in the condensation temperatures of some pesticides not being reached and therefore not being deposited.

Climate change forecasts have also predicted increased frequency and intensity of extreme weather events such as droughts, storms, cyclones and hurricanes. The preceding years weather conditions play a role in determining the runoff of water from rain. For example, under conditions of less than normal rainfall, the soil moisture content will be lower, and this decreases the permeability of soil to water. Less than normal rainfall may also decrease ground cover resulting in increased water runoff and increased soil erosion. An increase in the frequency and/or intensity of extreme rainfall events is likely to lead to increased surface run-off and the transportation of soil-bound pesticides.

Global warming could also change oceanic currents and thus the spatial distribution of pesticides. Global warming could lead to changes in the type and inten-

7

sity of pressure exerted by animal and plant pests to agriculture, and this could lead to increased or changed pesticide use (Noyes et al. 2009; Kattwinkel et al. 2011). Irrespective of how exactly climate change affects pesticide transport it is likely to have significant effects and is likely to cause many as yet unforeseen changes to pollutant transport.

Climate change could also affect the persistence and toxicity of pesticides (e.g. Noyes et al. 2009). The Q10 rule states that for every 10 °C increase in temperature, the rate of biological and chemical reactions will increase by two–threefold. Thus, degradation rates of chemicals could increase leading to decreased persistence, and this in turn could lead to increased or decreased toxicity depending on whether the degradates are more or less toxic than the parent compound. Increased ambient temperature is likely to decrease the risk that volatile pesticides will cause harm to aquatic organisms due to increased volatilisation. Temperature is a stressor to organisms, and they all have a thermal tolerance range. Ambient temperature outside an organism's tolerance range places an additional stress on the organism and is likely to increase the toxicity of pesticides (e.g. Negri et al. 2011, 2019a).

7.7 Marine Biocides

Within hours of a structure's submergence, a fine biofilm, which is comprised of microorganisms (mainly bacteria), develops on its surface (Steinberg et al. 2002). This initial layer then facilitates the attachment of macro-organisms such as barnacles, serpulid worms, bryozoans, ascidians and algae by providing biochemical cues that trigger larval settlement (◘ Figure 7.7). These biofouling communities are natural but they interfere with marine structures and shipping. Marine and estuarine aquaculture facilities, uptake and discharge pipes, and boats and ships of all sizes are susceptible to biofouling that significantly increases fuel consumption and infrastructure maintenance (Sonak et al. 2015). As a general rule of thumb, each 10 μm of surface roughness increases fuel consumption by approximately 1% (Lackneby 1962 in Westergaard 2007). Antifoulants are reported to save hundreds of millions of dollars in fuel costs to the shipping industry and result in significant reductions of carbon dioxide emissions (Champ 1999; Sonak et al. 2015), but they can cause harmful environmental effects.

Antifoulant paints are applied directly to structures that are placed in the marine environment. The most widely known antifoulants are tributyltin (TBT), a class of organotin compounds (OGTCs). Copper is also a well-known antifoulant paint additive and has been used in various forms for many hundreds of years. ◘ Figure 7.8

◘ **Figure 7.7** Examples of biofilms—**a** "Boat fouling organisms" *Photo*: Doug Beckers CC BY-SA 2.0 and **b** fouling community in pearl oyster cages. *Photo*: A. Reichelt-Brushett

shows a historical use of copper sheeting to protect wooden pylons from attack by marine borers. The photo reveals that even though the coverage of copper sheeting is now far from complete it is still exerting its biocidal properties—as indicated by the state of the pylon.

As a function of its lipophilic character and low water solubility, up to 90% of TBT introduced into water readily adsorbs to particulate matter. The level of adsorption is dependent on the sediment characteristics, salinity, pH, oxygen levels, temperature and the presence of dissolved organic matter (e.g. Burton et al. 2004). It is also thought that the biocidal action of TBT may inhibit normal biodegrading processes. Estimates of the half-life of TBT in the environment vary significantly. The half-life of TBT in the water column ranges from a few days to weeks, whereas once deposited in benthic sediments, TBT can last unaltered for decades, particularly in anoxic conditions (Svavarrson 2001).

High concentrations of TBT are normally associated with commercial ports, dockyards and marinas

Figure 7.8 Copper sheeting was used to protect pylons prior to the invention of antifouling paints. This metal remains a legacy pollutant (Richmond River, NSW, Australia 2020). *Photo*: A. Reichelt-Brushett

and along coastal shipping routes, particularly if these sites have fine silt-like sediment. TBT reservoirs in sediment create a long-term legacy, which extends decades past their actual use dates and their legal restrictions and/or ban. There is growing evidence that since the banning of TBT impacted ecosystems are recovering (Gibson and Wilson 2003; Kim et al. 2017). However, research in Arcachon Bay (one of the first TBT sites identified) showed that sediment contamination was still affecting non-target species after 20 years (Ruiz et al. 1996), and it is suggested that TBT will be a legacy problem into the future (Langston et al. 2015). A recent review by de Oliveira et al. (2020) also showed that the biological impacts of TBT continued to be reported globally between 2000 and 2019.

TBT reservoirs are particularly difficult to manage in areas such as ports and docks, where remobilisation of TBT is ongoing because of maritime traffic and/or dredging. To handle ever-larger vessels, dredging activities at ports are increasing worldwide. These dredging activities are responsible for frequent re-suspension, and re-loading of TBT from/to the sediment and through the disposal of dredge spoil can also impact sites previously unaffected by TBT (Svavarrson 2001).

7.7.1 Impacts of TBT Use and Regulation

Within two decades of the widespread use of TBT as antifouling agents, the toxicological problems associated with their use became clear, when the commercial oyster production in France almost completely collapsed in the late 1970s (Alzieu et al. 1989). Further established toxicological effects include imposex in gastropods (a phenomenon of masculinisation of females) (▶ Box 7.2) and immuno-suppression in seabirds and marine mammals, both of which result in significant population declines (Tester et al. 1996; Kannan et al. 1998; Tanabe et al. 1998). Frouin et al. (2010) provide a detailed review of the behaviour and toxicity of organotin compounds in marine environments and highlight the toxicological effects of extremely low concentrations (ng/L–µg/L range). The review by Frouin et al. (2010) is highly recommended for further reading if this topic interests you.

The detrimental effects of TBT have resulted in many legislative changes in developed countries and have culminated in the global ban of organotin compounds as antifouling agents since 2008. Countries in which the use of TBT had been regulated since the 1980s have shown a significant decrease in concentration in both waters and organisms (Evans et al. 1996; Tester et al. 1996). In July 1987, the UK Government banned the use of TBT-based antifouling paints for vessels under 25 m in length and also its use in aquaculture. Similarly, USA, Europe, Canada, South Africa restricted the use of TBT in antifouling boat bottom paints by vessel size (less than 25 m in length) and restricted the release rates of TBT from co-polymer paints. Japan and New Zealand banned TBT on all vessels. Most Australian States and Territories contributed to international efforts by banning the application to vessels less than 25 m in length (late 1980s–90 s). The IMO Convention on the Control of Harmful Anti-fouling Systems on Ships (AFS Convention) was adopted in 2001 and came into force in September 2008. A complete ban on the application of TBT on vessels under 25 m length came into force on 1 January 2003, with a complete global prohibition by 1 January 2008. After initial banning on small boats, the use of TBT on ocean-going vessels was still permitted on the grounds that these vessels do not sit for long periods of time in near-shore waters and are therefore unlikely to affect local shellfisheries. Legislation banning, the use of TBT on vessels over 25 m, is coming into effect in some countries.

Box 7.3: The Effects of TBT on Non-target Organisms

The main objective in the development of antifouling paints has been to inhibit the settlement of marine organisms and barnacles. TBT is toxic to a large range of organisms from plankton to marine mammals and potentially humans. There is considerable evidence that TBT is more toxic to marine organisms than freshwater species (Leung et al. 2007) and effects have been recorded in the ng/L range (Antizar-Ladislao 2008; Frouin et al. 2010). TBT exhibits the strongest (known) toxic effects to gastropods, with at least 200 species known to be susceptible to TBT-induced endocrine disruption. TBT interferes with the gastropod endocrine systems by inhibiting the P450 cytochrome system, which is linked to the conversion of male hormones (androgens) to female hormones (estrogens) in females. The ecological effect is a masculinisation of females, which is termed **imposex**. Affected females grow a vas deferens and in severe cases a penis which eventually blocks the oviducts. Although egg production continues during several gradual stages of penis growth, eventually the organism is unable to maintain a constant production of eggs (◘ Figure 7.9). Advanced stages of imposex result in sterility, premature death of the female and subsequent population decline.

TBT also reduces growth rates in adult oysters, thickens oyster shells and reduces the meat content. The shell thickening is caused by an enzyme disfunction in shell deposition and results in making the oysters unmarketable. Oysters affected in this way are often referred to as **golf-ball oysters**.

The ecotoxicological impact of TBT on fish, birds, seals and other marine mammals has been less well studied, although there is considerable evidence that TBT causes immuno-suppression in these taxonomic groups, thereby acting as a co-factor in causing increased mortality of, for example, sea otters (Kannan et al. 1998). Based on a significant association between TBT concentrations and parasitic infection rates of lung nematodes in porpoises, Nakayama et al. (2009) suggest that OGTCs are responsible for increasing susceptibility of infectious disease for this species. Likewise, mass mortality of coastal bottle-nose dolphins in the USA has been attributed to immune suppression in these inshore dolphins which have significantly higher TBT tissue levels than their offshore counterparts (Tanabe et al. 1998).

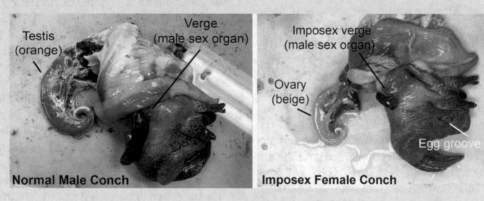

◘ **Figure 7.9** ▶ Box 7.2: An imposex impacted female conch (right) with normal male conch (left). *Images*: FWC Research CC BY-NC-ND 2.0

7.7.2 Advancing Technologies

Initially, the antifoulant paints that included TBT were designed so that the TBT would passively leach from the paint. However, the release rate was unpredictable and inconsistent, and therefore its antifoulant properties were not optimal. Self-polishing co-polymer paints were introduced in the 1970s. The advantage of this technology is that the biocide is slowly released as a result of wave action or the forward motion of the ship. Once the outer surface is worn away, the next layer begins to release the biocide. Thus, the biocidal activity of the antifoulant paint is consistent throughout its 5-year lifespan, which is twice that of the first TBT paints.

Since the ban on TBT, copper has again become the most widely used antifoulant, but because of its well-established toxicity to non-target organisms, it has also come under increasing scrutiny. It is subject to restrictions in a number of countries such as Denmark and Canada and is entirely banned for the use on small boats on the Baltic Coast of Sweden. The US EPA is currently reviewing its regulation, and it is expected that restrictions on the use of copper will result from this review in the future (Blossom et al. 2018).

Other widely used additives to antifoulant paints are cybutryne (commonly referred to as Irgarol 1051) and diuron—both Photosystem II inhibiting herbicides, and chlorothalonil (a fungicide). These additives diffuse out of the paint and have been detected in marine waters and sediment throughout Europe, North and Central America and Asia (Harino et al. 2009; Thomas and Langford 2009). Since around 1990, there has been considerable interest in developing natural antifoulants that are synthesised by marine organisms. These include many different types of chemicals: toxins, anaesthetics, surfactants, attachment and metamorphosis inhibitors and repellents. To date, no natural antifoulants have been commercially released.

Perhaps the search for better or safe biocides is fundamentally flawed because the very properties that are required to be an effective biocide (toxic to algae, crustaceans and molluscs and reasonable persistence) are also the properties that mean the biocide is likely to exert harmful effects on non-target organisms. It is also complicated by the fact that there are multiple organisms in biofilms. In spite of these challenges, many novel approaches are being researched. **Non-stick** foul-release coatings, such as Intersleek 700, Sealion and Bioclean, which incorporate silicone elastomers, waxes or silicone oils, and **natural** coatings that utilise secondary metabolites with biocidal properties derived from soft corals, sponges and microorganisms, present the two main lines of research into non-toxic alternatives (e.g. Michalek and Bowden 1997). Non-stick technology, while entirely non-toxic, has one major disadvantage, they only self-clean effectively at high velocity (Dafforn et al. 2011), and therefore they are only suitable for high speed vessels. Other examples of novel approaches to inhibit fouling include electrical currents (Hong et al. 2008) and intelligent paints that take up copper from seawater (Elmas et al. 2018).

7.8 Effects of Pesticides in Marine Environments

Early investigations of pesticides in the marine environment focused on the biomagnifying persistent organic compounds and their transfer through food chains into higher order species including cetaceans, sharks and large fish such as tuna and marlin. The health of sea birds, particularly large species such as pelicans and sea eagles, which are at the top of the food chain, was also of early concern. Research on the effects of pesticide on marine organisms is growing, yet many national guideline values for single pesticides in marine waters are considered to be of only limited reliability for ensuring the protection of marine ecosystems. In essence, marine ecotoxicological data are insufficient to determine reliable guideline values for many pesticides and of the guideline values that have been determined they are generally lower for marine waters compared to freshwaters.

The effects of pesticides can be measured according to responses of organisms from the subcellular level to whole organism level to the ecosystem level. ◘ Table 7.8 shows the examples of the current literature of a range of pesticides on their toxicity to marine species. Most studies have been completed on commercially important species. Other studies have shown the effects of atrazine on corals, seagrass, macro algae, algae, coralline algae, diatoms, phytoplankton and cyanobacteria (e.g. Howe et al. 2017 and citations therein). Negri et al. (2019b) recently conducted a large study that determined the toxicity of 21 pesticides on 16 tropical aquatic species. ◘ Table 7.10 highlights the lack of sublethal EC50 data and highly variable exposure times for different studies. More ecotoxicological research needs to be conducted on pesticides relevant to the marine environment.

Very few studies have investigated the effects of multiple concurrent pesticide exposures nor has much research been completed on effects of pulse exposure that are more consistent with natural environmental flows to the ocean. The study of the toxicity of mixtures is essential to understanding the actual risk that pesticides pose to marine organisms. The multiple stressors that organisms may need to respond to may reduce their overall fitness which further reduces the resilience of the population and long-term intergenerational survivorship. A combination of pesticides, like those that are present in catchment runoff (Brodie and Landos 2019; Warne et al. 2020a; Spilsbury et al. 2020), may each only need to be present at low concentrations in order to see measurable responses. This is illustrated in work by Thai et al. (2020) who monitored the Great Barrier Reef in-shore marine waters at nine sites for 30 pesticides. As the Great Barrier Reef lagoon is a National Marine Park and a World Heritage Site, it is recommended that at least 99% of species should be protected (Australian Government and Queensland Government 2018). In monitoring for 2018/2019, they reported that no individual pesticide exceeded their corresponding Australian and New Zealand water quality guideline values (Thai et al. 2020). However, when the combined toxicity of pesticide mixtures was accounted for using the method developed by Warne et al. (2020b), it was estimated that between ~2 and 10% of aquatic species would experience adverse sub-lethal effects at three of the nine monitoring sites. Research continues to develop guideline approaches that include the consideration of mixtures, their varying concentrations, mode of actions and relative toxicity for a range of representative organisms and keystone species.

7.9 Summary

This chapter introduced you to the various types of pesticides and biocides, how they are transported to marine waters and the risks they pose to the marine

Table 7.10 Examples of ecotoxicological studies on the effects of a range of pesticides on marine organisms (*Note* TBT and antifouling products are not included in this table as they have been discussed earlier)

Pesticide	Species affected	Exposure effect	Endpoint	Source
Azamethiphos (organophosphate) used for the control of sea lice	*Homarus Gammarus* European lobster	1-h LC50 8.5–75.7 µg/L, 1-h EC50 9.2–15.5 µg/L	Larval survival Immobility	Parsons et al. (2020)
	Metacarcinus edwardsii crab	30-min exposure caused significant mortality at 10–500 µg/L	Larval survival	Gebauer et al. (2017)
Chlorpyrifos (organophosphate)	*Litopenaeus vannamei* prawn	4-day LC50 2.1 µg/L Elevated catalase activity (CAT) and Lipid oxidation levels (LPO) in muscle, hepatopancreas and gills. Decreased acetyl-cholinesterase (AChE) activity in the brain and an increased glutathion-S-transferase (GST) activity in the hepatopancreas	Adult survival and biochemical function	Duarte-Restrepo et al. (2020) See also review by Huang et al. (2020)
Diazinon (organophosphate)	*Acropora tenuis* scleractinian coral	48-h EC50 54.7 µg/L	Larval metamorphosis	Flores et al. (2020)
Deltamethrin (pyrethroid) insecticide used for the control of sea lice	*Homarus Gammarus* European lobster	1-h LC50 2.6–2.9 ng/L, 1-h EC50 0.4–0.6 ng/L	Larval survival Immobility	Parsons et al. (2020)
Imidacloprid (IMI) (neonicotinoid) insecticide	*Metapenaeus macleaya* Eastern school prawn	8-day exposure 1–4 µg/L alterations in metabolic homeostasis and more frequent moulting	Juvenile metabolic homeostasis and moulting	McLuckie et al. (2020)
	Acropora tenuis scleractinian coral	48-h EC50 347 µg/L	Larval metamorphosis	Flores et al. (2020)
	Saccostrea glomerate Sydney rock oyster	1 mg/L significant effect on gill AChE activity 4-day exposure to 2 mg/L resulted in significant changes in gene expression and 14 gene ontology terms overexpressed	Adult biochemical function Genotoxicity	Ewere et al. (2019)
Calypso 480 SC (neonicotinoid) insecticide	*Mytilus galloprovincialis* Bivalve mollusc	96-h LC50 7.77 g/L LC50	Adult survival	Stara et al. (2020)
Chlorothalonil fungicide	*Acropora tenuis* scleractinian coral	48-h EC50 6.0 µg/L	Larval metamorphosis	Flores et al. (2020)
Fipronil (phenylpyrazole) insecticide	*Acropora tenuis* scleractinian coral	48-h EC50 29.1 µg/L	Larval metamorphosis	Flores et al. (2020)
Propiconazole fungicide	*Acropora tenuis* scleractinian coral	48-h EC50 1008 µg/L	Larval metamorphosis	Flores et al. (2020)
DDT (organochlorine)	*Mercenaria mercenaria* hard clam	24-h LC50 0.61 mg/L	Juvenile survival	Chung et al. (2007)

(continued)

◻ Table 7.10 (continued)

Pesticide	Species affected	Exposure effect	Endpoint	Source
Atrazine (triazine) herbicide	*Mercenaria mercenaria* Hard clam	96-h LC50 5608 μg/L	Juvenile survival	Lawton et al. (2010)
	Crassostrea gigas Pacific oyster	Significant differences in aneuploidy after treatments of 46.5 nM and 465 nM in comparison to control	Juvenile and adult (genotoxicity: abnormal number of chromosomes)	Bouilly et al. (2003)
Lindane (gamma-HCH)	*Crassostrea gigas* Pacific oyster	12-day LC50 2.22 mg/L 1.9 mg/L cell viability significantly decreased	Adult survival Cytotoxicity	Anguiano et al. (2006)
Diuron (organochlorine) herbicide	*Crassostrea gigas* Pacific oyster	Induction of aneuploidy after treatments of 300 and 3000 ng/L Significant aneuploidy in offspring	Adult (genotoxicity: abnormal number of chromosomes)	Bouilly et al. (2007)
	Exaiptasia pallida and in-hospite[a] *Symbiodinium* spp. anemone	48-h EC50 8 μg/L	Adult Photosynthesis reduction in effective quantum yield (EQY)	Howe et al. (2017)
	Acropora millepora scleractinain coral	24-h EC50 2.98 μg/L	Adult (EQY)	Negri et al. (2011)
Endosulfan (organochlorine) pesticide	*Crassostrea gigas* Pacific oyster	DNA strand breaks at 150 and 300 nM	Embryos	Wessel et al. (2007)

[a]In-hospite means living within a host cell

environment. Because of their very purpose, pesticides are toxic to target organisms but sometimes also to non-target organisms and occupational exposure mainly of agricultural workers. There are further concerns about potential impacts on human health in relation to the consumption of contaminated food. The most common source of pesticides in marine ecosystems is application on land although some pesticides are used in marine situations (e.g. anti-fouling agents). Transport of pesticides to the ocean is predominantly via the atmosphere and surface waters. The physicochemical properties of the pesticides and the physicochemical conditions and biochemical processes in air and fresh and estuarine systems influence the pathway of pesticides to the marine environment.

Pesticide solubility in water can vary according to their chemical characteristics, and this, in turn, affects the exposure to organisms via water. Once in water, pesticides are distributed (via a process called partitioning) into various compartments of the environment (e.g. water, suspended particles, sediment, plant and animal tissue and the air). For example, pesticides will bind, in varying degrees, to organic and inorganic particles and/or dissolved forms of organic matter. Binding to particles leads to a decrease in exposure of water column dwelling organisms but increased exposure to sediment-dwelling organisms (meiobenthos). Binding to dissolved organic matter influences solubility and subsequent distribution and availability to biota. Different types of pesticides have different rates of degradation or half-lives and this defines how long they may persist in the environment and pose a toxic threat. There are relatively few studies that investigate the toxic effects of pesticides to marine organisms but measurable responses at environmental relevant concentrations have been reported.

7.10 Study Questions and Activities

1. Determine the logKow of a new pesticide 'A' that has solubility in water of 0.67 mg/L and a solubility in octanol of 576,850 mg/L. Is this pesticide likely to accumulate in fatty tissue, explain your answer?
2. If the concentration of pesticide 'B' in fish tissue was 15 mg/kg and water concentration is 0.0028 mg/L, what is the BCF?
3. If the concentration of Pesticide 'C' in polychaete tissue was 12 mg/kg and the sediment concentration is 43 mg/kg, what is the BSAF?
4. Describe 'global distillation' in your own words.
5. Look up the FAO website: ▶ http://www.fao.org/faostat/en/#data/RP find the country you live in and using the FAOSTAT tool report the data available on organophosphate pesticide use in the country you live. Explore the website and the FAOSTAT tool.

References

Alzieu Cl, Sanjuan J, Michel P, Borel M, Dreno JP (1989) Monitoring and assessment of butyltins in Atlantic coastal waters. Mar Pollut Bull 20(1):22–26

Anguiano G, Llera-Herrera R, Rojas E, Vazquez-Boucard C (2006) Subchronic organismal toxicity, cytotoxicity, genotoxicity, and feeding response of Pacific oyster (*Crassostrea gigas*) to lindane (gamma-HCH) exposure under experimental conditions. Ecotoxicol Environ 65(3):388–394

Antizar-Ladislo B (2008) Environmental levels, toxicity and human exposure to tributyltin (TBT)-contaminated marine environment. A review. Environ Int 43:292–308

Australian Government and Queensland Government (2018) *Reef 2050 Water quality improvement plan: 2017–2022*. Reef Water Quality Protection Plan Secretariat. Available at: ▶ https://www.reefplan.qld.gov.au/__data/assets/pdf_file/0017/46115/reef-2050-water-quality-improvement-plan-2017-22.pdf. [Accessed 3 Oct 2020]

Bailey HC, Krassoi R, Elphick JR, Mulhall A-M, Hunt P, Tedmanson L, Lovell A (2000) Whole effluent toxicity of sewage treatment plants in the Hawkesbury-Nepean watershed, New South Wales, Australia, to *Ceriodaphia dubia* and *Selenastrum capricornutum*. Environ Toxicol Chem 19(1):72–81

Baird C, Cann M (2012) Environmental chemistry, 5th ed. New York, Freeman, p 736

Blossom N, Szafranski F, Lotz A (2018) Use of copper-based antifouling paint: A U.S. Regulatory Update. Coatings Technol 15(3). Available at: ▶ https://www.paint.org/coatingstech-magazine/articles/use-copper-based-antifouling-paint-u-s-regulatory-update/ [Accessed 29 Nov 2020]

Bouilly K, Leitão A, McCombie H, Lapègue S (2003) Impact of atrazine on aneuploidy in pacific oysters, *Crassostrea gigas*. Environ Toxicol Chem 1:219–223

Bouilly K, Bonnard M, Gagnaire B, Renault T, Lapègue S (2007) Impact of diuron on aneuploidy and hemocyte parameters in Pacific oyster, *Crassostrea gigas*. Arch Environ Contaminant Toxicol 52(1):58–63

Brodie J, Landos M (2019) Pesticides in Queensland and Great Barrier Reef waterways -potential impacts on aquatic ecosystems and the failure of national management. Estuar Coast Shelf Sci 230:106447

Burton ED, Phillips IR, Hawker DW (2004) Sorption and desorption behavior of tributyltin with natural sediments. Environ Sci Technol 38(24):6694–6700

Butcherine P, Benkendorff K, Kelaher B, Barkla BJ (2019) The risk of neonicotinoid exposure to shrimp aquaculture. Chemosphere 217:329–348

Cagnazzi D, Harrison PL, Parra JG, Reichelt-Brushett A, Marsilib L (2020) Geographic and temporal variation in persistent pollutants in Australian humpback and snubfin dolphins. Ecol Indicators 111:105990

Carson R (1962) *Silent spring*. Houghton Mifflin, p 368

Champ MA (1999) The need for the formation of an independent, international marine coatings board. Mar Pollut Bull 38(4):239–246

Chen C, Guo W, Ngo HH (2019) Pesticides in stormwater runoff–a mini review. Frontiers of Environ Sci Eng 13:72. ▶ https://doi.org/10.1007/s11783-019-1150-3

Chung KW, Fulton MH, Scott GI (2007) Use of the juvenile clam, *Mercenaria mercenaria*, as a sensitive indicator of aqueous and sediment toxicity. Ecotoxicol Environ Safety 67(3):333–340

Claeys RR, Caldwell RS, Cutshall NH, Holton R (1975) Chlorinated pesticides and polychlorinated biphenyls in marine species, Oregon/Washington Coast, 1972. Pesticide Monitor J 9(1):2–10

Coburn DR, Treichler R (1946) Experiments on toxicity of DDT to wildlife. J Wildl Manag 10(3):208–216

Commendatore M, Yori P, Scenna L, Ondarza PM, Suárez N, Marinao C, Miglioranza KSB (2018) Persistent organic pollutants in sediments, intertidal crabs, and the threatened Olrog's gull in a northern Patagonia salt marsh Argentina. Mar Pollut Bull 136:5330546

Connell DW, Hawker DW (1988) Use of polynomial expressions to describe the bioconcentration of hydrophobic chemicals in fish. Ecotoxicol Environ Safety 16:242–257

Corsolini S, Romeo T, Ademollo N, Greco S, Focardi S (2002) POPs in key species of marine Antarctic ecosystems. Microchem J 73:187–193

Corsolini S, Covaci A, Ademollo N, Focardi S, Schepens P (2003) Occurrence of organochlorine pesticides (OCPs) and their enantiomeric signatures, and concentrations of polybrominated diphenyl ethers (PBDEs) in the Adelie penguin food web, Antarctica. Environ Pollut 140:371–382

Dafforn KA, Lewis JA, Johnstone EL (2011) Antifouling strategies: history and regulation, ecological impacts and mitigation. Mar Pollut Bull 62(3):453–465

Das K, Malarvannan G, Dirtu A, Dulau V, Dumont M, Lepoint G, Mongin P, Covaci A (2017) Linking pollutant exposure of humpback whales breeding in the Indian Ocean to their feeding habits and feeding areas off Antarctica. Environ Pollut 220 (Part B):1090–1099.

Davis A, Lewis SE, Bainbridge ZT, Brodie J, Shannon E (2012) Pesticide residues in waterways of the lower Burdekin region: Challenges in ecotoxicological interpretation of monitoring data. Australas J Ecotoxicol 1:89–108

Day KE (1990) Pesticide residues in freshwater and marine zooplankton: a review. Environ Pollut 67:205–222

de Oliveira DD, Rojas EG, dos Santos Fernandez MA (2020) Should TBT continue to be considered an issue in dredging port areas? a brief review of the global evidence. Ocean Coastal Manag 197:105303

DEPA (The Danish Environmental Protection Agency) (2015) The EU list of potential endocrine disruptors. Ministry of Environment and Food, The Danish Environmental Protection Agency, København K, Denmark. Available from: ▶ https://eng.mst.dk/chemicals/chemicals-in-products/focus-on-specific-substances/endocrine-disruptors/the-eu-list-of-potential-endocrine-disruptors/ [Accessed 3 Oct 2021]

Devlin M, Waterhouse J, Taylor J, Brodie J (2001) Flood plumes in the Great Barrier Reef: spatial and temporal patterns in composition and distribution. Research Publication No. 68. Great Barrier Reef Marine Park Authority, Townsville, p 122

Duarte-Restrepo E, Jaramillo-Colorado B, Duarte-Jaramillo L (2020) Effects of chlorpyrifos on the crustacean *Litopenaeus vannamei*. PloS one, 15(4): e0231310, 1–16

Elmas S, Skipper K, Salehifar N, Jamieson T, Andersson G, Nydén M, Leterme S, Andersson M (2018) Cyclic copper uptake and release from natural seawater—A fully sustainable antifouling technique to prevent marine growth. Environ Sci Technol 55(1):757–766

Evans SM, Evans PM, Leksono T (1996) Widespread recovery of dogwhelks, *Nucella lapillus*, from tributyltin contamination in the North Sea and Clyde Sea. Mar Pollut Bull 32(3):263–269

Ewere EE, Powell D, Rudd D, Reichelt-Brushett A, Mouatt P, Voelcker N, Benkendorff K (2019) Uptake, depuration and sublethal effects of the neonicotinoid, imidacloprid, exposure in Sydney rock oysters. Chemosphere 230:1–13

Flores F, Kaserzon S, Esisei G, Ricardo G, Negri A (2020) Toxicity thresholds of three insecticides and two fungicides to larvae of the coral *Acropora tenuis*. PeerJ 8:e9615

FOA (Food and Agriculture Organization of the United Nations) (2021) Pesticides Use, pesticides trade and pesticides indicators 1990–2019—FOASTAT analytical brief 29. Available at: ▶ https://www.fao.org/faostat/en/#data/RP [Accessed 1 Feb2022]

FRAC (Fungicide Resistance Action Group) (2020) FRAC classification of fungicides. Fungal control agents by cross resistance pattern and mode of action 2020. Available at: ▶ https://www.frac.info/docs/default-source/publications/frac-mode-of-action-poster/frac-moa-poster-2020v2.pdf?sfvrsn=a48499a_2 [Accessed 6 Aug 2020]

Frouin H, Pelletier E, Lebeuf M, Saint-Louis R, Fournier M (2010) Toxicology of organotins in marine organisms: A review. In: Chin H (ed) Organometallic compounds: preparation, structure and properties. Nova Science Pub Inc, pp 1–47

Gallen C, Thai P, Paxman C, Prasad P, Elisei G, Reeks T, Eaglesham G, Yeh R, Tracey D, Grant S, Mueller J (2019) Marine monitoring program: annual report for inshore pesticide monitoring 2017–18. Report for the Great Barrier Reef Marine Park Authority, Great Barrier Reef Marine Park Authority, p 118

Gao Y, Zheng H, Xia Y, Chen M, Meng X-Z, Cai M (2019) Spatial distributions and seasonal changes of current-use pesticides from the North Pacific to the Arctic Oceans. J Geophys Res: Atmos 124(16):9716–9729

Gebauer P, Paschke K, Vera C, Toro J, Pardo M, Urbina M (2017) Lethal and sub-lethal effects of commonly used anti-sea lice formulations on non-target crab *Metacarcinus edwardsii* larvae. Chemosphere 185:1019–1029

Gibson CP, Wilson SP (2003) Imposex still evident in eastern Australia 10 years after tributyltin restrictions. Mar Environ Res 55(2):101–112

Glynn PW, Rumbold DG, Snedaker SC (1995) Organochlorine pesticide residues in marine sediment and biota from the Northern Florida reef tract. Mar Pollut Bull 30(6):397–402

Goldberg ED (1975) Synthetic organohalides in the sea. Proc Royal Soc London B 189:277–289

Greenhalgh R (1980) Definition of persistence in pesticide chemistry. Pure Appl Chem 52:2563–3566

Harino H, Arai T, Ohjii M, Miyazaki N (2009) Monitoring of alternative biocides: Asia. In: Arai T, Harino H, Ohji M, Langston W (eds) Ecotoxicology of antifouling biocides. Springer, pp 345–364

Haynes D, Müller J, Carter S (2000) Pesticide and herbicide residues in sediments and seagrasses from the Great Barrier Reef World Heritage Area and Queensland coast. Mar Pollut Bull 41(7–12):279–287

Heffernan AL, Gómes-Ramos MM, Gaus C, Vijayasarathy S, Bell I, Hof C, Mueller JF, Gómez-Ramos MJ (2017) Non-targeted, high resolution mass spectrometry strategy for simultaneous monitoring of xenobiotics and endogenous compounds in green sea turtles on the Great Barrier Reef. Sci Total Environ 599–600:1251–1262

Hellweger FL, Gordon AL (2002) Tracing Amazon River water into the Caribbean Sea. J Mar Res 60(4):537–549

Hermens JLM, de Bruijn JHM, Brooke DN (2013) The octanol-water partition coefficient: strengths and limitations. Environ Toxicol Chem 32(4):32–733

Hong S, Jeong J, Shim S (2008) Effect of electric currents on bacterial detachment and inactivation. Biotechnol Bioeng 100(2):379–386

Hook SE, Doan H, Gonzago D, Musson D, Du J, Kookana R, Sellars MJ, Kumar A (2018) The impacts of modern-use pesticides on shrimp aquaculture: an assessment for north eastern Australia. Ecotoxicol Environ Saf 148:770–780

Howe P, Reichelt-Brushett AJ, Clark M, Seery C (2017) Toxicity effects of the herbicides diuron and atrazine on the tropical marine cnidarian *Exaiptasia pallida* and in-hospite *Symbiodinium* spp. using PAM chlorophyll-a fluorometry. J Photochem Photobiol, B 171:125–132

HRAC (Herbicide Resistance Action Committee) (2020) HRAC mode of action classification 2020. HRAC, CropLife. Available at: ▶ https://hracglobal.com/files/HRAC_Revised_MOA_Classification_Herbicides_Poster.pdf [Accessed 6 Aug 2020]

Huang X, Cui H, Duan W (2020) Ecotoxicity of chlorpyrifos to aquatic organisms: a review. Ecotoxicol Environ Saf 200:110731

IbisWorld (2016) Available at: ▶ https://www.ibisworld.com/au/market-size/pesticide-manufacturing/ [Accessed 2 Feb 2022].

IRAC (Insecticide Resistance Action Committee) (2019) Mode of action classification. Poster Version 7, August 2019. IRAC, CropLife. Available at: file:///C:/Users/ac2458/Downloads/Eng_moa_structure_poster_Ed-7_4_18Aug19.pdf [Accessed 8 Aug 2020]

Iwata H, Tanabe S, Sakai N, Nishimura A, Tatsukawa R (1994) Geographic distribution of persistent organochlorines in air, water and sediments from Asia and Oceania, and their implications for global redistribution from lower latitudes. Environ Pollut 85:15–33

Jaga K, Dharmani C (2003) Global surveillance of DDT and DDE levels in human tissues. Int J Occup Med Environ Health 16(1):7–20

Jeschke P, Nauen R, Schindler M, Elbert A (2011) Overview of the status and global strategy for neonicotinoids. J Agric Food Chem 59(7):2897–2908

Jin M, Fu J, Xue B, Zhou S, Zhang L, Li A (2017) Distribution and enantiomeric profiles of organochlorine pesticides in surface sediments from the Bering Sea, Chukchi Sea and adjacent Arctic areas. Environ Pollut 222:109–117

Kannan K, Guruge KS, Thomas NJ, Tanabe S, Giesy JP (1998) Butyltin residues in southern sea otters Enhydra lutris nereis found dead along California coastal waters. Environ Sci Technol 32:1169–1175

Kattwinkel M, Kuhne JV, Foit K, Liess M (2011) Climate change, agricultural insecticide exposure, and risk for freshwater communities. Ecol Appl 21:2068–2081

Kim S-L (2020) Trophic transfer of organochlorine pesticides through food-chain in coastal marine ecosystem. Environ Eng Res 25(1):43–51

Kim NS, Hong SH, Shin K-H, Shim WJ (2017) Imposex in Reishia clavigera as an indicator to assess recovery of TBT pollution after a ban in South Korea. Arch Environ Contam Toxicol 73:301–309

Klinčić D, Romanić SH, Klaković-Gašpić Z, Tićina V (2020) Legacy persistent organic pollutants (POPs) in archive samples of wild Bluefin tuna from the Mediterranean Sea. Mar Pollut Bull 155:111086

Klumpp DW, von Westernhagen H (1995) Biological effects of pollutants in Australian tropical coastal waters: Embryonic malformations and chromosomal aberrations in developing fish eggs. Mar Pollut Bull 30(2):158–165

Lackenby H (1962) Resistance of ships, with special reference to skin friction and hull surface condition. Proc Inst Mech Eng 176:981–1014

Langston WJ, Pope ND, Davey M, Langston KM, O-Hara SCM, Gibbs PL, Pascoe P (2015) Recovery from TBT pollution in English channel environments: a problem solved? Mar Pollut Bull 95:551–564.

Lawton JC, Pennington PL, Chung KW, Scott GI (2010) Toxicity of atrazine to the juvenile hard clam, Mercenaria mercenaria. J Exp Biol 213:4010–4017

Leung KMY, Grist EPM, Morley NJ, Morritt D, Crane M (2007) Chronic toxicity of tributyltin to development and reproduction of the European freshwater snail Lymnaea stagnalis (L.). Chemosphere 66:1358–1366

Lewis SE, Brodie JE, Bainbridge ZT, Rohde KW, Davis AM, Masters BL, Maughan M, Devlin MJ, Mueller JF, Schaffelke B (2009) Herbicides: a new threat to the Great Barrier Reef. Environ Pollut 157(8):2470–2484

Lodeman EG (1896) The spraying of plants: a succinct account of the history, principles and practice of the application of liquids and powders to plants, for the purpose of destroying insects and fungi. New York, Macmillan, p 422

MacBean C (ed) (2012) The pesticide manual [OP]: a World compendium. 16th edn. Alton, British Crop Production Council (BCPC), p 1439

Mansouri A, Cregut M, Abbes C, Durand M-J, Landoulsi A, Thouand G (2017) The environmental issues of DDT pollution and bioremediation: a multidisciplinary review. Appl Biochem Biotechnol 181:309–339

Matthews GA (2018) A history of pesticides. Oxfordshire, CABI, p 310

McLuckie CM, Moltschaniwskyj N, Gaston TF, Dunstan RH, Crompton M, Butcherine P, Benkendorff K, Taylor MD (2020) Lethal and sub-lethal effects of environmentally relevant levels of imidacloprid pesticide to Eastern School Prawn, Metapenaeus macleaya. Sci Total Environ 742:140449

Menzies R, Quinete NS, Gardinali P, Seba D (2013) Baseline occurrence of organochlorine pesticides and other xenobiotics in the marine environment: Caribbean and Pacific collections. Mar Pollut Bull 70:289–295

Mercurio P, Mueller JF, Eaglesham G, Flores F, Negri AP (2015) Herbicide persistence in seawater simulation experiments. PLoS ONE 10(8):e0136391

Michalek K, Bowden BF (1997) A natural algicide from the soft coral Sinularia flexibilis (Coelenterata, Octocorallia, Alcyonacea). J Chem Ecol 23(2):1–16

Minggang C, Canong Q, Yuan S, MingHong C, ShuiYing H, biHau Q, JioungHui S, XiaoYan L (2010) Concentration and distribution of 17 organochlorine pesticides (OCPs) in seawater from the Japan sea northward to the Arctic Ocean. Sci China Chem 53(5):1033–1047

Miranda DA, Yogui GT (2016) Polychlorinated biphenyl and chlorinated pesticides in king mackerel caught off the coast of Pernambuco, northeastern Brazil: occurrence, contaminant profile, biological parameters and human intake. Sci Total Environ 569–570:1510–1516

Mitchell RT (1946) Effects of DDT spray on eggs and nestlings of birds. J Wildl Manag 10(3):192–194

Mnif W, Hassine AIH, Bouaziz A, Bartegi A, Thomas O, Roig B (2011) Effect of endocrine disruptor pesticides: a review. Int J Environ Res Public Health 8:2265–2303

Monirith I, Ueno D, Takahashi S, Nakata H, Sudaryanto A, Subramanian A, Karuppiah S, Ismail A, Muchtar M, Zheng J, Richardson BJ, Prudente M, Duc Hue N, Tana T-S, Tkalin AV, Tanabe S (2003) Asia-Pacific mussel watch: monitoring contamination of persistent organochlorine compounds in coastal waters of Asian countries. Mar Pollut Bull 46(30):281–300

Moreno-González R, León VM (2017) Presence and distribution of current-use pesticides in surface marine sediments from a Mediterranean coastal lagoon (SE Spain). Environ Sci Pollut Res 24(9):8033–8048

Muralidharan S, Dhananjayan V, Jayanthi P (2009) Organochlorine pesticides in commercial marine fishes of Coimbatore, India and their suitability for human consumption. Environ Res 109:15–21

Muzyed SKI, Kucuksezgin F, Tuzman M (2017) Persistent organochlorine residues in fish and sediments collected from Easter Aegean coast: levels, occurrence and ecological risk. Mar Pollut Bull 119:247–252

Mwevura H, Bouwman H, Kylin H, Vogt T, Issa MA (2020) Organochlorine pesticides and polycyclic aromatic hydrocarbons in marine sediments on polychaete worms from the west coast of Unguja island Tanzania. Reg Stud Mar Sci 36:101287

Nakayama K, Matsudaira C, Tajima Y, Yamada TK, Yoshioka M, Tanabe S (2009) Temporal and spatial trends of organotin contamination in the livers of finless porpoises (Neophocaena phocaenoides) and their association with parasitic infection status. Sci Total Environ 407(24):6173–6178

Negri A, Floes F, Röthig T, Uthicke S (2011) Herbicides increase the vulnerability of corals to rising sea surface temperature. Limnol Oceanogr 56(2):471–485

7

Negri AP, Smith RA, King O, Frangos J, Warne MStJ, Uthicke S (2019a) Adjusting tropical marine water quality guideline values for elevated ocean temperatures. Environ Sci Technol 54(2):1102–1110

Negri AP, Templeman S, Flores F, van Dam J, Thomas M, McKenzie M, Stapp LS, Kaserzon S, Mann RM, Smith RA, Warne MStJ, Mueller J (2019b) Ecotoxicology of pesticides on the Great Barrier Reef for guideline development and risk assessments. Final report to the National Environmental Science Program. Reef and Rainforest Research Centre Limited, Cairns (P 120). Available from: ► https://nesptropical.edu.au/wp-content/uploads/2020/04/NESP-TWQ-Project-3.1.5-Final-Report.pdf. [Accessed 3 Oct 2020]

Nhan DD, Am NM, Carvalho FP, Villeneuve J-P, Cattini C (1999) Organochlorine pesticides and PCBs along the coast of north Vietnam. Sci Total Environ 237–238:363–371

Nödler K, Licha T, Voutsa D (2013) Twenty years later-Atrazine concentrations in selected coastal waters of the Mediterranean and the Baltic Sea. Mar Pollut Bull 70:112–118

Noyes PD, McElwee MK, Miller HD, Clark BW, Van Tiem LA, Walcott KC, Erwin KN, Levin ED (2009) The toxicology of climate change: environmental contaminants in a warming world. Environ Int 35:971–986

OECD (Organisation of Economic and Cultural Development) (2012) OECD guidelines for testing of chemicals. Test no. 305. Bioaccumulation in fish: aqueous and dietary exposure. Available from: ► https://www.oecd-ilibrary.org/docserver/9789264185296-en.pdf?expires=1606616216&id=id&accname=guest&checksum=EE67909F9EC595773CF6FC0A-184FE8A6. [Accessed 3 Oct 2020]

Ojemaye CY, Onwordi CT, Pampanin DM, Sydnes MO, Petrik L (2020) Presence and risk assessment of herbicides in the marine environment of Camps Bay (Cape Town, South Africa). Sci Total Environ 738:140346

Ouyang W, Zhang Y, Gu X, Tysklind M, Lin C, Wang B, Xin M (2019) Occurrence, transportation, and distribution difference of typical herbicides from estuary to bay. Environ Int 130:104858

Packett R (2014) Project RRRD038 pesticide dynamics in the great Barrier Reef catchment and Lagoon: management practices (grazing, bananas and grain) and risk assessments. Dissolved and particulate herbicide transport in central Great Barrier Reef catchments (Subproject 3). Report to the Reef Rescue Water Quality Research and Development Program. Reef and Rainforest Research Centre Limited, Cairns, p 29. Available at: ► https://www.reefrescueresearch.com.au/research/all-projects/23-final-reports/177-rrrd038-final-report-1.html [Accessed 6 Feb 2022]

Parsons AE, Escobar-Lux RH, Sævick PN, Samuelsen OB (2020) The impact of anti-sea lice pesticides, azamethiphos and deltamethrin, on European lobster (*Homarus Gammarus*) larvae in the Norwegian marine environment. Environ Pollut 264:114725

Pedro S, Boba C, Deitz R, Sonne C, Rosing-Asvid A, Hansen M, Provatas A, McKinney MA (2017) Blubber-depth distribution and bioaccumulation of PCBs and organochlorine pesticides in Arctic-invading killer whales. Sci Total Environ 601–602:237–246

Phillips MWA (2020) Agrochemical industry development, trends in R&D and the impact of regulation. Pesticide Management Science 76(10):3348–3356

Ponce-Vélez G, de la Lanza-Espino G (2019) Organophosphate pesticides in coastal lagoon of the Gulf of Mexico. J Environ Prot 10(02):103–117

Prekker M (1992) The effects of the 1991 central Queensland flood waters around Heron Island, Great Barrier Reef, In: Byron G (ed) Workshop on the Impacts of Flooding. Workshop Series No. 17. Great Barrier Reef Marine Park Authority, Townsville, p 75

Qu C, Sun Y, Albanese S, Lima A, Sun W, Di Bonito M, Qi S, De Vivo B (2018) Organochlorine pesticides in sediments from Gulfs of Naples and Salerno, Southern Italy. J Geochem Explor 195:87–96

REACH (Registration, Evaluation, Authorisation and Restriction of Chemicals) (2021) Annex XIII: criteria for the identification of Persistent, bioaccumulative and toxic substances, and very persistent and very bioaccumulative substances. European Union. Available from: ► https://reachonline.eu/reach/en/annex-xiii.html?kw=substance#anch--substance [Accessed 3 Oct 2020]

Richardson BJ, Zheng GJ (1999) Chlorinated hydrocarbon contaminants in Hong Kong surficial sediments. Chemosphere 39(6):913–923

Riley JP, Wahby S (1977) Concentrations of PCBs, dieldrin and DDT residues in marine animals from Liverpool Bay. Mar Pollut Bull 8(1):9–11

Romero MB, Polizzi P, Chiodi L, Medici S, Blando M, Gerpe M (2018) Preliminary assessment of legacy and current-use pesticides in Franciscana dolphins from Argentina. Bull Environ Contam Toxicol 101(1):14–19

Ruiz JM, Bachelet G, Caumette P, Donard OFX (1996) Three decades of tributyltin in the coastal environment with emphasis on Arcachon Bay France. Environ Pollut 93(2):195–203

Sadler R, Connell D (2012) Global distillation in an era of climate change. In: Puzyn T, Mostrag A (eds) Organic pollutants ten years after the stockholm convention—environmental and analytical update, IntechOpen. Available from: ► https://www.intechopen.com/books/organic-pollutants-ten-years-after-the-stockholm-convention-environmental-and-analytical-update/global-distillation-in-an-era-of-climate-change [Accessed 3 Oct 2020]

Semeena VS (2005) Long-range atmospheric transport and total environment fate of persistent organic pollutants: a study using a general circulation model. Ph.D. Thesis, University of Hamburg, Hamburg. Available at ► https://mpimet.mpg.de/en/science/publications/2005 [Accessed 3 Oct 2020]

Sharma A, Kumar V, Shahzad B, Tanveer M, Sidhu GPS, Handa N, Kohli SK, Yadav P, Bali AS, Parihar RD, Dar OI, Singh K, Jasrotia S, Bakshi P, Ramakrishnan M, Bhardwaj SKR, Thukral AK (2019) Worldwide pesticide usage and its impacts on ecosystem. SN Appl Sci 1:1446

Simon-Delso N, Amaral-Rogers V, Belzunces LP, Bonmatin JM, Chagnon M, Downs C, Furlan L, Gibbons DW, Giorio C, Girolami V, Goulson D, Kreutzweiser DP, Krupke CH, Liess M, Long E, McField M, Mineau P, Mitchell EAD, Morrissey CA, Noome DA, Pisa L, Settele J, Stark JD, Tapparo A, Van Dyck H, Van Praagh J, Van der Sluijs JP, Whitehorn PR, Wiemers M (2015) Systemic insecticides (neonicotinoids and fipronil): trends, uses, mode of action and metabolites. Environ Sci Pollut Res 22(1):5–34

Sonak S, Giriyan A, Pangam P (2015) A method for analysis of costs and benefits of antifouling systems applied on ship's hull. J Ship Technol 6(1):73–83

Sonne C, Letcher RJ, Jenssen BM, Desforges J-P, Eulaers I, Andersen-Ranberg E, Gustavson K, Styrishave B, Dietz R (2017) A veterinary perspective on one health in the Arctic. Acta Vet Scand 59(84):1–11

Spilsbury FD, Warne MStJ, Backhaus T (2020) Risk assessment of pesticide mixtures in Australian rivers discharging to the Great Barrier Reef. Environ Sci Technol 54:14361–14371

Stara A, Pagano M, Capillo G, Faberello J, Sandova M, Vazzana I, Zuskova E, Velisek J, Matozza V, Faggio C (2020) Assessing the effects of neonicotinoid insecticide on the bivalve mollusc *Mytilus galloprovincialis*. Sci Total Environ 700:134914

Steinberg PD, De Nys R, Kjelleberg S (2002) Chemical cues for surface colonization. J Chem Ecol 28:1935–1951

7

Sundhar S, Shakila RJ, Jeyasekaran G, Aanand S, Shalini R, Arise-kar U, Surya T, Malini NAH, Boda S (2020) Risk assessment of organochlorine pesticides in seaweeds along the Gulf of Mannar, Southeast India. Mar Pollut Bull 161(B):111709

Svavarsson J, Granmo A, Ekelund R, Szpunars J (2001) Occurrence and effects of organotins on adult common whelk (*Buccinum undatum*) (Mollusca, Gastropoda) in harbours and in a simulated dredging situation. Mar Pollut Bull 42(5):370–376

Tanabe S, Prudente M, Mizuno T, Hasegawa J, Iwata H, Miyazaki N (1998) Butyltin contamination in marine mammals from North Pacific and Asian waters. Environ Sci Technol 32:192–198

Tester M, Ellis DV, Thompson JAJ (1996) Neogastropod imposex for monitoring recovery from marine TBT contamination. Environ Toxicol Chem 15(4):560–567

Thai P, Paxman C, Prasad P, Elisei G, Reeks T, Eaglesham G, Yeh R, Tracey D, Grant S, Mueller J (2020) Marine monitoring program: annual report for inshore pesticide monitoring 2018–19, p 69. Report to the Great Barrier Reef Marine Park Authority, Great Barrier Reef Marine Park Authority. Available from: ► http://elibrary.gbrmpa.gov.au/jspui/bitstream/11017/3666/4/Pesticides-Final-2018-19.pdf. [Accessed 3 Oct 2020]

Thomas KV, Langford KH (2009) Monitoring of alternate biocides: Europe and USA. In: Arai T, Harino H, Ohji M, Langston WJ (eds) Ecotoxicology of antifouling biocides. Springer, pp 331–344

Tsygankov VY, Lukyanova ON, Boyarova MD (2018) Organochlorine pesticide accumulation in seabirds and marine mammals from the Northwest Pacific. Mar Pollut Bull 128:208–213

University of Hertfordshire (2013) The Pesticide Properties DataBase (PPDB). Developed by the agriculture and environment research unit (AERU), University of Hertfordshire, 2006 (2013). Available from: ► http://sitem.herts.ac.uk/aeru/ppdb/ [Accessed 3 Jan 2022]

US EPA (United States Environmental Protection Authority). (undated). Estimating persistence, bioaccumulation, and toxicity using the PBT profiler. Available from: ► https://www.epa.gov/sites/default/files/2015-05/documents/07.pdf [Accessed 3 Oct 2020]

Vandergragt ML, Warne MStJ, Borschmann G, Johns CV (2020) Pervasive pesticide contamination of wetlands in the Great Barrier Reef Catchment Area. Integr Environ Assess Manag 16(6):968–982

Vass I (2011) Role of charge recombination processes in photodamage and photoprotection of the photosystem II complex. Physiol Plant 142(1):6–16

Vijayasarathy S, Baduel C, Hof C, Bell I, del Gómes-Ramos MM, Gómes-Ramos MJ, Kock M, Gaus C (2019) Multi-residue screening of non-polar hazardous chemicals in green turtle blood from different foraging regions of the Great Barrier Reef. Sci Total Environ 652:862–868

Wang X, Ren J, Gong P, Wang C, Xue Y, Yao T, Lohmann R (2016) Spatial distribution of the persistent organic pollutants across the Tibetan Plateau and its linkage with the climate systems: a 5-year air monitoring study. Atmos Chem Phys 16(11):6901–6911

Wania F (2003) Assessing the potential of persistent organic chemicals for long-range transport and accumulation in polar regions. Environ Sci Technol 37:1344–1351

Warne MStJ, Smith RA, Turner RDR (2020a) Analysis of pesticide mixtures discharged to the lagoon of the Great Barrier Reef Australia. Environ Pollut 265:114088

Warne MStJ, Neelamraju C, Strauss J, Smith RA, Turner RDR, Mann RM (2020b) Development of a method for estimating the toxicity of pesticide mixtures and a pesticide risk baseline for the reef 2050 water quality improvement plan. Report to the Queensland Department of Environment and Science for the Reef 2050 Water Quality Improvement Plan. Available at: ► https://www.publications.qld.gov.au/dataset/method-development-pesticide-risk-metric-baseline-condition-of-waterways-to-gbr/resource/c65858f9-d7ba-4aef-aa4f-e148f950220f [Accessed 3 Oct 2020b]

Waterhouse J, Schaffelke B, Bartley R, Eberhard R, Brodie J, Star M, Thorburn P, Rolfe J, Ronan M, Taylor B, Kroon F (2017) 2017 scientific consensus statement. Land use impacts on great barrier reef water quality and ecosystem condition. Queensland Government. Available from: ► https://www.reefplan.qld.gov.au/__data/assets/pdf_file/0029/45992/2017-scientific-consensus-statement-summary.pdf [Accessed 3 Oct 2020]

Weber J, Halsall CJ, Muir D, Teixeira C, Smail J, Solomon K, Hermanson M, Hung H, Bidleman T (2010) Endosulfan, a global pesticide: A review of its fate in the environment and occurrence in the Arctic. Sci Total Environ 408:2966–2984

Wessel N, Rousseau S, Caisey X, Quiniou F, Akcha F (2007) Investigating the relationship between embryotoxic and genotoxic effects of benzo[a]pyrene,17alpha- ethinylestradiol and endosulfan on *Crassostrea gigas* embryos. Aquatic Toxicol 85(2):133–142

Westergaard CH (2007) Comparison of fouling control coating performance to ship propulsion efficiency FORCE technology report no. 107- 24111, Part 2. Made for Hempel Marine Paints A/S on 11–06–2007

Whitford F, Pike D, Burroughs F, Hanger G, Johnson B, Brassard D, Blessing A (2006) The pesticide marketplace-discovering and developing new products. Purdue Extension PPP-71: Purdue University, West Lafayette, p 63

WHO (World Health Organisation) (2020) The WHO recommended classification of pesticides by hazard and guidelines to classification, 2019 edition. Available at: ► https://www.who.int/publications/i/item/9789240005662 [Accessed 19 Aug 2020]

Worldatlas (2018) ► https://www.worldatlas.com/articles/top-pesticide-consuming-countries-of-the-world.html. [Accessed 1 Sept 2020]

Wurl O, Obbard JP (2005) Organochlorine pesticides, polychlorinated biphenyls and polybrominated diphenyl ethers in Singapore's coastal marine sediments. Chemosphere 58:925–933

Zhang WJ (2018) Global pesticide use: Profile, trend, cost/benefit and more. Proc Int Acad Ecol Environ Sci 8(1):1–27

Zhang Q, Chen Z, Li Y, Wang P, Zhu C, Gao G, Xiao K, Sun H, Zheng S, Liang Y, Jiang G (2015) Occurrence of organochlorine pesticides in the environmental matrices from King George Island, west Antarctica. Environ Pollut 206:142–149

Zhang Y, Qin P, Lu S, Liu X, Zhai J, Xu J, Wang Y, Zhang G, Liu X, Wan Z (2020) Occurrence and risk evaluation of organophosphorus pesticides in typical water bodies of Beijing China. Environ Sci Pollut Res 28(2):1454–1463

Persistent Organic Pollutants (POPs)

Munro Mortimer and Amanda Reichelt-Brushett

Contents

8.1   Introduction – 186

8.2   History of POPs – 187

8.3   The Stockholm Convention – 188
8.3.1   Overview of the Convention – 188
8.3.2   Annexes and Exemptions for Some POPs – 189
8.3.3   The Original Set of 12 POPs Covered by the Stockholm Convention – 189
8.3.4   Additional POPs Now Covered by the Stockholm Convention – 189

8.4   Naming Conventions for Individual PCCD, PCDF, and PCB Compounds – 196

8.5   Assessment of Toxicity and Quantifying Exposure Risks for POPS – 197
8.5.1   Assessment of Toxicity and Exposure Risks for Dioxins, Furans, and Dioxin-Like PCBs – 198
8.5.2   The Meaning and Use of the Terms TEF and TEQ – 198
8.5.3   Use of Homologues and Congener Profiles in Forensic Investigations – 198

8.6   Case Studies – 199
8.6.1   Case Study 1—Dioxins, Furans, and Dioxin-Like PCBs in the Australian Aquatic Environment. – 199
8.6.2   Case Study 2—Spatial and Temporal Trends in Concentrations of Brominated Fire-Retardant POPs in Arctic Marine Mammal Tissues – 200

8.7   Summary – 203

8.8   Study Questions and Activities – 204

References – 204

© The Author(s) 2023
A. Reichelt-Brushett (ed.), *Marine Pollution—Monitoring, Management and Mitigation*,
Springer Textbooks in Earth Sciences, Geography and Environment,
https://doi.org/10.1007/978-3-031-10127-4_8

Abbreviations

AAS	Australian Academy of Science
ACS	American Chemical Society
BDE	Bromodiphenyl ether
BHC	An acronym for "benzene hexachloride"—a name sometimes used inappropriately for lindane (gamma-hexachlorocyclohexane). Using the acronym BHC for lindane is misleading because the 6 carbon ring chemical structure in the lindane molecule is cyclohexane, not a benzene ring. A more appropriate acronym for lindane is gamma HCH (HexaChlorocycloHexane).
CB	Chlorinated biphenyl
CDD	Chlorinated dibenzodioxin
CDF	Chlorinated dibenzofuran
DDD	1,1-Dichloro-2,2-bis(p-chlorophenyl)ethane
DDE	1,1-Dichloro-2,2-bis(p-chlorophenyl)ethylene
DDT	DichloroDiphenylTrichloroethane, a shortened version of a former name used for 1,1'-(2,2,2-trichloroethane-1,1-diyl)bis(4-chlorobenzene)
HBB	Hexabromobiphenyl
HBCD	Hexabromocyclododecane
HBCDD	Hexabromocyclododecane
HCBD	Hexachlorobutadiene
HCH	Hexachlorocyclohexane
NDP	National Dioxin Program
OCDD	Octachlorodibenzodioxin
OCDF	Octochlorodibenzofuran
PBB	Polybrominated biphenyl
PBDE	Polybrominated diphenyl ether
PCB	Polychlorinated biphenyl
PCDD	Polychlorinated dibenzo-p-dioxin
PCDF	Polychlorinated dibenzofuran
PCN	Polychlorinated naphthalene
PCP	Pentachlorophenol
PeCB	Pentachlorobenzene
PFOA	Perfluorooctanoic acid
PFHxS	Perfluorohexane sulfonic acid
PFOS	Perflurooctanoic acid
POP	Persistent Organic Pollutant
POPRC	Persistent Organic Pollutants Review Committee
SCCP	Short-chain chlorinated paraffin
TCDD	2,3,7,8-Tetrachlorodibenzo-p-dioxin
TEF	Toxic equivalent factor
TEQ	Toxic equivalent
UNEP	United Nations Environment Programme
USA	United States of America (USA)
WHO	World Health Organization
WHO-TEF	A TEF (toxic equivalent factor) published by the World Health Organisation

8.1 Introduction

Anyone interested in the history of environmental chemistry, and in particular the management and remediation of pollution, soon encounters the term **POP** along with the **Stockholm Convention** and **the dirty dozen**. This chapter gives an overview of this topic. It provides an important expansion of ▶ Chapter 7 on pesticides and biocides further addressing those pesticides that are also classified as POPs along with POPs that are derived from other sources.

The acronym **POP** is used for each member of a group of compounds called Persistent **Organic Pollutants**. The Stockholm Convention is an international agreement endorsed by most nations with the intent of managing (limiting the generation and usage, and where possible eliminating) further environmental contamination by this problematic group of chemicals. Some countries have not yet ratified (officially adopted) the terms of the Convention, (for example, the United States of America (USA)), or only partly ratified. The term **the dirty dozen** (a common descriptor for a group comprising 12 individuals of ill-repute) was used for the original 12 POPs listed in the Stockholm Convention in 2001.

The POPs are of concern because not only are they toxic contaminants but they also have a strong potential to bioaccumulate and persist (▶ Box 8.1). This aspect of their environmental fate and impact is related to their physicochemical properties.

> **Box 8.1: Definition of POPs and Their Problematic Properties**
>
> As defined by the United Nations Environmental Programme (UNEP) Stockholm Convention (UNEP 2018), Persistent Organic Pollutants (POPs) are organic (i.e. carbon-based) compounds possessing a particular combination of physical and chemical properties such that, once released into the environment, they:
>
> — remain intact for exceptionally long periods of time (many years);
> — become widely distributed throughout the environment as a result of natural processes involving soil, water and, most notably, air;
> — accumulate in the fatty tissue of living organisms including humans, and are found at higher concentrations at higher levels in the food chain; and
> — are toxic to both humans and wildlife.

8.2 History of POPs

As highlighted in ▶ Chapter 7, the problems of bioaccumulation and toxic effects associated with pesticides that are POPs became widely recognised with the publication of *Silent Spring* (Carson, 1962), which publi-cised the unintended environmental consequences associated with the use of organochlorine insecticides including DDT(▢ Figure 8.1), lindane (also known as BHC), heptachlor, and chlordane introduced during the 1940s and in widespread use by the 1950s. *Silent Spring* and the media publicity and public debate sent a

▢ **Figure 8.1** Spraying DDT in 1955 to manage western spruce budworm control, Powder River control unit, Oregon, United States of America (USA). *Source*: Wikimedia commons, *Photo*: R. B. Pope Date: July 1955 *Credit*: USDA Forest Service, Pacific Northwest Region, State and Private Forestry, Forest Health Protection. Collection: Portland Station Collection; La Grande, Oregon. *Image*: PS-1428

strong message to the community, and to governments and chemical regulators on a global scale (note: this happened before the internet or real-time social media communication was even conceived!).

Although the environmental impacts associated with POPs are now recognised by scientists and governments, the claims made in *Silent Spring* were hotly debated for many years following publication (ACS 2012). For example, in Australia in 1972 (a decade after the book was published), an enquiry established by the Australian Parliament was still reviewing and debating whether ongoing DDT use was problematic for wildlife and human health (Parliament of Australia, 1972). In 1972, the use of DDT in Australia and New Zealand was estimated as 900 tonnes per year (AAS 1972) despite restrictions introduced during the 1960s.

8.3   The Stockholm Convention

The focused attention by government agencies at an international scale on the environmental issues associated with POPs led to the establishment in 2001 of a global treaty among nations, formally known as the *United Nations Environment Programme (UNEP) Stockholm Convention on Persistent Organic Pollutants*. The convention, with the objective to "*protect human health and the environment from persistent organic pollutants*" came into force in 2004 and was ratified by 152 UN member states. By mid-2018, the number of ratifying states (including the European Union) had increased to 182 (◘ Figure 8.2).

The history, scope, amendments, and signatories can be found online at the United Nations Treaties website:
▸ https://treaties.un.org/Pages/ViewDetails.aspx?src=TREATY&mtdsg_no=XXVII-15&chapter=27&clang=_en

Full details of all aspects of the convention are available at the website of the Secretariat of the Stockholm Convention:
▸ http://www.pops.int/TheConvention/ThePOPs/tabid/673/Default.aspx

8.3.1   Overview of the Convention

The initial version of the convention covered 12 POP compounds known as **the dirty dozen**. The signatories to the original version of the convention agreed to:
− outlaw nine intentionally-manufactured POP chemicals;
− limit the use of DDT for malaria control; and
− reduce inadvertent production of dioxins and furans.

The convention committed developed countries to provide financial resources and measures to eliminate production and use of intentionally-produced POPs, eliminate unintentionally produced POPs where feasible, and manage and dispose of POP waste in an environmentally sound manner.

The convention includes an agreement to follow procedures to identify additional POPs and outlines the

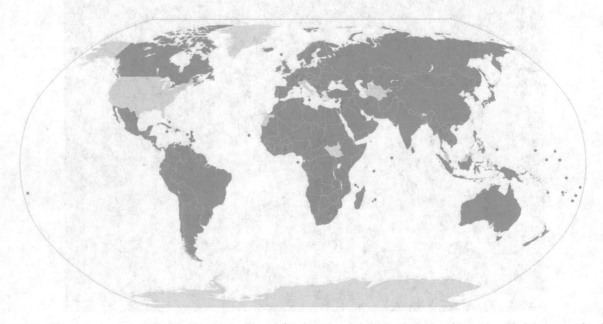

◘ **Figure 8.2**   Countries that have ratified the Stockholm Convention (green) and those that have signed but not ratified the convention (tan). *Source*: Wikimedia commons. *Author*: Canuckguy

criteria applied in doing so. Under the convention, reviews of chemicals as potential new POPs are led by a Persistent **Organic Pollutants Review Committee (POPRC)** comprising government-designated experts in chemical assessment and/or management. The members are nominated by and represent the signatory nations on a regional basis. The review process accounts for persistence, bioaccumulation, potential for long-range environmental transport, and toxicity of potential new POPs.

A series of review reports on individual POPs, published by the POPRC are available online, for example, UNEP (2015) in respect of the brominated flame retardant c-Decabromodiphenyl ether (c-decaBDE).

8.3.2 Annexes and Exemptions for Some POPs

Under the Stockholm Convention, each of the POPs is assigned an agreed **Annex** status. These are **A**—elimination, **B**—restriction, and **C**—unintended production:

- Annex A POPs (elimination) are prohibited from production and use (except for specific usage exemptions allowed only for convention parties that register for the exemption) and may be imported or exported only under specific restrictive conditions.
- Annex B POPs (restriction) are restricted from production and use (except for registered acceptable purposes) and imported and exported only under specific restrictive conditions. An example is that DDT use is still permitted for disease vector control for serious diseases such as malaria.
- Annex C POPs (unintentionally produced) are subject to release reduction and elimination, and the use of the best available techniques and best environmental practices for preventing releases of POPs into the environment are promoted.

Exemptions for some POPs and their uses are allowed in cases where replacement technologies do not yet exist, or viable replacement technologies are not readily available. This enables signatories to the Convention to take measures to reduce or eliminate releases from intentional production and use. Exemptions are limited to a specific time period.

In addition, the convention requires that signatory nations ensure stockpiles and wastes consisting of, containing, or contaminated with POPs, are managed safely and in an environmentally sound manner. This requires that such stockpiles and wastes be identified and managed to reduce or eliminate POPs releases from these sources, and that transport of wastes containing POPs across international boundaries is done

according to international rules, standards and guidelines.

8.3.3 The Original Set of 12 POPs Covered by the Stockholm Convention

The original **dirty dozen** POPs are listed in ▣ Table 8.1 with their structures and historical main uses, along with some general characteristics and toxicity. It is useful to note that:

- all are **chlorinated hydrocarbons** (also known as organohalogens) and nine are pesticides;
- all are chemically stable and **highly lipophilic** compounds that do not degrade readily even under extreme environmental conditions; and
- **some partially degrade into compounds of similar stability and toxicity**.

Furthermore, all have an established history of environmental persistence, bioaccumulation, and capacity for global distribution, and their use and release are associated with unintended adverse environmental consequences.

8.3.4 Additional POPs Now Covered by the Stockholm Convention

Since 2009, a further 16 compounds have been added to the listed Stockholm Convention POPs. In addition to several additional chlorinated hydrocarbons, the additions to the POPs list include several **polybrominated hydrocarbons** widely used as fire-retardants. The most recent **new POP** was added in 2015.

These new POPs are shown in ▣ Table 8.2 with their structures, uses, and general features of characteristics and toxicity. Like the original **dirty dozen** POPs, many of the additional POPs are chlorinated hydrocarbon pesticides, but also included are non-pesticide chlorinated hydrocarbons, several **brominated fire-retardant compounds**, and a **per-fluorinated compound** (known generally as a PFOS). Like the original POPs, these chemicals also have a demonstrated capacity for environmental persistence, for bioaccumulation, and for global distribution and most have been detected in the marine environment and various marine species. Studies of toxicity to marine species are limited for some POPs and consideration needs to be made about potential sources and risk to the marine environment. Atmospheric transport is an important pathway to the marine environment for some POPs. Their use and release are associated with unintended adverse environmental consequences.

◻ Table 8.1   The original 12 POPs listed in the Stockholm Convention (the dirty dozen)

POP name	Structure, uses, characteristics and toxicity
Aldrin (listed in Annex A)	 Formerly used as a soil treatment to kill termites, grasshoppers, corn rootworm, and other insect pests. After release to the environment, aldrin rapidly converts to dieldrin (see below, this table) which is also toxic. Can cause sublethal and lethal effects on marine crustaceans, molluscs, fish and birds. Also toxic to humans. Half-life in soil: approximately 110 days.
Chlordane (listed in Annex A)	 Formerly used extensively to control termites and as a broad-spectrum insecticide on a range of agricultural crops. Can cause lethal and sublethal effects in marine crustaceans, molluscs, fish and birds. Impacts the immune system in humans and possible carcinogen. Half-life in soil: variable approximately 1–4 years.
DDT (listed in Annex B)	 Widely used during World War II to protect soldiers and civilians from malaria, typhus, and other diseases spread by insects. Subsequently, widespread use of DDT continued for disease control, and use increased rapidly for pest control on a variety of agricultural crops, especially cotton. DDT is still used against mosquitoes in several countries to control malaria. DDT degrades to DDD and DDE, both of which are extremely persistent and have similar toxicity as DDT. Potential for trophic transfer and also to offspring of suckling mammals established. Causes egg shell thinning in birds, particularly high trophic order birds. Toxic to marine crustaceans, molluscs and fish. Half-life in soil: 10–15 years.
Dieldrin (listed in Annex A)	 Degradation product of aldrin (see above, this table), consequently concentrations of dieldrin measured in the environment may in part reflect a history of both dieldrin and aldrin usage. Formerly used widely to control termites and textile pests, dieldrin was also used to control insect-borne diseases and insect pests living in agricultural soils. Highly toxic to fish and other aquatic animals. Half-life in soil: approximately 5 years.
Endrin (listed in Annex A)	 Formerly used to control pests on crops such as cotton and grains, and rodents such as mice and voles. Animals can metabolise endrin, so it does not accumulate in their fatty tissue to the extent as structurally similar chemicals. Toxic to marine fish, crustaceans, echinoderms and algae. Half-life in soil: up to 12 years.

(continued)

□ Table 8.1 (continued)

POP name	Structure, uses, characteristics and toxicity
Heptachlor (listed in Annex A)	Formerly used to kill soil insects and termites including cotton insects, grasshoppers, other crop pests, and malaria-carrying mosquitoes. Believed to be responsible for the decline of wild bird populations including Canadian Geese and American Kestrels in the Columbia River basin in the USA. Toxic to amphipods, crustaceans, molluscs, and fish. Half-life in soil: 6–9 months.
Hexachlorobenzene (listed in Annex A) (listed in Annex C)	First introduced in 1945 to treat seeds, HCB kills fungi that affect food crops. It is also a by-product of the manufacture of some industrial chemicals and exists as an impurity in several pesticide formulations. Potential for trophic transfer and also to offspring of suckling mammals established. Limited marine toxicity data available. Half-life in soil: Highly variable depending on climate 0.6–6.3 years.
Mirex (listed in Annex A)	Used mainly to combat fire ants, and also against other types of ants and termites. One of the most stable and persistent pesticides. Also used as a fire retardant in plastics, rubber, and electrical goods. Limited marine toxicity data available. Half-life in soil: 10–12 years.
Toxaphene (listed in Annex A)	Formerly used as a pesticide on cotton, cereal grains, fruits, nuts, and vegetables. It was also used to control ticks and mites in livestock. Toxic to marine crustaceans, molluscs, and fish. Long-term exposure highly toxic to fish, with effects including reduced reproductivity. Half-life in soil: Variable, from 3 months up to 12 years.
Polychlorinated biphenyls (PCBs) (listed in Annex A with exemptions) (listed in Annex C)	These compounds were used in industry as heat exchange fluids, as fluids in electricity transformers and capacitors (□ Figure 8.3), and as additives in paint, carbonless copy paper, and plastics. Commercial mixtures of PCBs known as Aroclors were manufactured in large quantities (see ▶ Box 8.2). Of the 209 different types of PCBs, 13 exhibit a dioxin-like toxicity. Some studies show selected PCBs are toxic to marine crustaceans and molluscs, but research is limited. Half-life in soil: Their persistence in the environment corresponds to the degree of chlorination, and half-lives can vary from 10 days to 1.5 years.

(continued)

□ Table 8.1   (continued)

POP name	Structure, uses, characteristics and toxicity
Polychlorinated dibenzo-*p*-dioxins (PCDDs) and Polychlorinated dibenzo-furans (PCDFs) (listed in Annex C)	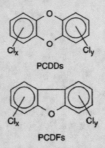 Produced unintentionally due to incomplete combustion, and also during the manufacture of pesticides and other chlorinated substances. Commonly emitted from the burning of hospital waste, municipal waste, and hazardous waste, and also from automobile emissions, peat, coal, and wood. There are 75 different dioxins, of which seven are considered to be of concern. Half-life in soil: Variable, 2.5–6.0 years.

□ Figure 8.3   Liquid Aroclors (PCB mixtures) had worldwide use as dielectric fluid in transformers common throughout power supply grids. In practice, the dielectric is an oily liquid that surrounds the electricity-conducting copper wire coils inside the transformer. Its roles are cooling and electrical insulation, and both the transformer body and the cooling pipes visible on the outside of the transformer shown here are filled with dielectric fluid. The disposal of these PCB mixtures from obsolete and redundant equipment is an ongoing concern. Poor disposal practices in the past (e.g. to unlined landfill) caused serious widespread contamination of waterways and aquatic ecosystems. *Photo*: M. Mortimer

□ Table 8.2   Additional POPs listed in the Stockholm Convention since 2009

POP name	Structure, uses, properties and toxicity
Hexachlorocyclohexane (alpha-isomer HCH and beta-isomer HCH) (listed in Annex A without exemptions)	alpha-HCH beta-HCH gamma-HCH The technical grade (meaning 'as manufactured') of hexachlorocyclohexane (HCH) contains five isomers, namely alpha-, beta-, gamma-, delta- and epsilon HCH. Use of alpha- and beta-HCH as insecticides was phased out years ago, but these chemicals were produced as by-products in the production of the gamma isomer (lindane). At many chemical plants an estimated 6–10 tons of alpha- and beta-HCH were produced for each ton of lindane manufactured (a ton is 0.9071 tonne). This resulted in large stockpiles of unwanted byproduct and site contamination (□ Fig. 8.4). Highly persistent in water, particularly in colder regions and may bioaccumulate and biomagnify in biota and arctic food webs, thus are often found in seals and polar bears. They are subject to long-range transport, classified as potentially carcinogenic to humans, and adversely affect wildlife and human health in contaminated regions. Half-life in soil: alpha-HCH 20 weeks; beta-HCH 7–10 years.

(continued)

◻ Table 8.2 (continued)

POP name	Structure, uses, properties and toxicity
Lindane (gamma-isomer HCH) (listed in Annex A with exemptions)	Gamma-hexachlorocyclohexane (lindane) formerly used as a broad-spectrum insecticide for seed and soil treatment, foliar applications, tree and wood treatment and against ectoparasites in both veterinary and human applications. Persistent, bioaccumulates easily in the food chain and bioconcentrates rapidly. Evidence of long-range transport and toxic effects (immunotoxic, reproductive and developmental effects) in laboratory animals and aquatic organisms. Toxic to marine rotifers, crustaceans, molluscs and fish. Half-life in soil: Variable, 3 months to over 3 years.
Chlordecone (listed in Annex A without exemptions)	Similar compound to Mirex and used as an agricultural pesticide. Chlordecone (Kepone) was first produced in 1951 and commercially introduced in 1958. Currently, no use or production of the chemical is reported, as many countries have already banned its sale and use. Highly persistent, and with high potential for bioaccumulation and biomagnification. Physico-chemical properties and modelling data show that chlordecone can be transported for long distances. It is classified as a possible human carcinogen and is extremely toxic to aquatic organisms. Likely exposure pathway is via food. Limited study on toxicity to marine species. Half-life in soil: 1–2 years.
Decabromodiphenyl ether (Commercial mixture, c-decaBDE) (listed in Annex A with exemptions)	Deca-BDE is used as an additive flame retardant, with many applications including in plastics/polymers/composites, textiles, adhesives, sealants, coatings and inks. DecaBDE-containing plastics are used in computer and TV casings, wires and cables, pipes and carpets. Commercially available decaBDE (c-decaBDE) contains small percentages of octa- and nona-BCE. Usage peaked in the early 2000s, but it is still extensively used worldwide. Adverse effects on fish and terrestrial species. Half-life in soil: Highly variable depending on conditions > 6 months up to 50 years.
Hexa- and hepta bromodiphenyl ether (Commercial octaBDE) (listed in Annex A with exemptions)	Hexabromodiphenyl ether and heptabromodiphenyl ether are the main components of commercial octabromodiphenyl ether. The commercial mixture of octaBDE is highly persistent. The only degradation pathway is through debromination and producing other bromodiphenyl ethers. Many products and materials in use still contain these compounds. Highly persistent is highly bioaccumulative and has a high potential for long-range environmental transport. Marine toxicity poorly understood. Half-life in soil: Highly variably between compounds and environmental conditions, from several weeks up to a decade.
Tetrabromodiphenyl ether (tetraBDE) and pentabromodiphenyl ether (pentaBDE) (listed in Annex A with exemptions)	Used as additive flame retardants. TetraBDE and pentaBDE are the main components of commercial pentabromodiphenyl ether. The commercial mixture of pentaBDE is highly persistent in the environment, bioaccumulative and has a potential for long-range environmental transport (it has been detected in humans throughout all regions). Considered toxic to fish and birds. Half-life in soil: Variable, aerobic soils 10 days to 1 year.

(continued)

■ **Table 8.2**   (continued)

POP name	Structure, uses, properties and toxicity
Hexabromobiphenyl (HBB) (listed in Annex A without exemptions)	 Formerly used as a flame retardant, mainly in the 1970s. No longer produced or used in most countries due to restrictions under national and international regulations. Highly persistent in the environment, highly bioaccumulative and with a high potential for long-range environmental transport. Half-life in soil: Variable.
Hexabromocyclododecane (HBCD or HBCDD) (listed in Annex A with exemptions)	 Formerly used as a flame retardant additive to polystyrene materials in the 1980s as a part of safety regulation for articles, vehicles, and buildings. Highly toxic to marine diatoms and considered very toxic to other marine biota. Half-life in soil: ~8 months.
Hexachlorobutadiene (HCBD) (listed in Annex A without exemptions) (Also listed in Annex C)	 Hexachlorobutadiene is commonly used as a solvent for other chlorine-containing compounds. It is no longer intentionally produced but is a by-product in the manufacture of chlorinated aliphatic compounds such as carbon tetrachloride and tetrachloroethene (both produced on a large scale). Highly persistent in the environment, highly bioaccumulative and with a strong potential to bioaccumulate and biomagnify, and has a potential for long-range environmental transport. It is extremely toxic to aquatic organisms. Toxic to marine invertebrates and fish. Half-life in soil: 4–26 weeks.
Pentachlorobenzene (PeCB) (listed in Annex A without exemptions) (Also listed in Annex C)	 Formerly used in PCB products, in dyestuff carriers, as a fungicide and a flame retardant. May still be used as a chemical intermediate (e.g. for the production of quintozene). Also produced unintentionally during combustion, thermal and industrial processes, and present as an impurity in products such as solvents or pesticides. Persistent in the environment, highly bioaccumulative and has a potential for long-range environmental transport. It is moderately toxic to humans and very toxic to aquatic organisms. Toxic to marine molluscs, crustaceans and fish. Half-life in soil: 45 days.
Pentachlorophenol and its salts and esters (PCP) (listed in Annex A with exemptions)	 Former uses as herbicide, insecticide, fungicide, algaecide, disinfectant and in antifouling paint. Applications included agricultural seeds, leather, wood preservation, cooling tower water, rope and in paper manufacture. Contaminants can include other polychlorinated phenols, PCDDs, and PCDFs. Toxic to algae, invertebrates and fish. Half-life in soil: weeks to months ~45 days.

(continued)

◪ Table 8.2 (continued)

POP name	Structure, uses, properties and toxicity
Perfluorooctane sulfonic acid, its salts and perfluorooctane sulfonyl fluoride (PFOS) (listed in Annex B with exemptions)	 PFOS is both intentionally produced and is a degradation product of related anthropogenic chemicals. The current intentional use of PFOS is widespread and includes: electric and electronic parts, fire-fighting foam, photo imaging, hydraulic fluids and textiles. PFOS is still produced in several countries. PFOS is extremely persistent and has bioaccumulation and biomagnifying properties. However, PFOS does not follow the classic pattern of other POPs by partitioning into fatty tissues, but instead binds to proteins in the blood and the liver. It has a capacity to undergo long-range transport and also meets the toxicity criteria of the Stockholm Convention. Limited study of toxicity to marine species, impacts measure on molluscs and fish. Half-life in soil: Highly variable: years to decades.
Polychlorinated naphthalenes (PCNs) (listed in Annex A with exemptions) (also listed Annex C)	 Commercial PCNs are mixtures of up to 75 chlorinated naphthalene congeners plus byproducts, used for insulated coatings on electrical wires, and as wood preservatives, rubber and plastic additives, for capacitor dielectrics and in lubricants. PCNs are unintentionally generated during high-temperature industrial processes in the presence of chlorine. While some PCNs can be broken down by sunlight and, at slow rates, by certain microorganisms, many PCNs persist in the environment. Bioaccumulation is confirmed for tetra- to hepta- CNs. Toxic to marine algae and crustaceans. Limited studies. Half-life in soil: >1 year.
Short-chain chlorinated paraffins (SCCPs) (listed in Annex A with exemptions)	 Used as a plasticiser in rubber, paints, adhesives, flame retardants for plastics as well as an extreme pressure lubricant in metal working fluids. SCCPs are produced by chlorination of straight-chained paraffin fractions. Appear to be hydrolytically stable and sufficiently persistent in air for long-range transport to occur and lead to significant adverse environmental and human health effects. Many SCCPs bioaccumulate. Toxic to marine algae, invertebrates but limited other studies. Half-life in soil: Variable.
Technical endosulfan and its related isomers (listed in Annex A with exemptions)	 Used as an insecticide/acaricide since the 1950s to control crop pests, tsetse flies and ectoparasites of cattle, and as a wood preservative. Currently used as a broadspectrum insecticide to control a wide range of pests on a variety of crops including coffee, cotton, rice, sorghum and soy. Endosulfan occurs as two isomers: alpha (α)- and beta (β)-endosulfan. Technical endosulfan is found as a mixture of conformational stereoisomers, typically in a roughly 70:30 mix of -: β- isomers. The α- isomer, or endosulfan-I, is more thermally stable, and the β- isomer, endosulfan-II, will irreversibly convert to the α- isomer, though the process is slow. The primary degradation product of endosulfan is endosulfan sulfate. Endosulfan bioaccumulate has the potential for long-range transport and is toxic to humans and a wide range of aquatic and terrestrial organisms. Toxic to marine algae, annelids, echnioderms, crustaceans, molluscs and fish. Half-life in soil: 2 months to 2.5 years.

Figure 8.4   Dumping of residual HCH isomers at a former lindane factory in the 1950s. *Source*: Vijgen (2006)

The addition of new POPs to the convention is a dynamic process, with an established protocol for expert review of changes (see ▶ Section 8.3.1). Very recently, in June 2022, the Conference of the Parties to the Stockholm Convention amended Annex A to list "*perfluorohexane sulfonic acid (PFHxS), its salts and PFHxS-related compounds in Annex A without specific exemptions (decision SC-10/13)*". For specific details see ▶ http://www.pops.int/TheConvention/ThePOPs/TheNewPOPs/tabid/2511/Default.aspx. The pesticide Methoxychlor, the flame retardant Declorane Plus, and a plastic stabiliser UV-238 are being reviewed for listing as POPs.

Box 8.2: Aroclors and Other Commercially Manufactured PCB Mixtures

PCBs were produced on a commercial scale as specific PCB congener mixtures formulated to obtain chemical properties desired for specific industrial applications. Most of these mixtures were produced for use as dielectric fluid in electrical transformers and capacitors, with other uses including hydraulic fluids, printing inks, and carbonless copy paper.

PCBs were produced commercially in the USA from 1929 through to 1977 by the Monsanto Chemical Company and marketed as mixtures called Aroclors. Monsanto reportedly produced from 500,000 to 600,000 tonnes of PCBs (about half the worldwide total) during its almost 50 years of production. PCB mixtures manufactured outside the USA by others had trade names such as Clophen (Germany), Prodolec (France), and Phenoclor (Japan). Although such PCB mixtures are no longer used, large quantities of Aroclors are present in old equipment still in use and as a legacy of poor disposal practices (e.g. into landfills) and potentially available for release to the environment.

The Aroclors manufactured in the USA each have a four-digit identification number. The last two digits (e.g. 60 in Aroclor 1260) indicate the average percentage of chlorine by weight in the mixture. Most Aroclor names have 12 as the first two digits, but this is not related to chemical structure. Aroclors 1016, 1242, 1254, and 1260 together comprised more than 90% of the PCBs that were produced in the USA. Note that **Aroclor** (a registered trade name of the Monsanto Company) is the correct spelling. However, it often appears **misspelt as Arochlor**.

8.4   Naming Conventions for Individual PCCD, PCDF, and PCB Compounds

The naming of the individual structural congeners is based on the number and positions of the chlorine atoms (▶ Box 8.3). The general formulae for PCCD (dioxins), PCDF (furans) and PCBs (polychlorinated biphenyls), along with two example structures are illustrated at ▶ Figure 8.5. For PCCD and PCDF, the numbers 1–9 indicate the possible positions of the chlorine atoms. For PCB, the numbers 2–6 (2′–6′) indicate the possible positions of the chlorine atoms at ortho(o), meta(m), and para(p) positions, respectively.

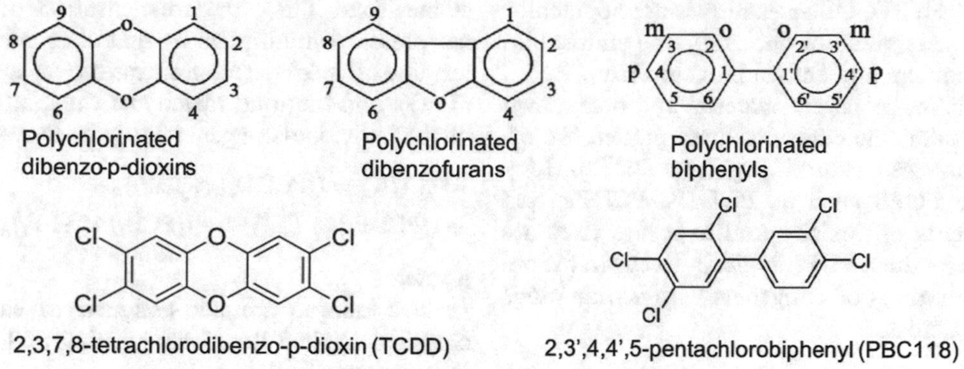

Figure 8.5   Chemical structures and naming conventions for PCDDs, PCDFs and PCBs

---

Box 8.3: The Meaning and Use of the Terms 'Congener', 'Congener Number', 'Homologue', and 'Homologous Series'

In chemistry, **congeners** are chemical substances **related to each other by origin, structure, or function**, and a **homologous series** is a set of compounds with the same functional group and similar chemical properties in which the members differ by the number of repeating units they contain. An example is the homologous series of straight-chained alkanes: i.e. methane (CH_4), ethane (C_2H_6), propane (C_3H_8), butane (C_4H_{10}), pentane (C_5H_{12}), etc. In that series, each successive member differs in structure by the addition of an extra methylene bridge (the $-CH_2-$ unit) inserted in the carbon chain.

A **homologue** (alternative spelling: homolog) is a compound belonging to a homologous series.

In the context of PCBs, polychlorinated dibenzo-*p*-dioxins (dioxins), and furans, the term **congener** is used to refer to an individual PCB, or dioxin, or furan compound within the series. Each of the different possible PBC structures is a **congener**, and PCBs occur in 209 congeners, with each structure unique in terms of the location of chlorine atoms within the PCB molecule. For example, the single chlorine atom in chlorobiphenyl can be in one of three locations, and the two chlorine atoms in dichlorobiphenyl can be located in any two of 12 locations in the PCB molecule.

The 209 possible PCB congeners can be named by their IUPAC or **BZ congener number** (for example, 2,3',4,4',5-penta polychlorinated biphenyl is **PCB 118** on the list of 1 through 209 possible congeners). The **BZ** is an acronym for the names of the authors (Ballschmiter and Zell) who proposed this identification system in 1980.

See ► https://www.epa.gov/sites/production/files/2015-09/documents/congenertable.pdf for the list and structure of all 209 PCB congeners.

Similarly, there are also 209 possible congeners of both PBBs (Polybrominated biphenyls) and PBDEs (polybrominated diphenyl ethers) (since like PCB, they are based on a double benzene structure), whereas CDDs (chlorinated dibenzodioxins) and CDFs (chlorinated dibenzofurans) have 75 and 135, possible congeners, respectively. BZ congener numbers are also assigned to each series.

In the context of polychlorinated dibenzo-*p*-dioxins (dioxins), the term **homologous series** means a group of dioxins with the same degree of chlorination (that is, with the same number of chlorine atoms in their structure, regardless of the positions of the chlorines in the dioxin structure). For example, the set of dioxins with five chlorines (the penta-CDDs) together comprise a homologous series, and the set of dioxins with six chlorines (the hexa-CDDs) is together another homologous series. Similarly, with the PCBs, PBBs and PBDEs and with the furans—the group of PCBs (or group of PBBs, PBDEs and furans) with the same number of chlorines (or bromines for PBBs and PBDEs) is a **homologous series**.

8.5 Assessment of Toxicity and Quantifying Exposure Risks for POPS

The toxicity of POPs can be expressed in standard terms such as LC50, EC50, etc. (see ► Chapter 3). However, POPs are commonly present in an environmental sample as a mixture of related compounds (see ► Chapter 14), for which the individual concentrations are measured and reported. Such **mixtures** may comprise both the parent compound and its breakdown products and metabolites—each with their own specific toxicities (e.g. DDT is commonly found with its breakdown products DDE and DDD). Some POPs are found and measured in the environment in one or more isometric forms (for example, hexachlorocyclohexane—HCH). Other examples are heptachlor (mostly found as heptachlor epoxide), and endosulfan (both parent isomers and endosulfan sulphate).

Where toxicities of parent material and breakdown product are similar, the concentrations present are often listed and assessed as totals (e.g. total DDTs). However, for some POPS such as PCDDs, PCDFs, and PCBs assessments of toxicity and exposure risks are more complex, as these POPS typically occur in the environment as mixtures of congeners with a wide range of potential toxicities.

8.5.1 Assessment of Toxicity and Exposure Risks for Dioxins, Furans, and Dioxin-Like PCBs

Dioxins, and furans have similar chemical structures, properties, and toxicities. By convention, when discussing toxicity and exposure risks, the term **dioxins** or **total dioxins** is generally taken to include PCDDs, PCDFs and **dioxin-like PCBs**. The **dioxin-like PCBs** are those PCB compounds exhibiting similar toxicity to dioxins, with toxic responses expressed including dermal toxicity, immunotoxicity, reproductive deficits, teratogenicity, endocrine toxicity and carcinogenicity/tumour promotion similar to those observed for 2,3,7,8-tetrachlorodibenzo-p-dioxin (TCDD).

8.5.2 The Meaning and Use of the Terms TEF and TEQ

In the environment, PCDDs, PCDFs, and **dioxin-like PCBs** are usually present in complex mixtures. For example, with PCBs, the proportion of congeners present in environmental samples is often related to the composition of the commercial formulation that is the source of contamination (e.g. Aroclor).

The World Health Organization (WHO) identifies 29 closely-related PCDD, PCDF, and **dioxin-like PCB** congeners as having a common mechanism of toxicity to PCBs. However, since their toxicities differ, WHO has adopted a **toxic equivalent factor** (TEF) for each of the 29 congeners to calculate a human risk assessment for individual components of the chemical mixture. TEFs are weighting factors assigned to specific congeners to reflect their toxicity relative to that of the most toxic dioxin, TCDD with TEF = 1.

The WHO-adopted TEFs are listed in ◻ Table 8.3. The original set of TEFs was adopted in 1998, but new TEF values were adopted following a review in 2005. The rationale and methods used in the TEF review process are reported in Van den Berg et al. (2006).

The toxicity of a mixture of PCDDs, PCDFs, and dioxin-like PCBs is quantified as its **Total TEQ**. To calculate Total TEQ, the concentration of each congener present is multiplied by that congener's TEF to determine a **weighted** concentration or **toxic equivalent** (TEQ), and the total toxicity of the mixture is the sum of the individual congener TEQs as shown in Eq. 8.1.

$$
\text{Total TEQ} = ([\text{PCDD}_i] \times \text{TEF}_i)_n \\
+ ([\text{PCDF}_i] \times \text{TEF}_i)n + ([\text{PCB}_i] \times \text{TEF}_i)_n \tag{8.1}
$$

where:

i = a unique number assigned to each individual chemical applied iteratively starting at 1 and proceeding sequentially.

n = the total number of unique pollutants.

Such results are expressed in concentration units such as picogram TEQ per gram.

It is important when comparing contamination assessments between sites, and/or for assessing historical changes in POP concentrations and distribution, that TEQ assessment dates are checked to ensure that all TEQs and conclusions are based on the current (2005) WHO-TEQs, and to recalculate any Total TEQs that are based on out-of-date TEQs (e.g. assessments dated earlier than 2005).

8.5.3 Use of Homologues and Congener Profiles in Forensic Investigations

Knowledge of the contaminant constituents in a sample containing a mixture of dioxins, furans, PCBs and/or PBBs, and PBDEs can be useful in developing a characteristic profile of contaminants present. For example, patterns of similarity in the presence and relative concentrations of members of a homologue set can be used to describe similarities and differences in contamination patterns between sites or across environmental media (e.g. EPHC 2005).

Frequently, similarities in **homologue and congener profiles** can be used as a **fingerprint** to make associations between PCDD/PCDF/PCB and/or PBB and PBDE contamination at a site and historical activities associated with the site. Similarly, PCDD/PCDF/PCB and/or PBB and PBDE profiles in samples from exposed persons/fish/livestock/food can be compared with profiles in samples taken from potential sources of the contamination. However, congener composition in a mixture released to the environment can change over time, making **fingerprinting** a complex task (Saba and Boehm, 2011).

Air, sediment, biota, and water samples are the most likely to have had their congener composition changed by environmental conditions. For example, composition changes can occur because PCDD/PCDF/PCB congeners with fewer chlorine atoms tend to partition into air and water more readily than those with more chlorine atoms. The relative partitioning potentials can be assessed from the partition coefficients

◼ **Table 8.3** Summary of WHO 1998 and WHO 2005 TEF values. Bold values show changes from 1998 to 2005. Adapted from Van den Berg, et al. 2006; see also ▶ https://www.who.int/ipcs/assessment/tef_values.pdf

Compound	WHO 1998 TEF	WHO 2005 TEF
Chlorinated dibenzo-*p*-dioxins		
2,3,7,8-TCDD	1	1
1,2,3,7,8-PeCDD	1	1
1,2,3,4,7,8-HxCDD	0.1	0.1
1,2,3,6,7,8-HxCDD	0.1	0.1
1,2,3,7,8,9-HxCDD	0.1	0.1
1,2,3,4,6,7,8-HpCDD	0.01	0.01
OCDD	0.0001	**0.0003**
Chlorinated dibenzofurans		
2,3,7,8-TCDF	0.1	0.1
1,2,3,7,8-PeCDF	0.05	**0.03**
2,3,4,7,8-PeCDF	0.5	**0.3**
1,2,3,4,7,8-HxCDF	0.1	0.1
1,2,3,6,7,8-HxCDF	0.1	0.1
1,2,3,7,8,9-HxCDF	0.1	0.1
2,3,4,6,7,8-HxCDF	0.1	0.1
1,2,3,4,6,7,8-HpCDF	0.01	0.01
1,2,3,4,7,8,9-HpCDF	0.01	0.01
OCDF	0.0001	**0.0003**
Non-*ortho*-substituted PCBs		
3,3′,4,4′-tetraCB (PCB 77)	0.0001	0.0001
3,4,4′,5-tetraCB (PCB 81)	0.0001	**0.0003**
3,3′,4,4′,5-pentaCB (PCB 126)	0.1	0.1
3,3′,4,4′,5,5′-hexaCB (PCB 169)	0.01	**0.03**
Mono-*ortho*-substituted PCBs		
2,3,3′,4,4′-pentaCB (PCB 105)	0.0001	**0.00003**
2,3,4,4′,5-pentaCB (PCB 114)	0.0005	**0.00003**
2,3′,4,4′,5-pentaCB (PCB 118)	0.0001	**0.00003**
2′,3,4,4′,5-pentaCB (PCB 123)	0.0001	**0.00003**
2,3,3′,4,4′,5-hexaCB (PCB 156)	0.0005	**0.00003**
2,3,3′,4,4′,5′-hexaCB (PCB 157)	0.0005	**0.00003**
2,3′,4,4′,5,5′-hexaCB (PCB 167)	0.00001	**0.00003**
2,3,3′,4,4′,5,5′-heptaCB (PCB 189)	0.0001	**0.00003**

of the individual compounds. For this reason, air and water samples are likely to be **enriched** with congeners with fewer chlorine atoms. In addition, the concentrations of individual congeners present in biota samples can also be altered through bio-degradation, with some congeners being selectively reduced and others remaining constant. Similar biodegradation involving debromination can occur with brominated contaminants, in fact more readily since the carbon-bromine bond is weaker than the carbon-chlorine bond.

8.6 Case Studies

8.6.1 Case Study 1—Dioxins, Furans, and Dioxin-Like PCBs in the Australian Aquatic Environment.

A comprehensive overview of dioxins, furans, and dioxin-like PCBs in Australia is provided in the series of studies comprising the National Dioxin Program

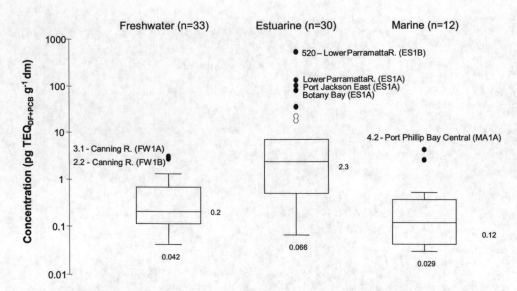

Figure 8.6   Concentrations of dioxin-like chemicals in aquatic sediments in Australia. *Source*: Müller et al. (2004)

(NDP) (EPHC 2005). The distributions of these POPs in the Australian aquatic environment are covered in Technical Report No. 6 (Müller et al. 2004). This study of Australian aquatic sediments showed that dioxins, furans, and dioxin-like PCBs were found in all samples, with middle-bound concentrations ranging from 0.002 to 520 pg TEQ g/dm (calculated using 1998 WHO-TEFs).

Aquatic sediments from urban/industrial sampling locations had significantly greater concentrations of dioxin-like chemicals than samples from remote and agricultural locations. The greatest concentrations occurred in sediments from the Parramatta River estuary (100 and 520 pg TEQ g/dm, ▣ Figure 8.6).

This pattern of POP distribution was similar in the case of bivalves and fish (▣ Figures 8.7 and 8.8). However, concentrations in the aquatic environment (sediments, bivalves and fish) were in most cases less than published levels for other industrialised countries.

Homologue and congener profiles for PCDD/PCDF were strongly dominated by OCDD, with the 1,2,3,4,6,7,8-heptachloro dibenzodioxin usually the congener with the second-highest concentration (▣ Figure 8.9).

8.6.2 Case Study 2—Spatial and Temporal Trends in Concentrations of Brominated Fire-Retardant POPs in Arctic Marine Mammal Tissues

Why are Arctic Marine Mammals at a High Risk of Bioaccumulating POPs?
Long-term **global transport pathways** for POPs result in the seas and **oceans being the ultimate sink for most**

of the POPs released to the environment. In particular, the major processes for global transport of POPs tend to accumulate them in the colder waters of the polar regions—for example, the Arctic is well recognised as a sink for many POPs due to their repeated deposition and remobilisation (Breivik et al. 2007; Burkow and Kallenborn, 2000) (see also Chapter 7, ▶ Section 7.6.2). As a consequence, marine animal life is exposed to the hazard of bioaccumulating POPs, and particularly marine mammals (seals, walruses, sea otters, polar bears, and whales) since they are relatively long-lived, air-breathing animals and are relatively high in the marine food chain—especially polar bears.

The major route of uptake of lipophilic POPs by air-breathing aquatic animals is through diet. This is in contrast to the importance of lipophilic POP uptake from the water column through gill membranes in aquatic animals such as fish and aquatic invertebrates.

Potential routes of excretion of accumulated lipophilic POPs in air-breathing aquatic animals are also different from the routes of excretion for lipophilic POPs available to marine fish and crustaceans. In marine fish and crustaceans, the membranes of gills or equivalent organs, provide an important route for excretion of lipophilic POP contaminants to the surrounding water. The availability of this important excretion route in marine fish and crustaceans tends to limit the accumulation of persistent lipophilic compounds to an equilibrium concentration in the tissue lipids that is mediated by the partition coefficients for the POPs between blood lipid and water. However, air-breathing aquatic animals lack this excretion pathway for lipophilic contaminants to the surrounding water. Instead of excreting lipophilic compounds across

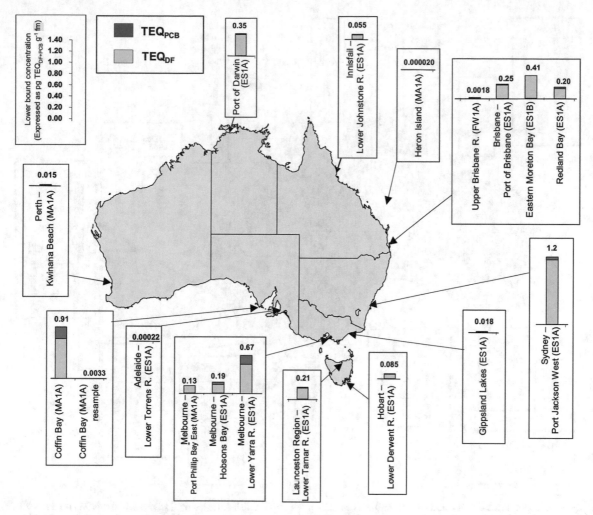

□ Figure 8.7   Geographical distribution of dioxin-like chemicals in Australian bivalve samples as TEQ values (from WHO 1998 TEFs) on a fresh mass basis. *Source*: Müller et al. (2004)

gill membranes to the surrounding water, air-breathing mammals are able to excrete such contaminants to air subsequently exhaled from the lungs. However, partitioning of lipophilic POPs from the blood plasma lipids across lung membranes to air subsequently exhaled from the lungs is much less efficient due to the orders of magnitude differences between typical POPs partition coefficients between lipid and water (Kow) versus lipid and air (Koa). The octanol-water coefficient (Kow) is a recognised surrogate for the partition coefficient between biotic lipid and water (Gobas and Mackay 1987) (see also ▶ Section 7.5.3, Chapter 7). Since Kow values for POPs typically range from 10^4 to 10^7, but Koa values range from 10^6 to 10^{12} (Wania and Mackay 1999), it is likely that rates for POPs partitioning from blood lipid into air via lungs in air-breathing are 1 to 3 magnitudes lower than for partitioning from blood lipid to water via gills. This leads to the expectation that because marine mammals are air-breathers, dietary acquired lipophilic POPs will inexorably accumulate in their tissue lipids.

Confounding Factors Involved When Assessing Spatial and Temporal Trends in POP Concentrations in Arctic Marine Mammals.

Although many studies have measured POPs concentrations in Arctic marine mammals, valid spatial or temporal comparisons are challenged by several confounding factors—for example that historical measurement data may not specify proximity of sampled animals to human populations and also biological factors such as species, age, health, and sex. In addition, sampling and analytical methodology for many organic contaminants was not standardised, making the results not easily comparable between studies (Green and Larson 2016).

For all mammals, transfer of lipophilic contaminants from mother to offspring is likely during lactation, resulting in increased body burden of contaminant in offspring and reduced burden in lactating females, thus sampling may show a pattern of greater contaminant concentrations in males than in females. However, this may be confounded in sampling data by

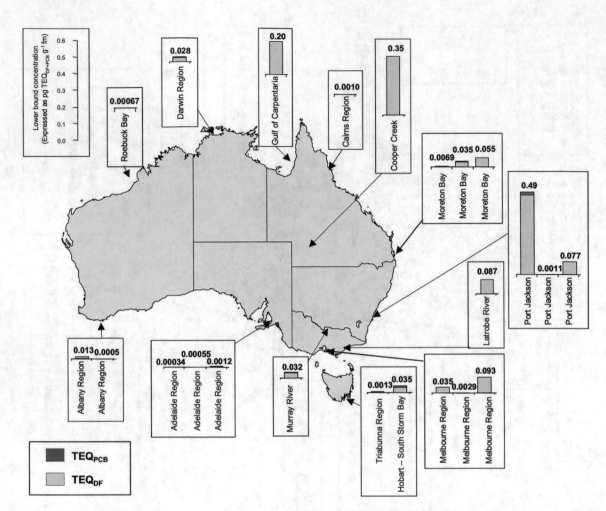

■ **Figure 8.8** Geographical distribution of dioxin-like chemicals in Australian fish samples as TEQ values (from WHO 1998 TEFs) on a fresh mass basis. *Source*: Müller et al. (2004)

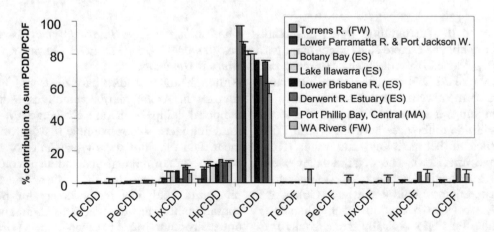

■ **Figure 8.9** PCCD/PCDF homologue profile for sediments sampled from various locations in Australia. *Source*: Müller et al. (2004)

other factors such as differences in diet and body condition at the time of sampling, for example, in polar bears (Lippold et al. 2019; Tartu et al. 2017). In respect to the influence of diet, it is suggested (Letcher et al. 2009) that changes in contaminant body burden of POPs can occur in polar bears in western Hudson Bay as a con-

sequence of the dietary shifts by these animals over the past several decades (Bentzen et al. 2007; Thiemann et al. 2008) from ice-associated to open-water-associated prey species when these animals responded to changes in sea ice cover. Regarding body condition, concentrations of lipophilic compounds in plasma and fat were 4–9 times higher in the thinnest compared to the fattest polar bears, whereas they were only 1.5–1.8 times higher in the individuals feeding at highest versus lowest trophic levels (Tartu et al. 2017). This means that in general, tissue concentrations of lipophilic POPs are highest when polar bears are at their thinnest from approximately November to April–May, depending on local sea ice conditions and food availability.

An additional factor affecting concentrations of brominated fire-retardant POPs in polar bears is their demonstrated capability to biotransform these compounds (McKinney et al. 2011; Krieger et al. 2015; Vetter et al. 2015).

Temporal Trends for Brominated Fire-retardant POPs in Polar Bears.
Temporal trends for body burden of brominated fire retardants have been studied in polar bear subpopulations in the Barents Sea, East Greenland and Hudson Bay.

In the Barents Sea, tetraBDE concentrations decreased by 3% per year in female polar bears from 1997 to 2017 (Lippold et al. 2019). However the same study showed that hexaBDE concentrations were stable.

A study of subadult polar bears in East Greenland (Dietz et al. 2013) showed that between 1983 and 2010 both tetraBDE and 2 congeners of pentaBDE body burdens increased by 6–8% per year until 2000–2004, after which tetraBDE declined by 31% per year, along with a non-significant decline in pentaBDE concentrations. In addition, this study showed that both hexaBDE and HBCDD concentrations increased by 3% per year from 1983 to 2010, although hexabromobiphenyl (HBB) body burdens did not change.

In Western Hudson Bay polar bears, there are reported (McKinney et al. 2011) increases in ΣPBDE (mostly tetra-, penta- and hexaBDE) of 13% per year from 1991 to 2007, but with HBCDD only detected since 2000, and no change in hexabromobiphenyl (HBB) concentrations, which is similar to the results in the East Greenland study (Dietz et al. 2013).

It is suggested (Routti et al. 2019) that the above-noted declines in tetraBDE concentrations in Barents Sea and East Greenland polar bears reflect the phase out of the production of the commercial pentaBDE mixture (a listed POP comprising both tetra- and pentaBDE as the main components) in Europe in the late 1990s, and in the USA in the mid-2000s. It is also suggested (Routti et al. 2019) that the stability of hexaBDE concentrations in all polar bear subpopulations may result from the biodegradation (debromination) of recent emissions of decaBDE, noting that although production (but not import) was phased out firstly in Europe (1999) and later in USA, in China—the largest producer of decaBDE mixtures (c-decaBDE) output remains constant.

The increase in PBDE concentrations in polar bear tissues from the 1970s to mid-1990s, followed by re-

ductions in some PBDEs since the late 1990s is similar to trends in marine mammals all over the world (Guo et al. 2012).

Recent (2012–2016) Spatial Trends for Brominated Fire-retardant POPs in Polar Bears.
A review (Routti et al. 2019) of available data on contaminant body burdens in polar bears (covering the period 2012–2016) shows that for brominated fire retardants, relative ΣPBDE concentrations for southern Hudson Bay > Barents Sea > East Greenland > western Hudson Bay > Chutki Sea and Kara Sea subpopulations (▪ Figure 8.10), and notes that this is consistent with a previous study that sampled bears in the period 1996–2002 (Muir et al. 2016). It is suggested (Routti et al. 2019) that this pattern of PBDE contamination reflects proximity to the sites of production, noting that in earlier years PBDE mixtures were produced in greater quantities in the USA than in Europe or Asia.

In respect of HBCDD, a recent review (Routti et al. 2019) shows that East Greenland and Barents Sea polar bear subpopulations carried greater concentrations than Hudson Bay bears (▪ Figure 8.10), and notes that is consistent with the higher use of HBCDD in Europe relative to the rest of the world.

8.7 Summary

POPS are organic compounds that are problematic for humans and the environment for many reasons. These compounds (and many of their degradation products) are highly toxic, accumulate in the fatty tissue of living organisms including humans, are found at higher concentrations at higher levels in the food chain, remain intact for exceptionally long periods of time, and become widely distributed throughout the environment as a result of natural processes involving soil, water and, most notably, air. Due to the high risks posed by POPs, the Stockholm Convention on Persistent Organic Pollutants was established in 2001 and by 2018 had been ratified by 182 nations. The original **dirty dozen** POPS were identified as posing significant risks to the humans and the environment. Since 2001 an additional 16 POPs have been added to the list of priority compounds in the Stockholm Convention and more are under consideration for listing.

POPs are often present in marine environments in complex mixtures of compounds and degradation products, and because many POPs (e.g. dioxins and furans) have similar chemical structures, properties, and toxicities, Toxicity Equivalent Factors (TEFs) and Toxic Equivalents (TEQs) are used to predict the toxicity of environmental samples.

The oceans are the ultimate sink for POPs after they are released into the environment. Because of the long-

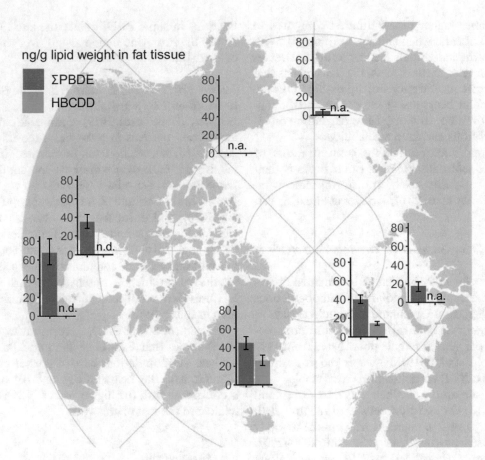

■ **Figure 8.10** Concentrations of brominated fire retardants measured in polar bear fat tissues between 2012 and 2016. The abbreviations n.a. and n.d. signify not analysed or not detected, respectively. *Source*: Routti et al. (2019)

range global transport of POPs, their persistence and tendency to bioaccumulate and biomagnify, marine organisms high in the food chain (e.g. large mammals) are exposed to high POP concentrations. POPs can be passed from mother to offspring, which often results in higher concentrations in males than in females.

8.8 Study Questions and Activities

1. Briefly explain the three different annexes placed on POPs in the Stockholm convention.
2. Explain why different types of POPs tend to be present in the tissues of large marine mammals, and why the concentrations may differ depending on an organism's age and sex.
3. Select three POPs that are still intentionally produced and do some research to determine their current levels of production, uses, distribution in the environment, and the potential threats they pose to marine ecosystems.
4. Define TEQs and TEFs and explain why they are used for predicting the toxicity of POPs.

References

AAS (Australian Academy of Science) (1972) The use of DDT in Australia. Report No. 14. Australian Academy of Science, Canberra ACT

ACS (American Chemical Society) (2012) The legacy of Rachel Carson's silent spring. American Chemical Society National Historic Chemical Landmarks. Available at ▶ http://www.acs.org/content/acs/en/education/whatischemistry/landmarks/rachel-carson-silent-spring.html. Accessed 6 Aug 2018

Bentzen T, Follmann E, Amstrup S, York G, Wooller M, O'Hara T (2007) Variation in winter diet of southern Beaufort Sea polar bears inferred from stable isotope analysis. Can J Zool 85(5):596–608

Breivik K, Sweetman A, Pacyna J, Jones K (2007) Towards a global historical emission inventory for selected PCB congeners—A mass balance approach. Sci Total Environ 377(2–3):296–307

Burkow I, Kallenborn R (2000) Sources and transport of persistent pollutants to the Arctic. Toxicol Lett 112–113:87–92

Carson R (1962) Silent spring. Houghton Mifflin, Boston, p 319

Dietz R, Rigét F, Sonne C, Born E, Bechshøft T, McKinney M, Drimmie R, Muir D, Letcher R (2013) Three decades (1983–2010) of contaminant trends in East Greenland polar bears (*Ursus maritimus*). Part 2: brominated flame retardants. Environ Int 59:494–500

EPHC (Environmental Protection and Heritage Council) (2005) National dioxins program: national action plan for addressing dioxins in Australia. Available at ▶ https://www.awe.gov.au/environment/protection/chemicals-management/dioxins. Accessed 6 Feb 2022

Gobas F, Mackay D (1987) Dynamics of hydrophobic organic chemical bioconcentration in fish. Environ Toxicol Chem 6:495–504

Green A, Larson S (2016) A review of organochlorine contaminants in nearshore marine mammal predators. J Environ Anal Toxicol 06(03):1000370

Guo Y, Shaw S, Kannan K (2012) Spatial and temporal trends of polybrominated diphenyl ethers. In: Loganathan B, Lam P (eds) Global contamination trends of persistent organic chemicals. CRC Press, Boca Raton, pp 33–72

Krieger L, Szeitz A, Bandiera S (2015) Evaluation of hepatic biotransformation of polybrominated diphenyl ethers in the polar bear (*Ursus maritimus*). Chemosphere 146:555–564

Letcher R, Gebbink W, Sonne C, Born E, McKinney M, Dietz R (2009) Bioaccumulation and biotransformation of brominated and chlorinated contaminants and their metabolites in ringed seals (*Pusa hispida*) and polar bears (*Ursus maritimus*) from East Greenland. Environ Int 35(8):1118–1124

Lippold A, Bourgeon S, Aars J, Andersen M, Polder A, Lyche J, Bytingsvik J, Jenssen B, Derocher A, Welker J, Routti H (2019) Temporal trends of persistent organic pollutants in Barents Sea polar bears (*Ursus maritimus*) in relation to changes in feeding habits and body condition. Environ Sci Technol 53(2):984–995

McKinney M, Dietz R, Sonne C, De Guise S, Skirnisson K, Karlsson K, Steingrímsson E, Letcher R (2011) Comparative hepatic microsomal biotransformation of selected PBDEs, including decabromodiphenyl ether, and decabromodiphenyl ethane flame retardants in Arctic marine-feeding mammals. Environ Toxicol Chem 30(7):1506–1514

Muir DCG, Backus S, Derocher AE, Dietz R, Evans TJ, Gabrielsen GW, Nagy J, Norstrom RJ, Sonne C, Stirling I, Taylor MK, Letcher R (2016) Brominated flame retardants in polar bears (*Ursus maritimus*) from Alaska, the Canadian Arctic, East Greenland, and Svalbard. Environ Sci Technol 40(2):449–455

Müller J, Muller R, Goudkamp K, Mortimer M, Haynes D, Paxman C, Hyne R, McTaggart A, Burniston D, Symons R, Moore M (2004) Dioxins in aquatic environments in Australia. National Dioxins Program Technical Report No. 6, Australian Government Department of the Environment and Heritage, Canberra. Available at ▶ http://www.environment.gov.au/protection/publications/dioxins-technical-report-06. Accessed 6 Feb 2022

Parliament of Australia (1972) Wildlife conservation. Report from the House of Representatives Select Committee, October 1972. The Parliament of the Commonwealth of Australia—1972 Parliamentary Paper No. 284. Available at ▶ https://www.aph.gov.au/Parliamentary_Business/Committees/House_of_Representatives_committees?url=/report_register/bycomlist.asp?id=108. Accessed 6 Aug 2018

Routti H, Atwood T, Bechshoft T, Andrei Boltunov A, Ciesielski T, Desforges J, Dietz R, Gabrielsen G, Jenssen B, Letcher R, McKinney MD, Morris A, Rigét F, Sonne C, Styrishave B, Tartu S (2019) State of knowledge on current exposure, fate and potential health effects of contaminants in polar bears from the circumpolar Arctic. Sci Total Environ 664:1063–1083

Saba T, Boehm P (2011) Quantitative polychlorinated biphenyl (PCB) congener and homologue profile comparisons. Environ Forensics 12(2):134–142

Tartu S, Bourgeon S, Aars J, Andersen M, Polder A, Thiemann G, Welker J, Routti H (2017) Sea ice-associated decline in body condition leads to increased concentrations of lipophilic pollutants in polar bears (*Ursus maritimus*) from Svalbard, Norway. Sci Total Environ 576:409–419

Thiemann G, Iverson S, Stirling I (2008) Polar bear diets and Arctic marine food webs: insights from fatty acid analysis. Ecol Monogr 78(4):591–613

UNEP (United Nations Environment Programme) (2018) United nations treaty collection. Environment Chapter XXVII. Stockholm convention on persistent organic pollutants. Available at ▶ https://treaties.un.org/Pages/ViewDetails.aspx?src=TREATY&mtdsg_no=XXVII-15&chapter=27&clang=_en. Accessed 6 Feb 2022

UNEP (United Nations Environment Programme) (2015) Persistent Organic Pollutants Review Committee, UNEP/POPS/POPRC.11/10/Add.1.Decabromodiphenyl Ether (Commercial Mixture, c-decaBDE): risk management evaluation. Stockholm Convention, Rome, pp 1–47. Available at ▶ http://chm.pops.int/Portals/0/download.aspx?d=UNEP-POPS-POPRC.11-10-Add.1.English.doc. Accessed 6 Feb 2022

Van den Berg M, Birnbaum L, Denison M, De Vito M, Farland W, Feeley M, Fiedler H, Hakansson H, Hanberg A, Haws L, Rose M, Safe S, Schrenk D, Tohyama C, Tritscher A, Tuomisto J, Tysklind M, Walker N, Peterson R (2006) The 2005 World Health Organization reevaluation of human and mammalian toxic equivalency factors for dioxins and dioxin-like compounds. Toxicol Sci 93(2):223–241

Vetter W, Gall V, Skirnisson K (2015) Polyhalogenated compounds (PCBs, chlordanes, HCB and BFRs) in four polar bears (*Ursus maritimus*) that swam malnourished from East Greenland to Iceland. Sci Total Environ 533:290–296

Vijgen J (2006) The legacy of Lindane HCH isomer production. Main report. A global overview of residue management, formulation and disposal. International HCH and Pesticides Association. Available at ▶ http://www.ihpa.info/docs/library/reports/Lindane%20Main%20Report%20DEF20JAN06.pdf. Accessed 6 Feb 2022

Wania F, Mackay D (1999) The evolution of mass balance models of persistent organic pollutant fate in the environment. Environ Pollut 100:223–240

Plastics

Kathryn L. E. Berry, Nora Hall, Kay Critchell, Kayi Chan, Beaudin Bennett, Munro Mortimer and Phoebe J. Lewis

Contents

9.1   Introduction – 208

9.2   Plastic Types and Characteristics – 208
9.2.1   Macroplastics – 210
9.2.2   Microplastics – 210

9.3   Sources – 210

9.4   Plastic Transport in the Marine Environment – 212
9.4.1   Modelling the Movements of Plastic – 213
9.4.2   Accumulation – 213
9.4.3   Plastics in Remote Environments – 214

9.5   Degrading Processes – 215
9.5.1   Complications of Measuring and Comparing Plastic Pollution – 216

9.6   Impacts of Plastic Debris – 216
9.6.1   Impacts Overview – 216
9.6.2   Physical Interactions with Wildlife – 216
9.6.3   Plastic as an Unnatural Substrate – 218
9.6.4   Chemical Effects of Microplastics – 219
9.6.5   Human Health Impacts – 220
9.6.6   Economic Impacts – 220

9.7   Actions to Drive Change – 220

9.8   Summary – 222

9.9   Questions and Activities – 222

        References – 222

© The Author(s) 2023
A. Reichelt-Brushett (ed.), *Marine Pollution—Monitoring, Management and Mitigation*,
Springer Textbooks in Earth Sciences, Geography and Environment,
https://doi.org/10.1007/978-3-031-10127-4_9

Acronyms and Abbreviations

BPA	Bisphenol A
EDCs	endocrine-disrupting chemicals
EU	European Union
FTIR	Fourier-transform infrared spectroscopy
GPGP	Great Pacific Garbage Patch
MARPOL	International Convention for the Prevention of Pollution from Ships
MOOC	Massive Open Online Course
POP	Persistent Organic Pollutant
PVC	polyvinyl chloride
SPI	Society of the Plastics Industry
UV	ultraviolet
USA	United States of America

9.1 Introduction

Plastic production has grown exponentially, from 1.5 million tonnes in the 1950s (Plastics Europe 2012) to 359 million tonnes in 2018 (Plastics Europe 2019). Valued for being versatile, durable, lightweight and inexpensive to produce, plastic is used in all aspects of our daily life. Plastic has shaped the development of modern society and has benefited many sectors, including healthcare, science and technology, agriculture, packaging, transportation, and construction (Napper and Thompson 2020; Plastics Europe 2017). The largest market demand of plastic is for single-use disposable packaging materials, with approximately 50% of all plastic production going towards single-use purposes (Hopewell et al. 2009; Xanthos and Walker 2017).

Plastic is extremely durable and non-biodegradable. Although plastic can break into pieces that are invisible to the naked eye, plastic longevity is estimated to range from hundreds to thousands of years (Barnes et al. 2009), making plastic waste management a global challenge. Plastic waste management is considered inadequate or non-existent in many parts of the world, despite high levels of plastic production and consumption (Bucci et al. 2020). Although most developed countries have invested in recycling technologies, there are many factors that impact recycling success, including the lack of technology to recycle all plastic types, lack of collection points, recycling feedstock contamination (which occurs when plastic food containers are not properly cleaned) and consumer apathy (Law 2017). Many developing countries lack the waste management practices, services, systems or infrastructure for garbage, let alone recycling. From 1950 to 2015, the cumulative waste generation of primary and recycled plastic amounted to 6300 million tonnes (6300 Mt), with only 9% recycled and 12% incinerated, while at least 60% persists in landfills or in the natural environment (Geyer et al. 2017).

In 2010, an estimated 4.8–12.7 million tonnes of plastic entered the world's oceans from land (Jambeck et al. 2015; Vince and Stoett 2018). Not surprisingly, plastics make up about 80% of all marine debris (defined as any persistent manufactured or processed solid material discarded, disposed of or abandoned in the marine environment and coastal environment) (UNEP 2016). Abundance estimates have predicted that tens of millions of metric tonnes of plastic debris is floating on global ocean surfaces (Lebreton et al. 2018), with microplastic estimates ranging between 15 and 51 trillion items (van Sebille et al. 2015), making plastic pollution internationally recognised (Rochman et al. 2013).

The longevity of plastic, large inputs into the ocean, and natural movement of the material via winds and currents have made plastic a persistent and ubiquitous pollutant throughout global coastal and marine environments, including in remote areas, such as the Arctic Ocean (Eriksen et al. 2020). Decades worth of evidence shows plastic pollution harms marine wildlife and habitats (Laist 1987; Gregory 2009; Baulch and Perry 2014; Beaumont et al. 2019) and human health (Thompson et al. 2009; Waring et al. 2018), with an associated economic loss and decline in ecosystem services (benefits people obtain from nature). This chapter explores the global issue of marine plastic pollution. In it we discuss topics such as plastic types and characteristics, sources of marine plastic pollution, transport and accumulation, impacts, challenges in governance, and initiatives aimed at reducing the use of plastics.

9.2 Plastic Types and Characteristics

The term **plastic** covers a wide range of synthetic or semi-synthetic materials that we use to help make life cleaner, easier, and safer (Andrady and Neal 2009; Plastics Europe 2020). They are produced from synthetic polymers, which are long, chain-like molecules of repeating chemical units (Napper and Thompson 2020). These units consist of hydrocarbons, usually sourced from fossil fuels such as coal, natural gas, and crude oil, but also from materials such as cellulose or salt (ACC 2020; Höfer and Selig 2012). Different plastic polymers are used for various product types (▶ Box 9.1), including polyethylene (clear food wrap, plastic bags, detergent bottles), polystyrene (Styrofoam packaging), polypropylene (packaging, industrial parts, textiles), and polyvinyl chloride (PVC, used for pipes and in the medical industry). To create durable plastic products, plastic polymers are combined with chemical additives such as fillers, plasticisers, flame retardants, and stabilisers (ultraviolet (UV) and thermal) (Andrady and Neal 2009). The coding of different types of plastics was developed by the Society of the Plastics Industry (SPI) and is used as the global standard (Wong 2010). Plastic polymers can be identified using laboratory techniques such as Fourier-transform infrared spectroscopy (FTIR) and Raman spectroscopy.

Box 9.1: Plastic Polymers: Recycling Numbers and Examples of Common Uses

Polymer name	Polyethylene terephthalate (PET or PETE)	High-density polyethylene (HDPE)	Polyvinyl chloride (PVC)	Low-density polyethylene (LDPE)	Polypropylene (PP)	Polystyrene (PS)	Other
Recycling Symbol	♺ 1	♺ 2	♺ 3	♺ 4	♺ 5	♺ 6	♺ 7
Common Uses	Soda bottles, water bottles, rope	Milk jugs, toys, snack food boxes	Plumbing pipes, credit cards, floor covering	Plastic wrap, bubble wrap, plastic grocery bags	Prescription bottles, most bottle tops, potato chip bags	Disposable foam cups, take-out food containers, plastic cutlery	Baby bottles, medical storage containers, eyeglasses

Adapted from Wong (2010)

In addition to polymer type, plastic debris is described using numerous characteristics including size, shape (e.g. beads, pellets, foams, fibres, fragments), colour, and original usage (e.g. fishing gear, food packaging) (Andrady 1994, 2017; Napper et al. 2015). The two most common size categories are macroplastic (>20 mm diameter) and microplastic (<5 mm), with further categorisations such as megaplastic (>1000 mm), mesoplastic (5–20 mm), and nanoplastic (<1000 nm) (▶ Box 9.2) (Barnes et al. 2009; Ivar do Sul and Costa 2014; Thompson et al. 2009). Plastic debris from all size categories are found throughout the marine environment, at beaches, on the water surface, in the water column, and on the seafloor (◘ Figure 9.1).

Box 9.2: Marine Plastic Debris: Examples of Debris in Different Size Categories

Nano (<1 μm)	Micro (<5 mm)	Meso (5–20 mm)	Macro (>20 mm)	Mega (>1000 mm)
• Fibres from clothing • Nano items in personal care products and pharmaceuticals	• Microbeads from personal care products • Fragments from larger existing plastic debris • Polystyrene balls from packaging	• Bottle caps • Cigarette filters and butts • Lighters • Candy wrappers	• Beverage bottles • Plastic bags • Cutlery • Beer-ties • Balloons • Fishing lines, floats, and buoys	• Abandoned fishing nets • Rope and rope conglomerates

Adapted from UNEP (2017)

◨ **Figure 9.1**   Plastic debris is ubiquitous in the marine environment, some examples include **a** plastic found washed up on beaches **b** floating on the water's surface **c** floating within the water column **d** and deposited on the seafloor. *Photos*: A. Malmgren (a), K. Berry (b–d)

9.2.1 Macroplastics

Macroplastic (>20 mm) debris commonly observed in the marine environment can include floating plastic bags and bottles and plastic beach debris (◨ Figure 9.2a). Significant levels of macroplastic debris can become a navigational hazard for both marine wildlife and vessels. Further significant impacts to the marine environment and organisms are numerous, for example, the smothering of coral reefs (Personal observation, K. Berry), seagrass beds (Kiessling et al. 2015) or mangroves (Martin et al. 2019), and the entanglement or ingestion of plastic debris by marine fauna (Gregory 2009; Wesch et al. 2016).

9.2.2 Microplastics

Microplastics (<5 mm) are sub-categorised into primary and secondary microplastics. Primary microplastics are intentionally manufactured to be small for various uses and include virgin plastic resin pellets, small items or spheres used in personal care products known as microbeads (e.g. for face washes, toothpaste, or cosmetics, ◨ Figure 9.2b, as well as abrasives in cleaning products (Cole et al. 2011; Derraik 2002). Secondary microplastics are created during the breakdown of larger plastic items. They commonly take the form of weathered and degraded plastic pieces (see ▶ Section 9.5) and microfi-

bres that are shed from synthetic and semi-synthetic fabrics during washing (◨ Figure 9.2c, d).

9.3 Sources

Plastic enters the marine environment from land and maritime sources, with a larger proportion (70–80%) entering from land (◨ Figure 9.3) (UNEP 2005, 2009). **Land-based sources** consist of mismanaged waste (e.g. uncovered garbage dumps or littered plastic, ◨ Figure 9.4), spillage of virgin plastic pellets, litter flowing into storm drains and rivers, treated and untreated sewage effluent, as well as aerial deposition (items or fibres that are emitted into the air from industrial facilities that are then deposited on the ocean) (Critchell et al. 2019). **Maritime sources** include shipping vessels, fishing and recreational boats, aquaculture facilities, offshore oil industry and tourism (Boucher and Friot 2017). Despite international regulations (see also ▶ Chapter 16) forbidding the discharge of waste at sea (International Convention for the Prevention of Pollution from Ships (MARPOL) 73/78), cargo loss during storms and intentional disposal of waste from ships does occur. Lost and abandoned fishing gear is also a major contributor to marine plastic pollution worldwide (Richardson et al. 2017). For example, fishing gear used for catching octopus accounts for 94% of larger plastic debris found in the Moroccan Southern Atlantic Ocean (Loulad et al. 2017).

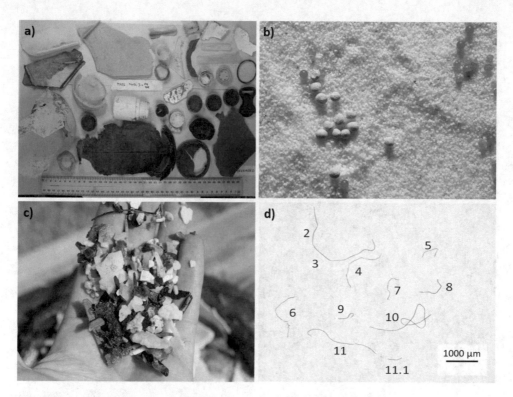

◻ Figure 9.2 Marine plastic pollution comes in all shapes and sizes and is categorised by size, shape, and colour. **a** Larger plastic debris is referred to as macroplastic (>20 mm) and is often observed floating **c** on the ocean surface or washed up on beaches. Smaller plastic debris, known as mesoplastic (5–20 mm) and **b** and **d** and microplastic (<5 mm) are harder or impossible to detect with the naked eye. Microplastics are sub-categorised into "primary", which are purposely manufactured to be small, such as microbeads used in exfoliating face cleansers (b), and "secondary" microplastics, formed from the breakdown of larger plastic items (c and d). *Photos*: K. Berry

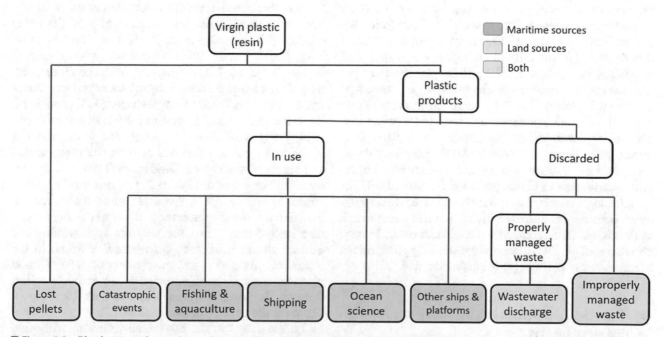

◻ Figure 9.3 Plastic enters the marine environment from land and maritime sources. Virgin plastic pellets may spill during manufacturing and transport, entering waterways. Manufactured plastic products may enter the marine environment due to degradation, accidental loss, or intentional disposal. Discarded waste, whether properly or improperly managed, may still enter the environment via numerous pathways such as wastewater effluent discharge, storm drains, and rivers. Adapted from Law (2017) by K. Berry

Many microplastic items, such as microbeads and microfibres, are washed down drains, entering waterways either directly or via wastewater management systems. More than 700,000 microplastic fibres can be re-leased from a typical six kilogram wash of synthetic clothing, such as polyester and nylon (Napper and Thompson 2016). Large quantities of microplastic items (up to 90%) can be removed from sewage during various

◘ Figure 9.4 Mismanaged waste is a major source of plastic debris to the marine environment. **a** and **b** Waste may be considered misman-aged due to lack of full containment, **c** and **d** that may result in accidental loss, or due to a lack of waste management infrastructure, which re-sults in plastic items being discarded directly into the environment. These photos, depicting mismanaged waste, were taken in Indonesia (a, b, c) and Myanmar (d). *Photos*: A. Reichelt-Brushett (a, b, c), K. Berry (d)

wastewater treatment stages (Carr et al. 2016), however, the capture of microplastic items is dependent on types of treatment processes. Due to their small size (microbe-ads are <50 μm) they are not always captured by filtra-tion devices. The quantity of microplastic items released in effluent can equate to 300 million plastic pieces per day, making wastewater discharge a major source of microplastic debris into the aquatic environment (Edo et al. 2019). The most commonly reported types of ma-rine microplastic debris worldwide are pellets, frag-ments, and fibres (GESAMP 2015), however, ropes, sponges, foams, rubber, and microbeads are also impor-tant contributors to plastic pollution (Auta et al. 2017).

Although most plastic enters the marine environ-ment because of human activity, natural events such as floods, earthquakes, and tsunamis can result in large quantities of plastic debris unintentionally entering the ocean (Murray et al. 2018; Veerasingam et al. 2016).

9.4 Plastic Transport in the Marine Environment

Plastics move through the marine environment via winds and ocean processes such as currents and eddies (Eriksen et al. 2014). Exactly how items move and how far is governed by the physical properties of the plas-tic object. The size, shape, and polymer density all in-fluence where the item will sit in the water column, and how easily it will move into another part of the water column (Chubarenko et al. 2016; Erni-Cassola et al. 2019; Lenaker et al. 2019). Ocean water has a density in the range of 1.02–1.03 g/cm^3 and therefore plas-tic polymers range from buoyant to negatively buoy-ant (e.g. PVC is denser than seawater [1.38 g/cm^3] and therefore tends to sink) (Andrady 2011; Plastics Europe 2014; Wang et al. 2016). Yet, where the plastic item sits in the water column depends also on the physical size and shape of the object. Despite PVC having a higher density than seawater (Syakti 2017), if a PVC object is large and hollow (e.g. a chemical drum), it may remain buoyant due to displacement. If it was a microplas-tic item (<5 mm), then the polymer type would have a much stronger influence on where it is found in the water column, and it will most likely sink. Very small plastic items such as microplastics can easily be mixed through the water column and can sink to different depths in the ocean (Reisser et al. 2015).

Plastic debris in the marine environment will often become substrate for sessile (immobile) marine organ-isms (this process is called bio-fouling), which can in-crease an item's density (e.g. Fazey and Ryan 2016; Kaiser et al. 2017). Smaller plastic debris and those with a density closer to that of sea water, which ex-perience bio-fouling, can have their density changed enough that the item will eventually sink to the seafloor (Kane and Clare 2019). Size is an important factor. Even microscopic sized pieces of plastic (known as nanoplastics, <1 μm), with a low polymer density making them very buoyant, can become tangled in marine snow (organic detritus in the water column) and sink (Porter et al. 2018). A similar process is thought to occur in the faeces of marine organisms that ingest and then excrete nanoplastics (Kvale et al. 2020). Larger plastic debris, instead, continues to drift in the ocean until it accumulates, either on beaches or in large ocean circulations, like the Great Pacific Garbage Patch (GPGP).

9.4.1 Modelling the Movements of Plastic

As with many ocean processes, it is not possible to study real-time **plastic** debris **dispersal and movement at ocean scales** in the field. The area is too large, the time scales are too long, and working on, or in, the ocean is expensive. Therefore, scientists use models to understand and predict plastic movement. Early studies modelled the movement of plastic debris at the scale of whole oceans (e.g. Law et al. 2010; Maximenko et al. 2012; van Sebille et al. 2012), while more recent studies focussed on the scales of seas and individual beaches (e.g. Cozar et al. 2014; Turrell 2018; Yabanlı et al. 2019). These models allow us to learn about the processes that transport and accumulate plastic debris in the environment.

9.4.2 Accumulation

Oceanic gyres are now infamous as large-scale debris accumulation areas for plastic pollution (Cozar et al. 2014; Eriksen et al. 2013). Gyres are large-scale eddies in the ocean, generated by oceanic currents and global wind patterns. The world's five major gyres (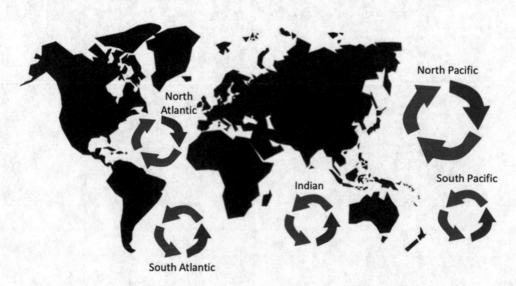 Figure 9.5) are found in the middle of the oceanic basins of the North and South Atlantic Ocean, the Indian Ocean, and the North and South Pacific Oceans. The largest gyre-associated "*floating garbage patch*" is the GPGP in the North Pacific Gyre (Lebreton et al. 2018), situated in the subtropical waters of the Pacific between California and Hawaii. In the GPGP, microplastic debris accounts for 94% of the plastic pieces floating in the area (Law et al. 2014). The micro- and meso-plastic debris concentrations in the GPGP are reported to be between 22,000 and 678,000 pieces/km^2, respectively (Lebreton et al. 2018).

Oceanic gyres are not a static accumulation of plastic debris, however, the time in which a plastic item remains within a gyre is very high (Howell et al. 2012). Accumulation can be defined as occurring when the supply (or input) to an area is larger than loss (or output). Each piece of plastic is perpetually moving, being mixed, and eventually leaving the gyre. Yet, this loss of plastic is very small when compared with the supply. Floating plastic debris has also accumulated in semi-enclosed regional seas globally, for example, the Caspian Sea (Nematollahi et al. 2020), the Mediterranean Sea (Suaria et al. 2016; Vianello et al. 2018), and Laizhou Bay in China (Teng et al. 2020).

The seabed is an accumulation zone that is only beginning to be understood (Woodall et al., 2014). Because degradation processes and bio-fouling can cause most categories of plastics to sink to the seafloor (Kowalski et al. 2016), microplastic debris has been found in sediments collected from the deepest parts of the ocean (Peng et al. 2020). This includes in deep sea sediments from the Great Australian Bight, the Southern Ocean, the North Atlantic Ocean, and the Mediterranean Sea (Van Cauwenberghe et al. 2013; Barrett et al. 2020;). A plastic bag was recently found in the world's deepest ocean trench, the 10,898 m deep Mariana Trench (Chiba et al. 2018).

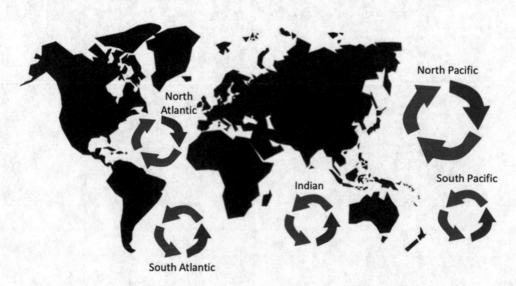

◻ Figure 9.5   The world's five major oceanic gyres. Adapted from what are the 7 continents (2020) by P. Lewis

9.4.3 Plastics in Remote Environments

Plastics have polluted remote terrestrial and marine environments, from the highest mountains to the depths of the ocean. These include the Arctic, Antarctic, and Southern Ocean, the Tibetan Plateau at 3000 m altitude, and the deep sea, at greater than 1000 m in depth (Wang et al. 2019a). Baseline pollution in remote polar regions, such as the Arctic and Antarctic, are considered indicators of global environmental health. The Arctic in particular is now being recognised as a global sink for anthropogenically derived particulates (Eriksen et al. 2020), with microplastics and microfibres being dispersed into the region from population centres by subsurface currents (Wichmann et al. 2019). Recent studies have also identified the atmosphere and snowfall as significant transport routes (Bergmann et al. 2019). Another significant source and transport vector within the region is the Arctic sea ice, which can trap between 38 to 234 plastic items per m^3 of ice (Obbard 2018), items that can then be re-released after the seasonal migration and melting of the ice in the North Atlantic (Peeken et al. 2018) (▶ Box 9.3).

At the other end of the world, the Antarctic Convergence current that surrounds Antarctica was thought to act as a potential barrier to flowing debris and pollutants from the north (Ainley et al. 1990). However, studies now show the presence of microplastics in sea ice, sediments, and surface waters of the Antarctic and Southern Ocean, as well as in the scat of seabirds from sub-Antarctic Islands and the Antarctic Peninsula (e.g. Isobe et al. 2014; Bessa et al. 2019; Kelly et al. 2020; Sfriso et al. 2020; Waluda et al. 2020).

Box 9.3: What is the Significance of Microplastic Items in Sea Ice?

Microplastic concentrations within Arctic sea ice can impact the absorption of incident solar radiation, which affects the light reflectance (albedo) of sea ice. Light reflectance is how the ice reflects solar energy and is one of its key properties, regulating heat exchange between the ocean and the atmosphere. High salinity sea ice has been associated with large concentrations of microplastic items, which could adversely affect albedo and how the ice melts, but also the brine volume content, which controls the permeability of sea ice. Microplastic impurities can be light-absorbing, affecting light penetration depth, potentially impacting algae that lives underneath, algae that forms the basis of the Arctic foodweb.[1] A total of 96 microplastic items from 14 types of polymers were discovered in sea ice samples collected near Casey Station in East Antarctica.[2] Local sources include clothing and equipment used by tourists and researchers, as well as varnishes and plastics commonly used by the fishing industry (◻ Figure 9.6).

For further reading:

The Guardian Australia ▶ https://mville.libguides.com/c.php?g=370027&p=5932225#:~:text=Structure%20of%20a%20citation%20for,Publisher%2C%20Publication%20date%2C%20URL.

◻ **Figure 9.6** ▶ Box 9.3: AWI scientists sample a melt pond on Arctic sea ice, discovering record levels of microplastics. *Photo*: Mar Fernandez/Alfred Wegener-Institute

---

1 The Conversation 2019
   ▶ https://theconversation.com/microplastics-may-affect-how-arctic-sea-ice-forms-and-melts-120721.
2 Ecowatch 2020
   ▶ https://www.ecowatch.com/antarctica-microplastics-sea-ice-2645809545.html?rebelltitem=2#rebelltitem2.

9.5 Degrading Processes

Durability is a valued property of plastic. Nonetheless, plastic items do not remain in their original form forever, and eventually degrade over time. Plastics can undergo different weathering and aging processes in the marine environment, due to a wide variety of environmental factors (▶ Box 9.4). These include photo-degradation from the sun, thermal aging, bio-film growth, and oxidation that results in the degradation of the plastic polymers (Andrady 1994; Min et al. 2020). The physical damage that results from this degradation can include cracking, surface erosion, and abrasion, all of which depends on the structure and chemical properties of the plastic polymer (Andrady 2011; Min et al. 2020). Photo-degradation, or the physical and chemical weathering by UV light, breaks polymer bonds, weakening the plastic structure and allowing the item to fragment, forming secondary microplastics (Efimova et al. 2018). Plastic that has sunk to ocean depths, or that is buried in sediment, does not experience exposure to UV light, therefore it will not undergo fragmentation processes, unless exposed to another mechanism of degradation (Andrady 2011). Mechanical forms of degradation are possible, particularly in the swash zone of high-energy beaches (Corcoran et al. 2009). The relentless battering of the plastic against sand grains, pebbles, and stones will cause it to break up, with previous UV light exposure exacerbating the process (◘ Fig. 9.7). These processes are believed to be the most common ways in which plastics become microplastics (described in ▶ Sect. 9.2.2).

Weathering processes can also release harmful additives from the plastic polymer matrix (Teuten et al. 2009). These can include plasticisers such as phthalates (Schrank et al. 2019), flame retardants (Fauser et al. 2020), and other endocrine-disrupting chemicals (EDCs) (Gallo et al. 2018). Biological degradation of plastic is also possible, through the bio-fouling of plastic surfaces (Fazey and Ryan 2016). Emerging research suggests that the cells of some microbes conform to the pits and grooves found on the surfaces of microplastics and may be degrading polymers in situ (Zettler et al. 2013; Reisser et al. 2014). Laboratory studies by McGivney et al. (2020) found physiochemical changes in microplastics exposed to bacterioplankton biofilms extracted from coastal waters in Sweden. Biofilm effects were dependent upon polymer type. Increases in crystallinity and maximum compression were observed in polyethylene and polystyrene items respectively, while polypropylene items decreased in stiffness when exposed to the biofilm (McGivney et al. 2020). Gene sequencing analyses found significantly higher abundances of *Sphingobium* spp., *Novosphingobium* ssp., and uncultured Planctomycetaceae on polyethylene, while polypropylene and polystyrene both had greater abundances of Sphingobacteriales and Alphaproteobacteria. These results provide evidence to support the hypothesis that bacteria are degrading microplastics and that different members of the bacterial community are responsible for this degradation, depending upon polymer type. More work is needed in order to determine how these biological modifications, in concert with the physical and chemical changes from abiotic factors, impact the fate of the various microplastic polymers in the marine environment.

Box 9.4: The Physical and Chemical Degradation Processes of Plastic

Type – Details

Biological – Microorganism actions cause degradation

Photo – UV light or photons, usually sunlight, cause degradation

Thermo-oxidative – Slow oxidative, molecular degradation at moderate temperatures

Hydrolysis – Chemical reaction with water causes degradation

Mechanical – Physical breakdown of plastics on high energy beaches

Adapted from Rochman et al. (2015).

Weathering Agents in Different Marine Zones

Weathering agent	Beach	Surface water	Deep water or sediment
Sunlight	Yes	Yes	No
Temperature	High	Moderate	Low
Oxygen levels	High	High/moderate	Low
Fouling (screens solar radiation)	No	Yes	Yes
Adapted from Andrady (2015)			

◼ Figure 9.7   Weathering (physical and chemical) contributes to the degradation of plastic items on beaches, **a** This blue plastic item was observed during the fragmentation process on a beach in Queensland, Australia, and **b** a sample was collected and taken back to the lab for imaging under a stereomicroscope, which revealed that the plastic pieces were fragmenting into more than 20 microplastic pieces, many of which were smaller than a grain of sand. If left on the beach, these new smaller fragments would continue to break into smaller and smaller pieces *Photos*: K. Berry

9.5.1   Complications of Measuring and Comparing Plastic Pollution

Scientific research on the **quantification** and environmental impacts of macro- and microplastic have increased drastically over recent years, providing critical information to scientists and policy makers (Forrest 2019). However, **discrepancies** in terminology, reporting units, and **inconsistencies** in methodologies make accurate geographical comparisons and summaries of this issue difficult (Provencher et al. 2020; Pittura et al. 2023).

Plastic quantification is presented by either (1) the number of plastic pieces per m², (2) the number of plastic items per litre of seawater, or (3) weight (Miller et al. 2017). The range of units makes it difficult to make accurate comparisons between study sites or obtain true estimates of total plastic contamination at the local, regional, or global level. Laboratory studies have quantified plastic ingestion by extracting plastics from animal tissue, yet all extraction methodologies have limitations (Miller et al. 2017). For example, digestion techniques using acid solutions can digest certain plastic polymers, resulting in the underestimation of plastics (Claessens et al. 2013; Li et al. 2015; Vandermeersch et al. 2015), while, methods using physical extractions may fragment plastic pieces, resulting in over-estimations (Kathryn Berry personal observation). These over- and underestimations can also occur when microplastic polymer types are not identified correctly, for example, many naturally derived materials can also resemble plastic, requiring these pieces to be validated as synthetic polymers (Lusher et al. 2020; Zhao et al. 2018). FTIR and Raman spectroscopy are the most commonly used methods for plastic polymer identification, however this equipment is expensive and the process is time consuming (Cozar et al. 2014; Lv et al. 2019). Consequently, any study that has not correctly validated microplastic polymers using one of these

techniques is likely overestimating microplastic contamination (Song et al. 2015; Provencher et al. 2020).

Lastly, procedural contamination by microfibres is a serious concern (Woodall et al. 2015; Torre et al. 2016), as is the ecological relevance of studies. Many studies investigating the potential effects of microplastics utilise concentrations and sizes of microplastics not commonly reported in the natural environment, meaning that the true implications of the results may be misinterpreted (Phuong et al. 2016). Increased baseline studies, standardised collection and quantification methods, and consistent reporting units will help provide accurate and comparable environmental data to inform management and policy decisions (Cowger et al. 2020; Pittura et al. 2023).

9.6   Impacts of Plastic Debris

9.6.1   Impacts Overview

The ubiquity of plastic debris and diversity of plastic debris characteristics (e.g. shape, size, density, chemical composition) results in many interaction pathways with marine wildlife (◼ Table 9.1) and humans. Plastic pollution is known to impact many trophic levels and can have physical and chemical effects. It is aesthetically unpleasing, creates human health concerns, and is an economic burden. In this section, we will discuss the impacts of plastic pollution on the environment, human health, and the economy.

9.6.2   Physical Interactions with Wildlife

Entanglement and ingestion are the most commonly reported interactions between marine plastic debris and wildlife (◼ Table 9.1) (Kühn and van Franeker 2020).

◻ Table 9.1 Summary of plastic debris impacts on marine wildlife related to encounter types (field and laboratory measurements)

Animal	Encounter type	Predominate debris type	Impact (response)	Study
Grey seals	ENT	Fishing line, net, rope	Constriction	Allen et al. (2012)
Manatees	ENT	Fishing line, bags, debris	Death	Beck and Barros (1991)
Elephant seals	ENT	Fishing line, fishing jibs	Dermal wound	Campagna et al. (2007)
Fur seals	ENT	Trawl net, packing bands	Death	Fowler (1987)
Invertebrates, fish, seabirds, marine mammals	ENT	Derelict gillnets	Death	Good et al. (2010)
Gorgonians	ENT	Fishing line	Damage/breakage	Pham et al. (2013)
Sea turtles	ENT	Fishing gear	Death	Vélez-Rubio et al. (2013)
Whales	ENT	Fishing line	Dermal wound	Winn et al. (2008)
Manatees	ING	Fishing line, bags, debris	Death	Beck and Barros (1991)
Penguins	ING	Plastic, fishing, debris	Perforated gut, death	Brandão et al. (2011)
Sea turtles	ING	Plastic bags, ropes	Gut obstruction, death	Bugoni et al. (2001)
Seabirds	ING	Plastic items, pellets	Perforated gut	Carey (2011)
Fish (L)	ING	Nano items	Biochemical/cellular	Cedervall et al. (2012)
Seabirds	ING	Plastic debris	Gut lesions	Fry et al. (1987)
Sperm whales	ING	Fishing gear, debris	Gastric tear, death	Jacobsen et al. (2010)
Copepods (L)	ING	Micro- and nanoplastics	Death	Lee et al. (2013)
Sea turtles	ING	Marine debris	Gut obstruction	Vélez-Rubio et al. (2013)
Seabirds	ING	Microplastics	Gut obstruction	Gilbert et al. (2015)
Mussels (L)	ING	Microplastics	Biochemical/cellular	von Moos et al. (2012)
Bivalves (L)	ING	Microplastics	Limited response	Bour et al. 2018
Marine larvae (L)	ING	Microplastics	Limited response	Kaposi et al. (2014)
Brine shrimp (L)	ING	Microplastics	Limited response	Wang et al. (2019b)
Marine fish (L)	ING	Microplastics	Limited response	Critchell and Hoogenboom (2018)
Copepods Zebrafish (L)	ING/CON	Microplastics	Trophic transfer, POP uptake	Batel et al. (2016)
Fish	ING	Microfibres	Limited response	Kroon et al. (2012)
Zebra Fish (L)	ING/CON	Microplastics	Pb (lead) bioavailable	Boyle et al. (2020)
Pearl oyster	ING/CON	Aquaculture gear	Leachate absorption, reproduction	Gardon et al. (2020)
Seabirds	ING/CON	Microplastics	PBDE body burden	Tanaka et al. (2013)
Coral reef	INT	Fishing gear	Tissue abrasion	Chiappone et al. (2005)
Seagrass	INT	Fishing gear, debris	Breakage, death	Uhrin and Schellinger (2011)
Coral (L)	INT	Microplastics	Limited response	Berry et al. (2019)

Abbreviations: *ENT* entanglement, *ING* ingestion, *INT* interaction, *(L)* laboratory experiment, *CON* contaminant

Wildlife is more likely to become entangled in certain shapes/types of plastic debris, such as ropes (◻ Figure 9.8a, b), bags, or circular plastic items, such as aluminium can six-pack rings. Entanglement can cause tissue abrasion, strangulation, reduced feeding efficiency, reduced growth and development, and death due to drowning (e.g. Allen et al. 2012). Plastic debris may cause additional **physical harm to marine habitats and** sessile benthic **organisms** (e.g. corals, seagrass, mangroves) via smothering (◻ Figure 9.8c), and when dragged along the seafloor. Fishing nets (referred to as ghost nets) that are lost, abandoned, or discarded at sea

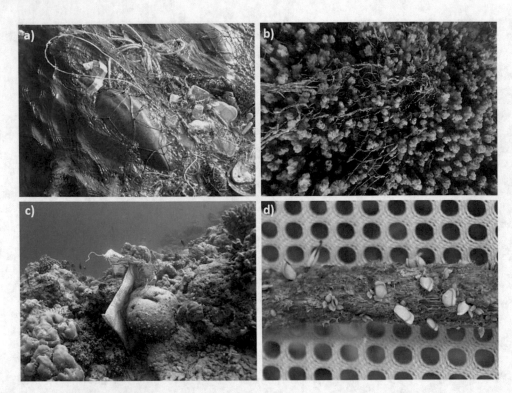

▣ Figure 9.8   Plastic debris interacts with the marine environment and wildlife in numerous ways: **a** and **b** Organisms, such as turtles and corals may become entangled in fishing nets/rope, **c** sunken plastic debris may smother sessile organisms such as corals, causing physical harm and blocking out essential light and **d** plastic can act as a platform to transport fouling organisms and microbes. *Photos*: A. Hassan (a), A. Reichelt-Brushett (b), K. Berry (c, d)

can continue to catch fish and other marine organisms such as rays and turtles for many years (Gunn et al. 2010). These environmental impacts may create economic loss associated with losses to fisheries due to depletion of fish stocks and gear replacement costs.

Many marine species are reported to ingest plastic debris, including the smallest marine animals at the bottom of the food chain, zooplankton (Cole et al. 2013), fish (Kroon et al. 2018), turtles (Caron et al. 2018), seabirds (Gilbert et al. 2015), whales and other large marine animals (Besseling et al. 2014; Germanov et al. 2019; Moore et al. 2020). Ingestion of plastic debris occurs due to an organism mistaking plastic debris for prey either by sight, for example, turtles mistaking plastic bags for jellyfish (Schuyler et al. 2014), or by smell, for example, some species of seabirds ingest microplastics after targeting zooplankton swarms (Gilbert et al. 2015; Savoca et al. 2016). Ingestion is influenced by the size and shape of the plastics, an organism's feeding behaviour, and feeding range (depth) within the water column (Fossi et al. 2012; Cole and Galloway 2015; Lusher et al. 2017). Impacts associated with ingestion are often related to size of the plastic debris, ranging from minimal effects (likely due to the animal simply passing the plastic debris through its digestive system) to obstruction of the intestinal tract and reduced stomach capacity (which can lead to malnutrition and reduced growth rates), internal injury, changes

in behaviour, reduced swimming performance, impaired reproduction, and oxidative stress (Sigler 2014; Cole et al. 2015; Gray and Weinstein 2017; Foley et al. 2018).

9.6.3   Plastic as an Unnatural Substrate

Micro- and macroplastics act as a platform for colonisation (▣ Figure 9.8d) by sessile organisms and microbes, including pathogens. Movement of colonised plastic debris may increase an organism's dispersal and transport of invasive species (Barnes et al. 2009; Gregory 2009). Colonisation of sunken plastic debris may alter habitat structure by providing sessile benthic organisms with alternative substrate to settle and grow upon. The long-term implications of plastic debris as a 3D habitat structure are unknown.

Plastic debris provides a novel habitat upon which microbes can flourish (Zettler et al. 2013). The **plastisphere** refers to the unique structure and taxonomy of the microbial community that forms on the surface of marine plastic debris, which differs significantly from the overall microbial community of the surrounding substrates (Bryant et al. 2016; Feng et al. 2020; Zettler et al. 2013). It has yet to be determined if the taxonomy of the plastisphere varies between polymer types, as other factors such as the age of the debris (i.e. virgin or weathered), season, and geographic location appear to also play a role (Erni-Cassola et al. 2019; Oberbeckmann et al. 2016; Zettler et al. 2013). The plastisphere has been shown to include pathogenic *Vibrio* and *Escherichia coli* species, antibiotic-resistant bacteria, harmful algal bloom species, and the fish disease causing bacteria *Aeromonas salmonicida* (Kirstein et al. 2016; Casabianca et al. 2019; Rodrigues et al. 2019; Silva et al. 2019; Laverty et al. 2020; Moore et al. 2020). Although this field of research is still novel, the likelihood of coral disease increased from 4 to 89% when corals (from 159 reefs in the Asia-Pacific region) were in contact with macroplastic debris. This suggests microbial colonisation of plastic by pathogens may contribute to disease outbreaks in the ocean (Lamb et al. 2018). Further research is required into the mechanisms of plastic as a vector for pathogens, trophic transfer of pathogens via plastic ingestion, and the potential for plastics to act as a vector for the long-distance dispersal of harmful microorganisms.

9.6.4 Chemical Effects of Microplastics

While ingestion of microplastics may cause physical harm to marine biota, there is also the potential for chemical impacts. The physical processes that weather plastic objects to microplastics can create a large specific surface area on the particles, causing the items to act as a **sponge** by taking up contaminants from sediments or the water column via adsorption (Fred-Ahmadu et al. 2020). Adsorbed contaminants can include Persistent Organic Pollutants (POPs) (Mato et al. 2001; Endo et al. 2005; Teuten et al. 2009), metals (Ashton et al. 2010), and EDCs (Hermabessiere et al. 2017).

The addition of chemical additives during the plastic manufacturing process can also pose a chemical threat (Rios et al. 2007; Oehlmann et al. 2009; Guo and Wang 2019). Additives include plasticisers such as phthalates or bisphenol A, flame retardants, and stabilisers such as lead (Pb) and other metals. These can leach into the marine environment as plastic weathers (Gardon et al. 2020; Lomonaco et al. 2020), or if ingested, into the tissues and guts of organisms (Teuten et al. 2007; Engler 2012; Tanaka et al. 2013). The harmful substances sorbed and leached are often persistent, enabling plastic objects to become vectors for contaminants or **biovectors** (◘ Figure 9.9) (Wang et al. 2020), resulting in bioaccumulation (Paul-Pont et al. 2016; Gallo et al. 2018).

The cumulative effects of microplastics and associated pollutants remain a developing field. Theoretical modelling has shown that the effect of adsorbed pollutants in organisms ingesting microplastics should be minor (Koelmans et al. 2016). However, laboratory-based studies have shown that once microplastics are ingested, the associated contaminants can be readily released into the bloodstream of marine organisms (Tanaka et al. 2013; Besseling et al. 2014). Further impacts may occur through biomagnification (e.g. Rochman et al. 2013; Batel et al. 2016). As these impacts can increase through the marine food chain, there are also implications for human health (Wang et al. 2019b; Enyoh et al. 2020).

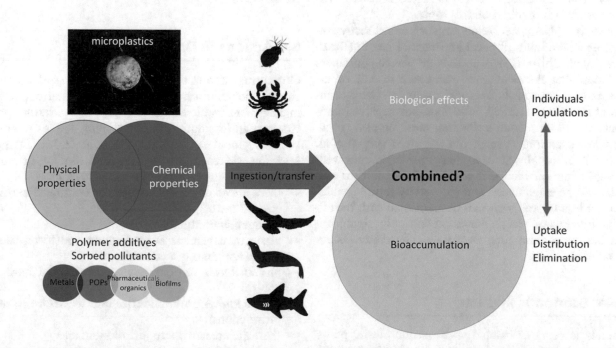

◘ **Figure 9.9** Through their physical and chemical properties, microplastics can act as biovectors of contaminants through marine food chains, with increasing biological effects and bioaccumulation through trophic levels. Yet the combined impacts of microplastic ingestion and transfer of chemicals on individuals or populations require further research. *Image*: P. Lewis

9.6.5 Human Health Impacts

Similar to marine wildlife, marine plastic pollution interacts with humans via numerous pathways. Humans may be exposed to plastic debris through seafood, as microplastic debris has been found in invertebrates, crustaceans, and fish harvested for human consumption (Van Cauwenberghe and Janssen 2014; Rochman et al. 2015; Carbery et al. 2018; Smith et al. 2018; Cox et al. 2019; Walkinshaw et al. 2020). Since most plastic remains in the digestive tract of an animal, the risk of ingestion by humans is higher when organisms are consumed whole, such as with small fish or bivalves (Rochman et al. 2015). There is concern that the chemical impacts associated with micro- and nanoplastic ingestion documented in marine wildlife are also a concern for human health, including adsorption across the gastrointestinal tract (Waring et al. 2018), chemical toxicity associated with leaching of plastic additives (e.g. BPA, heavy metals, EDCs) (Campanale et al. 2020), or sorbed contaminants (Smith et al. 2018), as well as hazards associated with microbial colonisation (Wright and Kelly 2017; Wang et al. 2019a, b). **Knowledge on the implications of microplastic consumption by humans is currently limited**, however the severity of impacts will be dependent on seafood contamination levels, exposure frequency, and effects of exposure. Similar to other organisms, it is possible humans simply ingest and then egest plastic pieces. While evidence is growing about the interactions between micro- and nanoplastic exposure, toxicology, and human health (Wright and Kelly 2017; Smith et al. 2018; De-la-Torre 2020), further research is required on this topic.

Marine plastic debris on beaches can directly impact an individual's physical and mental health (Beaumont et al. 2019). Sharp plastics or plastic containers that contains chemical waste can result in cuts or exposure to dangerous liquids and unsanitary items (Santos et al. 2005). Littered coastlines can negatively impact mood and mental wellbeing, resulting in a reduction in recreational use of littered areas (Wyles et al. 2016). Additionally, since some people experience wellbeing in the knowledge that culturally significant animals will be experienced and enjoyed by future generations, a loss of wellbeing can be associated with the adverse impacts of plastic debris on culturally significant marine megafauna such as turtles and whales (Beaumont et al. 2019).

9.6.6 Economic Impacts

Economic costs associated with marine plastic pollution can be either direct or indirect (McIlgorm et al. 2011). Marine plastic pollution negatively impacts tourism, fishing (subsistence, recreational, commercial), shipping, aquaculture, recreation, and other ecosystem services. Ecosystem service are the benefits people obtain from nature, including food, carbon storage and cultural benefits (Worm et al. 2006; Liquete et al. 2013) and evidence suggests that plastic pollution causes significant impacts to almost all global ecosystem service (Beaumont et al. 2019). In 2011, based on ecosystem service values and marine plastic abundance estimates, marine plastic's economic costs were conservatively estimated at between US$ 3300 and US$ 33,000 per tonne of marine plastic per year (Beaumont et al. 2019).

Renowned or frequently visited beaches that are littered may incur a range of economic costs including clean-up expenses and lost tourism revenue (Beaumont et al. 2019). Shipping, navy, coast guard, and fishing industries are impacted by direct damage and entanglement of fishing gear in propellers (Chen 2015). Fishing industries also suffer economic loss due to plastic debris negatively impacting fish habitats (e.g. sunken derelict fishing gear) and stocks (e.g. ghost fishing) (Kaiser et al. 2003; NOAA 2015). In Indonesia, local fisherfolk described the direct and indirect negative impacts of marine debris, including propeller entanglements, fouling of gill nets and hooks, damage to fishing gear, and injuries (Nash 1992). Such impacts can result in additional fishing time and modified fishing behaviour to attain the same yield compared to as if there were no waste associated losses. Some modified fishing behaviour includes the adoption of harmful fishing methods (Nash 1992).

9.7 Actions to Drive Change

Our current knowledge and understanding of the marine plastic pollution issue, including key sources, waste management inefficiencies, and gaps in legislation, provide a solid foundation for developing actions to combat this global issue (Rochman et al. 2016). Despite knowing where actions are required, finding effective solutions is a complex task for many reasons:

- there are economic incentives for continued and increased use of plastic;
- production continues to rise;
- waste management is inadequate and inconsistent within and amongst countries;
- plastic inputs are difficult to predict and hard to control;
- plastic knows no boundaries and will move to new jurisdictions;
- there are areas with no jurisdiction; and
- plastic debris accumulates in remote areas, or may sink out of sight.

As such, solutions require coordinated approaches by a range of stakeholders, including producers, consumers, scientists, and policy makers (local, regional, national, and international levels) (Löhr et al. 2017).

Global partnerships and commitments are being made to address marine plastic pollution (and other types of marine litter) at many major global fora (e.g. G7, G20, and the 2017 World Oceans Summit) (Vince and Hardesty 2018). International partnerships have led to instruments that regulate marine plastic pollution through conventions, strategies, action plans, agreements, and regulations (Chen 2015). For example, the EU Action Plan for a Circular Economy (a Europe-wide strategy committed to reducing plastic pollution impacts and increasing material value in the EU economy), MARPOL Convention (prevention of pollution from ships), the Honolulu Strategy (improving co-operation to prevent land-based plastic entering the oceans), and The Clean Seas Global Campaign on Marine Litter (worldwide elimination of single-use plastics and microplastics in cosmetics by 2022) (Ferraro and Failler 2020). In March 2022 Heads of State, Ministers of environment and other representatives from UN Member States endorsed a historic resolution at the UN Environment Assembly (UNEA-5) to

» *"End Plastic Pollution and forge an international legally binding agreement by 2024. The resolution addresses the full lifecycle of plastic, including its production, design and disposal"*.

Although international instruments are a step in the right direction, international policy framework can be fragmented, its focus can be limited, and laws are often soft (i.e. non-binding) (Vince and Hardesty 2018; Ferraro and Failler 2020). It is therefore imperative that these efforts coincide with actions taking place at local and national, levels, such as legislation and regulation creation.

A critical short-term action to reduce plastic inputs into the marine environment includes improvements to waste management regulations and infrastructure (Löhr et al. 2017). Around 4.8–12.7 million tonnes of marine plastic pollution enter the ocean from land-based sources annually, originating from 20 of 192 coastal countries (Jambeck et al. 2015). Highly polluting counties include China, Indonesia, Philippines, Vietnam, Sri Lanka, Thailand, Egypt, Malaysia, Nigeria, and Bangladesh (Jambeck et al. 2015). Many of the listed countries lack adequate waste management, making improvements to waste management (e.g. providing and improving collection infrastructure and technologies) critical for reducing plastic inputs into the ocean. Many new instruments are taking a hierarchical approach to waste management (Figure 9.10), which prioritises inhibiting waste generation and movement of litter into the marine environment, rather than cleaning up what is already in the ocean (Watkins et al. 2012). This is not to say that ocean and beach clean-ups are not important, but rather highlights how approaches that prioritise prevention rather than mitigation and curative measures are very important (Critchell et al. 2019; Watkins et al. 2012).

Large system changes such as behavioural changes and transitioning to a circular economy are suggested as longer-term solutions (Löhr et al. 2017). A circular economy focuses on purposeful design to incorporate end-use and reuse from the start of a product's life cycle (reduce, reuse, recycle, redesign, recover), encouraging supply chain investments that will ultimately reduce waste entering the ocean (Ellen MacArthur Foundation 2017). A circular economy approach is designed to not only benefit the environment, but also the economy, as it recaptures costs currently being lost (WEF 2016).

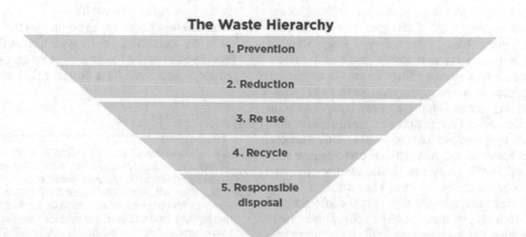

The Waste Hierarchy

1. Prevention

2. Reduction

3. Re use

4. Recycle

5. Responsible disposal

◻ **Figure 9.10**   Hierarchical approaches to waste management guide and rank waste management decisions. The preferred option is the prevention of waste generation, through limiting raw materials or acquiring used/recycled materials or materials that can be recycled. Waste disposal is unsustainable and can have long-term environmental impacts. Disposal is the least preferred option and should be carried out responsibly. *Image*: Wikibooks: CC-BY-SA-3.0

All pro-environmental movements require behavioural changes, which can be facilitated at numerous stakeholder levels using many strategies. For example, governments (regional and national) can create new policy and legislation aimed at reducing plastic product use or specific activities (e.g. bans on plastic grocery bags or microbead use in cosmetics, ten Brink et al. 2016). In 2015, the Canadian province of Ontario banned the production of microbeads (Legislative Assembly of Ontario 2015), and the United States of America (USA) passed a federal law banning microbead use in rinse off products by 2018 (Rochman et al. 2015; United States Congress 2015). Many other countries (e.g. Netherlands, United Kingdom) have expressed interest in creating similar laws (reviewed in Xanthos and Walker 2017). Economic incentives can target consumption, including the application of deposit refunds for plastic bottles, and charges/taxes to plastic bags and other one-use items (ten Brink et al. 2016).

Education and change-oriented public awareness is critical if we are to increase proper waste disposal rates and to promote consumers' refusal of products containing plastic and sustainable substitutes (ten Brink et al. 2016). Notable recent actions include the "*Beat The Microbead*" campaign (Plastic Soup Foundation 2020) and the Massive Open Online Course (MOOC) on Marine Litter, which is part of the "*Clean Seas*" campaign (Brown 2013). MOOC targets a range of sectors and stakeholders and aims to provide actionable and change-oriented open access learning on a global scale (Brown 2013; Leire et al. 2016).

9.8 Summary

Thirty-three billion tonnes of plastic will be created globally by 2050, if plastic production continues to increase at its current rate of 5% per year (Rochman et al. 2013). For context, compact cars weigh slightly over 1 tonne, while large cars weigh closer to 2 tonnes. It is imperative that strong efforts are made to reduce plastic production, use, and disposal as soon as possible. The natural environment is already inundated with plastic debris, and the ocean acts as a main sink for discarded and mismanaged plastic waste. This chapter highlighted some of the interactions and negative implications of plastic pollution in the marine environment, however, estimates of total plastic debris in the ocean are conservative, suggesting that the interactions with wildlife and potential adverse effects are much worse than what has been documented. As with any environmental stressor, it is also important to consider that the impacts of plastic may act in combination with other environmental stressors, such as over-exploitation, other types of pollution, and climate change, and that the cumulative effects of these stressors may be causing more damage than plastic pollution alone.

While there are many governance challenges and complexities influencing the success of plastic waste reduction and management, significant steps have been made. These include strategies to change consumer behaviour, transitioning to a circular economy, and the implementation and enforcement of policies and law (Löhr et al. 2017). The further success of initiatives will require actions from a range of stakeholders (e.g. producers, consumers, industry, and policy makers). Nonetheless, as is the case for most environmental issues, individuals can create positive change by staying informed, educating others, and changing their behaviour. Some easy actions to start with include: (1) read personal care product labels for plastic ingredients and don't purchase products that use microplastics; (2) carry reusable bottles/thermal cups and refuse single-use plastic items; (3) pick up and properly discard plastic litter; (4) educate yourself on local recycling policies and ensure you're recycling plastics properly; (5) read clothing labels and only purchase clothes that made from natural fibres, such as cotton, wool, hemp, and bamboo. These seemingly minor actions will contribute greatly to the positive changes occurring worldwide.

9.9 Questions and Activities

1. Take the time to monitor how much plastic waste you create each week. What activities result in the most plastic consumption?
2. What are three actions you can take regularly to reduce your plastic use?
3. What types of marine wildlife are most at risk from plastic floating in the ocean?
4. What characteristics make an animal more vulnerable to the impacts of plastic pollution in the ocean?
5. Provide examples of how a circular economy could reduce plastic waste from entering the ocean.

References

ACC (Americal Chemical Society) (2020) How plastics are made [WWW Document]. Available at: ► https://plastics.american-chemistry.com/How-Plastics-Are-Made/. Accessed 25 Aug 2020

Ainley D, Fraser WR, Spear LB, Beach S (1990) The incidence of plastic in the diets of Antarctic seabirds. In: Shomura R, Godfrey M (eds) Proceedings of the workshop on the fate and impact of marine debris, Honululu, Hawaii, pp 682–691

Allen R, Jarvis D, Sayer S, Mills C (2012) Entanglement of grey seals *Halichoerus grypus* at a haul out site in Cornwall, UK. Mar Pollut Bull 64:2815–2819

Andrady AL (2015) Persistence of plastic litter in the oceans. In: Bergmann M, Gutow L, Klages M (eds) Marine anthropogenic litter. Springer, Berlin, pp 57–72

Andrady AL (2017) The plastic in microplastics: a review. Mar Pollut Bull 119:12–22

Andrady AL (2011) Microplastics in the marine environment. Mar Pollut Bull 62:1596–1605

Andrady AL (1994) Assessment of environmental biodegradation of synthetic polymers. J Macromol Sci Part C: Polym Rev 34:25–76

Andrady AL, Neal MA (2009) Applications and societal benefits of plastics. Philos Trans R Soc B 364:1977–1984

Ashton K, Holmes L, Turner A (2010) Association of metals with plastic production pellets in the marine environment. Mar Pollut Bull 60:2050–2055

Auta HS, Emenike CU, Fauziah SH (2017) Distribution and importance of microplastics in the marine environment: a review of the sources, fate, effects, and potential solutions. Environ Int 102:165–176

Barnes DKA, Galgani F, Thompson RC, Barlaz M (2009) Accumulation and fragmentation of plastic debris in global environments. Philos Trans R Soc B 364:1985–1998

Barrett J, Chase Z, Zhang J, Holl MMB, Willis K, Williams A, Hardesty BD, Wilcox C (2020) Microplastic pollution in deep-sea sediments from the Great Australian Bight. Front Mar Sci 7:576170

Batel A, Linti F, Scherer M, Erdinger L, Braunbeck T (2016) Transfer of benzo[a]pyrene from microplastics to *Artemia* nauplii and further to zebrafish via a trophic food web experiment: CYP1A induction and visual tracking of persistent organic pollutants: trophic transfer of microplastics and associated POPs. Environ Toxicol Chem 35:1656–1666

Baulch S, Perry C (2014) Evaluating the impacts of marine debris on cetaceans. Mar Pollut Bull 80:210–221

Beaumont NJ, Aanesen M, Austen MC, Börger T, Clark JR, Cole M, Hooper T, Lindeque PK, Pascoe C, Wyles KJ (2019) Global ecological, social and economic impacts of marine plastic. Mar Pollut Bull 142:189–195

Beck CA, Barros NB (1991) The impact of debris on the Florida manatee. Mar Pollut Bull 22:508–510

Bergmann M, Mützel S, Primpke S, Tekman MB, Trachsel J, Gerdts G (2019) White and wonderful? Microplastics prevail in snow from the Alps to the Arctic. Sci Adv 5:eaax1157

Berry KLE, Epstein HE, Lewis PJ, Hall NM, Negri AP (2019) Microplastic contamination has limited effects on coral fertilisation and larvae. Diversity 11:228

Bessa F, Ratcliffe N, Otero V, Sobral P, Marques JC, Waluda CM, Trathan PN, Xavier JC (2019) Microplastics in gentoo penguins from the Antarctic region. Sci Rep 9:14191

Besseling E, Wang B, Lürling M, Koelmans AA (2014) Nanoplastic affects growth of *S. obliquus* and reproduction of *D. magna*. Environ Sci Technol 48:12336–12343

Boucher J, Friot D (2017) Primary microplastics in the oceans: a global evaluation of sources. IUCN, Gland Switzerland, p 43. Available at: ► https://portals.iucn.org/library/sites/library/files/documents/2017-002-En.pdf. Accessed 7 Feb 2022

Bour A, Haarr A, Keiter S, Hylland K (2018) Environmentally relevant microplastic exposure affects sediment-dwelling bivalves. Environ Pollut 236:652–660

Boyle D, Catarino AI, Clark NJ, Henry TB (2020) Polyvinyl chloride (PVC) plastic fragments release Pb additives that are bioavailable in zebrafish. Environ Pollut 263:114422

Brandão ML, Braga KM, Luque JL (2011) Marine debris ingestion by Magellanic penguins, *Spheniscus magellanicus* (Aves: Sphenisciformes), from the Brazilian coastal zone. Mar Pollut Bull 62:2246–2249

Brown S (2013) Back to the future with MOOCs. Conference Paper 3. Available at: ► http://www.icicte.org/Proceedings2013/Papers%202013/06-3-Brown.pdf. Accessed 7 Feb 2022

Bryant JA, Clemente TM, Viviani DA, Fong AA, Thomas KA, Kemp P, Karl DM, White AE, DeLong EF (2016) Diversity and activity of communities inhabiting plastic debris in the north Pacific gyre. Ecol Evol Sci 1(3):e00024-16

Bucci K, Tulio M, Rochman CM (2020) What is known and unknown about the effects of plastic pollution: a meta-analysis and systematic review. Ecol Appl 30(2):e02044

Bugoni L, Krause L, Virgínia Petry M (2001) Marine debris and human impacts on sea turtles in southern Brazil. Mar Pollut Bull 42:1330–1334

Campagna C, Falabella V, Lewis M (2007) Entanglement of southern elephant seals in squid fishing gear. Mar Mamm Sci 23:414–418

Campanale C, Massarelli C, Savino I, Locaputo V, Uricchio VF (2020) A detailed review study on potential effects of microplastics and additives of concern on human health. Int J Environ Res Public Health 17:1212

Carbery M, O'Connor W, Palanisami T (2018) Trophic transfer of microplastics and mixed contaminants in the marine food web and implications for human health. Environ Int 115:400–409

Carey MJ (2011) Intergenerational transfer of plastic debris by short-tailed shearwaters *Ardenna tenuirostris*. Emu—Austral Ornithol 111:229–234

Caron AGM, Thomas CR, Berry KLE, Motti CA, Ariel E, Brodie JE (2018) Ingestion of microplastic debris by green sea turtles (*Chelonia mydas*) in the Great Barrier Reef: validation of a sequential extraction protocol. Mar Pollut Bull 127:743–751

Carr SA, Liu J, Tesoro AG (2016) Transport and fate of microplastic particles in wastewater treatment plants. Water Res 91:174–182

Casabianca S, Capellacci S, Giacobbe MG, Dell'Aversano C, Tartaglione L, Varriale F, Narizzano R, Risso F, Moretto P, Dagnino A, Bertolotto R, Barbone E, Ungaro N, Penna A (2019) Plastic-associated harmful microalgal assemblages in marine environment. Environ Pollut 244:617–626

Cedervall T, Hansson L-A, Lard M, Frohm B, Linse S (2012) Food chain transport of nanoparticles affects behaviour and fat metabolism in fish. PLoS ONE 7:6

Chen C-L (2015) Regulation and management of marine litter. In: Bergmann M, Gutow L, Klages M (eds) Marine anthropogenic litter. Springer International Publishing, Cham, pp 395–428

Chiappone M, Dienes H, Swanson DW, Miller SL (2005) Impacts of lost fishing gear on coral reef sessile invertebrates in the Florida Keys National Marine Sanctuary. Biol Cons 121:221–230

Chiba S, Saito H, Fletcher R, Yogi T, Kayo M, Miyagi S, Ogido M, Fujikura K (2018) Human footprint in the abyss: 30 year records of deep-sea plastic debris. Mar Policy 96:204–212

Chubarenko I, Bagaev A, Zobkov M, Esiukova E (2016) On some physical and dynamical properties of microplastic particles in marine environment. Mar Pollut Bull 108:105–112

Claessens M, Van Cauwenberghe L, Vandegehuchte MB, Janssen CR (2013) New techniques for the detection of microplastics in sediments and field collected organisms. Mar Pollut Bull 70:227–233

Cole M, Galloway TS (2015) Ingestion of nanoplastics and microplastics by Pacific Oyster larvae. Environ Sci Technol 49:14625–14632

Cole M, Lindeque P, Fileman E, Halsband C, Galloway TS (2015) The impact of polystyrene microplastics on feeding, function and fecundity in the marine copepod *Calanus helgolandicus*. Environ Sci Technol 49:1130–1137

Cole M, Lindeque P, Fileman E, Halsband C, Goodhead R, Moger J, Galloway TS (2013) Microplastic ingestion by zooplankton. Environ Sci Technol 47:6646–6655

Cole M, Lindeque P, Halsband C, Galloway TS (2011) Microplastics as contaminants in the marine environment: a review. Mar Pollut Bull 62:2588–2597

Corcoran PL, Biesinger MC, Grifi M (2009) Plastics and beaches: a degrading relationship. Mar Pollut Bull 58:80–84

Cowger W, Booth AM, Hamilton BM, Thaysen C, Primpke S, Munno K, Lusher AL, Dehaut A, Vaz VP, Liboiron M, Devriese LI, Hermabessiere L, Rochman C, Athey SN, Lynch JM, De Frond H, Gray A, Jones OAH, Brander S, Steele C, Moore S, Sanchez A, Nel H (2020) Reporting guidelines to increase the reproducibility and comparability of research on microplastics. Appl Spectrosc 74:1066–1077

Cox KD, Covernton GA, Davies HL, Dower JF, Juanes F, Dudas SE (2019) Human consumption of microplastics. Environ Sci Technol 53:7068–7074

Cozar A, Echevarria F, Gonzalez-Gordillo JI, Irigoien X, Ubeda B, Hernandez-Leon S, Palma AT, Navarro S, Garcia-de-Lomas J, Ruiz A, Fernandez-de-Puelles ML, Duarte CM (2014) Plastic debris in the open ocean. Proc Natl Acad Sci 111:10239–10244

Critchell K, Bauer-Civiello A, Benham C, Berry K, Eagle L, Hamann M, Hussey K, Ridgway T (2019) Plastic pollution in the coastal environment: current challenges and future solutions. In: Wolanski E, Day J, Elliott M, Ramachandran R (eds) Coasts and estuaries. Elsevier, Amsterdam, pp 595–609

Critchell K, Hoogenboom MO (2018) Effects of microplastic exposure on the body condition and behaviour of planktivorous reef fish (*Acanthochromis polyacanthus*). PLoS ONE 13:e0193308

De-la-Torre GE (2020) Microplastics: an emerging threat to food security and human health. J Food Sci Technol 57:1601–1608

Derraik JGB (2002) The pollution of the marine environment by plastic debris: a review. Mar Pollut Bull 44:842–852

Edo C, Gonzalez-Pleiter M, Leganes F, Fernandez-Pinas F, Rosal R (2019) Fate of microplastics in wastewater treatment plants and their environmental dispersion with effluent sludge. Environ Pollut, 113837

Efimova I, Bagaeva M, Bagaev A, Kileso A, Chubarenko IP (2018) Secondary microplastics generation in the sea swash zone with coarse bottom sediments: laboratory experiments. Front Mar Sci 5:313

Ellen MacArthur Foundation (2017) The new plastics economy: rethinking the future of plastics and catalysing action. Available at: ▶ https://ellenmacarthurfoundation.org/the-new-plastics-economy-rethinking-the-future-of-plastics-and-catalysing. Accessed 7 Feb 2022

Endo S, Takizawa R, Okuda K, Takada H, Chiba K, Kanehiro H, Ogi H, Yamashita R, Date T (2005) Concentration of polychlorinated biphenyls (PCBs) in beached resin pellets: variability among individual particles and regional differences. Mar Pollut Bull 50:1103–1114

Engler RE (2012) The complex interaction between marine debris and toxic chemicals in the ocean. Environ Sci Technol 46:12302–12315

Enyoh CE, Shafea L, Verla AW, Verla EN, Qingyue W, Chowdhury T, Paredes M (2020) Microplastics exposure routes and toxicity studies to ecosystems: an overview. Environ Anal Health Toxicol 35(1):e2020004

Eriksen M, Borgogno F, Villarrubia-Gómez P, Anderson E, Box C, Trenholm N (2020) Mitigation strategies to reverse the rising trend of plastics in polar regions. Environ Int 139:105704

Eriksen M, Lebreton LCM, Carson HS, Thiel M, Moore CJ, Borerro JC, Galgani F, Ryan PG, Reisser J (2014) Plastic pollution in the world's oceans: more than 5 trillion plastic pieces weighing over 250,000 tons afloat at sea. PLoS ONE 9:e111913

Eriksen M, Maximenko N, Thiel M, Cummins A, Lattin G, Wilson S, Hafner J, Zellers A, Rifman S (2013) Plastic pollution in the South Pacific subtropical gyre. Mar Pollut Bull 68:71–76

Erni-Cassola G, Zadjelovic V, Gibson MI, Christie-Oleza JA (2019) Distribution of plastic polymer types in the marine environment; A meta-analysis. J Hazard Mater 369:691–698

Fauser P, Strand J, Vorkamp K (2020) Risk assessment of added chemicals in plastics in the Danish marine environment. Mar Pollut Bull 157:111298

Fazey FMC, Ryan PG (2016) Biofouling on buoyant marine plastics: an experimental study into the effect of size on surface longevity. Environ Pollut 210:354–360

Feng L, He L, Jiang S, Chen J, Zhou C, Qian Z-J, Hong P, Sun S, Li C (2020) Investigating the composition and distribution of microplastics surface biofilms in coral areas. Chemosphere 252:126565

Ferraro G, Failler P (2020) Governing plastic pollution in the oceans: institutional challenges and areas for action. Environ Sci Policy 112:453–460

Foley CJ, Feiner ZS, Malinich TD, Höök TO (2018) A meta-analysis of the effects of exposure to microplastics on fish and aquatic invertebrates. Sci Total Environ 631–632:550–559

Forrest A (2019) Eliminating plastic pollution: how a voluntary contribution from industry will drive the circular plastics economy. Front Mar Sci 6:11

Fossi MC, Panti C, Guerranti C, Coppola D, Giannetti M, Marsili L, Minutoli R (2012) Are baleen whales exposed to the threat of microplastics? A case study of the Mediterranean fin whale (*Balaenoptera physalus*). Mar Pollut Bull 64:2374–2379

Fowler CW (1987) Marine debris and northern fur seals: a case study. Mar Pollut Bull 18:326–335

Fred-Ahmadu OH, Bhagwat G, Oluyoye I, Benson NU, Ayejuyo OO, Palanisami T (2020) Interaction of chemical contaminants with microplastics: principles and perspectives. Sci Total Environ 706:135978

Fry M, Fefer S, Sileo L (1987) Ingestion of plastic debris by Laysan albatross and wedge-tailed shearwaters in the Hawaiian Islands. Mar Pollut Bull 18:339–343

Gallo F, Fossi C, Weber R, Santillo D, Sousa J, Ingram I, Nadal A, Romano D (2018) Marine litter plastics and microplastics and their toxic chemicals components: the need for urgent preventive measures. Environ Sci Eur 30:13

Gardon T, Huvet A, Paul-Pont I, Cassone A-L, Sham Koua M, Soyez C, Jezequel R, Receveur J, Le Moullac G (2020) Toxic effects of leachates from plastic pearl-farming gear on embryo-larval development in the pearl oyster *Pinctada margaritifera*. Water Res 179:115890

Germanov ES, Marshall AD, Hendrawan IG, Admiraal R, Rohner CA, Argeswara J, Wulandari R, Himawan MR, Loneragan NR (2019) Microplastics on the menu: plastics pollute Indonesian manta ray and whale shark feeding grounds. Front Mar Sci 6:00679

GESAMP (Joint Group of Experts in the Scientific Aspects of Marine Environmental Protection) (2015) Sources, fate and effects of microplastics in the marine environment: a global assessment. In: Kershaw PJ (ed) No. IMO/FAO/UNESCO-IOC/UNIDO/WMO/IAEA/UN/UNEP/UNDP. Rep. Study GESAMP No. 90, p 96

Geyer R, Jambeck JR, Law KL (2017) Production, use, and fate of all plastics ever made. Sci Adv 3:e1700782

Gilbert J, Reichelt-Brushett A, Bowling A, Christidis L (2015) Plastic ingestion in marine and coastal bird species of Southeastern Australia. Mar Ornithol 44:21–26

Good TP, June JA, Etnier MA, Broadhurst G (2010) Derelict fishing nets in Puget Sound and the Northwest Straits: patterns and threats to marine fauna. Mar Pollut Bull 60:39–50

Gray AD, Weinstein JE (2017) Size- and shape-dependent effects of microplastic particles on adult daggerblade grass shrimp (*Palaemonetes pugio*): uptake and retention of microplastics in grass shrimp. Environ Toxicol Chem 36:3074–3080

Gregory MR (2009) Environmental implications of plastic debris in marine settings—entanglement, ingestion, smothering, hangers-on, hitch-hiking and alien invasions. Philos Trans R Soc B 364:2013–2025

Gunn R, Hardesty BD, Butler J (2010) Tackling 'ghost nets': local solutions to a global issue in northern Australia: FEATURE. Ecol Manag Restor 11:88–98

Guo X, Wang J (2019) The chemical behaviors of microplastics in marine environment: a review. Mar Pollut Bull 142:1–14

Hermabessiere L, Dehaut A, Paul-Pont I, Lacroix C, Jezequel R, Soudant P, Duflos G (2017) Occurrence and effects of plastic additives on marine environments and organisms: a review. Chemosphere 182:781–793

Höfer R, Selig M (2012) Green chemistry and green polymer chemistry. In: Matyjaszewski K, Möller M, McGrath JE, Hickner MA, Höfer R (eds) Polymer science: a comprehensive reference, vol 10. Elsevier Science, Amsterdam, pp 5–14

Hopewell J, Dvorak R, Kosior E (2009) Plastics recycling: challenges and opportunities. Philos Trans R Soc B 364:2115–2126

Howell EA, Bograd SJ, Morishige C, Seki MP, Polovina JJ (2012) On North Pacific circulation and associated marine debris concentration. Mar Pollut Bull 65:16–22

Isobe A, Kubo K, Tamura Y, Kako S, Nakashima E, Fujii N (2014) Selective transport of microplastics and mesoplastics by drifting in coastal waters. Mar Pollut Bull 89:324–330

Ivar do Sul JA, Costa MF (2014) The present and future of microplastic pollution in the marine environment. Environ Pollut 185:352–364

Jacobsen JK, Massey L, Gulland F (2010) Fatal ingestion of floating net debris by two sperm whales (*Physeter macrocephalus*). Mar Pollut Bull 60:765–767

Jambeck JR, Geyer R, Wilcox C, Siegler TR, Perryman M, Andrady A, Narayan R, Law KL (2015) Plastic waste inputs from land into the ocean. Science 347:768–771

Kaiser D, Kowalski N, Waniek JJ (2017) Effects of biofouling on the sinking behavior of microplastics. Environ Res Lett 12:124003

Kaiser MJ, Collie JS, Hall SJ, Jennings S, Poiner IR (2003) Impacts of fishing gear on marine benthic habitats. In: Sinclair M, Valdimarsson G (eds) Responsible Fisheries in the Marine Ecosystem. CABI, Wallingford, pp 197–217

Kane IA, Clare MA (2019) Dispersion, accumulation, and the ultimate fate of microplastics in deep-marine environments: a review and future directions. Front Earth Sci 7:00080

Kaposi KL, Mos B, Kelaher BP, Dworjanyn SA (2014) Ingestion of microplastic has limited impact on a marine larva. Environ Sci Technol 48:1638–1645

Kelly A, Lannuzel D, Rodemann T, Meiners KM, Auman HJ (2020) Microplastic contamination in east Antarctic sea ice. Mar Pollut Bull 154:111130

Kiessling T, Gutow L, Thiel M (2015) Marine litter as habitat and dispersal vector. In: Bergmann M, Gutow L, Klages M (eds) Marine anthropogenic litter. Springer International Publishing, Cham, pp 141–181

Kirstein IV, Kirmizi S, Wichels A, Garin-Fernandez A, Erler R, Löder M, Gerdts G (2016) Dangerous hitchhikers? Evidence for potentially pathogenic *Vibrio* spp. on microplastic particles. Mar Environ Res 120:1–8

Koelmans AA, Bakir A, Burton GA, Janssen CR (2016) Microplastic as a vector for chemicals in the aquatic environment: critical review and model-supported reinterpretation of empirical studies. Environ Sci Technol 50:3315–3326

Kowalski N, Reichardt AM, Waniek JJ (2016) Sinking rates of microplastics and potential implications of their alteration by physical, biological, and chemical factors. Mar Pollut Bull 109:310–319

Kroon FJ, Kuhnert PM, Henderson BL, Wilkinson SN, Kinsey-Henderson A, Abbott B, Brodie JE, Turner RDR (2012) River loads of suspended solids, nitrogen, phosphorus and herbicides delivered to the Great Barrier Reef lagoon. Mar Pollut Bull 65:167–181

Kroon FJ, Motti CE, Jensen LH, Berry KLE (2018) Classification of marine microdebris: a review and case study on fish from the Great Barrier Reef, Australia. Sci Rep 8:16422

Kühn S, van Franeker JA (2020) Quantitative overview of marine debris ingested by marine megafauna. Mar Pollut Bull 151:110858

Kvale KF, Friederike Prowe AE, Oschlies A (2020) A critical examination of the role of marine snow and zooplankton fecal pellets in removing ocean surface microplastic. Front Mar Sci 6:00808

Laist DW (1987) Overview of the biological effects of lost and discarded plastic debris in the marine environment. Mar Pollut Bull 18:319–326

Lamb JB, Willis BL, Fiorenza EA, Couch CS, Howard R, Rader DN, True JD, Kelly LA, Ahmad A, Jompa J, Harvell CD (2018) Plastic waste associated with disease on coral reefs. Science 359:460–462

Laverty AL, Primpke S, Lorenz C, Gerdts G, Dobbs FC (2020) Bacterial biofilms colonizing plastics in estuarine waters, with an emphasis on *Vibrio* spp. and their antibacterial resistance. PLoS One 15:e0237704.

Law KL (2017) Plastics in the marine environment. Ann Rev Mar Sci 9:205–229

Law KL, Moret-Ferguson S, Maximenko NA, Proskurowski G, Peacock EE, Hafner J, Reddy CM (2010) Plastic accumulation in the North Atlantic Subtropical Gyre. Science 329:1185–1188

Law KL, Morét-Ferguson SE, Goodwin DS, Zettler ER, DeForce E, Kukulka T, Proskurowski G (2014) Distribution of surface plastic debris in the Eastern Pacific Ocean from an 11-year data set. Environ Sci Technol 48:4732–4738

Lebreton L, Slat B, Ferrari F, Sainte-Rose B, Aitken J, Marthouse R, Hajbane S, Cunsolo S, Schwarz A, Levivier A, Noble K, Debeljak P, Maral H, Schoeneich-Argent R, Brambini R, Reisser J (2018) Evidence that the Great Pacific Garbage Patch is rapidly accumulating plastic. Sci Rep 8:4666

Lee K-W, Shim WJ, Kwon OY, Kang J-H (2013) Size-Dependent effects of micro polystyrene particles in the marine copepod *Tigriopus japonicus*. Environ Sci Technol 47:11278–11283

Legislative Assembly of Ontario (2015) Microbead elimination and monitoring act. Available at: ► https://www.ola.org/en/legislative-business/bills/parliament-41/session-1/bill-75. Accessed 7 Feb 2022

Leire C, McCormick K, Richter JL, Arnfalk P, Rodhe H (2016) Online teaching going massive: input and outcomes. J Clean Prod 123:230–233

Lenaker PL, Baldwin AK, Corsi SR, Mason SA, Reneau PC, Scott JW (2019) Vertical distribution of microplastics in the water column and surficial sediment from the Milwaukee River Basin to Lake Michigan. Environ Sci Technol 53:12227–12237

Li J, Yang D, Li L, Jabeen K, Shi H (2015) Microplastics in commercial bivalves from China. Environ Pollut 207:190–195

Liquete C, Piroddi C, Drakou EG, Gurney L, Katsanevakis S, Charef A, Egoh B (2013) Current status and future prospects for the assessment of marine and coastal ecosystem services: a systematic review. PLoS ONE 8:e67737

Löhr A, Savelli H, Beunen R, Kalz M, Ragas A, Van Belleghem F (2017) Solutions for global marine litter pollution. Curr Opin Environ Sustain 28:90–99

Lomonaco T, Manco E, Corti A, La Nasa J, Ghimenti S, Biagini D, Di Francesco F, Modugno F, Ceccarini A, Fuoco R, Castelvetro V (2020) Release of harmful volatile organic compounds (VOCs) from photo-degraded plastic debris: a neglected source of environmental pollution. J Hazard Mater 394:122596

Loulad S, Houssa R, Rhinane H, Boumaaz A, Benazzouz A (2017) Spatial distribution of marine debris on the seafloor of Moroccan waters. Mar Pollut Bull 124:303–313

Lusher AL, Munno K, Hermabessiere L, Carr S (2020) Isolation and extraction of microplastics from environmental samples: an evaluation of practical approaches and recommendations for further harmonization. Appl Spectrosc 74:1049–1065

Lusher AL, Welden NA, Sobral P, Cole M (2017) Sampling, isolating and identifying microplastics ingested by fish and invertebrates. Anal Methods 9

Lv L, Feng L, Jiang S, Lu Z, XIe H, Sun S, Chen J, Li C (2019) Challenge for the detection of microplastics in the environment. Water Environ Res 93(1):5–15

Martin C, Almahasheer H, Duarte CM (2019) Mangrove forests as traps for marine litter. Environ Pollut 247:499–508

Mato Y, Isobe T, Takada H, Kanehiro H, Ohtake C, Kaminuma T (2001) Plastic resin pellets as a transport medium for toxic chemicals in the marine environment. Environ Sci Technol 35:318–324

Maximenko N, Hafner J, Niiler P (2012) Pathways of marine debris derived from trajectories of Lagrangian drifters. Mar Pollut Bull 65:51–62

McGivney E, Cederholm L, Barth A, Hakkarainen M, Hamacher-Barth E, Ogonowski M, Gorokhova E (2020) Rapid physicochemical changes in microplastic induced by biofilm formation. Front Bioeng Biotechnol 8:00205

McIlgorm A, Campbell HF, Rule MJ (2011) The economic cost and control of marine debris damage in the Asia-Pacific region. Ocean Coast Manag 54:643–651

Miller ME, Kroon FJ, Motti CA (2017) Recovering microplastics from marine samples: a review of current practices. Mar Pollut Bull 123:6–18

Min K, Cuiffi JD, Mathers RT (2020) Ranking environmental degradation trends of plastic marine debris based on physical properties and molecular structure. Nat Commun 11:727

Moore RC, Loseto L, Noel M, Etemadifar A, Brewster JD, MacPhee S, Bendell L, Ross PS (2020) Microplastics in beluga whales (*Delphinapterus leucas*) from the Eastern Beaufort Sea. Mar Pollut Bull 150:110723

Murray CC, Maximenko N, Lippiatt S (2018) The influx of marine debris from the Great Japan Tsunami of 2011 to North American shorelines. Mar Pollut Bull 132:26–32

Napper IE, Bakir A, Rowland SJ, Thompson RC (2015) Characterisation, quantity and sorptive properties of microplastics extracted from cosmetics. Mar Pollut Bull 99:178–185

Napper IE, Thompson RC (2020) Plastic debris in the marine environment: history and future challenges. Global Chall 4:1900081

Napper IE, Thompson RC (2016) Release of synthetic microplastic plastic fibres from domestic washing machines: effects of fabric type and washing conditions. Mar Pollut Bull 112:39–45

Nash AD (1992) Impacts of marine debris on subsistence fishermen: an exploratory study. Mar Pollut Bull 24:150–156

Nematollahi MJ, Moore F, Keshavarzi B, Vogt RD, Nasrollahzadeh Saravi H, Busquets R (2020) Microplastic particles in sediments and waters, south of Caspian Sea: frequency, distribution, characteristics, and chemical composition. Ecotoxicol Environ Saf 206:111137

NOAA (National Oceanic and Atmospheric Administration) (2015) Report on the impacts of "ghost fishing" via derelict fishing gear (Marine Debris Program Report). Available at: ▶ https://marinedebris.noaa.gov/sites/default/files/publications-files/Ghostfishing_DFG.pdf. Accessed 7 Feb 2022

Obbard RW (2018) Microplastics in polar regions: the role of long range transport. Curr Opin Environ Sci Health 1:24–29

Oberbeckmann S, Osborn AM, Duhaime MB (2016) Microbes on a bottle: substrate, season and geography influence community composition of microbes colonizing marine plastic debris. PLoS ONE 11:e0159289

Oehlmann J, Schulte-Oehlmann U, Kloas W, Jagnytsch O, Lutz I, Kusk KO, Wollenberger L, Santos EM, Paull GC, Van Look KJW, Tyler CR (2009) A critical analysis of the biological impacts of plasticizers on wildlife. Philos Trans R Soc B 364:2047–2062

Paul-Pont I, Lacroix C, González Fernández C, Hégaret H, Lambert C, Le Goïc N, Frère L, Cassone A-L, Sussarellu R, Fabioux C, Guyomarch J, Albentosa M, Huvet A, Soudant P (2016) Exposure of marine mussels *Mytilus* spp. to polystyrene microplastics: toxicity and influence on fluoranthene bioaccumulation. Environ Pollut 216:724–737

Peeken I, Primpke S, Beyer B, Gütermann J, Katlein C, Krumpen T, Bergmann M, Hehemann L, Gerdts G (2018) Arctic sea ice is an important temporal sink and means of transport for microplastic. Nat Commun 9:1505

Peng G, Bellerby R, Zhang F, Sun X, Li D (2020) The ocean's ultimate trashcan: hadal trenches as major depositories for plastic pollution. Water Res 168:115121

Pham CK, Gomes-Pereira JN, Isidro EJ, Santos RS, Morato T (2013) Abundance of litter on Condor seamount (Azores, Portugal, Northeast Atlantic). Deep Sea Res Part II 98:204–208

Phuong NN, Zalouk-Vergnoux A, Poirier L, Kamari A, Châtel A, Mouneyrac C, Lagarde F (2016) Is there any consistency between the microplastics found in the field and those used in laboratory experiments? Environ Pollut 211:111–123

Pittura L, Gorbi S, Mazzoli C, Nardi A, Benedetti M, Regoli F (2023) Microplastics and Nanoplastics. In: Blasco J and Tovar-Sanchez A (eds.) Marine Analytical Chemistry. Springer Nature, Cham, pp 323–348

Plastic Soup Foundation (2020) Beat the microbead. Available at: ▶ https://www.beatthemicrobead.org/about-us/. Accessed 7 Feb 2022

Plastics Europe (2012) Plastics—the Facts 2012. An analysis of European plastic production, demand. Available at: ▶ https://plasticseurope.org/wp-content/uploads/2021/10/2012-Plastics-the-facts.pdf. Accessed 7 Feb 2022

Plastics Europe (2014) Plastics—the facts 2014/2015. An analysis of European plastic production, demand and waste data. Available at: ▶ https://issuu.com/plasticseuropeebook/docs/final_plastics_the_facts_2014_19122. Accessed 7 Feb 2022

Plastics Europe (2017) Plastics—the facts 2017. Available at: ▶ https://plasticseurope.org/wp-content/uploads/2021/10/2017-Plastics-the-facts.pdf. Accessed 7 Feb 2022

Plastics Europe (2019) Plastics—the facts 2019. An analysis of European plastic production, demand and waste data. Available at: ▶ https://plasticseurope.org/wp-content/uploads/2021/10/2019-Plastics-the-facts.pdf. Accessed 7 Feb 2022

Plastics Europe (2020) What are plastics? Plastics Europe, Association of Plastics Manufacturers. Available at: ▶ https://www.plasticseurope.org/en/about-plastics/what-are-plastics. Accessed 20 June 2020

Porter A, Lyons BP, Galloway TS, Lewis C (2018) Role of marine snows in microplastic fate and bioavailability. Environ Sci Technol 52:7111–7119

Provencher JF, Covernton GA, Moore RC, Horn DA, Conkle JL, Lusher AL (2020) Proceed with caution: the need to raise the publication bar for microplastics research. Sci Total Environ 748:141426

Reisser J, Shaw J, Hallegraeff G, Proietti M, Barnes DKA, Thums M, Wilcox C, Hardesty BD, Pattiaratchi C (2014) Millimeter-sized marine plastics: a new pelagic habitat for microorganisms and invertebrates. PLoS ONE 9:e100289

Reisser J, Slat B, Noble K, du Plessis K, Epp M, Proietti M, de Sonneville J, Becker T, Pattiaratchi C (2015) The vertical distribution of buoyant plastics at sea: an observational study in the North Atlantic Gyre. Biogeosciences 12:1249–1256

Richardson K, Haynes D, Talouli A, Donoghue M (2017) Marine pollution originating from purse seine and longline fishing vessel operations in the Western and Central Pacific Ocean, 2003–2015. Ambio 46:190–200

Rios LM, Moore C, Jones PR (2007) Persistent organic pollutants carried by synthetic polymers in the ocean environment. Mar Pollut Bull 54:1230–1237

Rochman CM, Browne MA, Halpern BS, Hentschel BT, Hoh E, Karapanagioti HK, Rios-Mendoza LM, Takada H, Teh S, Thompson RC (2013) Classify plastic waste as hazardous. Nature 494:169–171

Rochman CM, Browne MA, Underwood AJ, van Franeker JA, Thompson RC, Amaral-Zettler LA (2016) The ecological impacts of marine debris: unraveling the demonstrated evidence from what is perceived. Ecology 97:302–312

Rochman CM, Tahir A, Williams SL, Baxa DV, Lam R, Miller JT, Teh F-C, Werorilangi S, Teh SJ (2015) Anthropogenic debris in seafood: plastic debris and fibers from textiles in fish and bivalves sold for human consumption. Sci Rep 5:14340

Rodrigues A, Oliver DM, McCarron A, Quilliam RS (2019) Colonisation of plastic pellets (nurdles) by *E. coli* at public bathing beaches. Mar Pollut Bull 139:376–380

Santos IR, Friedrich AC, Wallner-Kersanach M, Fillmann G (2005) Influence of socio-economic characteristics of beach users on litter generation. Ocean Coast Manag 48:742–752

Savoca MS, Wohlfeil ME, Ebeler SE, Nevitt GA (2016) Marine plastic debris emits a keystone infochemical for olfactory foraging seabirds. Sci Adv 2:e1600395

Schrank I, Trotter B, Dummert J, Scholz-Böttcher BM, Löder MGJ, Laforsch C (2019) Effects of microplastic particles and leaching additive on the life history and morphology of *Daphnia magna*. Environ Pollut 255:113233

Schuyler QA, Wilcox C, Townsend K, Hardesty B, Marshall N (2014) Mistaken identity? Visual similarities of marine debris to natural prey items of sea turtles. BMC Ecol 14:14

Sfriso AA, Tomio Y, Rosso B, Gambaro A, Sfriso A, Corami F, Rastelli E, Corinaldesi C, Mistri M, Munari C (2020) Microplastic accumulation in benthic invertebrates in Terra Nova Bay (Ross Sea, Antarctica). Environ Int 137:105587

Sigler M (2014) The effects of plastic pollution on aquatic wildlife: current situations and future solutions. Water Air Soil Pollut 225:2184

Silva MM, Maldonado GC, Castro RO, de Sá Felizardo J, Cardoso RP, dos Anjos RM, de Araújo FV (2019) Dispersal of potentially pathogenic bacteria by plastic debris in Guanabara Bay, RJ, Brazil. Mar Pollut Bull 141:561–568

Smith M, Love DC, Rochman CM, Neff RA (2018) Microplastics in seafood and the implications for human health. Current Environmental Health Report 5:375–386

Song YK, Hong SH, Jang M, Han GM, Rani M, Lee J, Shim WJ (2015) A comparison of microscopic and spectroscopic identification methods for analysis of microplastics in environmental samples. Mar Pollut Bull 93:202–209

Suaria G, Avio CG, Mineo A, Lattin GL, Magaldi MG, Belmonte G, Moore CJ, Regoli F, Aliani S (2016) The Mediterranean plastic soup: synthetic polymers in Mediterranean surface waters. Sci Rep 6:37551

Syakti AD (2017) Microplastics monitoring in marine environment. Omni-Akuatika J Fisheries Mar Res 13(2):1–6

Tanaka K, Takada H, Yamashita R, Mizukawa K, Fukuwaka M, Watanuki Y (2013) Accumulation of plastic-derived chemicals in tissues of seabirds ingesting marine plastics. Mar Pollut Bull 69:219–222

ten Brink P, Schweitzer J-P, Watkins E, Howe M (2016) Plastics, marine litter and the circular economy. Institute for European Environmental Policy (IEEP). Available at: ► https://ieep.eu/uploads/articles/attachments/15301621-5286-43e3-88bd-bd9a3f4b849a/IEEP_ACES_Plastics_Marine_Litter_Circular_Economy_briefing_final_April_2017.pdf?v=63664509972. Accessed 7 Feb 2022

Teng J, Zhao J, Zhang C, Cheng B, Koelmans AA, Wu D, Gao M, Sun X, Liu Y, Wang Q (2020) A systems analysis of microplastic pollution in Laizhou Bay, China. Sci Total Environ 745:140815

Teuten EL, Rowland SJ, Galloway TS, Thompson RC (2007) Potential for plastics to transport hydrophobic contaminants. Environ Sci Technol 41:7759–7764

Teuten EL, Saquing JM, Knappe DRU, Barlaz MA, Jonsson S, Björn A, Rowland SJ, Thompson RC, Galloway TS, Yamashita R, Ochi D, Watanuki Y, Moore C, Viet PH, Tana TS, Prudente M, Boonyatumanond R, Zakaria MP, Akkhavong K, Ogata Y, Hirai H, Iwasa S, Mizukawa K, Hagino Y, Imamura A, Saha M, Takada H (2009) Transport and release of chemicals from plastics to the environment and to wildlife. Philos Trans R Soc B 364:2027–2045

Thompson RC, Moore CJ, vom Saal FS, Swan SH (2009) Plastics, the environment and human health: current consensus and future trends. Philos Trans R Soc B 364:2153–2166

Torre M, Digka N, Anastasopoulou A, Tsangaris C, Mytilineou C (2016) Anthropogenic microfibres pollution in marine biota. A new and simple methodology to minimize airborne contamination. Mar Pollut Bull 113:55–61

Turrell WR (2018) A simple model of wind-blown tidal strandlines: how marine litter is deposited on a mid-latitude, macro-tidal shelf sea beach. Mar Pollut Bull 137:315–330

Uhrin AV, Schellinger J (2011) Marine debris impacts to a tidal fringing-marsh in North Carolina. Mar Pollut Bull 62:2605–2610

UNEP (United Nations Environment Programme) (2005) Marine Litter, an Analytical Overview. UNEP, Nairobi, p 47. Available at: ► https://wedocs.unep.org/handle/20.500.11822/8348. Accessed 7 Feb 2022

UNEP (United Nations Environment Programme) (2009) Marine litter, a global challenge. UNEP, Nairobi, p 232. Available at: ► https://wedocs.unep.org/handle/20.500.11822/7787;jsessionid=3685E6D5D2185DE0D914E3DE2A2D6991. Accessed 7 Feb 2022

UNEP (United Nations Environment Programme) (2016) Marine debris: understanding, preventing and mitigating the significant adverse impacts on marine and coastal biodiversity (Technical Series No.83. Secretariat of the Convention on Biological Diversity), p 78. Available at: ► https://www.cbd.int/doc/publications/cbd-ts-83-en.pdf. Accessed 7 Feb 2022

UNEP (United Nations Environment Programme) (2017) Marine litter socio economic study. UNEP, Nairobi, p 114. Available at: ► https://wedocs.unep.org/bitstream/handle/20.500.11822/26014/Marinelitter_socioeco_study.pdf?sequence. Accessed 8 July 2020

United States Congress (2015) H.R.1321—Microbead-Free Waters Act of 2015. Available at: ► https://www.congress.gov/bill/114th-congress/house-bill/1321/text. Accessed 8 Feb 2022

Van Cauwenberghe L, Vanreusel A, Mees J, Janssen CR (2013) Microplastic pollution in deep-sea sediments. Environ Pollut 182:495–499

Van Cauwenberghe L, Janssen CR (2014) Microplastics in bivalves cultured for human consumption. Environ Pollut 193:65–70

van Sebille E, England MH, Froyland G (2012) Origin, dynamics and evolution of ocean garbage patches from observed surface drifters. Environ Res Lett 7:044040

van Sebille E, Wilcox C, Lebreton L, Maximenko N, Hardesty BD, van Franeker JA, Eriksen M, Siegel D, Galgani F, Law KL (2015) A global inventory of small floating plastic debris. Environ Res Lett 10:124006

Vandermeersch G, Van Cauwenberghe L, Janssen CR, Marques A, Granby K, Fait G, Kotterman MJJ, Diogène J, Bekaert K, Robbens J, Devriese L (2015) A critical view on microplastic quantification in aquatic organisms. Environ Res 143:46–55

Veerasingam S, Mugilarasan M, Venkatachalapathy R, Vethamony P (2016) Influence of 2015 flood on the distribution and occurrence of microplastic pellets along the Chennai coast, India. Mar Pollut Bull 109:196–204

Vélez-Rubio GM, Estrades A, Fallabrino A, Tomás J (2013) Marine turtle threats in Uruguayan waters: insights from 12 years of stranding data. Mar Biol 160:2797–2811

Vianello A, Da Ros L, Boldrin A, Marceta T, Moschino V (2018) First evaluation of floating microplastics in the Northwestern Adriatic Sea. Environ Sci Pollut Res 25:28546–28561

Vince J, Hardesty BD (2018) Governance solutions to the tragedy of the commons that marine plastics have become. Front Mar Sci 5:214

Vince J, Stoett P (2018) From problem to crisis to interdisciplinary solutions: plastic marine debris. Mar Policy 96:200–203

von Moos N, Burkhardt-Holm P, Köhler A (2012) Uptake and effects of microplastics on cells and tissue of the blue mussel *Mytilus edulis* L. after an experimental exposure. Environ Sci Technol 46:11327–11335

Walkinshaw C, Lindeque PK, Thompson R, Tolhurst T, Cole M (2020) Microplastics and seafood: lower trophic organisms at highest risk of contamination. Ecotoxicol Environ Saf 190:110066

Waluda CM, Staniland IJ, Dunn MJ, Thorpe SE, Grilly E, Whitelaw M, Hughes KA (2020) Thirty years of marine debris in the Southern Ocean: annual surveys of two island shores in the Scotia Sea. Environ Int 136:105460

Wang J, Tan Z, Peng J, Qiu Q, Li M (2016) The behaviors of microplastics in the marine environment. Mar Environ Res 113:7–17

Wang W, Gao H, Jin S, Li R, Na G (2019a) The ecotoxicological effects of microplastics on aquatic food web, from primary producer to human: a review. Ecotoxicol Environ Saf 173:110–117

Wang X, Wang C, Zhu T, Gong P, Fu J, Cong Z (2019b) Persistent organic pollutants in the polar regions and the Tibetan Plateau: a review of current knowledge and future prospects. Environ Pollut 248:191–208

Wang Z, Dong H, Wang Y, Ren R, Qin X, Wang S (2020) Effects of microplastics and their adsorption of cadmium as vectors on the cladoceran *Moina monogolica* Daday: implications for plastic-ingesting organisms. J Hazard Mater 400:123239

Waring RH, Harris RM, Mitchell SC (2018) Plastic contamination of the food chain: a threat to human health? Maturitas 115:64–68

Watkins E, Hogg D, Mitsios A, Mudgal S, Neubauer A, Reisinger H, Troeltzsch, J, Acoleyen MV (2012) Use of economic instruments and waste management performances. Final report prepared for the European Commission—DG Environment 181. Available at: ► https://www.ecologic.eu/684. Accessed 8 Feb 2022

WEF (World Economic Forum) (2016) Circular economy and material value chains. Available at: ► https://www.weforum.org/projects/circular-economy. Accessed 2 Aug 2020

Wesch C, Bredimus K, Paulus M, Klein R (2016) Towards the suitable monitoring of ingestion of microplastics by marine biota: a review. Environ Pollut 218:1200–1208. ► https://doi.org/10.1016/j.envpol.2016.08.076

whatarethe7continents (2020) Ocean Gyres—formation, maps & more. Available at: ► https://www.whatarethe7continents.com/ocean-gyres-formation-maps-more/. Accessed 10 Feb 2021

Wichmann D, Delandmeter P, van Sebille E (2019) Influence of near-surface currents on the global dispersal of marine microplastic. J Geophys Res Oceans 124:6086–6096

Winn JP, Woodward BL, Moore MJ, Peterson ML, Riley JG (2008) Modeling whale entanglement injuries: an experimental study of tissue compliance, line tension, and draw-length. Mar Mamm Sci 24:326–340

Wong C (2010) A study of plastic recycling supply chain. University of Hull Business School and Logistics Institute, UK, The Chartered Institute of Logistics and Transport. Available at: ► https://ciltuk.org.uk/portals/0/documents/pd/seedcornwong.pdf. Accessed 8 Feb 2022

Woodall L, Sanchez-Vidal A, Canals M, Paterson GLJ, Coppock R, Sleight V, Calafat A, Rogers AD, Narayanaswamy BE, Thompson RC (2014) The deep sea is a major sink for microplastic debris. R Soc Open Sci 1:140317

Woodall LC, Gwinnett C, Packer M, Thompson RC, Robinson LF, Paterson GLJ (2015) Using a forensic science approach to minimize environmental contamination and to identify microfibres in marine sediments. Mar Pollut Bull 95:40–46

Worm B, Barbier EB, Beaumont N, Duffy JE, Folke C, Halpern BS, Jackson JBC, Lotze HK, Micheli F, Palumbi SR, Sala E, Selkoe KA, Stachowicz JJ, Watson R (2006) Impacts of biodiversity loss on ocean ecosystem services. Science 314:787

Wright SL, Kelly FJ (2017) Plastic and human health: a micro issue? Environ Sci Technol 51:6634–6647

Wyles KJ, Pahl S, Thomas K, Thompson RC (2016) Factors that can undermine the psychological benefits of coastal environments: exploring the effect of tidal state, presence, and type of litter. Environ Behav 48:1095–1126

Xanthos D, Walker TR (2017) International policies to reduce plastic marine pollution from single-use plastics (plastic bags and microbeads): a review. Mar Pollut Bull 118:17–26

Yabanlı M, Yozukmaz A, Şener İ, Ölmez ÖT (2019) Microplastic pollution at the intersection of the Aegean and Mediterranean Seas: a study of the Datça Peninsula (Turkey). Mar Pollut Bull 145:47–55

Zettler ER, Mincer TJ, Amaral-Zettler LA (2013) Life in the "Plastisphere": microbial communities on plastic marine debris. Environ Sci Technol 47:7137–7146

Zhao S, Zhu L, Gao L, Li D (2018) Limitations for microplastic quantification in the ocean and recommendations for improvement and standardization. In: Zeng E (ed) Microplastic contamination in aquatic environments. Elsevier, pp 27–49

Radioactivity

Amanda Reichelt-Brushett and Joanne M. Oakes

Contents

10.1   **Introduction – 230**

10.2   **Understanding Radioactivity and Units
of Measurement – 231**
10.2.1   Radioactivity and Radioactive Decay – 231
10.2.2   Alpha, Beta and Gamma Decay – 231
10.2.3   Developing a Measurable Unit – 231
10.2.4   Half-Lives – 233

10.3   **Sources of Radioactivity – 234**
10.3.1   Natural Radioactivity – 234
10.3.2   Anthropogenic Radioactivity – 236
10.3.3   Radioactive Waste Management – 239

10.4   **Effects on Marine Biota – 240**

10.5   **Summary – 242**

10.6   **Study Questions And Activities – 243**

          References – 243

© The Author(s) 2023
A. Reichelt-Brushett (ed.), *Marine Pollution—Monitoring, Management and Mitigation*,
Springer Textbooks in Earth Sciences, Geography and Environment,
https://doi.org/10.1007/978-3-031-10127-4_10

Acronyms and Abbreviations

ERICA	European project Environmental Risk from Ionising Contaminants: Assessment and Management
EW	Exempt waste
HLW	High level waste
IAEA	International Atomic Energy Association
ILW	Intermediate level waste
LLW	Low level waste
NOAA	National Oceanic and Atmospheric Association
NORM	Naturally Occurring Radioactive Materials
TENORM	Technologically Enhanced Naturally Occurring Radioactive Materials
UNSCEAR	United Nations Scientific Committee on the Effects of Atomic Radiation
USA	United States of America
VLLW	Very low level waste
VSLW	Very short lived waste

10.1 Introduction

Co-author Amanda shares an experience from her youth:

"When I was in high school we had a soap box event in the court yard once a week. The soap box was essentially an up-side-down milk crate that you could stand on and talk about anything. This was mostly a student led activity but teachers would sometimes become involved. One of the common topics discussed week-in-week-out was the threat of nuclear war and a following nuclear winter. This threat felt very real to us and after organising a lunchtime viewing of the video The Day After, a fictional story about nuclear war and post war life, it felt even more real and more frightening. It seemed to our young minds at the time that war games being played by Mikhail Gorbachev, president of the Union of Soviet Socialist Republics (USSR), and Ronald Reagan, president of the United States of America (USA), threatened the existence of the world as we knew it. The idea of Cold War resulting in nuclear winter was both literally and figuratively chilling and confronting. Certainly, there were tensions, but the 1980s was also a time of considerable negotiation and over 1987-1988 the Intermediate-Range Nuclear Force Treaty was signed, approved and ratified."

In 2018 Donald Trump, president of the USA at the time, announced that the USA was withdrawing from the Intermediate-Range Nuclear Force Treaty, due to Russian non-compliance and amidst the continuing growth of China's missile forces. The USA formally withdrew from the Treaty on the 2nd of August 2019. Today, in 2023, we face the threat of nuclear force being used in the Russian invasion of Ukraine. With the dissolution of the Soviet Union, Ukraine became an independent nation and gave up its sizeable nuclear arsenal in return for security guarantees offered by the United States and Russia. Ukraine is now vulnerable if those guarantees are not kept. Are we at another crossroads?

The devastating impacts of nuclear warfare were realized with the detonation of two nuclear weapons over Hiroshima and Nagasaki, on August 6th and August 9th, 1945, respectively, in effect ending World War II. The first nuclear weapons test had only occurred in July 1945 in New Mexico, USA. Many tests that followed relied upon the remoteness of uninhabited islands and atolls such as Enewetak and Bikini Atolls in the Marshall Islands, Johnston Atoll near Hawaii, Kiritimati in Kiribati, the archipelago Novaya Zemlya in the Arctic Ocean, Montebello Islands off the northwest coast of Australia, Mururoa and Fangataufa Atolls in French Polynesia, and in the open Pacific and South Atlantic Oceans. Many other tests occurred in remote mountainous areas and underground. Today, nine sovereign states (political entities with one centralized government) are considered to have nuclear weapons capabilities.

Similar technologies are required to make both nuclear weapons and nuclear power. Nuclear fission reactions are slower in a power plant compared to a weapon; however, they both use plutonium-239 and uranium-235, both produce waste, and both have responsibilities for various accidents that have resulted in radioactive pollution in the marine environment. But if we are to gain a rational understanding of nuclear science we must also consider natural sources of radioactivity and other uses of nuclear chemistry, such as for scientific research and medical therapy and diagnosis, all of which result in radioactive waste. There is also the matter of managing waste generated from the nuclear industry. This chapter introduces nuclear chemistry and radioactive pollution in the marine environment from intentional and accidental human activities. It describes how radioactivity is measured, what it is, natural and anthropogenic sources, legacy waste and current waste management practices, and discusses the effects of radioactivity on marine biota.

10.2 Understanding Radioactivity and Units of Measurement

10.2.1 Radioactivity and Radioactive Decay

After the discovery of X-rays by Wilhelm Roentgen (1845–1923) a new field of science emerged. Henri Becquerel (1852–1908) became interested in substances that became luminous after exposure to sunlight. One of these substances was uranium ore and as a result Becquerel discovered another type of radiation. Following from this work, Marie Curie (1867–1934) and her husband Pierre Curie (1859–1906) discovered polonium and radium and named the radiation they produced **radioactivity**.

Radioactivity results from the degradation of unstable atoms to achieve a more stable form. All matter around us is made of atoms, each of which has a nucleus made up of protons and neutrons. Whereas the number of protons (atomic number) is what defines an element (e.g. all atoms of carbon have nuclei containing six protons), the number of neutrons within the atoms of a given element can vary. Some combinations of protons and neutrons result in a nucleus that is unstable, with excess energy stored within it. Different forms of a given element are called **isotopes**. Isotopes of any given element have the same number of protons, but different numbers of neutrons. Isotopes of an element that have a combination of neutrons and protons that is stable are called stable isotopes and do not decay. Isotopes that have an unstable combination of neutrons and protons are called radioactive isotopes, radionuclides, or **radioisotopes**. Over time, these radioisotopes spontaneously lose nuclear material and energy (protons, neutrons, and/or electrons) to achieve a more stable state. This emission of radiation is measured as **radioactivity**.

The loss of nuclear material associated with radiation emission is called **radioactive decay** and results in a new atom, which may be a different element (due to a change in the number of protons) or a different isotope of the same element (due to a change in atomic mass). Often this new atom will have a stable nucleus and no further decay will occur. However, depending on the initial radioisotope and the form of radioactive decay that occurs, the new atom may be another radioisotope. In this circumstance, the atom will undergo further radioactive decay until the nucleus reaches a stable state (i.e. the atom becomes a stable isotope).

10.2.2 Alpha, Beta and Gamma Decay

In the years following the discovery of radioactivity, researchers investigated the properties of radiation. Some of the early experiments were conducted by Ernest Ru-

therford (1871–1937). It was in 1898, while Rutherford was still a student, that he noted two forms of radioactive rays with different abilities to penetrate matter. He named these rays alpha (α) and beta (β) rays, after the first two letters of the Greek alphabet. By mid-1902, this naming scheme had been extended to include gamma (γ) rays, named after the third Greek letter. Alpha, beta and gamma radiation represent the three most common forms of radioactive decay and vary in their properties and characteristics ◘ Fig 10.1.

Alpha rays, or alpha particles, are the most easily absorbed, with the lowest power to penetrate matter. Atoms undergo alpha decay through the loss of two protons and two neutrons from the nucleus. Alpha particles are therefore helium nuclei without electrons and are positively charged. Because they are relatively large, heavy, and strongly charged, alpha particles have a strong tendency to interact and collide with the molecules in matter. This results in their low penetrative power. Alpha rays travel only a few centimetres in air and can be stopped by a sheet of paper.

Beta rays, or beta particles, have a moderate ability to penetrate matter and can be produced through either negative beta decay or positive beta decay. Both of these beta decay processes result in the charge of an atom increasing or decreasing by one unit whilst the atomic mass number remains unchanged. In negative beta decay (electron emission), a neutron within the nucleus of an unstable atom decays into a proton and electron. Whereas the proton from this decay remains in the nucleus, the electron is emitted at high speed and is a negatively charged beta particle. In positive beta decay (positron emission), a proton in the nucleus decays into a neutron, which remains in the nucleus, and a positron is emitted. Positrons have similar properties to electrons, except that they have a positive charge. Beta particles are therefore high-energy, charged, fast-moving, and relatively small, with essentially no mass. These properties allow beta rays to travel some metres through air and mean that beta rays are able to pass through paper but can be absorbed and stopped by human tissue, or around a 0.5 mm sheet of aluminium.

Gamma rays have the greatest penetrative power but, unlike alpha and beta rays, do not consist of particles. Rather, gamma rays consist of photons, which are packets of high-frequency electromagnetic radiation that move in waves. Gamma rays have no mass and can travel indefinitely through air. Thick sheets of lead or metres of concrete are required to stop gamma rays (◘ Figure 10.1).

10.2.3 Developing a Measurable Unit

Many of the units for measuring radiation and radioactivity (◘ Table 10.1) are named after the pioneering scientists of the field—Wilhelm Roentgen (1845–1923),

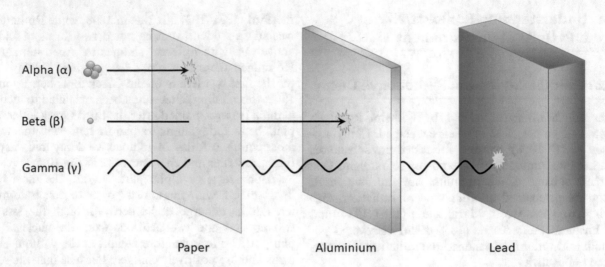

◘ Figure 10.1   Penetrative power of alpha radiation (helium nuclei), beta radiation (high-energy electrons), and gamma rays (photons moving in waves). *Image*: J. Oakes

◘ Table 10.1   Units of radiation and radioactivity

	Measure	Quantity	Unit
Radiation source	Energy of ionising radiation	Radiation energy	Electron volts Joules Ergs
	Amount of radioactivity (number of particles or photons emitted per second)	Activity	Becquerel (Bq)[a] Rutherford (Rd) = 1.0×10^6 Bq Curie (Ci) = 3.7×10^{10} Bq
	Amount of radioactivity per unit mass of a radionuclide	Relative or specific radioactivity	e.g. Bq/mmol or Ci/mmol
Received radiation	Ionisation in air	Exposure	Coulombs per kilogram (C/kg)[a] Roentgen (R)
	Absorbed energy per mass	Absorbed dose	Gray (Gy)[a] Radiation absorbed dose (rad)
	Absorbed dose weighted by type of radiation (measure of effective biological damage)	Equivalent dose	Sievert (Sv)[a] Roentgen equivalent man (rem)

[a] SI unit (unit specified by International System of Units)
Note: 1 Sv = 100 Roentgens = 100 rem, 1 Gy = 100 rad

Henri Becquerel (1852–1908), Marie Curie (1867–1934) and her husband Pierre Curie (1859–1906), and Ernest Rutherford (1871–1937).

The original unit for measuring the amount of radioactivity was the curie (Ci)–first defined to correspond to radioactive decay of one gram of radium-226 but more recently defined as:

1 curie = 3.7×10^{10} radioactive decays per second (exactly)

The International System of Units (SI) has replaced the curie with the becquerel (Bq), where:

1 becquerel = 1 radioactive decay per second = 2.703×10^{-11} Ci

and is the number of nuclei that decay per unit time (◘ Table 10.1). The specific radioactivity or relative radioactivity can be determined as the radioactivity per unit mass of a substance.

Ionising radiation (radiation with enough energy to ionise [or remove electrons from] other atoms) is measured using electron-volts, joules and ergs. The electron-volt (eV) is the energy gained by an electron when it moves from rest through a potential difference of one volt (e.g. the energy an electron gains as it moves from a negative plate to a positive plate with a 1-V higher potential). Electron-volts are a useful unit for expressing very small amounts of energy. One joule (J) is equal to 6.242×10^{18} electron-volts and is equivalent to the amount of energy used by a one-watt light bulb lit for one second. The erg is a unit of energy equal to 6.242×10^{11} electron-volts or 1×10^{-7} J.

There are also other interrelated ways to consider radiation based on the objective of a study. For example, **exposure** describes the amount of radiation traveling through the air and is used in monitoring exposure. The units for exposure are the roentgen (R) and coulomb/kilogram (C/kg). Sometimes we might be interested in the **absorbed dose,** which is the amount of radiation absorbed by a living organism or an object. The units for absorbed dose are the radiation absorbed dose (rad) and gray (Gy). If we are interested in **effective dose** or dose equivalent, we consider both the amount of radiation absorbed and the effect of that radiation. Units for dose equivalent are the roentgen equivalent man (rem) and sievert (Sv). Biological dose equivalents are commonly measured in 1/1000th of a rem (known as a millirem or mrem). They are influenced by the penetrating power of alpha, beta and gamma radiation.

10.2.4 Half-Lives

In addition to quantifying and describing radioactivity and dose, as described in ▶ Section 10.2.3, it is useful to be able to express how slowly or rapidly radioactive material decays.

All unstable atoms will undergo radioactive decay at some point, but this decay is a random event; it is impossible to predict at what point in time any given atom will decay. However, for a very large number of atoms, the number of nuclei that will decay in a given period of time is predictable. The proportion of atoms decaying in a given period of time remains constant, (i.e. the number of atoms of a given radioisotope remaining within a sample reduces exponentially over time). This radioactive decay is expressed in terms of **half-life**.

The half-life of a radioisotope is defined as the time taken for half of the radioactive nuclei within a sample of that isotope to decay or the time taken for the activity (the number of decays per unit of time) to halve (▶ Box 10.1). Each radioisotope has a specific half-life, which may be anywhere from microseconds to hundreds of years, or even longer. In fact, the half-life of some radioisotopes is so long that they have remained in their current state since before the Earth was formed, and some isotopes have half-lives that are longer than the age of the universe. For example, bismuth-209 has recently been found to have a half-life of 1.9×10^{19} years, whereas the universe is estimated to have an age of only 1.38×10^{10} years.

Knowing the **decay rate** of radioisotopes has a number of practical applications. For example, decay rates can allow us to determine how long an environment, plant, or animal contaminated by radioactive waste will remain hazardous and can allow us to determine the age of various materials including archaeological artefacts, sediment, etc. (e.g. ^{14}C dating; ▶ Box 10.2).

Box 10.1: Understanding Half-lives

The half-life of a radioisotope is the amount of time that it takes for one-half of the original number of atoms to undergo radioactive decay to form a new element. For example, lead-210 decays to Bi-210 according to the nuclear equation below.

$$^{210}_{82}\text{Pb} \longrightarrow \, ^{210}_{83}\text{Bi} + \, ^{0}_{-1}\text{e}$$

The half-life of lead-210 is 22.2 years so the radioactivity halves every 22.2 years, as shown in ◻ Figure 10.2. You can calculate the remaining radioactivity for any given time period using a simple equation, demonstrated here:

You have 150 g of lead-210. How much lead-210 remains after 92 years?

$$\frac{92}{22.2} = 4.14 \text{ half lives}$$

Fraction remaining after 4.14 half-lives:

$$\frac{1}{2^n} = \frac{1}{2^{4.14}} = \frac{1}{18}$$

$n =$ number of half-lives

The amount of lead-210 remaining $\left(\frac{1}{18}\right)(150)$ g $= 8.33$ g

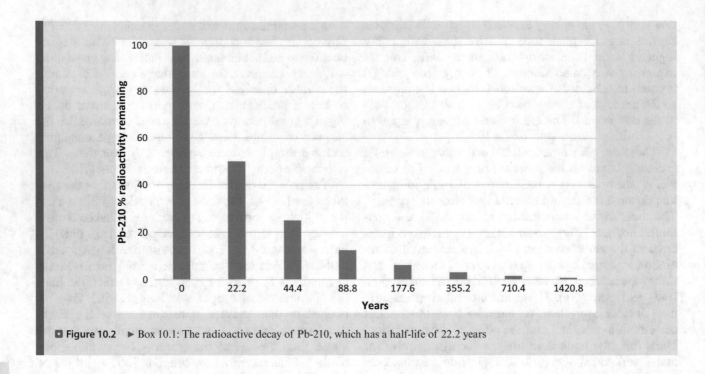

Figure 10.2   ► Box 10.1: The radioactive decay of Pb-210, which has a half-life of 22.2 years

10.3 Sources of Radioactivity

Radioactive substances occur naturally across the whole biosphere, and life has evolved in this radioactive environment. The natural background levels provide a reference for acceptable levels and are important to understand before we attempt to measure anthropogenic increases. Radioactive elements may be found in differing concentrations around the world as a result of natural and anthropogenic processes. To date, around 3000 natural and artificial radioisotopes have been identified.

10.3.1 Natural Radioactivity

Radioisotopes of naturally occurring elements comprise **Naturally Occurring Radioactive Materials (NORM)** and are ubiquitous in the environment, occurring in soil, sand, clay, rocks, air, water, and the tis-

sues of plants and animals. These radioisotopes undergo radioactive decay that results in one or more types of radiation. Cosmic rays from the sun and outer space are referred to as ionising radiation and constantly bombard the Earth. Most naturally occurring radioactive substances (predominantly radium and radon) are the result of uranium and thorium decay. They may be mobilised, redistributed and concentrated by human activities such as fossil fuel mining and burning and fertiliser mining. When NORM are concentrated, or the potential for exposure has been enhanced, due to human activities they are termed **Technologically Enhanced Naturally Occurring Radioactive Materials (TENORM)** (Ojovan et al. 2019).

Naturally occurring radioisotopes have been used in the environmental sciences for over 150 years and enable the study of processes from a cellular level to broad oceanic scales (see ► Box 10.2). They can be applied both in field and laboratory studies (◻ Table 10.2). The

◻ **Table 10.2**   Selected naturally-occurring radioisotopes that are used in industry and science

Radioisotope	Half-life	Uses
Radon-222	3.82 days	Detecting and quantifying groundwater input to estuaries
Beryllium-7	53.22 days	Determining age of water and sediment
Lead-210	22.2 years	Dating layers of sand and soil laid down up to 80 years ago
Carbon-14	5700 years	Measuring the age of organic material up to 50,000 years old
Chlorine-36	301,000 years	Measuring sources of chloride and the age of water up to 2 million years old
Beryllium-10	1.39 mill years	Investigating soil formation and erosion rates, time of rock exposure (exposure dating), and dating of layers within ice cores
Uranium-235	704 mill years	Used in nuclear reactors, nuclear weapons, nuclear powered submarines

evolving field of **radioecology** has had a strong focus on marine research since the 1970s, and new applications are expanding the research scope (Cresswell et al. 2020). Studies using naturally occurring radioisotopes are useful for understanding the chronological formation of the Earth, sedimentology, contaminant behaviour, nutrient transport through food chains, global element cycles, defining natural and anthropogenic sources of nutrients, industry compliance, ecotoxicology, and remediation success, among others (e.g. Vandecasteele 2004; Call et al. 2015; Riekenberg et al. 2020; Cresswell et al. 2020). Indeed, the Journal of Environmental Radioactivity, established in 1984, is dedicated to this research field.

Box 10.2: Radioisotopes in Environmental Science: Nutrients Release 6000 Year Old Carbon from Coastal Sediment

The carbon-14 radioisotope is commonly used to date artefacts of biological origin from up to about 50,000 years ago. This technique, called radiocarbon dating or carbon-14 dating, can also be used to determine the age of organic matter in marine sediment. Radiocarbon dating uses the ratio of carbon-14 (^{14}C) to the common, stable form of carbon (^{12}C). In living plants and animals, which are constantly taking in new carbon, the ^{12}C:^{14}C ratio is relatively constant. However, once a plant or animal dies, and no new carbon is taken in, the amount of ^{14}C in its tissues begins to decline due to radioactive decay. Because the amount of ^{12}C remains unchanged, there is a shift in the ^{12}C:^{14}C ratio. Based on this shift, and the known half-life of ^{14}C (5700 years) it is possible to estimate how much time has elapsed since organic matter was part of a living thing.

Radiocarbon dating was used in a recent study looking at the impact on coastal systems of nutrients, which are increasing in coastal and marine systems globally due to human activities (Rockström et al. 2009). Riekenberg et al. (2020) observed that coastal sediments subjected to high concentrations of nutrients lost more carbon to the overlying water than unaffected sediments (Figure 10.3). This is concerning, given that coastal sediments are increasingly recognised as important sites for storage of excess carbon. Release of stored carbon from coastal sediments could increase atmospheric carbon dioxide concentrations, contributing to climate change. However, the source of the extra carbon lost from the nutrient-impacted sediments was unknown; was it stored (old) carbon, or new carbon (e.g. produced in the sediment by microalgae)?

Radiocarbon dating showed that the carbon lost from nutrient-impacted sediment was around 6000 years old, confirming that increasing nutrient inputs to coastal systems may cause the loss of old, stored carbon from sediments. This use of radiocarbon dating highlighted the potential for coastal nutrient inputs to shift carbon budgets locally and, if nutrient inputs increase more broadly, possibly impact climate change by altering atmospheric carbon concentrations.

 Figure 10.3 ► Box 10.2: Mud flat in the Richmond River, NSW, Australia, where radiocarbon dating showed that excess nutrients cause loss from the sediment of 6000 year old stored carbon (Riekenberg et al. 2020). *Photo*: J. Oakes

---

Mobilisation and Distribution from Agriculture

The agricultural industry is an important potential source of TENORM in the marine environment. Agricultural TENORM are associated with the production of phosphorus-containing fertilisers, which are applied to soil to enhance the growth and production of crops and pastures.

Phosphorus-containing fertilisers are derived from phosphate ore, which **naturally contains small amounts**

of radioisotopes, including uranium, radium, and thorium, and the radioisotopes produced through their decay. During the treatment of phosphate ore to produce fertiliser, some of these radioisotopes transfer to the fertiliser. Phosphorus-containing fertiliser added to fields can therefore increase the concentration of radioisotopes in soils (Pfister et al. 1976; Hameed et al. 2014), sometimes over many decades of application, although in some instances there is no or negligible increase in radioisotope concentration (e.g. Saueia et al. 2006). Radioisotopes in soil have the **potential to transfer to crops** that are consumed by humans, and the potential to enter adjacent waterways via erosion and/or groundwater and ultimately accumulate in the marine environment. Whether or not fertiliser application causes harmful levels of radioisotopes in the environment will depend on the radioactivity of the fertiliser used, its application rate, and biogeochemical characteristics of the soil and receiving environment.

A more significant source of TENORM associated with agriculture is phosphogypsum (hydrated calcium sulphate), which is a solid by-product of phosphorus fertiliser production. Around 100–280 megatonnes of phosphogypsum are produced globally per year (Yang et al. 2009; Parreira et al. 2003), with around 5 tonnes produced per 1 tonne of phosphorus fertiliser (Rutherford et al. 1994). Phosphogypsum has potential application in agriculture as a readily available source of gypsum, which adds calcium and sulfur to the soil, thereby enhancing root penetration (Nisti et al. 2015). However, during fertiliser production, up to 90% of the radioisotopes in phosphate ore, particularly radium (^{226}Ra), selectively transfer to phosphogypsum (Mazzilli et al. 2000). Due to this elevated radioactivity, the use of phosphogypsum in agriculture is restricted and phosphogypsum is typically treated as waste.

Where phosphogypsum is treated as waste, it may be directly discharged to the marine environment (El Kateb et al. 2018; Belahbib et al. 2021) and can cause substantial radioisotope contamination (e.g. Martínez-Aguirre and García-León 1994; Villa et al. 2009). However, this practise has become less common in recent times. Instead, vast quantities of phosphogypsum are stored in large stacks around the world, including in Europe, China, and the USA. These stacks are often near or in the coastal zone (Papaslioti et al. 2020) and leaching from the stacks has the potential to contaminate groundwater, transferring radionuclides to coastal and marine sediment and water (Tayibi et al. 2009). This is particularly the case for older stacks that were constructed and operational in the 1990s and earlier, before practises were improved to minimise environmental contamination. Even in these older stacks, however, leached radionuclides can be rapidly attenuated within the underlying sediment due to reactions with, and adsorption to, reactive coastal sediment (e.g.

within 50 cm; Guerrero et al. 2020). Accordingly, for ^{226}Ra in phosphogypsum stacks there is often an initial pulse to the environment, with the remaining ^{226}Ra only slowly dissolved thereafter (Haridasan et al. 2002).

For radioisotopes that enter the marine environment, their distribution and impact, and whether they reach a level that is harmful to humans and ecological communities, is determined by their interaction with salinity and tidal movement (Martínez-Aguirre and García-León 1994), as well as redox conditions and the presence or absence of ion exchangers within the sediment.

10.3.2 Anthropogenic Radioactivity

Of more than 3000 known radioactive isotopes, only around 84 occur naturally. Most radioactive isotopes are artificially produced in reactors and accelerators for the purposes of research, energy generation, and/or medical treatments and diagnosis, or result from radioactive decay of these isotopes (□ Table 10.3). Anthropogenic emissions of radioactive isotopes add to the natural background levels of radioactivity. Much research has explored the risk to the marine environment from the production and distribution of anthropogenic radioactivity throughout the world (see review by Livingston and Povinec 2000), including in the Barents Sea and Arctic Ocean (e.g. Klungsøyr et al. 1995; Macdonald and Brewers 1996), the Western Sea on the Swedish west coast (Lindahl et al. 2003), the Pacific Ocean (e.g. Eigl et al. 2017; Buesseler et al. 2018), the North Atlantic Ocean (Villa-Alfageme et al. 2018), and the Flores Sea and Lombok Strait (Suseno and Wahono 2018). These studies vary not only in location but also in the source of the radioactive risk.

Nuclear Weapons
There are currently nine sovereign states considered to have nuclear capabilities: Russia, USA, France, China, United Kingdom, Israel, Pakistan, India and North Korea. Weapons testing is the predominant form of intentional nuclear emissions, including that arising from their use in war. There are around 13,000 nuclear weapons in the world, primarily in Russia (6255) and the USA (5550), with as few as 40–50 in North Korea (SIPRI 2021). Both the USA and Russia also have the highest stock piles of enriched uranium and separated plutonium (SIPRI 2021). The main nuclear weapon test sites that have resulted in marine contamination are in Novaya Zemlya, the Marshall Islands, Christmas Island, French Polynesia, and Lop Nop (Livingston and Povinec 2000).

Nuclear Energy
The International Atomic Energy Association (IAEA) is an international organization with 171 member states, founded in July 1957, that seeks to promote the

■ Table 10.3   Selected artificially produced radioisotopes that are used in industry and science

Radioisotope	Half-life	Uses
Technetium-99 m	6.01 h	Studying sewage and liquid waste movements. Also used in medical imaging. Produced in 'generators' from the decay of molybdenum-99, which is in turn produced in reactors
Gold-198	2.70 days	Tracing sand movement in river beds and on ocean floors, and studying coastal erosion. Also used to trace factory waste causing ocean pollution, and to study sewage and liquid waste movements
Chromium-51	27.7 days	Tracing sand to study coastal erosion
Ytterbium-169	32.03 days	Used in gamma radiography
Iridium-192	73.83 days	Used in gamma radiography. Also used to trace sand to study coastal erosion
Zinc-65	243.66 days	Predicting the behaviour of heavy metal components in effluents from mining waste water
Manganese-54	312.12 days	Predicting the behaviour of heavy metal components in effluents from mining waste water
Cobalt-60	5.27 years	Used in gamma radiography, gauging, commercial medical equipment sterilisation, and cancer treatment. Also used to irradiate fruit fly larvae in order to contain and eradicate outbreaks, as an alternative to the use of toxic pesticides. Used to irradiate some foods to extend shelf-life
Hydrogen-3 (tritium)	12.32 years	Used as a tracer in tritiated water to study sewage and liquid wastes, animal metabolism, and in biochemical research. Also used for luminous (glow in the dark) dials
Cesium-137	30.08 years	Radiotracer to identify sources of soil erosion and depositing, and also used for thickness gauging. Also a marker for sediment deposited in the mid-1960s (which had high Cs-137 levels due to nuclear bomb fallout) contributing to dating of sediment layers and quantification of subsequent rates of sedimentation
Americium-241	432.5 years	Used in neutron gauging and smoke detectors
Sodium-24	15 h	Detection of leaks in pipes
Sulphur-35	87.5 days	Determining sulphate reduction rates in coastal sediments
Fluorine-18	109.7 min	Used in medical imaging as a positron source for positron emission tomography (PET) scans
Calcium-47	4.5 days	Investigating bone metabolism
Californium-252	2.6 years	Used in cancer treatment, detection of gold and silver ore, portable metal detectors, detection of metal fatigue and stress
Iodine-131	8.04 days	Treatment of overactive thyroid and thyroid cancer. Also used in diagnostic imaging. Also used as an industrial tracer
Gadolinium-153	241.6 days	Used as a contrast agent in Magnetic Resonance Imaging (MRI)

peaceful use of nuclear energy, and to inhibit its use for any military purpose, including nuclear weapons. According to the IAEA (2021) Annual Report there were 437 operational nuclear power reactors in the world. The global use of nuclear energy is growing with 56 reactors currently under construction. The IAEA (2019) predicted that nuclear power capacity will increase by 12–25% by 2030 and up to 80% by 2050. Still, there is not a commonly agreed solution to the growing nuclear waste problem (Choudri and Baawain 2016). Furthermore, in the context of the marine environment **nuclear reactors require large volumes of water in their cooling towers** and ocean water is often used as a cheap and suitable source, avoiding the consumption of freshwater resources. A consequence of this is that many reactors are located on coastlines; there is even some discussion in the literature regarding floating nuclear power reactors (Srandring et al. 2009).

Nuclear energy provides a carbon free energy source and countries with the lowest carbon emissions are those that have a higher dependence on nuclear energy. But this type of energy does not come without its own risks. Unintentional release of radioactive materials can occur because of human and mechanical error (e.g. 1979—Three Mile Island, USA and 1986 -Chernobyl, Ukraine), and due to extreme natural events (e.g. 2011—damage caused by a tsunami generated from an earthquake at Fukushima, Japan ; see ▶ Box 10.3).

10

The production of nuclear energy is controversial and accidents create emotive responses from the public (◘ Figure 10.4). Of course, there is much to be concerned about with the long-term global effects of nuclear accidents, global distribution of fall out and impacts on marine and terrestrial food chains. This concern has resulted in long-term research studies related to accident sites. However sometimes there is a misrepresentation of facts in the media, which leads to heightened public concern and enhanced public anxiety. One outstanding example of this misrepresentation is how the image in ◘ Figure 10.5 was promoted in the media and widely used to show the radiation leakage from the Fukushima accident. However, the National Oceanic and Atmospheric Association (NOAA) actually produced this map to show the maximum wave heights of the tsunami generated by the Japan earthquake on March 11, 2011. That being said, research studies have shown enhanced levels of radioactivity derived from the Fukushima accident in locally sourced seafood (Buesseler et al. 2012), and seafood from the North Pacific (e.g. Azouz and Dulai 2017), with some radioactivity transported via fish migration (Madigan et al. 2012). The levels detected have predominantly been below various limits of concern (e.g. Buesseler et al. 2012; Fisher et al. 2013; Azouz and Dulai 2017).

Box 10.3: Radioactive Pollution in the Marine Environment from the Fukushima Dai-ichi Nuclear Power Plant Accident

An earthquake generated tsunami wave seriously damaged the reactors at the Dai-ichi Nuclear Power Plant, Fukushima, Japan, on the 11th March 2011. Like other reactors, the Fukushima reactors have many radioactive elements, but three radioactive isotopes were of particular concern for marine ecosystems after the accident: iodine-131, cesium-137, and cesium-134. Iodine-131 has a half-life of 8 days, which means it is highly radioactive in the short term and was of concern immediately after the accident. Cesium-137 and -134 were released in the largest amounts. Levels 50 million times higher than before the accident were recorded in the ocean, posing a direct threat to marine life at the site. Levels dropped sharply after the first month but ongoing leaks have been indicated (e.g. Inoue 2018). Cesium-137 has a relatively long half-life (30.08 years), but is present in the ocean due to nuclear weapons testing in the 1950s and 1960s. Cesium-134 is much shorter-lived (2.06 years) and therefore, if present in seawater samples, it most likely comes from Fukushima. Tritium-2 (12.3 year half-life) was also measured throughout the western North Pacific at very low concentrations (Kaizer et al. 2018). Although of relatively low concern in regards to health impacts, it is found in stored water at the site even after decontamination processes—management considerations continue.

Most Japanese fisheries were unaffected by the accident, but coastal fisheries nearest the reactors were closed because of concern that some species, particularly those that are benthic and sessile, would be exposed. Biota testing to date still occurs on a regular basis and is compared against Japan's limits for radiation in seafood (which are more stringent than USA regulations). If seafood exceeds these regulations it cannot be sold. Fortunately, the contamination is very localised but, in light of the high consumption of seafood in the Japanese diet, there has been much concern raised within Japan about seafood safety as a result of the accident. A questionnaire, exploring factors affecting consumer behaviour towards seafood from regions near the accident with uncertain risks, highlighted that the consumer class perceiving the highest risk and greatest negativity towards this seafood were parents of young children and of higher academic achievement. Interestingly, environmental awareness and higher age range categories showed a more positive response to seafood from this location indicating that the desire to support the economic recovery of the seafood industry outweighed the risk concerns (Aruga and Wakamatsu 2018). No studies have been published that show consumption of seafood from the impacted area causes serious human health risks.

10.3.3 Radioactive Waste Management

The first sea dumping of radioactive waste took place in 1946 at a site in the North East Pacific Ocean, about 80 km off the coast of California (Calmet 1989).

Dumping of low level waste (LLW) continued for 36 years and occurred as late as 1982, at a site about 550 km off the European continental shelf in the Atlantic Ocean (Calmet 1989). An estimated 63 PBq (1.7 MCi) of radioactive waste coming from research, med-

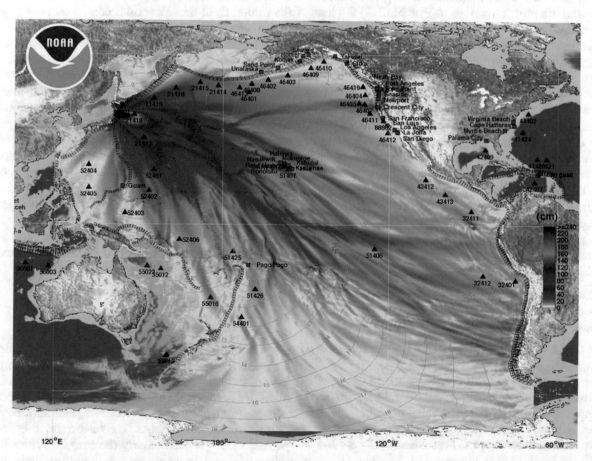

□ **Figure 10.5** This is not radioactive leakage from the Fukushima nuclear accident spreading across the Pacific Ocean. *Image:* created by NOAA's Center for Tsunami Research and graphically shows maximum wave heights of the tsunami generated by the Japan earthquake on March 11, 2011

◘ Table 10.4   Classification of radioactive waste according to the IAEA (2009)

Classification code	General criteria
Exempt waste (EW)	Classification explained in IAEA (2004)
Very short lived waste (VSLW)	Contains only radionuclides with very short half-lives, can be stored until the activity has fallen below the levels for clearance[a]
Very low level waste (VLLW)	Waste arising from decommissioning of nuclear facilities with levels only slightly above specified levels, other waste containing naturally occurring radionuclides[a]
Low level waste (LLW)	Radioactive waste that does not need shielding during normal handling, suitable for near surface disposal[a]
Intermediate level waste (ILW)	Contains long lived radionuclides in quantities that need a greater degree of containment and isolation from the biosphere than provided in near surface disposal[a]
High level waste (HLW)	Contains high concentrations of both short and long lived radionuclides where long term safety needs to be ensured. These are heat generating wastes arising from spent fuel from nuclear reactors. Requires deep geologic disposal[a]

[a] Further details in IAEA (2009)

icine, and other nuclear industry activities were packaged, usually in metal drums lined with a concrete and bitumen matrix, and disposed of at sea (Calmet 1989). Over 50 dump sites are recorded across the northern Atlantic and Pacific Oceans (Calmet 1989).

All human uses of radioactive material generate waste, and as humans are inclined to do, we collect our waste and attempt to manage it. Pritchard (1960) highlighted the production of radioactive waste as an unavoidable consequence of utilising atomic energy. He recognised the responsibilities to assure that the atomic energy industry did not endanger humans or the resources of the sea. After considering the analysis he suggested permissible concentrations and disposal conditions for radioisotope disposal to the sea. Similarly, Vilks (1976) provided insight into a workshop held at Woods Hole in 1976 to discuss the disposal of high level waste (HLW) in oceans, showing that concern was there but serious consideration was not yet being given to ocean disposal. Due to the concerns raised, novel approaches were proposed to manage some forms of radioactive waste. Krutenat (1978) proposed that plutonium-239 waste, with a half-life of 25,000 years, should:

» *"Be disposed of in the basement rock of an oceanic plate at the edge of its subduction zone [to] allow the crustal movement to carry the waste to the centre of the earth".*

The IAEA was no doubt considering the results and recommendations of these and many similar studies at the time. Yet, even with this engaging and long-lasting discussion, there has been considerable disposal of nuclear wastes to the oceans over the years which remains as legacy waste in ageing storage containments. **Today** radioactive waste **has been classified by the IAEA (2009)** (◘ Table 10.4; ◘ Figure 10.6). This classification

scheme considers the type of waste based on half-life as well as its state (solid, aqueous, organic, liquid) and provides detailed direction for appropriate disposal.

Site specific **legacy nuclear waste and radioactivity** from both intentional activities (e.g. waste dumping, weapons testings) and accidents (e.g. power plants, nuclear submarines) will need to be managed long into the future. Even though serious consideration was given to disposing of nuclear waste in the sea in the 1960s and 1970s (e.g. Pritchard 1960; Vilks 1976) it is now not considered an option. There are, however, current sources of anthropogenically derived radioactivity that enter the marine environment, including global fall out, and low level release from nuclear power plants and nuclear fuel reprocessing plants as part of normal operating procedures (Livingston and Povinec 2000). Potential for accidents exists wherever radioactive material is used or when it is transported. Most LLW can be managed safely on land in most places and there are now dedicated facilities for reprocessing and/or storage of intermediate and high level nuclear waste. Interestingly, there is a trade in nuclear waste management and this results in radioactive material being transported from the site of production to the site of disposal, usually by shipping transport.

10.4   Effects on Marine Biota

The enrichment of radioactive material in the marine environment causes risks to marine organisms and to human populations that consume these organisms. Radiation causes changes in living cells as it interferes with normal chemical processes within and between cells. Water within cells can be transformed to hydrogen peroxide. This is particularly the case for white blood cells and impacts an organism's ability to fight infec-

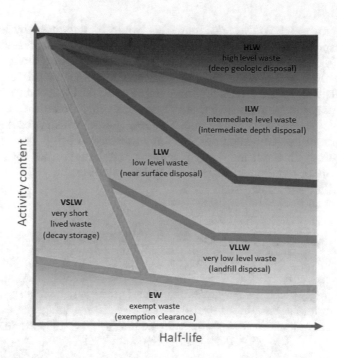

▣ Figure 10.6 Conceptual diagram of the radioactive waste classification scheme of the International Atomic Energy Agency. Adapted from IAEA 2009 by A. Reichelt-Brushett

tion. Radiation has also been shown to induce cancer-like diseases like leukaemia in blood forming organs. It may also cause mutations that impact on heredity. Interestingly, the global background radiation may have influenced the current genetic evolution of species.

The fate of radionuclides depends largely on ocean geochemistry, physical processes and biological uptake, and these characterise exposure in marine environments. Radioactive elements bioaccumulate in a similar manner to other pollutants that contain similar chemical characteristics. The accumulation of radioactive elements is dependent upon chemical behaviour, physical chemistry, and how organisms interact with their environment. As with many metals and organic compounds some contaminant will disperse, but much will bind to particulates, accumulate in benthic environments, and have the potential for remobilisation (Batlle et al. 2011; Buesseler et al. 2017).

The study of in situ exposure is challenging (Buesseler et al. 2017), and exposure doses are highly variable given dispersal by currents and dilution (Batlle et al. 2014). There are increasing numbers of studies that document concentrations of radionuclides in marine species but few investigate effects; rather they rely on recommended values. Recent studies by Men et al. (2020a) showed that ^{134}Cs, ^{137}Cs and ^{110m}Ag accumulated in dolphin fishes after the Fukushima Daiichi Nuclear Power Plant Accident, but decreased with time. It was concluded from these radiation dose as-

sessments that the released radiation would not have caused harm to dolphin fishes in the open ocean of the Northwest Pacific. Similar results were found for neon flying squid (Men et al. 2020b).

The European project Environmental Risk from Ionising Contaminants: Assessment and Management (ERICA) proposed a benchmark at the ecosystem level of 10 μGy/h. Another benchmark from the United Nations Scientific Committee on the Effects of Atomic Radiation (UNSCEAR 1996) concluded that dose rates of up to 400 μGy/h to a small proportion of individuals in aquatic populations would not result in adverse effects at the population level.

There is limited evidence of radioactivity having caused impacts to marine biota (e.g. Batlle 2011; Hosseini et al. 2012). The biological impacts of radiation on chronically exposed organisms are limited to a few laboratory studies, and it is suggested that there is a broad range of species' sensitivities (Batlle et al. 2014) (see ▶ Box 10.4). As with all environmental pollution studies, there is a multitude of endpoints that could be considered to indicate biological impacts of radiation. The selection of an endpoint is usually considered in respect to the likely or possible response to the contaminant of interest. In the case of radionuclides, cellular dysfunction and abnormalities are probable choices. There has been limited targeted analysis to determine genotoxic effects in studies on the impacts of radiation on marine biota. Jha et al. (2005) completed laboratory assays to assess the genotoxic effects of tritium on the adult life stage of the mussel *Mytilus edulis*; they found a dose dependent response in micronuclei, and DNA single strand breaks (Comet assay), highlighting genetic damage. This study further suggested that the generic dose limits recommended by the IAEA for the protection of aquatic biota may not be applicable to all aquatic organisms.

Batlle et al. (2018) have made recommendations for the field of marine radioecology through the development of models to better predict radionuclide transfer to biota in non-equilibrium situations. Such mod-

els would be enhanced by an increased understanding of biogeochemical processes and their influence on radionuclide dispersion along with antagonistic and synergistic interactions related to uptake. Furthermore, Batlle et al. (2018) recommended a more integrated approach to marine radioecology that includes oceanography, radiochemistry, ecology, ecotoxicology and climate science to bring more ecological thinking into the discipline, with further focus on food chains and ecosystem processes.

Box 10.4: Bikini Atoll Five Decades On

A vast number of nuclear weapons tests occurred in waters and coral reef areas of the Marshall Islands and specifically Bikini Atoll, which was physically decimated by 23 surface and subsurface thermonuclear experiments (◼ Figure 10.7). Five decades later, Bikini Atoll was shown to be flourishing with coral reefs and plentiful fish (Richards et al. 2008). Richards et al. (2008) determined that overall species richness before nuclear testing and 50 years after testing was approximately the same, but the species mix was different, suggesting that 28 species were genuine losses, predominantly from the lagoon habitat. The presumed initial losses were mainly attributed to physical impacts, shock waves, temperature rises, and sediment and nutrient suspension.

◼ **Figure 10.7** ▶ Box 10.4: Operation crossroads, test Baker as seen from Bikini Atoll, July 25, 1946. *Photo*: x-ray delta one licensed under CC BY-SA 2.0

10.5 Summary

Radioactivity results from the degradation of unstable atoms to achieve a more stable form. The units it is measured in are unique, and understanding these provides for an enhanced understanding of the topic of radioactivity.

There are both natural and artificial sources of radioactivity. Human uses of both are wide ranging; it is a relatively common source of power, it is used as a weapon, and also in life saving medical science and other scientific investigations. It behaves in the environment in a similar way to some other contaminants that bioaccumulate. The fate and behaviour of radioisotopes is influenced by biogeochemical and physical processes, and the degradation of radioisotopes is dependent on their half-life.

The world's oceans have been exposed to anthropogenic radioactivity as a result of nuclear accidents, weapons testing and waste disposal, and they are considered slightly contaminated by anthropogenic radionuclides. Currently, global fallout and authorised release of low-level waste from nuclear reprocessing facilities and power plants are the main sources of radionuclides to the ocean. There are now global guidelines and restrictions for the management of radioactive waste and recommended safe exposure levels for humans and ecosystems. However, there are few studies that have investigated concentration and effect relationships of radiation on marine biota. A more integrated approach to marine radioecology would help by bringing more ecological thinking into the discipline.

10.6 Study Questions And Activities

1. Describe alpha, beta, and gamma radiation.
2. Using ▶ Box 10.1 determine how much lead-210 would remain after 135 years.
3. Does your home country need to manage nuclear waste from energy generation? If so, see if you can investigate how that waste is managed.
4. Find a journal article that explores the impact of radioactivity on a marine species. Report how the effect is being measured.
5. Explore the IAEA website and record two new facts that you learn.

References

Aruga K, Wakamatsu H (2018) Consumer perceptions towards seafood produced near the Fukushima nuclear plant. Mar Resour Econ 33(4):373–386

Azouz HR, Dulai H (2017) In the wake of Fukushima: radiocesium inventories of selected Northern Pacific fish. Pac Sci 71(20):107–115

Batlle J (2011) Impact of nuclear accidents on marine biota. Integr Environ Assess Manag 7(3):365–367

Batlle J, Aono T, Brown J, Hosseini A, Garier-Laplace J, Sazykina T, Steenhuisen F, Stand P (2014) The impact of the Fukushima nuclear accident on marine biota: retrospective assessment of the first year and perspectives. Sci Total Environ 487:143–153

Batlle J, Aoyama M, Bradshaw C, Brown C, Brown J, Busseler K, Casacuberta N, Christl M, Duffa C, Impens N, Iosjpe M, Masqué P, Nishikawa J (2018) Marine radioecology after the Fukushima Dai-ichi nuclear accident: are we better positioned to understand the impact of radionuclides in marine ecosystems? Sci Total Environ 618:80–92

Belahbib L, Arhouni FE, Boukhair A, Essadaoui A, Ouakkas S, Hakkar M, Abdo MAS, Benjelloun M, Bitar A, Nourreddine A (2021) Impact of phosphate industry on natural radioactivity in sediment, seawater, and coastal marine fauna of El Jadida Province, Morocco. J Hazard Toxic Radioactive Waste 25:04020010

Buesseler KO, Jayne SR, Fisher NS, Rypina II, Baumann H, Baumann Z, Brier CF, Douglass EM, George J, Macdonald AS, Miyamoto H, Nishikawa J, Pike SM, Yoshida S (2012) Fukushima-derived radionuclides in the ocean and biota off Japan. Proc Natl Acad Sci USA 109(16):5984–5988

Buesseler K, Dai M, Aoyama M, Benitez-Nelson C, Charmasson S, Higley K, Maderich V, Masqué P, Morris P, Oughton D, Smith J (2017) Fukushima Daiichi-derived radionuclides in the ocean: transport, fate and impacts. Ann Rev Mar Sci 9:173–203

Buesseler KO, Charette MA, Pike SM, Henderson PB, Kipp LE (2018) Lingering radioactivity at the Bikini and Enewetak Atolls. Sci Total Environ 621:1185–1198

Call M, Maher DT, Santos IR, Ruiz-Halpern S, Mangion P, Sanders CJ, Erler DV, Oakes JM, Rosentreter J, Murray R, Eyre BD (2015) Spatial and temporal variability of carbon dioxide and methane fluxes over semi-diurnal and spring-neap-spring timescales in a mangrove creek. Geochim Cosmochim Acta 150:211–225

Calmet DP (1989) Ocean disposal of radioactive waste: status report. IAEA Bull 4:47–50. Available at: ▶ https://www.iaea.org/sites/default/files/31404684750.pdf. Accessed 20 Feb 2022

Choudri BS, Baawain M (2016) Radioactive wastes. Water Environ Res 88(10):1486–1503

Cresswell T, Metian M, Fisher NS, Charmasson S, Hansman RL, Bam W, Bock C, Swarzenski PW (2020) Exploring new frontiers in marine isotope tracing -adapting to new opportunities and challenges. Front Mar Sci 7:00406

Eigl R, Steier P, Sakata K, Sakaguchi A (2017) Verticle distribution of ^{236}U in the North Pacific Ocean. J Environ Radioact 169–170:70–78

El Kateb A, Stalder C, Rüggeberg A, Neururer C, Spangenberg JE, Spezzaferri S (2018) Impact of industrial phosphate waste discharge on the marine environment in the Gulf of Gabes (Tunisia). PLoS ONE 13(5):e0197731

Fisher NS, Beaugelin-Seiller K, Hinton TG, Baumann Z, Madigan DJ, Garnier-Laplace J (2013) Evaluation of radiation doses and associated risk from the Fukushima nuclear accident to marine biota and human consumers of seafood. Proc Natl Acad Sci USA 110(26):10670–10675

Guerrero J, Gutierrez-Alvarez I, Mosqueda F, Gazquez MJ, García-Tenorio R, Olías M, Bolívar JP (2020) Evaluation of the radioactive pollution in salt-marshes under a phosphogypsum stack system. Environ Pollut 258:113729

Hameed PS, Pillai GS, Mathiyarasu R (2014) A study on the impact of phosphate fertilizers on the radioactivity profile of cultivated soils in Srirangam (Tamil Nadu, India). J Radiat Res Appl Sci 7:463–471

Haridasan PP, Maniyan CG, Pillai PMB, Khan AH (2002) Dissolution characteristics of ^{226}Ra from phosphogypsum. J Environ Radioact 62:287–294

Hosseini A, Brown J, Gwynn J, Dowdall M (2012) Review of research on impacts to biota of the discharges of naturally occurring radionuclides in produced water to the marine environment. Sci Total Environ 438:325–333

IAEA (International Atomic Energy Agency) (2004) Application of the concepts of exclusion, exemption and clearance. Safety Guide No. RS-G-1.7. IAEA, Vienna

IAEA (International Atomic Energy Agency) (2009) Classification of radioactive waste -general safety guide. No. GSG-1. IAEA, Vienna

IAEA (International Atomic Energy Agency) (2019) IAEA annual report 2019, 156. Available at: ▶ https://www.iaea.org/sites/default/files/publications/reports/2019/gc64-3.pdf. Accessed 9 Sept 2021

IAEA (International Atomic Energy Agency) (2021) IAEA annual report 2021, 185. Available at: ▶ https://www.iaea.org/sites/default/files/publications/reports/2021/gc66-4.pdf. Accessed 28 February 2023

Inoue M, Shirotani Y, Yamashita S, Takata H, Kofuji H, Ambe D, Honda N, Yagi Y, Nagao S (2018) Temporal and spatial variations of 134Cs and 137Cs levels in the Sea of Japan and Pacific coastal region: implications for dispersal of NDNPP-derived radiocesium. J Environ Radioact 182:142–150

Jha A, Dogra Y, Turner A, Millward G (2005) Impact of low doses of tritium on the marine mussel, *Mytilus edulis*: genotoxic effect and tissue-specific bioconcentration. Mar Res 586:47–57

Kaizer J, Aoyama M, Kumamoto Y, Molnár M, Palscu L, Povinec PP (2018) Tritium and radiocarbon in the western North Pacific waters: post-Fukushima situation. J Environ Radioact 184–185:83–94

Klungsøyr J, Sætre R, Føyn L, Loeng H (1995) Man's impact on the Barents Sea. Arctic 48(3):279–296

Krutenat RA (1978) A proposal for the permanent disposal of nuclear and other toxic wastes utilising offshore technology. In: 10th annual offshore technology conference, Houston, Texas, 8–11 May 1978

Lindahl P, Ellmark C, Gäfvert T, Mattsson S, Roos P, Holm E, Erlandsson B (2003) Long-term study of 99Tc in the marine environment on the Swedish west coast. J Environ Radioact 67:145–156

Livingston H, Povinec P (2000) Anthropogenic marine radioactivity. Ocean Coast Manag 43(8–9):689–712

Macdonald RW, Brewers JM (1996) Contaminants in the arctic marine environment: priorities for protection. ICES J Mar Sci 53(3):537–563

Madigan DJ, Baumann Z, Fisher NS (2012) Pacific bluefin tuna transport Fukushima-derived radionuclides from Japan to California. Proc Natl Acad Sci USA 109(24):9483–9486

Martínez-Aguirre A, García-León M (1994) The distribution of U, Th and 226Ra derived from the phosphate fertilizer industries on an estuarine system in southwest Spain. J Environ Radioact 22:155–177

Mazzilli B, Palmiro V, Saueia C, Nisti MB (2000) Radiochemical characterization of Brazilian phosphogypsum. J Environ Radioact 49:113–122

Men W, Wang W, Yu W, He J, Lin F, Deng F, Ma H, Zeng Z (2020a) Impact of the Fukushima Dai-ichi nuclear power plant accident on dolphin fishes in the Northwest Pacific. Chemosphere 257:12767

Men W, Wang F, Yu W, He J, Lin F, Deng F (2020b) Impact of the Fukushima Dai-ichi nuclear power plant accident on neon flying squids in the Northwest Pacific from 2011 to 2018. Environ Pollut 264:114647

Nisti MB, Saueia CR, Malheiro LH, Groppo GH, Mazzilli BP (2015) Lixiviation of natural radionuclides and heavy metals in tropical soils amended with phosphogypsum. J Environ Radioact 144:120–126

Ojovan MI, Lee WE, Kalmykov SN (2019) Naturally occurring radionuclides. In: Ojovan M, Lee W, Kalmykov S (eds) An introduction to nuclear waste immobilisation, 3rd edn. Elsevier, pp 35–46

Papaslioti EM, Perez-Lopez R, Parviainen A, Phan VTH, Marchesi C, Fernandez-Martinez A, Garrido CJ, Nieto JM, Charlet L (2020) Effects of redox oscillations on the phosphogypsum waste in an estuarine salt-marsh system. Chemosphere 242:125174

Parreira AB, Kobayashi ARKJ, Silvestre OB (2003) Influence of Portland cement type on unconfined compressive strength and linear expansion of cement-stabilized phosphogypsum. J Environ Eng 129:956–960

Pfister H, Philipp G, Pauly H (1976) Population dose from natural radionuclides in phosphate fertilizers. Radiat Environ Biophys 13(3):247–261

Pritchard DW (1960) The application of existing oceanographic knowledge to the problem of radioactive waste disposal into the sea. In: Proceedings of the IAEA conference on disposal of radioactive wastes, Monaco, Nov., 16–21, 1959, vol 2. International Atomic Energy Agency, Vienna, pp 229–248

Riekenberg PM, Oakes JM, Eyre BD (2020) Shining light on priming in euphotic sediments: nutrient enrichment stimulates export of stored organic matter. Environ Sci Technol 54(18):11165–11172

Richards Z, Beger M, Pinca S, Wallace C (2008) Bikini Atoll coral biodiversity resilience five decades after nuclear testing. Mar Pollut Bull 56:503–515

Rockström J, Steffen W, Noone K, Persson Å, Chapin FS III, Lambin E, Lenton TM, Scheffer M, Folke C, Schellnhuber H, Nykvist B, De Wit CA, Hughes T, van der Leeuw S, Rodhe H, Sörlin S, Snyder PK, Costanza R, Svedin U, Falkenmark M, Karlberg L, Corell RW, Fabry VJ, Hansen J, Walker B, Liverman D, Richardson K, Crutzen P, Foley J (2009) Planetary boundaries: exploring the safe operating space for humanity. Ecol Soc 14(2):32

Rutherford PM, Dudas MJ, Samek RA (1994) Environmental impacts of phosphogypsum. Sci Total Environ 149:1–38

Saueia CHR, Mazzilli BP (2006) Distribution of natural radionuclides in the production and use of phosphate fertilizers in Brazil. J Environ Radioact 89:229–239

SIPRI (Stockholm International Peace Research Institute) (2021) SIPRI Yearbook 2021 armaments, disarmament and international security-summary. Oxford University Press, Oxford, p 32

Standring WJF, Dowdall M, Amundsen I, Strand P (2009) Floating nuclear power plants: potential implications for radioactive pollution of the northern marine environment. Mar Pollut Bull 59:174–178

Suseno H, Wahono IB (2018) Present status of 137Cs in seawaters of the Lombok Strait and the Flores Sea at the Indonesia through flow (ITF) following the Fukushima accident. Mar Pollut Bull 127:458–462

Tayibi H, Choura M, Lopez FA, Alguacil FJ, Lopez-Delgado A (2009) Environmental impact and management of phosphogypsum. J Environ Manage 90:2377–2386

UNSCEAR (United Nations Scientific Committee on the Effects of Atomic Radiation) (1996) Sources and effects of ionising radiation. Available at: ▶ https://www.unscear.org/unscear/en/publications/1996.html. Accessed 10 Jan 2022

Vandecasteele CM (2004) Environmental monitoring and radioecology: a necessary synergy. J Environ Radioact 72:17–23

Villa M, Mosqueda F, Hurtado S, Mantero J, Manjón G, Periañez R, Vaca F, García-Tenorio R (2009) Contamination and restoration of an estuary affected by phosphogypsum releases. Sci Total Environ 408:69–77

Villa-Alfageme M, Chamizo E, Santos-Arévalo FJ, López-Gutierrez JM, Gómez-Martínez A, Hurtado-Bermúdez S (2018) Natural and artificial radionuclides in a marine core. First results of 236U in North Atlantic sediments. J Environ Radioact 186:152–160

Vilks G (1976) The disposal of high level nuclear waste in the oceans. Geosci Can 3(4):295–298

Yang J, Liu W, Zhang L, Xiao B (2009) Preparation of load-bearing building materials from autoclaved phosphogypsum. Constr Build Mater 23:687–693

10

Atmospheric Carbon Dioxide and Changing Ocean Chemistry

Kai G. Schulz and Damien T. Maher

Contents

11.1   Introduction – 248

11.2   The Global Carbon Cycle – 248

11.3   The Physical and Biological Carbon Pumps – 251

11.4   Human-Induced Changes to the Global Carbon Cycle – 252
11.4.1   Ocean Acidification – 252
11.4.2   Potential Effects of Ocean Acidification on Key Organisms and Processes of the Marine Carbon Cycle – 253
11.4.3   Potential Effects of Ocean Acidification on Biogeochemical Element Cycling – 256

11.5   Outlook – 256

11.6   Summary – 256

11.7   Study Questions and Activities – 257

References – 257

© The Author(s) 2023
A. Reichelt-Brushett (ed.), *Marine Pollution—Monitoring, Management and Mitigation*,
Springer Textbooks in Earth Sciences, Geography and Environment,
https://doi.org/10.1007/978-3-031-10127-4_11

Abbreviations

CA   Carbonate alkalinity
DIC   Dissolved inorganic carbon
IPCC   Intergovernmental Panel on Climate Change
Ω_{arag} Saturation state for aragonite
Ω_{calc} Saturation state for calcite
POC   Particulate organic carbon
ppm   Parts per million
Pg   Peta gram
TA   Total alkalinity
TOC   Total organic carbon

11.1 Introduction

"*They call it pollution, we call it life*" is an infamous quote which ignores many facts about why carbon dioxide (CO_2) poses a significant problem for the ocean. But before we get to this, let's start at the beginning.

All organisms on Earth require a particular set of elements for growth. In the case of plants, these elements are needed to synthesise organic matter in a process called primary production via photosynthesis, and in the case of animals, these elements are directly assimilated by either consuming plant material or by preying on other animals. In this respect, one of the key elements is carbon. Being the molecular backbone for a number of vital organic compounds such as sugars, proteins and nucleic acids (containing genetic information), carbon can be considered as the building block of life. Similar requirements also apply to bacteria, fungi and viruses, which stand outside the plant and animal kingdoms. However, carbon is not only exchanged between organisms. There are in fact much larger reservoirs on Earth and fluxes in between, especially in the marine environment (◘ Figure 11.1). In their entirety, they are referred to as the global carbon cycle.

11.2 The Global Carbon Cycle

The largest carbon reservoir on Earth are **sedimentary rocks** in its crust and upper mantle (i.e. the lithosphere). It is approximately three orders of magnitude ($\times 1000$) larger than all other reservoirs combined (Falkowski et al. 2000), however, it has the slowest exchange rates (◘ Figure 11.1). As a consequence, it requires relatively large changes to these fluxes and typically long geologic time scales (on the order of thousands to millions of years) to significantly affect the other reservoir sizes.

The second-largest carbon reservoir are the oceans, where about 900 Pg are found in the surface ocean (1 Pg equals 10^{15} g or 10^9, or 1 billion tonnes) and approximately 37,100 Pg in the intermediate and deep ocean, mostly in the form of dissolved inorganic carbon (DIC)

(► Box 11.1). Given the relatively large exchange rates between the surface and the deep ocean and also with the atmosphere, small changes can lead to a significant redistribution, for instance in the atmosphere, which is one of the smaller reservoirs. As an example, shutting down the organic carbon pump (see below) could easily double atmospheric CO_2 levels, based on a back-on-the-envelope calculation, assuming a mean deep ocean DIC enrichment of 50 μmol kg^{-1} (compare ◘ Figure 11.2a) at an overall volume of $\sim 1.3 \times 10^{18}$ m^3 of seawater and an average depth of 3.7 km (Eakins and Sharman 2010).

On land, it is the dead organic matter stored in soils which constitutes the largest carbon reservoir. On time scales on the order of years to millennia, an important connection between these three reservoirs of easily interchangeable carbon (the soils on land, the atmosphere, and the oceans) are the living organisms. In this respect plants play a key role by taking up CO_2 via photosynthesis and utilising light as an energy source to build up biomass, a process termed **primary production**. Interestingly, net primary production which is the balance between the overall amount of carbon fixed by plants during the day and respired again to CO_2 by these organisms throughout day and night (to produce metabolic energy-sustaining cellular functioning, especially in the dark) is similar on land and at sea ($\sim$50 Pg C per year).

However, in the marine environment, this amount of carbon is being fixed by plants containing about 100 times less carbon ($\sim$3 Pg C) than those on land (450–650 Pg C). This is because on land much of the carbon stored in plants is structural (e.g. tree trunks) and only a small fraction is actively participating in carbon fixation via photosynthesis (e.g. leaves). Structural carbon is required to fight gravity and gain a competitive advantage over other plants by getting as much exposure to light as possible (imagine, for instance, a tree overshadowing the grass underneath). **In the ocean, microscopic unicellular algae (termed phytoplankton) dominate primary production.** As these organisms, float in the sunlit surface ocean and are constantly being moved by winds and currents, no structural carbon is required (or would help) in maximising light exposure.

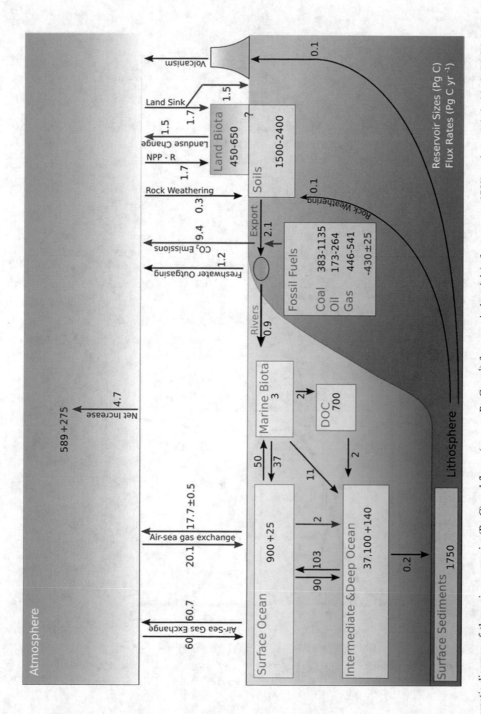

□ **Figure 11.1** Schematic diagram of the major reservoirs (Pg C) and fluxes (arrows–Pg C yr⁻¹) for a pre-industrial (reference year 1750) carbon cycle in black (IPCC 2013). In red are anthropogenic perturbations of this quasi-steady state, where changes in reservoir sizes and fluxes (period of 2008–2017) are taken from IPCC (2013) and Le Quéré et al. (2018). In this respect, the land sink has been divided into a biological and non-biological (see text for details) component (Kirschbaum et al. 2019), and the question mark highlights the unknowns concerning the overall amount of anthropogenic carbon taken up by the terrestrial biosphere. Uncertainties are only given for the anthropogenic air-sea gas exchange component and the amount of fossil fuels that have been extracted, but are representative for uncertainties in the overall budget. NPP-R refers to the balance of net primary production by plants and respiration by all other organisms. *Image*: K. Schulz

Box 11.1: Basic Carbonate Chemistry

In water carbon dioxide (CO_2) is being found as a dissolved gas ($CO_{2(aq)}$) and after hydration to H_2CO_3 and dissociation, as bicarbonate (HCO_3^-) and carbonate ions (CO_3^{2-}). As $CO_{2(aq)}$ and H_2CO_3 are chemically not separable, they are typically simply referred to as CO_2. The sum of the concentrations of all species is then dissolved inorganic carbon (DIC), defined as shown in Eq. 11.1

$$DIC = [CO_2] + [HCO_3^-] + [CO_3^{2-}] \tag{11.1}$$

The underlying acid–base equilibria are given by Eq. 11.2

$$CO_2 + H_2O \rightleftharpoons HCO_3^- + H^+ \rightleftharpoons CO_3^{2-} + 2H^+ \tag{11.2}$$

Another key concept in carbonate chemistry is that of total alkalinity (TA), which can be defined as the excess of proton acceptors over proton donors (Dickson 1981). Essentially, TA can be seen as the buffering capacity of the ocean to resist changes in pH. TA can be expressed as (Zeebe and Wolf-Gladrow 2001) Eq. 11.3:

$$\begin{aligned} TA = {} & [HCO_3^-] + 2[CO_3^{2-}] + [B(OH_4)^-] + [OH^-] \\ & + [HPO_4^{2-}] + 2[PO_4^{3-}] + [H_3SiO_4^-] \\ & + [NH_3] + [HS^-] - [H^+]_F - [HSO_4^-] \\ & - [HF^-] - [H_3PO_4] + \cdots - \cdots \end{aligned} \tag{11.3}$$

with the last two terms accounting for minor components of alkalinity such as conjugate bases of organic acids, and minor proton donors.

In this interlinked system, the relative contribution of the three DIC components to the overall concentration is controlled by pH (■ Figure 11.2). However, it worth noting that any change to the concentration of one of the DIC components, or TA, will equally lead to changes in pH.

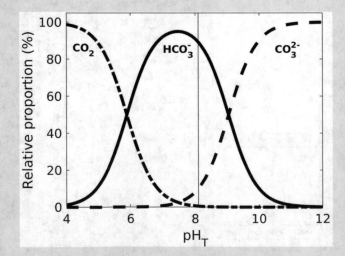

■ **Figure 11.2** ▶ Box 11.1 Relative contributions of CO_2, HCO_3^- and CO_3^{2-} to overall DIC as a function of pH (total hydrogen ion scale), also known as Bjerrum plot and calculated using stoichiometric equilibrium constants for carbonic acid from Mehrbach et al. (1973) and refitted by Dickson and Millero (1987)

Before the beginning of the Industrial Revolution, this global cycling of carbon was thought to have been in quasi-steady state (Pongratz et al. 2009), meaning that the flow of carbon in and out of a reservoir was roughly balanced (Figure 11.1). However, significant changes have occurred since then, but before we come to these we shall have a closer look at carbon cycling within the ocean.

11.3 The Physical and Biological Carbon Pumps

The so-called **carbon pumps** describe the flux of dissolved inorganic carbon and total alkalinity from the ocean surface to depth. They are operating against a concentration gradient, and their relative strengths drive the direction of CO_2 air-sea gas exchange (Figure 11.3b).

The physical carbon pump is based on increasing CO_2 solubility when warmer waters cool. Subsequently when sea ice forms at cold high latitudes in the North Atlantic and around Antarctica, the DIC-rich surface waters become particularly dense (due to high salinity) and sink (Figure 11.3b). Such deep-water formation at high latitudes is responsible for large-scale ocean circulation with water masses flowing at depth

from the North to the South Atlantic, and into the Indian and Pacific Oceans where they are eventually subject to upwelling and return as surface currents back to the North Atlantic. The **biological carbon pump** can be split into two; the organic carbon (or soft tissue) pump, and the carbonate counter pump. As the name suggests the latter operates in the opposite way to the former in terms of atmospheric CO_2 uptake of the surface ocean (Figure 11.3b).

The **organic carbon pump** is driven by marine phytoplankton (microscopic unicellular plants) which, as mentioned above, fix CO_2 into organic matter in the sun-lit surface ocean via photosynthesis. Most of the carbon will end up in the particulate rather than the dissolved organic fraction and rain down to the ocean floor when particles have a higher density than the surrounding seawater. During downward transport most of the particulate organic carbon (POC) will be converted back to CO_2 by bacterial respiration and only a fraction of the ~11 Pg exported each year will be buried in sediments (compare Figures 11.1 and 11.3b).

The **carbonate counter pump** is driven by calcium carbonate ($CaCO_3$) formation of open ocean organisms, mainly calcifying phytoplankton (coccolithophores), and zooplankton (foraminifera and pteropods). As the precipitation of $CaCO_3$ shifts the carbonate chemistry equilibrium towards lower pH and

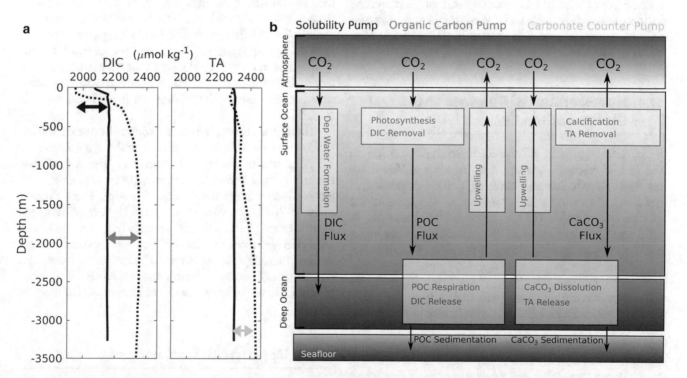

 Figure 11.3 **a** Depth profiles of dissolved inorganic carbon (DIC) and total alkalinity (TA) in the North Atlantic at 57.3° N/39.5° W (solid line) and the Tropical Pacific at 15° N/135° W (dotted line) taken from Alvarez (1997) and Goyet et al. (1997), respectively, and **b** a schematic of the two carbon pumps. They can be divided into the physical solubility pump (black arrow in **a**), and the organic carbon (green arrow in **a**) and carbonate counter pumps (yellow arrow in **a**). The reason for higher DIC and TA at depth in the Pacific compared to the Atlantic profiles is the amount of time without contact with the atmosphere and the water age. In general, the older the water mass the more particulate organic carbon (POC) respiration and calcium carbonate ($CaCO_3$) dissolution will have taken place at depth. *Image*: K. Schulz

hence higher CO_2 (▶ Box 11.2), the ocean's storage capacity for the latter is reduced. At present, $CaCO_3$ is a stable mineral in most of the surface ocean, an important fact as some organisms critically rely on these minerals (see ▶ Section 11.4.2 for details). Being denser than seawater, however, there is a steady rain of $CaCO_3$ to the sea floor, estimated at approximately 1 Pg C per year. About 90% of this will dissolve as the saturation state decreases with pressure at depth and waters eventually become undersaturated. Although $CaCO_3$ formation in the surface ocean has an immediate opposite effect on CO_2 air-sea gas exchange in comparison to photosynthesis, $CaCO_3$ is thought to strengthen the efficiency of the organic carbon pump as aggregates of POC and the much denser $CaCO_3$ will sink more rapidly (Klaas and Archer 2002; De La Rocha and Passow 2014). This would move respiration to deeper layers and keep the CO_2 produced away from atmospheric contact for longer time. As a consequence, this would increase the surface to depth DIC gradient and allow for more atmospheric CO_2 to be stored at depth.

Whether the positive effects of ballasting on oceanic CO_2 uptake outweigh the negative effects of $CaCO_3$ production is difficult to reconcile, however, models suggest that they may cancel each other out (e.g. Barker et al. 2003; Kvale et al. 2021). While the two carbon pumps govern biogeochemical element cycling in the ocean, they are part of a bigger picture, the **global carbon cycle**, which is currently perturbed by human activities.

11.4 Human-Induced Changes to the Global Carbon Cycle

Since the beginning of the Industrial Revolution in the late eighteenth century, the burning of fossil fuels such as oil, gas and coal, as well as cement production, has come with emissions of the greenhouse gas carbon dioxide (CO_2). Together with changes in land-use and deforestation, these anthropogenic emissions have significantly changed the flow of carbon between land, ocean and atmosphere, impacting respective reservoir sizes (◻ Figure 11.1). From the ~430 Pg of carbon stemming from fossil fuels and cement production, ~275 Pg have been accumulating in the atmosphere. With ~235 Pg estimated to originate from land-use changes and ~165 Pg having entered the oceans, the terrestrial carbon sink is estimated to be ~225 Pg. For a long time, this has been considered the result of CO_2 - enhanced photosynthesis coupled with vast amounts of nitrogen and phosphorus being released into the environment as fertiliser, enhancing growth in the last 100 years. Although this simple mass balance is similar to estimates from various dynamic vegetation models (Le Quéré et al. 2018), it has been proposed recently that almost half of the current annual land sink for anthropogenic carbon is not linked to an increase in terrestrial biomass storage. Rather, it comprises fossil fuels which are not turned into CO_2 (i.e. plastics and bitumen), harvested wood products and CO_2 being absorbed by cement carbonation in concrete structures. Hence, at the moment, it is difficult to put a number on the biological component of the terrestrial carbon sink.

Nevertheless, the human-induced changes to the global carbon cycle in general, and the atmosphere in particular are driving climate change by affecting Earth's radiative balance, causing mean surface temperatures to increase. While this is an accepted fact within the scientific community after decades of research–the first assessment report on climate change by the Intergovernmental Panel on Climate Change (IPCC) dates back to the 1990s and the most recent report was released in 2021 (IPCC 1990, 2021)–there are groups of activists campaigning against this scientific consensus. A memorable anecdote is the release of a television advertisement by the Competitive Enterprise Institute, a neoconservative think tank in May 2006. This video appeared to be targeting Al Gore's film *An Inconvenient Truth*, which was about to be released in cinemas worldwide at the time and highlighted the connections between a fossil fuel-based economy and climate change. In an attempt to dismiss the negative effects of increasing atmospheric CO_2 levels on Earth's climate system and referring to the fact that plants utilise CO_2 in photosynthesis for growth, the advertisement culminated in the claim "CO_2: They call it pollution, we call it life!".

Today we are in a situation where it appears highly unlikely to be able to stabilise global temperatures at 1.5 °C above pre-industrial levels until the end of this century (IPCC 2018). This was a goal adopted by 179 states that ratified the United Nations Paris Agreement in 2015, in order to keep the threats of temperature-driven global climate change (e.g. ice melt, sea level rise, and intensification of extreme weather events) within manageable boundaries. However, these are not the only issues with increasing levels of atmospheric CO_2, as it also has a direct impact on the world's oceans.

11.4.1 Ocean Acidification

In the last 250 years, atmospheric CO_2 levels have steadily increased from 280 ppm (parts per million) to 417 ppm in 2022 (Friedlingstein et al. 2022). As a result of equilibration at the air-sea interface, substan-

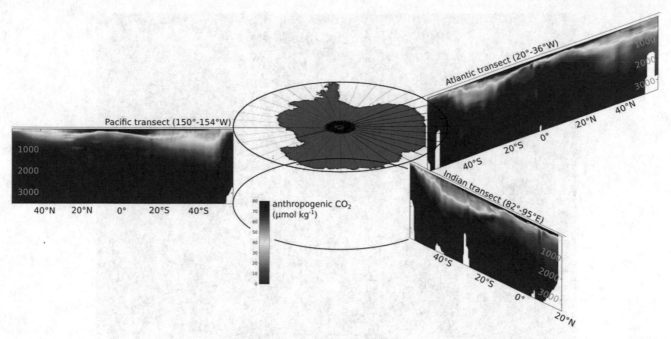

Figure 11.4   Concentrations of anthropogenic CO_2 found in all three major ocean basins, centred around the Antarctic continent (depth in metres). Calculations from measured oxygen, DIC, salinity and temperature follow the principles outlined in Sarmiento and Gruber (2006) and are described in detail in Schulz and Riebesell (2011). Data for the Pacific (Feely et al. 2013a; b), Atlantic (Wanninkhof et al. 2013; Peltola et al. 2013) and Indian ocean (Feely et al. 2013c; d) are from transects of the CLIVAR repeat hydrography program. *Image*: K. Schulz

tial amounts of this anthropogenic CO_2 have also made its way into the ocean (◨ Figure 11.4). When CO_2 dissolves in seawater it forms carbonic acid, which reduces its pH (Equation 11.2). This is known as **ocean acidification and has been coined the evil twin of climate change** (Pelejero et al. 2010). The pH of the ocean has already dropped by ~0.1 units since the Industrial Revolution, which is equivalent to a ~30% increase in acidity (note that the pH scale is logarithmic). This drop in pH has changed carbonate chemistry speciation (▶ Box 11.1), with potentially significant implications for a number of key species in the ocean.

11.4.2 Potential Effects of Ocean Acidification on Key Organisms and Processes of the Marine Carbon Cycle

With the realisation that ongoing ocean acidification might pose a threat to marine organisms and ecosystems, research into the potential biological responses has grown exponentially during the last two decades. Today the number of studies is probably in the thousands (compare Gattuso and Hansson 2011). Since it is beyond the scope of this Chapter to cover all of them, we will focus on key groups and processes in terms of impact on global carbon cycling.

One of the first groups suspected to be affected by ocean acidification was calcium **carbonate ($CaCO_3$) producing (or calcifying) organisms.** The reasoning was that the CO_2 - induced reduction in pH would reduce carbonate ion concentrations (▶ Box 11.2) and hence substrate availability for calcification, in turn negatively affecting $CaCO_3$ precipitation rates (Broecker and Takahashi 1966). Today, a majority of studies (including meta-analyses) have indeed found that calcification rates in most marine taxa are negatively impacted by ocean acidification. This includes hard corals, calcifying phytoplankton (a very important taxon for marine carbon cycling at the base of marine food webs and for marine carbon cycling ◨ Figure 11.5), algae, foraminifera, pteropods and molluscs (Gattuso and Hansson 2011; Kroeker et al. 2010, 2013; Schulz et al. 2017), although it has to be acknowledged that there are species-specific differences and sensitivities.

In this respect, it is interesting to note that there is still an ongoing debate on the physiological mechanisms underlying the susceptibility of calcification to ocean acidification. **Calcification** is a process under tight cellular control and involves diffusion and transport of ions across membranes. While there is a consensus that calcification rates are a function of the internal saturation state at the site of $CaCO_3$ precipitation, the key question is by which external bulk seawater carbonate chemistry parameter(s) it is gov-

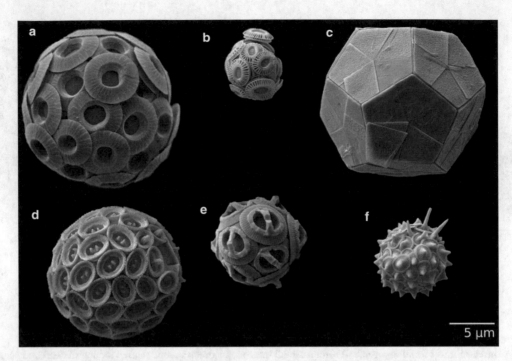

□ **Figure 11.5**   Calcifying phytoplankton divsersity in the modern ocean, exemplified by **a** *Umbilicosphaera sibogae*, **b** *Emiliania huxleyi*, **c** *Braarudosphaera bigelowii*, **d** *Coronosphaera binodata*, **e** *Gephyrocapsa oceanica* and **f** *Acanthoica quattrospina*. *Images*: M. Dawes. For details on coccolithophore-global carbon cycle interactions see Rost and Riebesell (2004)

11

erned. For coccolithophores, there is experimental evidence that it is decreasing seawater pH levels which are unfavourable for calcification (e.g. Bach et al. 2015). The underlying principle is a reduction in the internal to external electrochemical proton gradient, which is thought to increase the costs of removing the protons (e.g. Gafar et al. 2019 and references therein) that are generated by $CaCO_3$ precipitation (▶ Box 11.2). These basic principles could also be the mechanisms underlying the sensitivity of calcification in other marine organisms. While this idea has recently been advocated (Cyronak et al. 2016a, b), it has also been contested (Waldbusser et al. 2016).

The other key group of marine organisms (when it comes to global carbon cycling) are **photoautotrophs**. Since they harness the energy of light to fix CO_2 in photosynthesis, it might be expected that this process could profit from ocean acidification. Indeed, there is (with the exception of calcifying algae) a trend towards increased photosynthesis in response to ocean acidification. However, it appears that photosynthesising calcifiers (such as corals and coccolithophores) might benefit less than non-calcifying groups such as the main contributor to ocean primary productivity [i.e. diatoms, fleshy algae and seagrasses (Kroeker et al. 2010, 2013)]. Again, it is important to stress that sensitivities are species-specific.

While most of the experiments on the effects of **ocean acidification** on marine calcification and primary production have been conducted on single species, to predict future changes it is necessary to include important factors on the **ecosystem level** such as species interaction (e.g. predation, and competition for resources). This requires either larger-scale manipulative in-situ studies or manipulative incubations with entire communities. A good example of the former are experiments which showed that net calcification rates of an entire reef–including a variety of calcifiers ranging from corals to coralline algae and foraminifera in sediments–were decreasing with decreasing pH and vice versa (Albright et al. 2016, 2018). The future survival of coral reef ecosystems depends on this intricate budget, **taking into account all $CaCO_3$ precipitation and dissolution processes**. In addition, mass balance calculations including corals and sediments suggest that below a seawater aragonite saturation state of 2.55, reefs with 5% coral cover might become net dissolving (Eyre et al. 2018). Depending on the future CO_2 emission scenario, this threshold could be reached well before the turn of this century (IPCC 2013), in particular for reefs that experience upwelling of naturally CO_2-rich deep waters, such as documented along the East Australian coast, amplifying ocean acidification (Schulz et al. 2019).

Box 11.2: Calcium Carbonate

The formation of the mineral calcium carbonate ($CaCO_3$) leads to redistributions in carbonate chemistry speciation, decreasing pH and CO_3^{2-} ions, and increasing CO_2 concentrations. This can be understood by realising that CO_3^{2-} ion concentrations can be expressed as TA-DIC, upon simplifying Eq. 11.1 and 11.3 as shown in Eq. 11.4:

$$TA \approx CA = [HCO_3^-] + 2[CO_3^{2-}] \tag{11.4}$$

with CA denoting carbonate alkalinity and as shown in Eq. 11.5:

$$DIC \approx [HCO_3^-] + 2[CO_3^{2-}] \tag{11.5}$$

These are valid approximations in typical seawater as CA contributes about 95% to total alkalinity and CO_2 less than 1% to DIC. Since the formation of $CaCO_3$ reduces TA twice as much as DIC, the difference between them, i.e. the CO_3^{2-} ion concentration, decreases. This is equivalent to shifting carbonate chemistry speciation towards the left in a Bjerrum plot (◻ Figure 11.1), leading to decreasing pH and increasing CO_2 concentrations. It is acknowledged that the actual acid-base re-equilibration is more complex as overall DIC concentrations are changing (assumed constant in a Bjerrum plot), but the change of direction is the same (for details see Zeebe and Wolf-Gladrow 2001).

Ocean Acidification and Saturation State

Carbonate ion availability is also reduced by ocean acidification (Equation 11.2). This is crucial in governing the saturation state of the mineral calcium carbonate ($CaCO_3$) which has two main polymorphs, i.e. the more soluble aragonite and the less soluble calcite. Whether these minerals are stable in seawater is given by their respective saturation states, defined as shown in Eq. 11.6:

$$\Omega_{arag/calc} = \frac{[Ca^{2+}][CO_3^{2-}]}{K_{sp}} \tag{11.6}$$

with K_{sp} denoting the temperature and salinity dependent solubility product in equilibrium for aragonite or calcite, and $[Ca^{2+}]$ and $[CO_3^{2-}]$ denoting respective concentrations in seawater. Hence, seawater Ω levels above One are indicative of stable mineral conditions in the water column, while at levels below One $CaCO_3$ would start to dissolve.

When it comes to assessing the effects of ocean acidification on the main drivers of marine primary productivity, mesocosms have proven an invaluable tool. They are incubation units large enough to house a community of organisms on multiple trophic levels (e.g. primary producers, primary consumers, and secondary consumers). Most **mesocosm studies** to date have not found positive effects of ocean acidification on autotrophic biomass build-up (see Bach et al. 2017; Maugendre et al. 2017; Schulz et al. 2017, and references therein). This suggests that either the products of potentially enhanced photosynthesis (compare single-species experiments) are effectively transferred up the trophic ladder by grazing or are remineralised by bacterial respiration, or that a change in species composition is having a buffering effect. Although at this stage the underlying mechanisms are not fully understood, changes in species composition have been found in most experiments. A re-occurring pattern in this respect is that when effects are found, small picoeukaryote abundances are positively affected and coc-

colithophore abundances are negatively affected by ocean acidification (Schulz et al. 2017). In regard to diatoms, their abundances are mostly positively affected, with shifts towards larger species (Bach and Taucher 2019). These findings have the potential to significantly change marine carbon cycling with feedbacks to the climate system (see below). However, **changes in phytoplankton community composition** can also change the transfer efficiency to higher trophic levels and affect fish production, either through changes in food quality (Rossoll et al. 2012) or by disrupting the vital link between primary producers and fish (i.e. the zooplankton community through toxin-producing harmful algal blooms.

Last but not least, it is not only necessary to scale up experiments in space and time to include the required realism with species interaction, competition and natural variability, but also to assess the interactive effects of other environmental drivers which are changing in concert with ongoing ocean acidification (Riebesell and Gattuso 2014)–often referred to as anthropogenic stressors. These include, but are not limited to, increasing surface ocean temperatures which enhance stratification and hence reduce mixed layer depths, which is thought to increase light but decrease nutrient availability for primary production.

11.4.3 Potential Effects of Ocean Acidification on Biogeochemical Element Cycling

Changes to the efficiency and/or strength of the biological carbon pumps are bound to affect air-sea CO_2 gas exchange (compare ☐ Figure 11.3). Hence, ongoing ocean acidification-driven changes would constitute a feedback loop, either amplifying or dampening increasing atmospheric CO_2 levels (Riebesell et al. 2009). As described above, the mineral ballasting ($CaCO_3$ or biogenic silica) of particulate organic matter produced in the sun-lit surface ocean, which is otherwise nearly neutrally buoyant, enhances the transport of organically bound CO_2 into the deep ocean. Hence, reductions in $CaCO_3$ production there are thought to negatively impact this export flux (e.g. Armstrong et al. 2002; Boyd and Doney 2002). Experimental assessment, however, has proven difficult, in part due to the large distances involved (1000s of metres). However, on smaller scales (10s of metres) in mesocosm experiments, ocean acidification has been found to significantly reduce the bloom-forming potential of the cosmopolitan coccolithophore *Emiliania huxleyi* (☐ Figure 11.5) which in turn reduced the amount of sedimenting organic carbon by approximately a quarter (Riebesell et al. 2018). As this was directly related to reductions in sinking speed velocities with reduced $CaCO_3$ production (Bach et al. 2016), it might be reasonable to assume this mechanism is also operating in the real ocean. When it comes to biogenic silica production by diatoms, the other important ballast mineral, it is more complicated as of a more variable response to ocean acidification (Bach and Taucher 2019). Although, overall contributions of diatoms to primary production have recently been projected to decrease in the future ocean (Trèguer et al. 2018), with the exception of the Southern Ocean, in which however, the amount of cellular silica ballast produced by diatoms have been found to decrease with ocean acidification (Petrou et al. 2019).

The shift from larger to smaller sized picophytoplankton as a result of ocean acidification found in a number of experiments (see Schulz et al. 2017 and references therein), could have a similar effect on the flux of organic carbon to depth and hence atmospheric CO_2 levels. This is based on observations of lower export efficiencies at a number of open ocean sites with significant picophytoplankton community contributions, although considerable variability exists locally and seasonally (see De La Rocha and Passow 2014 for a review).

In summary, experiments on the effects of ocean acidification on surface ocean carbon cycling imply that positive could outweigh negative feedbacks, amplifying increasing levels of atmospheric carbon dioxide.

11.5 Outlook

Without a reduction in anthropogenic CO_2 emissions, the future ocean will continue to become more acidic. Depending on CO_2 emission scenario extrapolations to the year 2100 suggest a further decrease in pH of ~0.2–0.3 units (IPCC 2013). The direction and magnitude of all the feedback mechanisms associated with ocean acidification are currently difficult to project, but clearly there will be significant effects on certain organisms and ecosystems.

To successfully reduce anthropogenic CO_2 emissions will require a combination of legislative and technological advances. Several significant steps have been made in trying to curb the emissions of CO_2, such as the UN Paris Agreement and national arrangements such as Canada's Greenhouse Gas Pollution Price Act. However, despite these important steps, CO_2 emissions as of today are still increasing (Le Quéré et al. 2018), highlighting the need to further strengthen our efforts. In terms of technological advances, the efficiency of the leading multi-junction cell solar panels has reached ~46% (Green et al. 2018), and renewable energy output is increasing globally (as demand has unfortunately too), supplying almost 30% of global electricity demand in 2020 (IEA 2020). With continued advancement in technology, along with the implementation of legislative tools, a zero-emission future is achievable, but if it will be quick enough remains to be seen.

11.6 Summary

Carbon is considered the building block of life. It is exchanged between organisms, but there are much larger reservoirs and fluxes on Earth, especially in the marine environment. Carbon reservoirs include sedimentary rocks in the Earth's crust and upper mantle (primarily), the oceans, the soils on land, and to a lesser extent the atmosphere. Significant changes have occurred since the Industrial Revolution, when global cycling of carbon (the flow of carbon in and out of a reservoir) was near balanced in a quasi-steady state (Pongratz et al. 2009).

Carbon pumps are either biological or physical and operate against a concentration gradient. Their relative strengths drive the direction of CO_2 air-sea gas exchange, and govern biogeochemical element cycling in the ocean. The physical carbon pump is driven by increasing CO_2 solubility when warmer waters cool. When sea ice forms in the North Atlantic and around Antarctica, the DIC-rich surface waters increase in density and sink. This is driving large-scale ocean circulation.

The **biological carbon pump** includes the **organic carbon pump** and the **carbonate counter pump**, and these operate in an opposite direction in terms of atmospheric CO_2 uptake. The organic carbon pump is driven by the constant rain of particulate organic matter from the surface ocean, where it had been produced by photosynthesis. During sinking, most particulate organic carbon will be converted back to CO_2 by bacterial respiration. The carbonate counter pump is driven by $CaCO_3$ formation by marine organisms, whereby precipitation of $CaCO_3$ shifts the carbonate chemistry equilibrium towards lower pH and hence a higher CO_2 concentration.

Anthropogenic CO_2 emissions since the Industrial Revolution, along with changes in land-use and deforestation, have significantly altered the flow of carbon between land, oceans, and the atmosphere. These human-induced changes to the global carbon cycle in general, and the atmosphere in particular, are driving current climate and ocean change. In the last 250 years, atmospheric CO_2 levels have increased from 280 to 417 ppm (in 2022). Substantial amounts of CO_2 have made their way into the ocean causing ocean acidification, with a reduction in the pH of surface waters by ~0.1 units (equivalent to a 30% increase in acidity) as of today. This will **most likely have increasing implications for numerous marine keystone species.** Furthermore, experiments on the effects of ocean acidification on surface ocean carbon cycling suggest that positive feedback effects are likely to amplify increasing levels of atmospheric CO_2. A combination of legislative and technological advances is needed to reduce anthropogenic CO_2 emissions and mitigate global climate/ocean change.

11.7 Study Questions and Activities

1. Describe how pH influences the proportions of CO_2, CO_3^{2-} and HCO_3^- in water and vice versa.
2. Discuss the various carbon pumps and how they influence marine carbon cycling.
3. Describe how $CaCO_3$ formation in the surface ocean has the opposite effect to photosynthesis on CO_2 air-sea gas exchange.
4. Explain the meaning of "saturation state", and how this is relevant to calcifying marine organisms.
5. Summarise The United Nations Paris Agreement, and report up-to-date information regarding its implementation.

References

Albright R, Caldeira L, Hosfelt J, Kwiatkowski L, Maclaren JK, Mason BM, Nebuchina Y, Ninokawa A, Pongratz J, Ricke KL, Rivlin T (2016) Reversal of ocean acidification enhances net coral reef calcification. Nature 531:362–365

Albright R, Takeshita Y, Koweek DA, Ninokawa A, Wolfe K, Rivlin T, Nebucchina Y, Young J, Caldeira K (2018) Carbon dioxide addition to coral reef waters suppresses net community calcification. Nature 555:516–519

Alvarez MS (1997) Carbon dioxide, hydrographic, and chemical data obtained during the R/V discovery cruise in the North Atlantic Ocean during WOCE Section A25 (7 August–17 September 1997). US Department of Energy, Carbon Dioxide Information Analysis Center, Oak Ridge National Laboratory, Oak Ridge, Tennessee. ▸ https://doi.org/10.3334/CDIAC/otg.WOCE_A25. ▸ http://cdiac.ornl.gov/ftp/oceans/CARINA/Discovery/74DI230/

Armstrong RA, Lee C, Hedges JI, Honjo S, Wakeham SG (2002) A new mechanistic model for organic carbon fluxes in the ocean based on the quantitative association of POC with ballast minerals. Deep-Sea Res II 49:219–236

Bach LT, Taucher J (2019) CO_2 effects on diatoms: a synthesis of more than a decade of ocean acidification experiments with natural communities. Ocean Sci 15:1159–1175

Bach LT, Riebesell U, Gutowska MA, Federwisch L, Schulz KG (2015) A unifying concept of coccolithophore sensitivity to changing carbonate chemistry embedded in an ecological framework. Prog Oceanogr 135:125–138

Bach LT, Boxhammer T, Larsen A, Hildebrandt N, Schulz KG, Riebesell U (2016) Influence of plankton community structure on the sinking velocity of marine aggregates. Glob Biogeochem Cycles 30:1145–1165

Bach LT, Alvarez-Fernandez S, Hornick T, Stuhr A, Riebesell U (2017) Simulated ocean acidification reveals winners and losers in coastal phytoplankton. PLoS ONE 12(11):e0188198

Barker S, Higgins JA, Elderfield H (2003) The future of the carbon cycle: review, calcification response, ballast and feedback on atmospheric CO_2. Philos Trans Roy Soc Lond a: Math Phys Eng Sci 361(1810):1977–1999

Boyd PW, Doney SC (2002) Modelling regional responses by marine pelagic ecosystems to global climate change. Geophys Res Lett 29(16):53-2–53-4

Broecker WS, Takahashi T (1966) Calcium carbonate precipitation on the Bahama Banks. J Geophys Res 171:1575–1602

Cyronak T, Schulz KG, Jokiel PL (2016a) The Omega myth: what really drives lower calcification rates in an acidifying ocean. ICES J Mar Sci 73:558–562

Cyronak T, Schulz KG, Jokiel PL (2016b) Response to Waldbusser et al. (2016b): 'Calcium carbonate saturation state: on myths and this or that stories'. ICES J Marine Sci 7:569–571

Dickson AG (1981) An exact definition of total alkalinity and a procedure for the estimation of alkalinity and total inorganic carbon from titration data. Deep-Sea Res 28:609–623

Dickson AG, Millero FJ (1987) A comparison of the equilibrium constants for the dissociation of carbonic acid in seawater media. Deep-Sea Res 34:1733–1743

De La Rocha C, Passow U (2014) 8.4—The biological pump. In: Holland H, Turekian K (eds) Treatise on geochemistry, vol 8, 2nd edn. Elsevier, Amsterdam, pp 93–122

Eakins BW, Sharman GF (2010) Volumes of the world's oceans from ETOPO1. NOAA National Geophysical Data Center, Boulder, CO. Available at ▶ https://www.ngdc.noaa.gov/mgg/global/etopo1_ocean_volumes.html. Accessed 22 Feb 2022

Eyre BD, Cyronak T, Drupp P, De Carlo EH, Sachs JP, Andersson AJ (2018) Coral reefs will transition to net dissolving before end of century. Science 359:908–911

Falkowski P, Scholes RJ, Boyle E, Canadell J, Canfield D, Elser J, Gruber N, Hibbard K, Hoegberg P, Linder S, Mackenzie FT, Moore B III, Pedersen T, Rosenthal Y, Seitzinger S, Smetacek V, Steffen W (2000) The global carbon cycle: a test of our knowledge of Earth as a system. Science 290:291–296

Feely RA, Sabine CL, Millero FJ, Langdon C, Fine RA, Bullister JL, Hansell DA, Carlson CA, McNichol A, Key RM, Byrne RH, Wanninkhof R (2013a) Dissolved inorganic carbon (DIC), total alkalinity, dissolved organic carbon (DOC), chlorofluorocarbons, temperature, salinity and other hydrographic and chemical variables collected from discrete samples and profile observations during the R/V Roger Revelle Cruise CLIVAR_P16S_2005 (EXPOCODE 33RR20050109) in the South Pacific Ocean from 2005-01-09 to 2005-02-19 (NODC Accession 0108095). Version 3.3. National Oceanographic Data Center, NOAA. Dataset. ▶ https://doi.org/10.25921/09v7-bq08

Feely RA, Sabine CL, Millero FJ, Langdon C, Fine RA, Bullister JL Hansell DA, Carlson CA, McNichol A, Key RM, Byrne RH, Wanninkhof R (2013b) dissolved inorganic carbon (DIC), total alkalinity, pH on total scale, partial pressure of carbon dioxide (pCO_2), temperature, salinity and other hydrographic and chemical variables collected from discrete samples and profile observations during the R/V Thomas G. Thompson Cruise CLIVAR_P16N_2006 (EXPOCODE 325020060213) in the Pacific Ocean from 2006-02-13 to 2006-03-30 (NODC Accession 0108062). Version 2.2. National Oceanographic Data Center, NOAA. Dataset. ▶ https://doi.org/10.25921/bvaz-zm17

Feely RA, Sabine CL, Dickson AG, Wanninkhof R, Hansell DA (2013c) Dissolved inorganic carbon (DIC), total alkalinity, temperature, salinity and other variables collected from discrete sample and profile observations during the R/V Roger Revelle Cruise CLIVAR_I08S_2007 (EXPOCODE 33RR20070204) in the Indian Ocean from 2007-02-04 to 2007-03-17 (NODC Accession 0108119). Version 2.2. National Oceanographic Data Center, NOAA. Dataset. ▶ https://doi.org/10.3334/CDIAC/OTG.CLIVAR-I08S-2007

Feely RA, Sabine CL, Millero FJ, Wanninkhof R, Hansell DA (2013d) Dissolved inorganic carbon (DIC), total alkalinity, pH on sea water scale, temperature, salinity and other variables collected from discrete sample and profile observations during the R/V Roger Revelle Cruise CD139, CLIVAR I09N_2007 (EXPOCODE 33RR20070322) in the Indian Ocean from 2007-03-22 to 2007-05-01 (NODC Accession 0110791). Version 2.2. National Oceanographic Data Center, NOAA. Dataset. ▶ https://doi.org/10.3334/CDIAC/OTG.CLIVAR-I09N-2007

Friedlingstein P et al. (2022) Global Carbon Budget 2022. Earth Syst Sci Data 14:3811–4900. ▶ https://essd.copernicus.org/articles/14/4811/2022/essd-14-4811-2022-discussion.html

Gafar NA, Eyre BD, Schulz KG (2019) Particulate inorganic to organic carbon production as a predictor for coccolithophorid sensitivity to ongoing ocean acidification. Limnol Oceanogr Lett 4:62–70

Gattuso J-P, Hansson L (2011. Ocean acidification. Oxford University Press, Oxford, p 408

Goyet C, Key RM, Sullivan KF, Tsuchiya M (1997) Carbon dioxide, hydrographic, and chemical data obtained during the R/V Thomas Washington Cruise TUNES-1 in the equatorial Pacific Ocean (WOCE Section P17C). ORNL/CDIAC-99, NDP-062. Carbon Dioxide Information Analysis Center, Oak Ridge National Laboratory, Oak Ridge, Tennessee. ▶ https://doi.org/10.3334/CDIAC/otg.ndp062

Green MA, Hishikawa Y, Dunlop ED, Levi DH, Hohl-Ebinger J, Ho-Baillie AWY (2018) Solar cell efficiency tables (version 51). Prog Photovolt Res Appl 26(1):3–12

IEA (International Energy Agency) (2020) Global energy review 2020. ▶ https://www.iea.org/reports/global-energy-review-2020/renewables. Accessed 11 Jan 2022

IPCC (Intergovernmental Panel on Climate Change) (1990) Climate Change. In: Houghton J, Jenkins G, Ephraums J (eds) The IPCC Scientific Assessment, Working Group I. Cambridge University Press, Cambridge, p 351. Available at ▶ https://www.ipcc.ch/site/assets/uploads/2018/03/ipcc_far_wg_I_full_report.pdf. Accessed 7 Feb 2022

IPCC (Intergovernmental Panel on Climate Change) (2013) In: Stocker TF, Qin D, Plattner G-K, Tignor M, Allen SK, Boschung J, Nauels A, Xia Y, Bex V, Midgley PM (eds) Climate Change 2013: the physical science basis. Contribution of Working Group I to the Fifth Assessment Report of the Intergovernmental Panel on Climate Change. Cambridge University Press, Cambridge, p 1535. Available at ▶ https://www.ipcc.ch/report/ar5/wg1/. Accessed 7 Feb 2022

IPCC (Intergovernmental Panel on Climate Change) (2018) Summary for policymakers. In: Masson-Delmotte V, Zhai P, Pörtner HO, Roberts D, Skea J, Shukla PR, Pirani A, Moufouma-Okia W, Péan C, Pidcock R, Connors S, Matthews JBR, Chen Y, Zhou X, Gomis XI, Lonnoy E, Maycock T, Tignor M, Waterfield T (eds) Global warming of 1.5 °C. An IPCC special report on the impacts of global warming of 1.5 °C above pre-industrial levels and related global greenhouse gas emission pathways, in the context of strengthening the global response to the threat of climate change, sustainable development, and efforts to eradicate poverty. World Meteorological Organization, Geneva, p 24. Available at ▶ https://www.ipcc.ch/sr15/chapter/spm/. Accessed 7 Feb 2022

IPCC (Intergovernmental Panel on Climate Change) (2021) In: Masson-Delmotte V, Zhai P, Pirani A, Connors SL, Péan C, Berger S, Caud N, Chen Y, Goldfarb L, Gomis MI, Huang M, Leitzell K, Lonnoy E, Matthews JBR, Maycock TK, Waterfield T, Yelekçi O, Yu R, Zhou B (eds) Climate Change 2021: the physical science basis. Contribution of Working Group I to the Sixth Assessment Report of the Intergovernmental Panel on Climate Change. Cambridge University Press, Cambridge (in press)

Kirschbaum MUF, Zeng G, Ximenes F, Giltrap DL, Zeldis JR (2019) Towards a more complete quantification of the global carbon cycle. Biogeosciences 16:831–846

Klaas C, Archer DE (2002) Association of sinking organic matter with various types of mineral ballast in the deep sea: implications for the rain ratio. Glob Biogeoch Cycles 16(4):63-1–63-14

Kroeker KJ, Kordas RL, Crim RN, Singh GG (2010) Meta-analysis reveals negative yet variable effects of ocean acidification on marine organisms. Ecol Lett 13:1419–1434

Kroeker KJ, Kordas RL, Crim R, Hendriks IE, Ramajo L, Sing GS, Duarte CM, Gattuso J (2013) Impacts of ocean acidification on marine organisms: quantifying sensitivities and interaction with warming. Glob Change Biol 19:1884–1896

Kvale K, Koeve W, Mengis N (2021) Calcifying phytoplankton demonstrate an enhanced role in greenhouse atmospheric CO_2 regulation. Front Mar Sci 7:583989

Le Quéré C, Andrew RM, Friedlingstein P, Sitch S, Hauck J, Pongratz J, Pickers PA, Korsbakken JI, Peters GP, Canadell JG, Arneth A, Zheng B (2018) Global carbon budget 2018. Earth Syst Sci Data 10(4):2141–2194

Maugendre L, Guieu C, Gattuso J-P, Gazeau F (2017) Ocean acidification in the Mediterranean Sea: Pelagic mesocosm experiments. a synthesis. Estuarine Coastal Shelf Sci 186:1–10

Mehrbach C, Culberson CH, Hawley JE, Pytkowicz RM (1973) Measurement of the apparent dissociation constants of carbonic acid in seawater at atmospheric pressure. Limnol Oceanogr 18:897–907

Pelejero C, Calvo E, Hoegh-Guldberg O (2010) Paleo-perspectives on ocean acidification. Trends Ecol Evol 25(6):332–344

Peltola E, Wanninkhof R, Feely RA, Hansell DA, Castle RD, Greeley D, Zhang J-Z, Millero FJ, Gruber N, Bullister JL, Taylor G (2013) Partial pressure (or fugacity) of carbon dioxide, dissolved inorganic carbon, pH, alkalinity, temperature, salinity and other variables collected from discrete sample and profile observations using alkalinity titrator, CTD and other instruments from NOAA Ship RONALD H. BROWN in the North Atlantic Ocean and South Atlantic Ocean from 2003-06-04 to 2003-08-11 (NODC Accession 0108061). Version 3.3. National Oceanographic Data Center, NOAA. Dataset. ► https://doi.org/10.3334/CDIAC/OTG.NDP085

Petrou K, Baker KG, Nielsen DA, Hancock AM, Schulz KG, Davidson AT (2019) Acidification diminishes diatom silica production in the Southern Ocean. Nat Clim Chang 9:781–786

Pongratz J, Reick CH, Raddatz T, Claussen M (2009) Effects of anthropogenic land cover change on the carbon cycle of the last millennium. Glob Biogeochem Cycles 23:Gb4001

Riebesell U, Gattuso JP (2014) Lessons learned from ocean acidification research. Nat Clim Change 5:12–14

Riebesell U, Körtzinger A, Oschlies A (2009) Sensitivities of marine carbon fluxes to ocean change. Proc Natl Acad Sci 106(49):20602–20609

Riebesell U, Aberle-Malzahn N, Achterberg EP, AlguerólMuñiz M, Alvarez-Fernandez S, Arístegui J, Bach LT, Boersma M, Boxhammer T, Guan W, Haunost M (2018) Toxic algal bloom induced by ocean acidification disrupts the pelagic food web. Nat Clim Change 8:1082–1086

Rossoll D, Bermúdez R, Hauss H, Schulz KG, Riebesell U, Sommer U Winder M (2012) Ocean acidification-induced food quality deterioration constrains trophic transfer. PloS one 0034737

Rost B, Riebesell U (2004) Coccolithophores and the biological pump responses to environmental changes. In: Thierstein H, Young J (eds) Coccolithophores: from molecular processes to global impact. Springer, Berlin, pp 99–125

Sarmiento JL, Gruber N (2006) Ocean biogeochemical dynamics. Princeton University Press, Princeton, p 503

Schulz KG, Riebesell U (2011) Versauerung des Meerwassers durch anthropogenes CO₂. In: Lozán J, Graßl H, Reise K (eds) Warnsignal Klima: Die Meere - Änderungen und Risiken. Universität Hamburg, Hamburg, pp 160–163

Schulz KG, Bach LT, Bellerby RGJ, Bermúdez R, Büdenbender J, Boxhammer T, Czerny J, Engel A, Ludwig A, Meyerhöfer M, Larsen A, Paul AJ, Sswat M, Riebesell U (2017) Phytoplankton blooms at increasing levels of atmospheric carbon dioxide: experimental evidence for negative effects on prymnesiophytes and positive on small picoeukaryotes. Front Marine Sci 4:64

Schulz KG, Hartley S, Eyre BD (2019) Upwelling amplifies ocean acidification on the East Australian Shelf: implications for marine ecosystems. Front Mar Sci 6:636

Tréguer P, Bowler C, Moriceau B, Dutkiewicz S, Gehlen M, Aumont O, Bittner L, Dugdale R, Finkel Z, Iudicone D, Jahn O, Guidi L, Lasbleiz M, Leblanc K, Levy M, Pondaven P (2018) Influence of diatom diversity on the ocean biological carbon pump. Nat Geosci 11:27–37

Waldbusser G, Hales B, Haley BA (2016) Calcium carbonate saturation state: on myths and this or that stories. ICES J Mar Sci 73:569–571

Wanninkhof R, Doney SC, Castle RD, Millero FJ, Bullister JL, Hansell DA, Warner MJ, Langdon C, Johnson GC, Mordy C (2013) Partial pressure (or fugacity) of carbon dioxide, dissolved inorganic carbon, pH, alkalinity, temperature, salinity and other variables collected from discrete sample and profile observations using alkalinity titrator, barometric pressure sensor and other instruments from NOAA Ship RONALD H. BROWN in the South Atlantic Ocean, South Pacific Ocean and Southern Oceans from 2005-01-11 to 2005-02-24 (NODC Accession 0108153). Version 3.3. National Oceanographic Data Center, NOAA. Dataset. ► https://doi.org/10.3334/CDIAC/OTG.NDP087

Zeebe RE, Wolf-Gladrow D (2001) CO₂ in seawater: equilibrium, kinetics, isotopes. Elsevier oceanography series, vol 65. Elsevier Science BV, Amsterdam, p 346

Other Important Marine Pollutants

Amanda Reichelt-Brushett and Sofia B. Shah

Contents

12.1   Introduction – 262

12.2   Noise Pollution – 262
12.2.1 Natural Sources of Sound in the Sea – 263
12.2.2 Anthropogenic Sources of Sounds in the Sea – 263
12.2.3 Effects of Anthropogenic Noises – 266

12.3   Light Pollution – 266

12.4   Thermal Pollution – 268

12.5   Particulates – 269
12.5.1 Particulate Organic Matter – 269
12.5.2 Suspended Sediments – 270

12.6   Pathogens – 271
12.6.1 Sources of Marine Pathogens – 272

12.7   Personal Care Products (PCPs) – 272
12.7.1 Triclosan and Triclocarban – 273
12.7.2 Sunscreens – 273

12.8   Non-native Species – 275

12.9   Summary – 278

12.10   Study Questions and Activities – 278

        References – 278

© The Author(s) 2023
A. Reichelt-Brushett (ed.), *Marine Pollution—Monitoring, Management and Mitigation*,
Springer Textbooks in Earth Sciences, Geography and Environment,
https://doi.org/10.1007/978-3-031-10127-4_12

Acronyms and Abbreviations

ALAN   Artificial light at night
ELP   ecological light pollution
NNS   Non-native species
NRC   National Research Council
NOAA   National Oceanic and Atmospheric Administration
PCPs   personal care products
STD   submarine tailings disposal also known as deep sea tailings placement (DSTP)
TCC   triclocarban
TCS   triclosan

12.1 Introduction

"*Sorry! What did you say?*" Consider how easy it is to miss some conversation details when it is noisy. Communication is an important aspect of all social interactions for animals, many use sound as a means of communication particularly longer distance communication.

"*Is it day or night?*" Light deprivation can have dramatic effects, so can too much.

"*Too hot, too cold?*" We all have our preferences but there are critical points of temperature ranges and rates of change that are detrimental to us and to all organisms on Earth.

This chapter introduces you to some forms of marine pollution that you might not immediately consider pollution. The point that helps to provide clarity about these pollutants is if they cause adverse effects. Indeed, they do. Commonly, the less easily recognised marine pollutants are not chemically based, but rather mechanical, physical or biological.

12.2 Noise Pollution

Sound is constituted by mechanical disturbance (or vibration) that moves through a material (Bradley and Stern 2008; Penar et al. 2020) and is a fundamental constituent of the marine environment. Sound propagates energy through the ocean and, like with sound movement through air, it moves in waves. The knowledge of the feature of sound is essential to fully understand the impacts of sound on marine organisms; and parameters such as frequency, wavelength and intensity best describe the characteristics of sound (Peng et al. 2015) (▶ Box 12.1). Water being denser than air is a great medium of sound conduction as sound propagates faster in the sea than in the air. Being a liquid, water is less compressible than air and therefore transmits the sound wave faster when compared to air.

Sound is a part of the natural seascape. Oceans are naturally noisy with natural sounds originating from a great variety of sources.

The ocean is intimately coupled to the geosphere and the atmosphere and as such, most of the significant physical sources of natural sound occur at the interfaces of these media (NRC 2003). For instance, as described by the National Research Council (NRC 2003), many sounds originate in the atmosphere and enter the ocean surface; and elastic vibrations in the earth introduce sound into the underwater acoustic field.

Sound is regarded as an important feature of marine habitats, with most marine species relying on it for critical life functions (Hawkins and Popper 2017; Southall et al. 2020). Many marine organisms use sounds as a means of communication, thus overcoming the many complications that living in the sea implies. Communication is a vital aspect of all social interactions (Butler and Maruska 2020), and sound is an important part of communication. In fact, animals rely on sound signals that encode information about the sender's species, sex, motivation, reproductive state and identity (Butler and Maruska 2020).

In contrast to sound, **noise** is more specific and defined as any unwanted or disturbing sound (Kunc et al. 2016) and there are varieties of sources of underwater noise. Underwater ambient (sound) noise is a component of background noise, and it varies depending on depth, time and location. According to the National Oceanic and Atmospheric Administration (NOAA) ocean noise refers to sounds made by human activities, which can interfere with or obscure the ability of marine animals to hear natural sounds in the ocean. Excess noise affects both the anatomy and morphology of an organism, by mechanically damaging single cells as

Box 12.1: Characteristics of Sound

Sound is a form of energy, which enables us to hear. Sound travels in the form of waves; which are vibratory disturbances in a medium carrying energy from one point to another. Sound can be described by five characteristics, namely: speed (or velocity); frequency; wavelength; amplitude and time period. The frequency (f) of the wave is the number of oscillations in a second. The speed (v) of the wave is the distance travelled by the wave in one second. The wavelength (λ) is the minimum distance in which a sound wave repeats itself. Amplitude (A) is the maximum displacement of the particles from their original undisturbed positions. Time period (T) is the time required to produce 1 complete wave. Relationship between period and frequency Eq. 12.1:

$$T = 1/f \text{ or } f = 1/T \tag{12.1}$$

Relationship between speed, frequency and wavelength Eq. 12.2 and 12.3:

$$V = f \times \lambda \tag{12.2}$$

$$V = \lambda \times 1/T \tag{12.3}$$

well as entire organs (Kunc et al. 2016). Sources of underwater noise can pose local impacts or regional and global impacts.

12.2.1 Natural Sources of Sound in the Sea

The underwater marine environment consists of **biotic** and **abiotic sounds** that are closely related to the survival and reproduction of marine organisms (Slabbekoorn et al. 2010). These natural sounds are both localized and dispersed and include surface waves, turbulence(wind), rainfall, water flow, seismic disturbances, cracking polar ice, and subsea earthquakes and volcanoes and sounds of biological origin (Bradley and Stern 2008; Peng et al. 2015; Hawkins and Popper 2017; Erbe et al. 2018). Natural biological sounds include whale songs, dolphin clicks and fish vocalizations among many others (NRC 2003). Background or ambient sound describes naturally occurring sounds from distributed sources. The combination of sounds produced by an ecosystem shows eco-acoustic complexity, and it is suggested that the more complex the natural soundscape, the healthier an ecosystem is (e.g. Linke et al. 2018; Di Iorio et al. 2021).

Fish, marine mammals, invertebrates and other marine organisms produce natural biotic sound in the marine environment, and these are regarded as biotic or biological sources of sound (e.g. ◻ Figure 12.1). Biotic sounds can be produced in many different ways such as rubbing parts of the body such as bones, teeth or the valve of shells, mechanical flapping of teeth or plates; and compression and decompression of the bladder through muscle strength. Biotic sound in the sea is used to communicate, navigate, locate and avoid prey, mate detection, and orientation including locating appropriate habitats and locations (e.g. Lillis et al. 2014; Simp-

son et al. 2016; Hawkins and Popper 2017; Lecchini et al. 2018). Erbe et al. (2017) reported that marine mammals have evolved to use sound as their primary sensory modality-both actively (sound production) and passively (sound reception). A passive mode of sound is when an animal does not actively generate sound impulses but only responds to them with a particular behaviour and these include identification of predators, capture of prey and direction change. Through active sound animals can communicate during mating, search for food, navigate over long distances, fight for territory and social disputes, distract a predator to escape, stun and catch prey, and produce alarm signals. **Echolocation** is the ability to gain information from sounds produced by the animal that bounce off distant objects and return as echoes and is necessary for navigation. Examples of mammals that use echolocation include Odontocetes, sperm whales, finback whales and other dolphins ► Box 12.2.

Natural biotic sources of sound usually occur over an extremely broad frequency range; spatially very limited in extent and occur over a short time (Bradley and Stern 2008) and provide important information to marine organisms about their surrounding environments. Organisms vary in their complete sensitivity and spectral range of hearing (Peng et al. 2015) (◻ Table 12.1). Abiotic sounds usually occur over a broad frequency range but they generally have a wide distribution and are generated over a long time (hours/days) (Bradley and Stern 2008).

12.2.2 Anthropogenic Sources of Sounds in the Sea

Noise generated intentionally and unintentionally from human activities is usually regarded as anthropo-

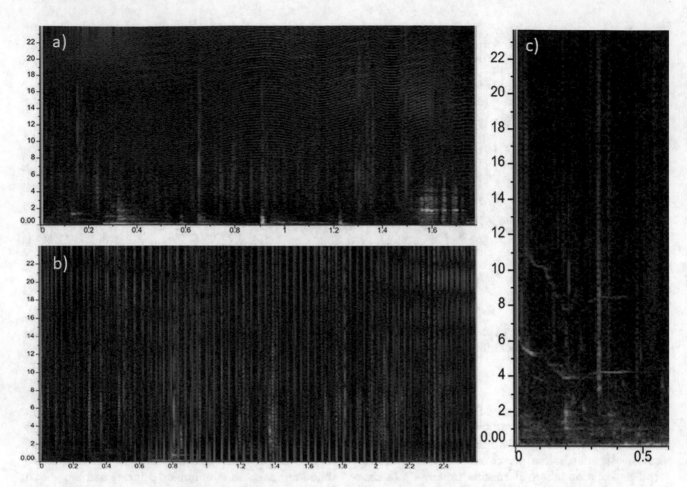

■ Figure 12.1 The spectrograms represent different sounds emitted by snubfin dolphins **a** burst pulse, **b** click train and **c** whistle. X axis represents seconds of the sound, y axis represents frequency (kHz) Click trains are mainly used for echolocation and are a unidirectional sound type. They are produced by directing clicks out of the melon of the dolphin towards a target. Whistles are omnidirectional and they are used for social activity and communication. There is not much knowledge on burst pulse but it is believed they are also used for hunting. *Images and caption text*: D. Cagnazzi

■ Table 12.1 Some examples of hearing range of marine species

Species	Hearing sensitivity	Source
Pacific bluefin tuna (*Thunnus orientalis*)	Most sensitive range 400–500 Hz	Dale et al. (2015)
Common prawn (*Palaemon serratus*)	Responsive to sound between 100–3000 Hz	Lovell et al. (2005)
Common octopus (*Octopus vulgaris*)	400–1000 Hz	Hu et al. (2009)

Box 12.2: Cetaceans, Seabirds and Ocean Noise

Cetacean and Ocean Noise

Marine mammals produce sound profusely for communication purposes. Odontocetes (toothed whales): emit echolocation clicks during foraging and navigation. Odontocete sounds are typically classified as whistles, burst-pulse sounds and clicks (Erbe et al. 2017) and produce mid and high frequency sounds around 1–150 kHz. Mysticetes (baleen whales): emits constant-wave tones, frequency-modulated sounds as well as broadband pulses (Erbe et al. 2017) and produce low-frequency sounds around 10–2000 Hz. Blue whales and humpback whales organsie sound into songs lasting for hours to days.

Marine mammals and other marine animals rely heavily on acoustics for navigation, hunting, reproduction and communication. Marine mammals such as whales and dolphins are highly adapted physiologically and behaviorally to utilize sound. Cetaceans are highly dependent on sound as their primary sense. Cetacean vocalizations cover a wide range of frequencies, from the infrasonic calls of the large Mysticetes (baleen whales) to the ultrasonic clicks of the Odontocetes (toothed whales) (Weilgart 2007) (see also ◘ Figure 12.1). The frequency of calls produced by the cetacean species is dependent on the body size. Larger body size correlates to lower frequency of calls; and cetacean calls and hearing span a broad range of frequencies because of highly sophisticated auditory systems. Mysticetes produce and use sound at the frequencies emitted by large ships, and are considered to be more sensitive at these low frequencies; whilst Odontocetes produce high frequency sounds as emitted by ships (see Erbe et al. 2019). It has also been shown that gray whales increased their vocalization rates and humpback whales have increased their vocalizations due to increased noise exposures from tourism vessels (Erbe et al. 2019).

Seabirds and Ocean Noise

Seabird families include Spheniscidae, Laridae, Stercorariidae, Procellariidae and Sulidae. Seabirds forage at sea but breed on land and hence use calls for communication to and from their colonies for their kin as well as partner recognition. Seabirds freely transit between air and water and enact key behaviours in both habitats (Mooney et al. 2019). However, with increasing human utilization of coastal areas, the soundscapes of these areas are changing. Anthropogenic noise seems to be a major stressor leading to the degradation of seabird habitat. Generally, noise pollution can affect birds by causing physical damage to ears, stress responses, changes in behavior, reproductive success, fright-flight responses, changes in vocal communications, habitat loss and changes in the ability to hear predators.

genic noise and more specifically in the marine environment as ocean or marine noise. Marine noise pollution is thus defined as any source of anthropogenic sound happening in the marine environment, which is capable of producing harmful effects on marine life.

Anthropogenic noise is a pervasive pollutant to almost all aquatic and terrestrial environments (Halfwerk and Slabbekoorn 2015). Many human activities generate noise within the hearing ranges of other animals, at sound levels above those found naturally and with different acoustic characteristics from natural sounds (Hildebrand 2009). The marine noise generated by human activities has amplified significantly since the industrial revolution (Frisk 2012); and hence the ocean is now reported to be 2–10 times louder compared to the preindustrial era (Hildebrand 2009; Frisk 2012). Escalating human population, coastal urbanisation, maritime traffic, oil extraction, civil and military sonars and ocean-based energy production systems (wind and wave energy farms) will continue to contribute to marine noise (di Franco et al. 2020).

Anthropogenic noises are multifaceted (Bradley and Stern 2008) and includes commercial shipping, oil and gas exploration, naval operations (e.g. military sonars, communications, and explosions), fishing (e.g. commercial/civilian sonars, acoustic deterrent, and harassment devices), dredging and drilling operations, marine renewable energy devices, research (e.g. air guns, sonars, telemetry, communication, and navigation) anti-predator devices, seismic surveys, cabling and other activities such as construction, icebreaking, and recreational boating (Hildebrand 2009; Jerem and Mathews 2020; Pieretti et al. 2020).

Underwater noise from shipping is a significant and pervasive pollutant with the potential to affect the marine ecosystems on a global scale (Clark et al. 2009; Williams et al. 2014; Merchant et al. 2015). In fact, ship noise is rising concomitantly with the increased use of shipping in transport and ships are becoming the most ubiquitous and pervasive source of anthropogenic noise in the oceans (Erbe et al. 2019; Vakili et al. 2020).

Anthropogenic noise can be categorised as either **high-intensity impulsive noise** or **low-frequency stationary noise** (Peng et al. 2015). High-intensity or acute marine noise pollution has a short duration and is often emitted repeatedly, over frequencies ranging from a few hertz (Hz) to hundreds of thousands of Hz. Low-intensity noise or chronic marine noise pollution has a longer duration with frequencies below 1 kHz (1000 Hz). Pile driving, underwater blasting, seismic exploration and active sonar application create high-intensity noise whilst ships and vessels generate low-frequency stationary noise (Codarin et al. 2009; Peng et al. 2015).

Chronic marine noise pollution is regarded as the main contributor to the increase in ocean background noise (Hildebrand 2009). Both acute and chronic marine noise pollution can co-occur and interact in producing their impact on marine life (di Franco et al. 2020).

12.2.3 Effects of Anthropogenic Noises

Increased anthropogenic noise imposes new **constraints on communication** (Vieira et al. 2021) such that it can interfere with the vocalizations emitted by many animals as well as the natural sounds that are used by animals for their routine behaviour. Biological sounds can be impaired by anthropogenic noise and possibly determine cascade effects at the population and community level (Kunc et al. 2016). In fact, noise exposure can change hearing capabilities.

The effects of anthropogenic underwater noise on aquatic life have become an important environmental issue (Thomsen et al. 2020) and a global concern which can cause auditory masking, behavioural disturbances, hearing damage and even death for marine animals (Peng et al. 2015; Halliday et al. 2020). Faulkner et al. (2018) further emphasised that underwater noise pollution poses a global threat to marine life and has become a growing concern for policy makers and environmental managers.

12.3 Light Pollution

Natural light at night is derived from the moon, the stars and the Milky Way (Ayalon et al. 2019; Duarte et al. 2019) whilst the day light is from the sun. The natural sources of light play a **fundamental role on the behavioural patterns of marine as well as terrestrial organisms** and the timing of the ecological processes (Ayalon et al. 2019; Duarte et al. 2019). The vast majority of species have evolved under natural and predictable regimes of moonlight, sunlight and starlight (Davies et al. 2014). Smyth et al. (2020) highlighted that photobiological life history adaptations to the moon and sun are near ubiquitous in the surface ocean (0–200 m), such that cycles and gradients of light intensity and spectra are major structuring factors in marine ecosystems.

With human population growth, the progress of energy supply for lighting and lighting technologies, **artificial light has steadily altered natural cycles** in many locations. Human population growth and migration to the coastal regions have led to an increase in the amount of lighting near coastal environments. Of emerging concern is the artificial light, which is now central to the functioning of modern society and concomitantly referred to as the **artificial light at night** (**ALAN**). ALAN is the alteration of natural light levels due to anthropogenic light sources (Cinzano et al. 2001; Falchi et al. 2016; Duarte et al. 2019) and is closely related to the rate of urban development, especially with the presence of outdoor night lights (Maggi and Serôdio 2020) (◻ Figure 12.2). Artificial light is an emerging threat to global biodiversity (Reid et al. 2019) and is escalating swiftly in coastal habitats due to rapid urbanisation, fisheries and aquaculture.

ALAN affects the adjacent abiotic environment both directly (through light sources of variable intensity) and indirectly (through the formation of a skyglow) (O'Connor et al. 2019; Maggi and Serôdio 2020). The skyglow is a diffuse light field of low intensity, and continuous lighting that is detectable as a glowing dome over built up areas such as coastal settlements and marine infrastructure, spreading its influence on sub-urban and rural sites (Gaston 2018; O'Connor et al. 2019) (◻ Figure 12.2b). Sources of ALAN include fixed lamps along the coastal streets, promenades, ports and marinas, lighthouses, oil platforms and from mobile sources such as commercial and tourist boats (O'Connor et al., 2019; Maggi and Serôdio 2020). The emission spectrum of the light sources creates vertical variability in the water column due to precise attenuation patterns amongst different wavelengths (Tamir et al. 2017).

Depledge et al. (2010) reported that light pollution occurs when organisms are exposed to light in the wrong place, at the wrong time or at the wrong intensity. Light pollution or **ecological light pollution (ELP)** describes all types of artificial light that alter the natural patterns of light and dark in ecosystems (Longcore and Rich 2004). Light pollution mainly affects nocturnal species by triggering unnatural processes that can result in important physiological and behavioral changes (Navara and Nelson 2007).

Artificial light disturbs a variety of fundamental biological processes such as the development of visual cells, pigmentation, growth, and development in the early life stages of fish (Boeuf and Le Bail 1999; O'Connor et al. 2019; Zapata et al. 2019); the structure and functions of invertebrate and fish communities in ecosystems (Davies et al. 2012; Zapata et al. 2019); harming biodiversity hotspots (Guette et al. 2018) spawning and settlement patterns of different species of corals and thus affecting the local and spatial community structure.

ALAN impacts species behaviour and inter-species interactions through the fluctuating visual surroundings (O'Connor et al. 2019). ALAN has prevalent effects on marine turtles (Tuxbury and Salmon 2005; Lorne and Salmon 2007; Dimitriadis et al. 2018), fish (Brüning et al. 2015; Pulgar et al. 2019), invertebrate communities (Jelassi et al. 2014; Underwood et al. 2017) and corals (Vermeij and Bak 2002; Gleason et al. 2006; Schlacher et al. 2007; Strader et al. 2015). Of concern is the light pollution on coral reefs since coral reef fishes depend on natural lunar cues to regulate reproductive periodicity in adults and the timing of reef-colonization by the larvae at the end of their pe-

lagic dispersal phase (Naylor 1999; Davies et al. 2013; Besson et al. 2017).

Several studies have highlighted the effect of ALAN on coastal organisms and habitats such as effects on settlement processes both in invertebrates and bacteria (Davies et al. 2015; Maggi and Benedetti-Cecchi 2018); changes in behaviour such as orientation and nesting of turtles, vertical migration of zooplankton and fish, antipredator and locomotor activities, trophic pressure (Witherington and Bjorndal 1991; Underwood et al. 2017; Ludvigsen et al. 2018; Duarte et al. 2019; Maggi et al. 2019) and the composition of assemblages (Garratt et al. 2019; Maggi et al. 2020). Furthermore, ALAN also has an effect on predator–prey interactions (Cravens et al. 2018; O'Connor et al. 2019; Yurk and Trites 2000); species phenology (Gaston et al. 2017; Bennie et al. 2018); foraging behaviour (Underwood et al. 2017; Farnworth et al. 2018); and orientation (Lorne and Salmon 2007). O'Connor et al. (2019) observed that light pollution causes changes in behaviour, physiological function and post-settlement survival in surgeonfish (*Acanthurus triostegus*) lar-

vae (◘ Table 12.2). The distance from a source that results in an insignificant effect of ALAN will vary between species.

Symbiotic corals are highly photosensitive and are likely to be susceptible to ecological light problems since they are often found in shallow, clear water with relatively high light levels (Rosenberg et al. 2019). Rosenberg et al. (2019) reported that human instigated ELP could alter the natural light regimes of coral reefs by causing persistence disturbance or chronic stress. Oxidative stress and physiological effects from exposure to ALAN had been observed for scleractinian corals, *Acropora eurystoma* and *Pocillopora damicornis*, from the Gulf of Eilat in the Red Sea, from exposure to blue LED and white LED lights (Ayalon et al. 2019).

Seabirds of the order Procellariiformes (such as shearwaters, petrels and albatrosses) are nocturnal (active at night) so as to avoid predation; exploit bioluminescent and vertically migrating prey and navigate the night sky. However, these seabirds are vulnerable to artificial light and easily get disoriented by intense

Table 12.2 Some examples of outcomes of research studies investigating the impacts of light pollution on larval fish

Species or population	Source of light	Impact of ALAN	Source
Acanthurus triostegus (coral reef fish)	Overhead light with dimmable smd5050 white LED strip lights (6500 k, λp=450 nm, placed 40 cm above the water surface, with a light intensity of 650–700 lx during the day (7 a.m.–7 p.m.). At night 20–25 lx from 7 p.m.–7a.m.	Larvae settled in dark areas. lowered thyroid hormone levels during metamorphosis; faster swimming rate, increased growth, decreased probability of survival, increased probability of predation	O'Connor et al. (2019)
Amphiprion ocellaris (anemonefish)	LED light programmed to 12:12 h light–dark photoperiod, measuring approximately 2000 lx during the day and 0 lx at night	ALAN at levels as low as ~15 lx resulted in significant negative effects on *A. ocellaris* reproductive success. Cool-white light (464 nm) had a greater impact on *A. ocellaris* hatching success than warm-white light (636 nm). Both light treatments resulted in smaller embryo sizes at the end of the developmental period	Fobert et al. (2021)

sources of artificial light. The vulnerability to artificial lighting varies between different species and age classes of birds and is influenced by season, lunar phase and weather conditions (Birdlife International 2012).

Sea turtles require regular intervals of natural diurnal and nocturnal light when they come ashore to lay their eggs (Silva et al. 2017). Light pollution tends to decrease the availability and suitability of sea turtle habitats and can become a crucial threat to entire sea turtle populations (Hopkins and Richardson 1984); especially nesting of adult marine turtles (Silva et al. 2017).

12.4 Thermal Pollution

Thermal pollution is the degradation of water quality due to changes in the ambient temperature of seawater, thus causing deleterious ecological effects. The influence of thermal discharges on aquatic ecosystems has become a significant issue in the field of marine and environmental protection. Thermal pollution can be caused by either hot or cold water discharges, and both the **rate** and **extent** of **temperature change** that deviates from normal conditions are important factors affecting marine organisms. Discharges from industrial activities are important sources of thermal pollution and in longer term, more subtle timeframes ocean warning from atmospheric change is considered a risk (e.g. Baag and Mandal 2022) and has been implicated in global coral reef bleaching events (Ainsworth et al. 2021). Community and ecosystem responses to thermal pollution include reduced species abundance, species richness and species diversity (e.g. benthic foraminiferal assemblages in Israel, Arieli et al. 2011) and may result in localised biological extinction (Dong et al. 2018).

A common cause of thermal pollution is the discharge of water, used as a coolant in industries, data storage centres and power plants (Abbaspour et al. 2005; Issakhov and Zhandaulet 2019; Mokhtari and Arabkoohsar 2021). Coolant waters may also contain contaminants such as metals through corrosion of infrastructure. When the coolant water is returned to the marine environment (usually at a higher temperature) it results in decreased availability of dissolved oxygen. Dissolved oxygen is essential for underwater life and if lacking may lead to deleterious effects such as fish kills (Speight 2020). An upsurge in seawater temperature also leads to seawater stratification (Huang et al. 2019). Seawater stratification occurs when isolated layers of water are formed with the upper warm layer (epilimnion) being separated by the cold layer (hypolimnion). Littlefair et al. (2020) stated that the hydrological layers give rise to distinct temperature and oxygen circumstances, thus creating different habitat niches for aquatic organisms which are adapted to particular temperature ranges (Speight 2020).

Many marine organisms have specific temperature needs, and hence rapid temperature changes can be deleterious (e.g. thermal shocks can result in reproduction difficulties and less resistance to diseases). Slower rates of temperature change can impact species if they exceed the upper (or lower) thermal tolerance level of species. Upper and lower **thermal tolerances** are not known for all species and research in this field highlights the complexity of the combination of responses to stressors including metabolic regulation, oxygen limitation and heat tolerance (e.g. Marshall and McQuaid 2020). Interestingly, the same species from different geographic locations can have different upper and lower thermal tolerances, highlighting population adaptability (Black et al. 2015). Furthermore, temperature change can result in increased sensitivity to other pollutants (Black et al. 2015) (see also ▶ Chapter 14). Notable consequences of artificial temperature rise include forced migration, massive fish kills as a result of slow-ing of metabolism, increased sensitivity to toxins, and loss of biodiversity.

Warming air temperatures over the past several decades have resulted in mass coral bleaching events in many parts of the world. The bleaching patterns vary spatially and temporally and are most common in tropical mid-latitudes (15°–20° north and south of the Equator) (Sully et al. 2019). Sully et al. (2019) further suggest that rates of change in sea surface temperatures are strong predictors of coral bleaching with faster rates of change correlating with higher levels of bleaching.

12.5 Particulates

Marine water quality is crucial for plants and animals that live in the sea; especially for marine species that rely on photosynthesis. **Water clarity** is an important water quality parameter and is a measure of how far light can penetrate through the water column. Light penetration is vital for the process of **photosynthesis** and contributes to the conditions that provide for the enormous diversity present in the ocean waters (◘ Figure 12.3). Kennicutt (2017) reported that access to sunlight is vital for the well-being of submerged aquatic vegetation, which aids as food and habitat for other biota. Water clarity is important as clear waters enable more sunlight to reach the photic zone, enabling the production of oxygen. Clear waters usually have low concentrations of suspended particles, and both natural and anthropogenic sources of suspended and dissolved solids affect water clarity (Kennicutt 2017). Dissolved substances as well as the productivity of phytoplankton also affect water clarity and colour. Floating

plastic particulates are also of concern and have been discussed specifically in ► Chapter 9.

12.5.1 Particulate Organic Matter

Marine phytoplankton; mostly single-celled algae and bacteria are extraordinarily diverse in morphology, evolutionary history, and biochemical behavior. They make up most of the **organic particulate matter** in seawater via the process of photosynthesis (Pilson 2013) and while essential to ecosystem structure they too can become a problem due to increased nutrient availability (► Chapter 4). The formation of organic matter from phytoplankton is referred to as **primary production**; where carbon dioxide, water and other nutrients in the presence of sunlight are converted to organic matter. Organic matter can also enter the marine environment from river discharges, from the atmosphere, from photosynthesis by larger fixed algae along the shores, and by bacterial chemosynthesis on parts of the ocean floor (Pilson 2013). Organic matter composed of algae, plants and other animals is regarded as **autocthonous** organic matter whilst those composed of terrestrial material are **allochthonous**. In the aquatic environment, organic matter can be present as particulate organic matter and dissolved organic matter.

The organic matter present in aquatic ecosystems is typically composed of proteins, lipids, carbohydrates, humic substances (e.g. humic and fulvic acids), plant tissues (rich in cellulose and hemicellulose) and animals and other acids of different molecular weights (Benner 2003). Organic matter can undergo transformations in the water column of aquatic environments and later become part of the sediments. In aquatic eco-

◘ **Figure 12.3** Scleractinian corals host dinoflgellates (Symbiodinium) that provide them with photosynthetically derived nutrients. These algae live within the coral and need access to light for photosynthesis. Picture **a** shows the high level of water clarity often associated with coral reefs, Pig Island, Papua New Guinea and **b** at times reefs are exposed to more turbid conditions during natural or anthropogenic disturbances. Species composition in consistently turbid waters can be markedly different from areas with high clarity waters. *Photos*: A. Reichelt-Brushett

systems, sediments may receive large amounts of organic matter and as it settles through the water it provides essential energy for the deep sea. Most deep sea ecosystems are **heterotrophic**, waiting for food to sink from the euphotic zone (<200 m) and the surface production can vary both temporally and spatially resulting in variable deposition of organic matter to the sea floor (Ramirez-Llodra et al. 2010 and authors there in).

12.5.2 Suspended Sediments

Sediments are principally unconsolidated materials, products of modification of rocks, soils, and organic matter that have undergone weathering, transportation, transformation and deposition near the Earth's surface or in water bodies (Cardoso et al. 2019). Sediments at the bottom of the oceans have formed from particulate matter that settles out of the water column and may consist of coarse gravel, sand, clay and organic ooze, together with contaminants. Sediments in the oceans are repositories for physical and biological debris, and serve as sinks for a wide variety of chemicals. In aquatic ecosystems, sediments provide habitat and substrate for a wide variety of benthic organisms and chemicals that bind to sediment particles can cause grave pollution problems (Chapters 1, 2, 4, 5 and 6).

Regardless of any chemicals associated with suspended sediments the particles themselves can result in deleterious impacts on organisms and communities. As a physical pollutant, suspended sediments cause:

- reduce water clarity and limit the depth sunlight can penetrate for photosynthesis to occur;
- excess fine sediments can injure gills of some types of fish and shell fish;
- reduced visibility from reduced water clarity causes a reduction in the number of organisms that use visual methods to seek prey and hide from predators; and
- when they eventually settle sediments may smother sessile species resulting in death (◻ Figure 12.4).

◻ **Figure 12.4** **a** and **b** sediment smothered coral around Henning Island, Whitsunday Islands, Australia, **c** after a rainfall event turbid river waters mixing with ocean water at the mouth of the Richmond River, NSW, Australia, **d** terrestrial inputs of suspended sediment to the ocean, Eastern Indonesia. *Photos*: A. Reichelt-Brushett

Figure 12.5 Capital dredging works in Platypus Channel, Cleveland Bay, Australia in 1991 (insert bottom right). In the main picture suspended sediments can be seen drifting from the dredge site to the shores of Magnetic Island. The trailing suction hopper dredge in action. *Photos*: Dredging Assessment Project Team, James Cook University, 1991

Enhanced turbidity can be generated naturally from storms and terrestrial runoff, however, vegetation clearing in catchments and poorly managed riparian zones enhance soil loss to waterways which is subsequently transported to the ocean (◧ Figure 12.4d). Port and harbour facilities for shipping require relatively deep water and entrances often need to be dredged to establish and maintain access. Dredging re-mobilises deposited sediment which can then be transported by tidal currents, settling in areas of low velocity. ◧ Figure 12.5 shows capital dredging works in 1991, in Cleveland Bay, Townsville, Australia. Sediments drifted from the dredging and dump sites and settled on coral surfaces around Magnetic Island offshore from Townsville (Reichelt and Jones 1994).

Throughout the world there are numerous examples of submarine tailings disposal (STD) also known as deep sea tailings placement (DSTP) (Vare et al. 2018). During STD operations tens to hundreds of thousands of tailings waste from terrestrial mining activities are discharged on a daily basis to the ocean at a depth between 80 and 150 m. The site of pipeline discharge is generally placed at the edge of a continental shelf, and tailings are meant to fall down the continental slope to rest in canyons (▶ Chapter 5, ◧ Figure 5.3). The process results in the smothering of marine species in the impact zone and contributes to turbidity in the water column due to plume sheering (Reichelt-Brushett 2012; Stauber et al. 2022).

12.6 Pathogens

Pathogens are organisms that cause disease to their host, with the severity of the disease symptoms referred to as virulence (Balloux and van Dorp 2017). They are widely diverse taxonomically and consist of bacteria, viruses, fungi and some parasites as well as unicellular and multicellular eukaryotes, potentially harmful to humans, marine species and ecosystems. Host–pathogen relationships are capable of influencing population dynamics, community structure, and biogeochemical cycles, and these are expected to shift in response to global climate change (Cohen et al. 2018).

Pathogens can be found in association with marine animals, phytoplankton, zooplankton, sediments and detritus (Stewart et al. 2008). Environmental factors such as salinity, temperature, nutrients and light influence the survival and sometimes the proliferation of pathogens (Stewart et al. 2008). Diseases have been identified as a major contributor to the decline of corals worldwide, particularly in the Western Atlantic (Bourne et al. 2009). The causes of disease are either from new pathogens or changed environmental conditions that affect the host, pathogen, environment relationship (◧ Figure 12.6).

Pathogens cause illness to their hosts in a variety of ways such as direct damage of tissues or cells during replication, or through the production of toxins. Bacterial toxins are among the deadliest poisons known

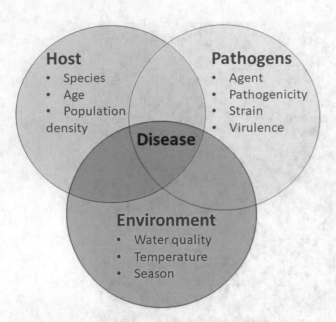

■ **Figure 12.6** Host, pathogen and environmental factors that contribute to the causes of disease. By managing the interacting factors well, the further the circles will separate, reducing the size of the disease risk. *Image*: A. Reichelt-Brushett

and include famous examples such as tetanus, anthrax or botulinum taoxin (Balloux and van Dorp 2017). The majority of the antibiotic classes are originally derived from bacteria and fungi (some of which are derived from the marine environment); with 64% of antibiotic classes being derived from filamentous actinomycetes (Gomes et al. 2021). Marine actinomycetes produce secondary metabolites that show a range of biological activities including antibacterial, antifungal, anticancer, insecticidal and enzyme inhibition. Marine pathogens and other parasites play important roles in composing the makeup, diversity, and health of natural marine communities (Baskin 2006). They may also be responsible for a broad spectrum of acute and chronic human diseases such as gastroenteritis, ocular and respiratory infections, hepatitis, myocarditis, meningitis, and neural paralysis (Brettar et al. 2007).

Pathogens are divided as facultative or obligate pathogens depending on how intimately their life cycle is tied to their host (Balloux and van Dorp 2017). Facultative pathogens are primarily environmental bacteria and fungi that can occasionally cause infection and include many of the hospital-acquired bacteria involved in the antimicrobial resistance pandemic (Balloux and van Dorp 2017). However, obligate pathogens necessitate a host to fulfil their life cycle. For instance, all viruses are obligate pathogens as they are dependent on the cellular machinery of their host for their reproduction (Balloux and van Dorp 2017).

12.6.1 Sources of Marine Pathogens

Pathogens artificially introduced to the marine environment get carried via sewage effluent, ship ballast water, agricultural runoff (defecation/urination/shedding from human or animal hosts), stormwater runoff, human recreational, industrial processes, introduction of exotic species and plastics (Baskin 2006). Non-host environments such as water, decaying organic matter and abiotic surfaces are important constituents of the lifespan of some pathogens since these environments provide habitats in which pathogens may replicate or survive; thus facilitating transmission (Lanzas et al. 2019).

Corals belong to the phylum Cnidaria, which consist of organisms including jellyfish, anemones, and hydra that form polyps with stinging cells and scleractinian corals which are the major reef building corals. Coral reefs are declining world-wide as a result of global changes; and one of the factors of concern include destructive diseases. Sharma and Ravindran (2020) mentioned that pathogens and parasites causing infectious diseases of scleractinian or stony corals especially in India include bacteria, fungi, viruses, and parasitic infections by protozoans, metazoans and parazoans; which leads to partial or entire-colony mortality. These infectious diseases cause lesions or bands of tissue loss on the coral colonies, thus affecting the entire reef ecosystem (Sokolow 2009). Diseases lead to significant alterations in coral reproduction and growth rates, thus changing community structure, species diversity and abundance of reef-associated organisms (Loya et al. 2001). White band disease on *Acropora palmate* and *Acropora cervicornis* in the 1980s caused an estimated 95% reduction in colonies (Vollmer and Kline 2008). White pox disease in the Florida Keys reduced the cover of *Acropora palmata* by up to 70% (Patterson et al. 2002).

12.7 Personal Care Products (PCPs)

Personal care products (PCPs) are intended for external application on the human body and generally enter the environment unaltered during water recreation, washing, showering or bathing and are considered as **emerging pollutants** (▶ Chapter 13). PCPs that usually reach the aquatic environment are bioactive, pseudo-persistent, exhibit a high degree of bioaccumulation in aquatic organisms (Cortez et al. 2012; Montesdeoca-Esponda et al. 2018) and have been shown to impact marine organisms (Câmara et al. 2021). The environmental fate of PCPs depends on their physico-chemical properties such as water solubility, adsorption behaviour, volatility and degradability (Montesdeoca-Esponda et al. 2018). Consequently little is known

Table 12.3 Some subgroups of personal care products (PCPs) and example compounds

Subgroups of PCPs	Example compounds
Antimicrobial agents/disinfectants	Triclosan Triclocarban
Synthetic musks/fragrances	Galaxolide (HHCB) Toxalide (AHTN)
Insect repellents	N,N-diethyl-m-toluamide (DEET)
Preservatives	Parabens (alkyl-p-hydroxybenzoates)
Sunscreen UV filters	2-ethyl-hexyl-4-trimethoxycinnamate (EHMC) 4-methyl-benzilidine-camphor (4MBC)

about the fate and the toxicity of PCPs introduced into the environment, hence, increasing attention is being placed on their occurrence, persistence, and potential threats to aquatic environment and human health.

PCPs include a large and diverse group of organic compounds used in disinfectants, soaps, shampoos, lotions, skin creams, toothpaste, fragrances/synthetic musks, sunscreens, insect repellants, and preservatives. The primary classes of personal care products include disinfectants (e.g. triclosan), fragrances (e.g. musks), insect repellants (e.g. DEET), preservatives (e.g. parabens) and UV filters (e.g. methylbenzylidene camphor) (Brausch and Rand 2011). UV filters are used to protect skin from UV solar radiation and usually contains chemicals of different chemical families such as benzimidazoles, camphor derivatives, triazines, benzotriazoles, cinnamates, salicylates, benzophenones, p-aminobenzoates (Câmara et al. 2021).

Many of these compounds are environmentally persistent, bioactive, potentially bioaccumulative and have lipophilic characteristics (Peck 2006; Mackay and Barnthouse 2010; Brausch and Rand 2011). Table 12.3 identifies some subgroups of PCPs and the characteristic compounds present in them see also ▶ Box 12.3.

12.7.1 Triclosan and Triclocarban

The PCPs Triclosan (TCS) and triclocarban (TCC) two distinctive **antimicrobial agents** used in soaps, deodorants, skin creams, toothpaste and plastics, among other things (see US EPA 2008); are frequently detected in seawater (McAvoy et al. 2002; Liu and Wong 2013); and are amongst the top 10 most commonly detected

organic wastewater compounds for frequency and concentration (Kolpin et al. 2002; Halden and Paull 2015). The effectiveness of TCS against gram-negative and gram-positive bacteria resulted in its widespread use (Cortez et al. 2012). TCS is regarded as an environmental concern due to its photodegradation into dioxins and furans; structural similarity to Bisphenol-A; biological methylation due to formation of more toxic compounds; and its bioaccumulative and toxic nature (see Cortez et al. 2012) (Table 12.4). The TCS molecule possesses both the phenol (5-chloro-2(2,4-dichlorophenoxy) phenol) and ether (2,4,4-trichloro-2-hydroxydiphenylether) functional groups (Olaniyan et al. 2016). TCS consists of multiple halogen atoms and is highly xenobiotic; and hence many microorganisms lack the necessary metabolic pathways and enzymes to degrade it (Abbot et al. 2020) and therefore it is highly persistent and bioaccumulates in the environment (Halden 2014). Cortez et al. (2012) demonstrated via laboratory assays that TCS caused acute and chronic toxicity to gametes and embryos of *Perna perna* at concentrations not yet reported in marine surface waters. TCC (3-(4-chlorophenyl)-1-(3,4-dichlorophenyl)) is used as a broad spectrum antibacterial and antifungal agent in many personal care products. It is a trichlorinated binuclear aromatic compound which has toxic, persistent and bioaccumulating properties (Halden 2014). TCC concentrations have been measured at 6.75 µg/L in raw wastewater (Halden and Paull 2015).

12.7.2 Sunscreens

Sunscreens are of emerging concern both to human and environmental health; however, their regulation is constantly evolving, largely due to the potential risks related to the ingredients they contain (Labille et al. 2020). Sunscreens typically consist of an oil–water emulsion in which the major active ingredients are UV filters, incorporated in high concentrations (Labille et al. 2020). Sunscreen products contain active constituents that protect human skin from UV radiation. Ramos et al. (2015) reported that UV filters and stabilizers are assimilated into a wide range of manufactured products to provide protection from UVA (315–400 nm) and UVB (280–315 nm) radiation. These include organic compounds that absorb UV rays (e.g. cinnamates, camphor derivatives, benzophenones) and/or inorganic compounds (e.g. TCC (3-(4-ch, TiO_2 and ZnO), which act as chemical or physical filters preventing or limiting UV penetration (Corinaldesi et al. 2017; Carve et al. 2021).

12

Table 12.4 Some examples to ecotoxicological studies that assess the effects of personal care products (PCPs) on marine Biota

Species/population	Source and type of PCP	Impact	Citation
Marine mussel (*Perna perna*)	Triclosan (TCS) Prepared in the laboratory by mixing TCS with dimethyl sulfoxide	Laboratory assay results: Fertilization assay: fertilization was inhibited at mean concentration of 0.49 ± 0.048 mg/L Embryo-larval development was inhibited by 0.135 ± 0.028 mg/L Cytotoxicological assays: adverse effects at concentrations of 1200 ng/L and 12,000 ng/L. laborator results indicate acute and chronic toxicity to gametes and embryos of *P.perna*	Cortez et al. (2012)
King mackerel (*Scomberomorus cavalla*), leatherjacket (*Oligoplites saurus*), Mullet (*Mugil incilis*), Gafftopsail catfish (*Bagre marinus*), White sea catfish (*Genidens barbus*), crevalle jack (*Caranx hippos*)	Triclosan (TCS)	Detected in seawater samples but not in fish muscles. (perhaps bioaccumulation occurred in the liver which was not tested)	Pemberthy et al. (2020)
Marine bivalve (*Ruditapes philippinarum*)	Triclosan (TCS) OTNE (octahydro-tetramethyl-naphthalenyl-ethanone) BP-3: benzophenone-3 OCR: octocrylene TiO$_2$: inorganic UV filter	Accumulation of inorganic and organic PCPs in the clam tissue was of the order: BP-3>TiO$_2$>TCS>OC>OTNE BP-3 and TCS accumulated in the tissue at higher concentartions	Sendra et al. (2017)
Coral planula (*Stylophora pistillata*)	BP-3	In both darkness and light, BP-3 transformed planulae from a motile state to a deformed sessile condition. Coral bleaching occurred with increasing doses of BP-3	Downs et al. (2016)
Corals (adult) *Acropora* spp. *Stylophora pistillata* *Millepora complanata*	BP-3 OCR OMC (ethylhexylmethoxycinnamate) EHS (ethylhexylsalicylate)	Addition of low quantities of sunscreen resulted in release of coral mucuos; thus resulting in coral bleaching	Danavaro et al. (2008)
Limpets (*Cymbula oculus* and *Cymbula granatina*), Mussels (*Mytilus galloprovincialis*), Sea snail (*Oxystele sinensis* and *Oxystele tigrina*), Sea urchin (*Parechinus angulosus*), Starfish (*Marthasterias glacialis*)	TCS	Not detected	Ojemaye and Petrik (2022)

Box 12.3: Some Personal Care Products (PCPs) of Concern

4-methylbenzylidene-camphor (4-MBC)

This compound is an organic UVB filter and is referred to as enzacamene. It is a high lipophilic component, easily absorbed through the human skin and exhibits a toxic activity as estrogenic endocrine disruptor. 4-MBC shows myriad effects on aquatic organisms, for instance, laboratory studies have shown that 4-MBC causes oxidative stress to an aquatic protozoan, *Tetrahymena thermophile,* resulting in inhibited growth and developmental defects in embryonic zebrafish (Li et al. 2016); toxicity to *Mytilus galloprovincialis* and *Paracentrotus lividus* (Paredes et al. 2014); and reduced growth, alterations on behaviour, imbalance of neurotransmission related endpoints and decreased enzyme activity were reported in Senegales Sole due to varied concentrations of 4-MBC (Araujo et al. 2018).

benzophenone-3

This compound is also referred to as oxybenzone or 2-hydroxy-4-methoxphenyl phenylmethanone and is a class of organic UV filter that is used in organic products to prevent burning of the skin by UVA and UVB radiation. Benzophenone-3 is known to cause a bleaching effect to coral, inhibiting growth and possibly killing the organism; and causing the mobile planulae to become deformed and trapped within its own calcium carbonate skeleton (Downs et al. 2016). Oxybenzone is also an active ingredient in PCPs including body fragrances, hair styling products, shampoos and conditioners, antiaging creams, insect repellants, as well as hand soaps (CIR 2005). In addition, the oxybenzone sunscreens can promote viral infections in corals, resulting in additional bleaching events (Danovaro et al. 2008). Oxybenzone can cause deformities in juvenile coral and damage their DNA and is a skeletal endocrine disrupter.

As the number of vacationers visiting the world's oceans increases, the rate of sunscreen inadvertently washed into these marine environments also rises. UV filters enter the environment directly from sloughing off while swimming and other recreational activities or indirectly via effluent from waste water treatment plants (Brausch and Rand 2011). Whilst these compounds have relatively short half-lives in seawater; they are continuously reintroduced via recreational activities and wastewater discharge, making them environmentally persistent (Horricks et al. 2019). UV filters are considered to be ubiquitous environmental contaminants of increasing concern, due to their bioaccumulation potential, and as endocrine disruptors (Ozáez et al. 2013).

Miller et al. (2021) mentioned that UV filters used in sunscreens and other PCPs may impact coral health on a local scale and also affect other marine species (► Box 12.3). Research studies have suggested that exposure of corals to several widely used UV filters have produced negative health effects including bleaching and mortality (see Miller et al. 2021). Research findings of Danovaro et al. (2008) and Downs et al. (2014, 2016) raised public concern and Hawaii became the first place to take legislative action to ban Benzophenone-3 (BP-3) and octinoxate (EHMC).

12.8 Non-native Species

Non-native species (NNS) (synonyms: exotic, alien taxa, non-indigenous, allochthonous, introduced) are species, sub-species or lower taxa introduced outside of their natural range and outside of their natural dispersal potential, dispersed by direct or indirect, intentional or unintentional human activities (Walther et al. 2009; Occhipinti-Ambrogi and Galil 2010; Rotter et al. 2020). NNS can be introduced and spread to waters through several different pathways (e.g. Alidoost Salimi et al. 2021). The major threat to indigenous species diversity and community structures occurs as a result of human-mediated introduction of a marine species outside their natural range of distribution (Rotter et al. 2020). NNS are a component of global change in all marine coastal ecosystems (Occhipinti-Ambrogi 2007) since they are a major threat to global biodiversity. Scientists and policy makers increasingly see the introduction of alien species as a major threat to marine biodiversity and a contributor to environmental change (Bax et al. 2003).

Harbours are known introduction foci of NNS, acting as recipients of new introductions and as sources for regional spread (Peters et al. 2017). Aquaculture is another primary pathway of the introduction of NNS (e.g., Wang et al. 2021), Additional to these, NNS can **hitch-hike** clinging to scuba gear between uses, attached to marine litter or debris, or in consignments of live organisms traded as live bait and plants and animals destined for the aquarium trade (Ruiz et al. 1997; Bax et al. 2003; Godwin 2003; Padilla and Williams 2004; Cagauan 2007; Molnar et al. 2008; De Silva et al. 2009; Anderson et al. 2015; Wang et al. 2021). Interestingly, NNS have been used in restoration programs and for biological control, sometimes with devastating consequences.

Table 12.5 Some examples of Non-Native Species (NNS) impacting locations of intentional or unintentional introduction

Species or population	Location (obtained from)	Introduced to	Impact	Citation
Red mangrove (*Rhizophora mangle*)	South Florida	Moloka 'i, Hawai 'i O'ahu	Introduced to stabilise shorelines Successful invasion on the island of Hawaii, obnoxious odors (Anoxic conditions) and clogging of tidal streams	Chimmer et al. (2006), Allen (1998)
Seagrass (*Halophila stipulacea*)	India	Eastern continent of Africa, Madagascar, the Red Sea and the Persian Gulf	Fast spreading, competes with native seagrasses	Den Hartog (1970), Short et al. (2007), Willette et al. (2014)
Orange keyhole sponge (*Mycale grandis*)	Indonesia	Hawaii	Threat to corals	Coles et al. (2007), Coles and Bolick (2007)
Encrusting sponge (*Chalinula nematifera*)	Indo-Pacific	Isla Isabel National Park	Overgrows corals, dead skeletons and coralline algae	Ávila and Carballo (2009), Turicchia et al. (2018)
Hard corals (*Tubastrea coccinea, Tubastrea tagusensis,* and *Tubastrea micranthu*)	Indo-Pacific	Brazil	Caused significant environmental, economic, and social impacts	Creed (2006), Lages et al. (2011)
Predatory seastar (*Asterias amurensis*)	native to the coasts of northern China, North Korea, South Korea, Russia and Japan	Tasmania, Australia	Threat to mariculture and wild shellfish fisheries	Ross et al. (2003)
Devil firefish (*Pterois miles*) and red lionfish (*P. volitans*)	Aquarium trade	Atlantic waters off south east USA, Gulf of Mexico and the Caribbean	Threat to biodiversity, outcompete native species	Ballew et al. (2016)

12

Some NNS can have slight impacts within their new habitat; whilst others can become invasive and pose serious threats affecting marine biodiversity, coastal economies, local cultures and livelihoods and human health. If NNS succeed in attaining high abundances, then they have the potential to displace native species, disturb ecosystem processes and function, change community structure, impact human health, decrease native biodiversity and cause substantial economic losses (Mack et al. 2000; Grosholz 2002; Bax et al. 2003; Simberloff 2005; Ojaveer et al. 2015). Furthermore, NNS may bring with them new diseases and parasites, and genetic modifications (e.g. aquaculture species) (Cook et al. 2016).

Davidson et al. (2015) reported that marine macroalgae are a major constituent of NNS worldwide, having current estimations of introductions in excess of 300 species. The NNS usually have fast growth rates, morphological plasticity, production of tetraspores in abundance and grow on other algae (Russell 1992; Smith et al. 2002). Red alga *Kappaphycus alvarezii* was widely farmed in the Philippines in the 1960s (Bixler 1996; Sulu et al. 2004); in Hawaii from 1970 (Conklin and Smith 2005); and in the Gulf of Mannar from 1990 (Kamalakannan et al. 2014); and established populations have spread outside the farmed areas in India, Tanzania, Panama, Venezuela,

Hawaii and Fiji (Rodgers and Cox 1999; Ask et al. 2003; Chandrasekaran et al. 2008; Sellers et al. 2015). *Kappaphycus alvarezii* is cultured in close proximity to coral reef ecosystems such as Kāneʻohe Bay, Hawaiʻi and the Gulf of Mannar, India (Rodgers and Cox 1999; Chandrasekaran et al. 2008) and tends to reduce the density and diversity of native fish and decreases the species richness and abundance of native macroalgae, coral and other benthic macrofauna (Neilson et al. 2018; HISC 2019).

Other examples of red algae including *Gracilaria Salicornia*, *Acanthophora spicifera* and *Hypnea musciformis* have been known to cause problems in Hawaii (Alidoost Salimi et al. 2021). *Gracilaria Salicornia* leads to the acidification of water, causing coral reef deterioration (Martinez et al. 2012) whilst *A. spicifera* and *H. musciformis* were observed to smother corals and algae (Smith et al. 2002). Further examples of other NNS that have had significant environmental impacts are highlighted in ◘ Table 12.5.

There are wide ranging action programs to deal with invasive marine species and websites dedicated to educating people about these and their impacts on biodiversity. Some examples include:

— Lionfish:　　　► http://lionfish.gcfi.org/index.php
　(◘ Figure 12.7)

◘ **Figure 12.7**　Lionfish, native to the Indo-Pacific are a pest species in the Atlantic Ocean off south east USA, the Gulf of Mexico and the Caribbean. *Photo*: A. Reichelt-Brushett

- Northern Pacific Seastar: ▶ https://dpipwe.tas.gov.au/conservation/the-marine-environment/marine-pests-and-diseases/pest-identification/northern-pacific-seastar
- Black Striped Mussel: ▶ https://nt.gov.au/marine/for-all-harbour-and-boat-users/biosecurity/aquatic-pests-marine-and-freshwater/black-striped-mussel
- Red Mangroves: ▶ https://malamaopuna.org/our-work/past-work/mangrove-removal-waiopae/.

It is helpful to tackle an invasive species problem when it is relatively small and when the reproductive effort of the introduced population is not at its maximum. There are basic frameworks to develop invasive marine species management plans that can help direct effort to ensure cost-effective returns. Some programs are government-funded, but others are driven by community organisations and partnerships.

12.9 Summary

This chapter has introduced you to some of the many other marine pollution problems that are being tackled in many different ways. It is a diverse chapter highlighting sources and impacts from noise, light, temperature, particulates, pathogens, personal care products, and non-native species. To understand each of these topics requires focused research activity, and this research highlights the need to develop solutions. Mitigating marine pollution is an essential research area now and in the future. ▶ Chapter 15 provides a helpful introduction to marine pollution mitigation and habitat restoration.

12.10 Study Questions and Activities

1. Investigate the upper and lower thermal tolerances of some marine species. Record your findings.
2. Find out about the biology of a non-native species (NNS) that has been introduced in your home country. What are the ecological and economic consequences of this introduction?
3. How can suspended sediment impact marine species? Provide an example of a species that may be impacted in each of the ways described in the dot points of ▶ Section 12.5.2.
4. This chapter is unlikely to have covered all the 'other' pollutants in the marine environment that have not had chapters dedicated to them in this textbook. Form a group and discuss pollutants that have not been covered in this textbook. Send your recommendation to the Editor as there might just be a second edition of the book.

References

Abbaspour M, Javid AH, Moghimi P, Kayhan K (2005) Modeling of thermal pollution in coastal area and its economical and environmental assessment. Int J Environ Sci Technol 2(1):13–26

Abbott T, Kor-Bicakci G, Islam MS, Eskicioglu C (2020) A review on the fate of legacy and alternative antimicrobials and their metabolites during wastewater and sludge treatment. Int J Mol Sci 21:9241

Ainsworth TD, Leggat W, Silliman BR, Lantz CA, Bergman JL, Fordyce AJ, Page CE, Renzi JJ, Morton J, Eakin CM, Heron SF (2021) Rebuilding relationships on coral reefs: coral bleaching knowledge-sharing to aid adaptation planning for reef users. BioEssays 43(9):2100048

Alidoost Salimi P, Creed JC, Esch MM, Fenner D, Jaafar Z, Levesque JC, Montgomery AD, Alidoost Salimi M, Patterson Edward JK, Diraviya Raj K, Sweet M (2021) A review of the diversity and impact of invasive non-native species in tropical marine ecosystems. Mar Biodivers Rec 14:11

Allen J (1998) Mangroves as alien species: the case of Hawai'i. Glob Ecol Biogeogr Lett 7(1):61–71

Anderson LG, Roclitte S, Haddaway NR, Dunn AM (2015) The role of tourism and recreation in the spread of non-native species: a systematic review and meta-analysis. PLoS ONE 10:0140833

Araujo MJ, Rocha RJM, Soares AMVM, Benede JL, Chisvert A, Monteiro MS (2018) Effects of UV filter 4-methylbenzylidene camphor during early development of Solea senegalensis Kaup, 1858. Sci Total Environ 628:1395–1404

Arieli RN, Almogi-Labin A, Abramovich S, Herut B (2011) The effect of thermal pollution on benthic foraminiferal assemblages in the Mediterranean shoreface adjacent to Hadera power plant (Israel). Mar Pollut Bull 62:1002–1012

Ask EI, Batibasaga A, Zertuche-Gonzalez JA, De San M (2003) Three decades of Kappaphycus alvarezii (Rhodophyta) introduction to non-endemic locations. In: Proceedings of the 17th international seaweed symposium, Cape Town, South Africa, vol 17, pp 49–57

Ávila E, Carballo JL (2009) A preliminary assessment of the invasiveness of the Indo- Pacific sponge Chalinula nematifera on coral communities from the tropical Eastern Pacific. Biol Invasions 11(2):257–264

Ayalon I, Marangoni LFD, Benichou JIC, Avisar D, Levy O (2019) Red Sea corals under artificial light pollution at night (ALAN) undergo oxidative stress and photosynthetic impairment. Glob Change Biol 25:4194–4207

Baag S, Mandal S (2022) Combined effects of ocean warming and acidification on marine fish and shellfish: a molecule to ecosystem perspective. Sci Total Environ 802:149807

Ballew NG, Bacheler NM, Kellison GT, Schueller AM (2016) Invasive lionfish reduce native fish abundance on a regional scale. Sci Rep 6:321169

Balloux F, van Dorp L (2017) Q&A: what are pathogens, and what have they done to and for us? BMC Biol 15:91

Baskin Y (2006) Sea sickness: the upsurge in marine diseases. Bioscience 56(6):464–469

Bax N, Williamson A, Aguero M, Gonzalez E, Geeves W (2003) Marine invasive alien species: a threat to global biodiversity. Mar Policy 27(4):313–323

Benner R (2003) Molecular indicators of the bioavailability of dissolved organic matter. In: Findlay S, Sinsabaugh R (eds) Aquatic ecosystems interactivity of dissolved organic matter. Academic Press, San Diego, pp 121–135

Bennie J, Davies TW, Cruse D, Bell F, Gaston KJ (2018) Artificial light at night alters grassland vegetation species composition and phenology. J Appl Ecol 55:442–450

Besson M, Gache C, Brooker RM, Moussa RM, Waqalevu VP, LeRohellec M, Jaouen V, Peyrusse K, Berthe C, Bertucci F (2017) Consistency in the supply of larval fishes among coral reefs in French Polynesia. PLoS ONE 12:e0178795

Black J, Reichelt-Brushett AJ, Clark M (2015) The effect of copper and temperature on juveniles of the eurybathic brittle star *Amphipholis squamata*. Chemosphere 124:32–39

Birdlife International (2012) Light pollution has a negative impact on many seabirds including several globally threatened species. Available at: ▶ www.birdligr.org. Accessed 29 Oct 2021

Bixler HJ (1996) Recent developments in manufacturing and marketing carrageenan. Hydrobiologia 326:35–57

Boeuf G, Le Bail PY (1999) Does light have an influence on fish growth? Aquaculture 177(1):129–152

Bourne DG, Garren M, Work TM, Rosenberg E, Smith GW, Harvell CD (2009) Microbial disease and the coral holobiont. Trends Microbiol 17(12):554–562

Bradley DL, Stern R (2008) Underwater sound and the marine mammal acoustic environment: a guide to fundamental principles prepared for the U.S. Marine Mammal Commission. Available at: ▶ https://www.mmc.gov/wp-content/uploads/sound_bklet.pdf. Accessed 17 Dec 2021

Brausch JM, Rand GM (2011) A review of personal care products in the aquatic environment: environmental concentrations and toxicity. Chemosphere 82:1518–1532

Brettar I, Guzman CA, Hofle MG (2007) Human pathogens in the marine environment—an ecological perspective. CIESM Workshop Monographs No. 31.: Marine Sciences and Public Health, CIESM, Geneva, pp 59–68

Brüning A, Hölker F, Franke S, Preuer T, Kloas W (2015) Spotlight on fish: light pollution affects circadian rhythms of European perch but does not cause stress. Sci Total Environ 511:516–522

Butler JM, Maruska KP (2020) Underwater noise impairs social communication during aggressive and reproductive encounters. Anim Behav 164:9–23

Cagauan AG (2007) Exotic aquatic species introduction in the Philippines for aquaculture—a threat to biodiversity or a boom to the economy. J Environ Sci Manage 10(1):48–62

Câmara JS, Montesdeoca-Esponda S, Freitas J, Guedes-Alonso R, Sosa-Ferrara Z, Perestrelo R (2021) Emerging contaminants in seafront zones. Environmental impact and analytical approaches. Separations 8(7):95

Cardoso SJ, Quadra GR, Resende NdS, Roland F (2019) The role of sediments in the carbon and pollutant cycles in aquatic ecosystems. Thematic Section: mini-reviews in applied limnology. Acta Limnologica Brasiliensia 31:e201

Carve M, Nugegoda D, Allinson G, Shimeta J (2021) A systematic review and ecological risk assessment for organic ultraviolet filters in aquatic environments. Environ Pollut 268(B):115894

Chandrasekaran S, Nagendran NA, Pandiaraja D, Krishnankutty N, Kamalakannan B (2008) Bioinvasion of *Kappaphycus alvarezii* on corals in the Gulf of Mannar, India. Curr Sci 94(9):1167–1172

Chimner RA, Fry B, Kaneshiro MY, Cormier N (2006) Current extent and historical expansion of introduced mangroves on Oʻahu, Hawaiʻi. Pac Sci 60(3):377–383

Cinzano P, Falchi F, Elvidge CD (2001) The first world atlas of the artificial night sky brightness. Mon Not R Astron Soc 328(2):689–707

Clark CW, Ellison WT, Southall B, Hatch LT, Van Parijs S, Frankel AS, Ponirakis D (2009) Acoustic masking in marine ecosystems: intuitions, analysis and implications. Mar Ecol Prog Ser 395:201–222

Codarin A, Wysicki LE, Ladich F, Picculin M (2009) Effects of ambient and boat noise on hearing and communication in three fish species living in a marine protected area (Miramare, Italy). Mar Pollut Bull 58:1880–1887

Cohen RE, James CC, Lee A, Martinelli MM, Muraoka WT, Ortega M, Sadowski R, Starkey L, Szesciorka AR, Timko SE, Weiss EL, Franks PJS (2018) Marine host-pathogen dynamics influences of global climate change. Oceanography 31(2 Special Issue):182–193

Coles SL, Bolick H (2007) Invasive introduced sponge *Mycale grandis* overgrows reef corals in Kāneʻohe Bay, Oʻahu, Hawaiʻi. Coral Reefs 26(4):911

Coles SL, Marchetti J, Bolick H, Montgomery A (2007) Assessment of invasiveness of the orange keyhole sponge *Mycale armata* in Kāneʻohe Bay, Oʻahu, Hawaiʻi (Final Report, 2). Bishop Museum Technical Report, 2006–2

Conklin EJ, Smith JE (2005) Abundance and spread of the invasive red algae, *Kappaphycus* spp in Kāneʻohe Bay, Hawaiʻi and an experimental assessment of management options. Biol Invasions 7(6):1029–1039

Cook EJ, Payne R, Macleod A, Brown S (2016) Marine biosecurity: protecting indigenous marine species. Res Rep Biodivers Stud 5:1–14

Corinaldesi C, Damiani E, Marcellini F, Falugi C, Tiano L, Brugè F, Danovaro R (2017) Sunscreen products impair the early developmental stages of the sea urchin *Paracentrotus lividus*. Sci Rep 7:7815

Cortez FS, Pereira CDS, Santos AR, Cesar A, Choueri RB, de Assis Martini G, Bohrer-Morel MB (2012) Biological effects of environmentally relevant concentrations of the pharmaceutical Triclosan in the marine mussel *Perna perna* (Linnaeus, 1758). Environ Pollut 168:145–150

CIR (Cosmetic Ingredient Review) (2005) Annual review of cosmetic ingredient safety assessments: 2003/2003. Int J Toxicol 24:1–102

Cravens ZM, Brown VA, Divill TJ, Boyles JG (2018) Illuminating prey selection in an insectivorous bat community exposed to artificial light at night. J Appl Ecol 55:705–713

Creed JC (2006) Two invasive alien azooxanthellate corals, *Tubastraea coccinea* and *Tubastraea tagusensis*, dominate the native zooxanthellate *Mussismilia hispida* in Brazil. Coral Reefs 25:350

Dale JJ, Gray MD, Popper AN, Rogers PH, Block BA (2015) Hearing thresholds of swimming Pacific Bluefin tuna *Thunnus orientalis*. J Comp Physiol A 201:441–454

Danovaro R, Bongiorni L, Corinaldesi C, Giovannelli D, Damiani E, Astolfi P, Greci L, Puseddu A (2008) Sunscreens cause coral bleaching by promoting viral infections. Environ Health Perspect 116(4):441–447

Davies TW, Bennie J, Gaston KJ (2012) Street lighting changes the composition of invertebrate communities. Biol Let 8(5):764–767

Davies TW, Bennie J, Inger R, Gaston KJ (2013) Artificial light alters natural regimes of night-time sky brightness. Sci Rep 3:1722

Davies TW, Duffy JP, Bennie J, Gaston KJ (2014) The nature, extent and ecological implications of marine light pollution. Front Ecol Environ 12(6):347–355

Davies TW, Coleman M, Griffith KM, Jenkins SR (2015) Night-time lighting alters the composition of marine epifaunal communities. Biol Let 11:20150080

Davidson A, Campbell ML, Hewitt CL, Schaffelke B (2015) Assessing the impacts of nonindigenous marine macroalgae: an update of current knowledge. J Botanica Marina 58(2):55–79

Depledge MH, Godard-Codding CAJ, Bowen RE (2010) Light pollution in the sea. Mar Pollut Bull 60:1383–1385

Den Hartog C (1970) The sea-grasses of the world. North-Holland, Amsterdam. P 21. Available at: ▶ https://pdf.usaid.gov/pdf_docs/PNAAM467.pdf. Accessed 26 Feb 2022

De Silva SS, Nguyen TT, Turchini GM, Amarasinghe US, Abery NW (2009) Alien species in aquaculture and biodiversity: a paradox in food production. Ambio 38(1):24–28, 19260343

di Franco E, Pierson P, Di Iorio L, Calò A, Cottalorda JM, Derijard B, Di Franco A, Galvé A, Guibbolini M, Lebrun J, Micheli F, Priouzeau F, Risso-de Faverney C, Rossi F, Sabourault C, Spennato G, Verrando P, Guidetti P (2020) Effects of marine noise pollution on Mediterranean fishes and invertebrates: a review. Mar Pollut Bull 159:111450

280 A. Reichelt-Brushett and S.B. Shah

12

Di Lorio L, Audax M, Deter J, Holon F, Lossent J, Gervaise C, Boissery P (2021) Biogeography of acoustic biodiversity of NW Mediterranean coralligenous reefs. Sci Rep 11:16991

Dimitriadis C, Fournari-Konstantinidou I, Sourbès L, Koutsoubas D, Mazaris AD (2018) Reduction of sea turtle population recruitment caused by nightlight: evidence from the Mediterranean region. Ocean Coastal Manage 153:108–115

Dong ZG, Chen YH, Ge HX, Li XY, Wu HL, Wang CH, Hu Z, Wu YJ, Fu GH, Lu JK, Che H (2018) Response of growth and development of the Pacific oyster (*Crassostrea gigas*) to thermal discharge from a nuclear power plant. BMC Ecol 18(1):31

Downs CA, Kramarsky E, Fauth JE, Segal R, Bronstein O, Jeger R, Lichtenfeld Y, Woodley CM, Pennington P, Kushmaro A, Loya Y (2014) Toxicological effects of the sunscreen UV filter, benzophenone-2, on planulae and in vitro cells of the coral, *Stylophora pistillata*. Ecotoxicology 23:175–191

Downs CA, Kramarsky-Winter E, Segal R, Fauth J, Knutson S, Bronstein O, Ciner FR, Jeger R, Lichtenfeld Y, Woodley CM, Pennington P, Cadenas K, Kushmaro A, Loya Y (2016) Toxicopathological effects of the sunscreen UV filter, oxybenzone (benzophenone-3), on coral planulae and cultured primary cells and its environmental contamination in Hawaii and the US Virgin Islands. Arch Environ Contam Toxicol 70(2):265–288

Duarte C, Quintanilla-Ahumada D, Anguita C, Manríquez PH, Widdicombe S, Pulgar J, Silva-Rodríguez EA, Miranda C, Manríquez K, Quijón PA (2019) Artificial light pollution at night (ALAN) disrupts the distribution and circadian rhythm of a sandy beach isopod. Environ Pollut 248:565–573

Erbe C, Dunlop R, Jenner KCS, Jenner MNM, McCauley RD, Parnum I, Parsons M, Rogers T, Salgado-Kent C (2017) Review of underwater and in-air sounds emitted by Australian and Antarctic marine mammals. Acoust Aust 45:179–241

Erbe C, Dunlop R, Dolman S (2018). Effects of noise on marine mammals. In: Slabbekoorn H, Dooling R, Popper A, Fay R (eds) Effects of anthropogenic noise on animals. Springer, New York, pp 277–309

Erbe C, Marley SA, Schoeman RP, Smith JN, Trigg LE, Embling C. (2019. The effects of ship noise on marine mammals—a review. Front Mar Sci 6(606)

Falchi F, Cinzano P, Duriscoe D, Kyba CCM, Elvidge CD, Baugh K, Portnov BA, Rybnikova NA, Furgoni R (2016) The new world atlas of artificial night sky brightness. Sci Adv 2(6):e1600377

Faulkner RC, Farcas A, Merchant ND (2018) Guiding principles for assessing the impact of underwater noise. J Appl Ecol 55(6):2531–2536

Fobert EK, Schubert KP, da Silva KB (2021) The influence of spectral composition of artificial light at night on clownfish reproductive success. J Exp Mar Biol Ecol 540:151559

Farnworth B, Innes J, Kelly C, Littler R, Waas JR (2018) Photons and foraging: artificial light at night generates avoidance behaviour in male, but not female, New Zealand weta. Environ Pollut 236:82–90

Frisk GV (2012) Noiseonomics: the relationship between noise levels in the sea and global economic trends. Sci Rep 2:437

Garratt MJ, Jenkins SR, Davies TW (2019) Mapping the consequences of artificial light at night for intertidal ecosystems. Sci Total Environ 691:760–768

Gaston KJ, Davies TW, Nedelec SL, Holt LA (2017) Impacts of artificial light at night on biological timings. Ann Rev Ecol Evol Syst 48:49–68

Gaston KJ (2018) Lighting up the nighttime. Science 362:8226

Gleason DF, Edmunds PJ, Gates RD (2006) Ultraviolet radiation effects on the behaviour and recruitment of larvae from the reef coral *Porites astreoides*. Mar Biol 148:503–512

Gomes NGM, Madureura-Carvalho A, Dias-da-Silva D, Valentão P, Andrade PB (2021) Biosynthetic versatility of marine-derived fungi on the delivery of novel antibacterial agents against priority pathogens. Biomed Pharmacother 140:111756

Godwin LS (2003) Hull fouling of maritime vessels as pathway for marine species invasions to the Hawai'ian Islands. Biofouling 19(S1):123–131

Grosholz E (2002) Ecological and evolutionary consequences of coastal invasions. Trends Ecol Evol 17(1):22–27

Guette A, Godet L, Juigner M, Robin M (2018) Worldwide increase in artificial light at night around protected areas and within biodiversity hotspots. Biol Cons 223:97–103

Halden RU (2014) On the need and speed of regulating triclosan and triclocarban in the US. Environ Sci Technol 48:3603–3611

Halden RU, Paull DH (2015) Co-occurrence of triclocarban and triclosan in US water resources. Environ Sci Technol 39:1420–1426

Halfwerk W, Slabbekoorn H (2015) Pollution going multimodal: the complex impact of the human-altered sensory environment on animal perception and performance. Biol Let 11:20141051

Halliday WD, Pine MK, Insley SJ (2020) Underwater noise and Arctic marine mammals: review and policy recommendations. Environ Rev 28(4):438–448

Hawkins AD, Popper AN (2017) A sound approach to assessing the impact of underwater noise on marine fishes and invertebrates. ICES J Mar Sci 74(3):635–651

Hildebrand JA (2009) Anthropogenic and natural sources of ambient noise in the ocean. Mar Ecol Prog Ser 395:5–20

HISC (Hawai'i Invasive Species Council) (2019) Kappaphycus algae. Available at: ▶ https://dlnr.hawaii.gov/hisc/info/invasive-species-profiles/kappapchyus-algae/. Accessed 17 Dec 2021

Hopkins SR, Richardson JI (eds) (1984) A recovery plan for marine turtles, p 355. Available at: ▶ https://www.fws.gov/oregonfwo/documents/RecoveryPlans/Marine_Turtles_RP.pdf. Accessed 19 Feb 2022

Hu MY, Yan H-Y, Chung W-S, Hwang P-P (2009) Acoustically evoked potentials in two cephalopods inferred using the auditory brainstem response (ABR) approach. Comp Biochem Physiol A Physiol 153(3):278–283

Huang F, Lin J, Zheng B (2019) Effects of thermal discharge from coastal nuclear power plants and thermal power plants on the thermocline characteristics in sea areas with different tidal dynamics. Water 11:2577

Horricks RA, Tabin SK, Edwards JJ, Lumsden JS, Marancik DP (2019) Organic ultraviolet filters in nearshore waters and in the invasive lionfish (*Pterois volitans*) in Grenada West Indies. PLoS One, 0220280

Issakhov A, Zhandaulet Y (2019) Numerical simulation of thermal pollution zones' formations in the water environment from the activities of the power plant. Eng Appl Comput Fluid Mech 13(1):279–299

Jelassi R, Ayari A, Nasri-Ammar K (2014) Effect of light intensity on the locomotor activity rhythm of *Orchestia montagui* and *Orchestia gammarellus* from the supralittoral zone of Bizerte lagoon (North of Tunisia). Biol Rhythm Res 45(5):817–829

Jerem P, Mathews F (2020) Trends and knowledge gaps in field research investigating effects of anthropogenic noise. Conserv Biol 35(1):115–129

Kamalakannan B, Jeevamani JJ, Nagendran NA, Pandiaraja D, Chandrasekaran S (2014) Impact of removal of invasive species *Kappaphycus alvarezii* from coral reef ecosystem in Gulf of Mannar, India. Curr Sci 106(10):1401–1408

Kennicutt MC (2017) Water quality of the Gulf of Mexico. In: Ward C (ed) Habitats and Biota of the Gulf of Mexico: Before the Deepwater Horizon Oil Spill. Springer, New York, pp 55–164

Kolpin DW, Furlong ET, Meyer MT, Thurman EM, Zaugg SD, Barber LB, Buxton HT (2002) Pharmaceuticals, hormones and other organic wastewater contaminants in US streams, 1999–2000: a national reconnaissance. Environ Sci Technol 36:1202–1211

Kunc HP, McLaughlin KE, Schmidt R (2016) Aquatic noise pollution: implications for individuals, populations, and ecosystems. Proc R Soc B: Biol Sci 283:20160839

Labille J, Slomberg D, Catalano R, Robert S, Apers-Tremelo ML, Boudenne JL, Manasfi T, Radakovitch O (2020) Assessing UV filter inputs into beach waters during recreational activity: A field study of three French Mediterranean beaches from consumer survey to water analysis. Sci Total Environ 706:136010

Lages BG, Fleury BG, Menegola C, Creed JC (2011) Change in tropical rocky shore communities due to an alien coral invasion. Mar Ecol Prog Ser 438:85–96

Lanzas C, Davies K, Erwin S, Dawson D (2019) On modelling environmentally transmitted pathogens. Interface Focus 10:20190056

Lecchini D, Bertucci F, Gache C, Khalife A, Besson M, Roux N, Berthe C, Singh S, Parmentier E, Nugues MM, Brooker RM, Dixson DL, Hédouin L (2018) Boat noise prevents soundscape-based habitat selection by coral planulae. Sci Rep 8:9283

Li VW, Tsui MP, Chen X, Hui MNY, Jin L, Lam RHW, Yu RMK, Murphy MB, Cheng J, Lam PKS, Cheng SH (2016) Effects of 4-methylbenzylidene camphor (4-MBC) on neuronal and muscular development in zebrafish (*Danio rerio*) embryos. Environ Sci Pollut Res (int) 23:8275–8285

Lillis A, Eggleston D, Bohnenstiehl D (2014) Soundscape variation from a larval perspective: the case for habitat-associated sound as a settlement cue for weakly swimming estuarine larvae. Mar Ecol Prog Ser 509:57–70

Linke S, Gifford T, Desjonquères C, Tonolla D, Aubin T, Barclay L, Karaconstaintis C, Kennard MJ, Ryback F, Sueur J (2018) Freshwater ecoacoustics as a tool for continuous ecosystem monitoring. Front Ecol Environ 16:231–238

Littlefair JE, Hrenchuck LE, Blanchfield PJ, Rennie MD, Cristescue ME (2020) Thermal stratification and fish thermal preference explain vertical eDNA distributions in lakes. Mol Ecol 30913:3083–3096

Liu JL, Wong MH (2013) Pharmaceuticals and personal care products (PPCPs): a review on environmental contamination in China. Environ Int 59:208–224

Longcore T, Rich C (2004) Ecological light pollution. Front Ecol Environ 2:191–198

Lorne J, Salmon M (2007) Effects of exposure to artificial lighting on orientation of hatchling sea turtles on the beach and in the ocean. Endangered Species Res 3:23–30

Lovell JM, Findlay MM, Maote RM, Yan HY (2005) The hearing abilities of the prawn *Palaemon serratus*. Comp Biochem Physiol A 140(1):89–100

Loya Y, Sakai K, Yamazato K, Nakano Y, Sambali H, van Woesik R (2001) Coral bleaching: the winners and the losers. Ecol Lett 4:122–131

Ludvigsen M, Berge J, Geoffroy M, Cohen JH, De La Torre PR, Nornes SM, Singh H, Sørensen AJ, Daase M, Johnsen G (2018) Use of an autonomous surface vehicle reveals small-scale diel vertical migrations of zooplankton and susceptibility to light pollution under low solar irradiance. Sci Adv 4:9887

Mack RN, Simberloff D, Mark Lonsdale W, Evans H, Clout M, Bazzaz FA (2000) Biotic invasions: causes, epidemiology, global consequences, and control. Ecol Appl 10(3):689–710

Mackay D, Barnthouse L (2010) Integrated risk assessment of household chemicals and consumer products: addressing concerns about triclosan. Integr Environ Assess Manag 6:390–392

Maggi E, Benedetti-Cecchi L (2018) Trophic compensation stabilizes marine primary producers exposed to artificial light at night. Mar Ecol Prog Ser 606:1–5

Maggi E, Bertocci I, Benedetti-Cecchi L (2019) Light pollution enhances temporal variability of photosynthetic activity in mature and developing biofilm. Hydrobiologia 847:1793–1802

Maggi E, Bongiorni L, Fontanini D, Capocchi A, Dal Bello M, Giacomelli A, Benedetti-Cecchi L (2020) Artificial light at night erases positive interactions across trophic levels. Funct Ecol 34:694–706

Maggi E, Serôdio J (2020) Artificial light at night: a new challenge in microphytobenthos research. Front Mar Sci 7:329

Marshall DJ, McQuaid CD (2020) Metabolic regulation, oxygen limitation and heat tolerance in a subtidal marine gastropod reveal the complexity of predicting climate change vulnerability. Front Physiol 11:1106

Martinez JA, Smith CM, Richmond RH (2012) Invasive algal mats degrade coral reef physical habitat quality. Estuarine Coastal Shelf Sci 99:42–49

McAvoy DC, Schatowitz B, Jacob M, Hauk A, Eckhoff WS (2002) Measurement of triclosan in wastewater treatment systems. Environ Toxicol Chem 21:1323–1329

Merchant ND, Fristup KM, Johnson MP, Tyack PL, Witt MJ, Blondel P, Parks SE (2015) Measuring acoustic habitats. Methods Ecol Evol 6:257–265

Miller IB, Pawlowski S, Kellerman MY, Petersen-Thiery M, Moeller M, Nietzer S, Schupp PJ (2021) Toxic effects of UV filters from sunscreens on coral reefs revisited: regulatory aspects for "reef safe" products. Environ Sci Eur 33:74

Mokhtari R, Arabkoohsar A (2021) Feasibility study and multi-objective optimization of seawater cooling systems for data centers: a case study of Caspian Sea. Sustain Energy Technol Assess 47:101528

Molnar JL, Gamboa RL, Revenga C, Spalding MD (2008) Assessing the global threat of invasive species to marine biodiversity. Front Ecol Environ 6(9):485–492

Montesdeoca-Esponda S, Checchini L, Del Bubba M, Sosa-Ferrera Z, Santana-Rodriguez J (2018) Analytical approaches for the determination of personal care products and evaluation of their occurrence in marine organisms. Sci Total Environ 633:405–425

Mooney TA, Smith A, Larsen ON, Hansen KA, Wahlberg M, Rasmussen MH (2019) Field-based hearing measurements of two seabird species. The company of biologists. J Exp Biol 222(4):1–7

NRC (National Research Council) (2003) Sources of sound in the ocean and long-term trends in ocean noise. In: Ocean Noise and Marine Mammals. Washington: National Academies Press (US) Committee on potential impacts of ambient noise in the ocean on marine mammals. Available at: ► https://www.ncbi.nlm.nih.gov/books/NBK221253/. Accessed 17 Dec 2021

Navara KJ, Nelson RJ (2007) The dark side of light at night: physiological, epidemiological, and ecological consequences. J Pineal Res 43(3):215–224

Naylor E (1999) Marine animal behaviour in relation to lunar phase. Earth Moon Planet 85:291–302

Neilson BJ, Wall CB, Mancini FT, Gewecke CA (2018) Herbivore biocontrol and manual removal successfully reduce invasive macroalgae on coral reefs. PeerJ 6:E5332

Occhipinti-Ambrogi A (2007) Global change and marine communities: alien species and climate change. Mar Pollut Bull 55:342–352

Occhipinti-Ambrogi A, Galil B (2010) Marine alien species as an aspect of global change. Adv Oceanogr Limnol 1(1):199–218

O'Connor JJ, Fobert EK, Besson M, Jacob H, Lecchini D (2019) Live fast, die young: behavioural and physiological impacts of light pollution on a marine fish during larval recruitment. Mar Pollut Bull 146:908–914

Ojaveer H, Galil BS, Campbell ML, Carlton JT, Canning-Clode J, Cook EJ, Davidson AD, Hewitt CL, Jelmert A, Marchini A, McKenzie CH, Minchin D, Occhipinti-Ambrogi A, Olenin S Ruiz G (2015) Classification of non-Indigenous species based on their impacts: considerations for application in marine management. PLoS Biol 13(4): e1002130

Ojemaye CY, Petrik L (2022) Pharmaceuticals and personal care-products in the marine environment around False BAY, Cape-Town, South Africa: occurrence and risk-assessment study. EnvironToxicol Chem, 1–21

Olaniyan LWB, Mkwetshana N, Okoh AI (2016) Triclosan in water, implications for human and environmental health. Springerplus 5:1639

Ozáez I, Martínez-Guitarte JL, Morcillo G (2013) Effects of in vivo exposure to UV filters (4-MBC, OMC, BP-3, 4-HB, OC, OD-PABA) on endocrine signaling genes in the insect Chironomus riparius. Sci Total Environ 456–457:120–126

Padilla DK, Williams SL (2004) Beyond ballast water: aquarium and ornamental trades as sources of invasive species in aquatic ecosystems. Front Ecol Environ 2(3):131–138

Paredes E, Perez S, Rodil R, Quintana JB, Beiras R (2014) Ecotoxicological evaluation of four UV filters using marine organisms from different trophic levels Isochrysis galbana, Mytilus galloprovincialis, Paracentrotus lividus, and Siriella armata. Chemosphere 104:44–50

Patterson KL, Porter JW, Ritchie KB, Polson SW, Mueller E, Pesters EC, Santavy DL, Smith GW (2002) The etiology of white pox, a lethal disease of the Carribbean elkhorn coral Acropora palmata. Proc Natl Acad Sci USA 99:8725–8730

Peck AM (2006) Analytical methods for the determination of persistent ingredients of personal care products in environmental matrices. Anal Bioanal Chem 386:907–939

Pemberthy D, Padilla Y, Echeverri A, Penuela GA (2020) Monitoring pharmaceuticals and personal care products in water and fish from Gulf of Uraba, Columbia. Heliyon 6:e04215

Penar W, Magiera A, Klocek C (2020) Applications of bioacoustics in animal ecology. Ecol Complex 43:100847

Peng C, Zhao X, Liu G (2015) Noise in the sea and its impacts on marine organisms. Int J Environ Res Public Health 12:12304–12323

Peters K, Sink K, Robinson TB (2017) Raising the flag on marine alien fouling species. Manage Biol Invasions 8(1):1–11

Pieretti N, Lo Martire M, Corinaldesi C, Musco L, Dell'Anno A, Danovaro R (2020) Anthropogenic noise and biological sounds in a heavily industrialised coastal area (Gulf of Naples, Mediterranean Sea). Mar Environ Res 159:105002

Pilson MEQ (2013) An introduction to the chemistry of the sea, 2nd edn. Cambridge University Press, Cambridge, p 539

Pulgar J, Zeballos D, Vargas J, Aldana M, Manriquez , Manriquez K, Quijon PA, Widdicombe S, Anguita C, Quintanilla D, Duarte C (2019) Endogenous cycles, activity patterns and energy expenditure of an intertidal fish is modified by artificial light pollution at night (ALAN). Environ Pollut 244:361–366

Ramirez-Llodra E, Brandt A, Danovaro R, De Mol B, Escobar E, German CR, Levin LA, Martinez Arbizu P, Menot L, Buhl-Mortensen P, Narayanaswamy BE, Smith CR, Tittensor DP, Tyler PA, Vanreusel A, Vecchione M (2010) Deep, diverse and definitely different: unique attributes of the world's largest ecosystem. Biogeosciences 7(9):2851–2899

Ramos S, Homem V, Alves A, Santos L (2015) Advances in analytical methods and occurrence of organic UV-filters in the environment—a review. Sci Total Environ 526:278–311

Reichelt AJ, Jones GB (1994) Trace metals as tracers of dredging activities in Cleveland Bay-field and laboratory study. Aust J Mar Freshw Res 45:1237–1257

Reichelt-Brushett AJ (2012) Risk assessment and ecotoxicology—limitations and recommendations in the case of ocean disposal of mine waste. Oceanography 25(4):40–51

Reid AJ, Carlson AK, Creed IF, Eliason EJ, Gell PA, Johnson PTJ, Kidd KA, MacCormack TJ, Olden JD, Ormerod SJ, Smol JP, Taylor WW, Tockner K, Vermaire JC, Dudgeon D, Cooke SJ (2019) Emerging threats and persistent conservation challenges for freshwater biodiversity. Biol Rev 94(3):849–873

Rodgers S, Cox EF (1999) Rate of spread of introduced Rhodophytes Kappaphycus alvarezii, Kappaphycus striatum, and Gracilaria Salicornia and their current distribution in Kāneʻohe Bay, Oʻahu Hawaiʻi. Pac Sci 53(3):232–241

Rosenberg Y, Doniger T, Levy O (2019) Sustainability of coral reefs are affected by ecological light pollution in the Gulf of Aqaba/Eilat. Commun Biol 2:289

Ross DJ, Johnson CR, Hewitt CL (2003) Assessing the ecological impacts of an introduced seastar: the importance of multiple methods. Biol Invasions 5(1–2):3–21

Rotter A, Klun K, Francé J, Mozetič P, Orlando-Bonaca M (2020) Non-indigenous species in the Mediterranean Sea: turning from pest to source by developing the 8Rs model, a new paradigm in pollution mitigation. Front Mar Sci 7:178

Ruiz GM, Carlton JT, Grosholz ED, Hines AH (1997) Global invasions of marine and estuarine habitats by non-indigenous species: mechanisms, extent and consequences. Am Zool 37(6):621–632

Russell DJ (1992) The ecological invasion of Hawaiian reefs by two marine red algae, Acanthophora spicifera (Vahl) Boerg. and Hypnea musciformis (Wulfen) J. Ag., and their association with two native species, Laurencia nidifica J. Ag. and Hypnea cervicornis. ICES Mar Sci Symp 194:110–125

Schlacher TA, Stark J, Fischer AB (2007) Evaluation of artificial light regimes and substrate types for aquaria propagation of the staghorn coral Acropora solitaryensis. Aquaculture 269:278–289

Sellers AJ, Saltonstall K, Davidson TM (2015) The introduced alga Kappaphycus alvarezii (Doty ex PC Silva, 1996) in abandoned cultivation sites in Bocas del Toro, Panama. BioInvasions Records 4:1–7

Sendra M, Pintado-Herrera MG, Aguirre-Martínez GV, Monero-Garrido I, Martin-Díaz LM, Lara-Martin PA, Blasco J (2017) Are the TiO$_2$ NPs a "Trojan horse" for personal care products (PCPs) in the clam Ruditapes philippinarum? Chemosphere 185:192–204

Sharma D, Ravindran C (2020) Diseases and pathogens of marine invertebrate corals in Indian reefs. J Invertebr Pathol 173:107373

Short F, Carruthers T, Dennison W, Waycott M (2007) Global seagrass distribution and diversity: a bioregional model. J Exp Mar Biol Ecol 350(1–2):3–20

Silva E, Marco A, da Graca J, Perez H, Abella E, Patino-Martinez J, Martins S, Almeida C (2017) Light pollution affects nesting behaviour of loggerhead turtles and predation risk of nests and hatchlings. J Photochem Photobiol, B 173:240–249

Simberloff D (2005) Non-native species do threaten the natural environment. J Agric Environ Ethics 18(6):595–607

Simpson SD, Radford AN, Holles S, Ferarri MCO, Chivers DP, McCormick MI, Meekan MG (2016) Small-boat noise impacts natural settlement behaviour of coral reef fish larvae. In: Popper A, Hawkins A (eds) The effects of noise on aquatic life II. Springer, New York, pp 1041–1048

Slabbekoorn H, Bouton N, van Opzeeland I, Coers A, Cate C, Popper AN (2010) A noisy spring: the impact of globally rising underwater sound levels on fish. Trends Ecol Evol 25:419–427

Smith JE, Hunter CL, Smith CM (2002) Distribution and reproductive characteristics of nonindigenous and invasive marine algae in the Hawaiʻian Islands. Pac Sci 56(3):299–315

Smyth T, Davies T, McKee D (2020) The global extent of artificial light pollution in the marine environment. 22nd EGU General Assembly (online): 4–8 May 2020. Available at: ▸ https://meetingorganizer.copernicus.org/EGU2020/EGU2020-2415.html. Accessed 17 Dec 2021

Sokolow S (2009) Effects of a changing climate on the dynamics of coral infectious disease: a review of the evidence. Dis Aquatic Organ 87:5–18

Southhall BL, Southall H, Antunes R, Nichols R, Rouse A, Stafford KM, Robards M, Rosenbaum HC (2020) Seasonal trends in underwater ambient noise near St. Lawrence Island and the Bering Strait. Mar Pollut Bull 157:111283

Speight JG (2020) Sources of water pollution. In: Speight J (ed) Natural water remediation chemistry and technology. Elsevier, Amsterdam, pp 165–198

Stewart JR, Gast RJ, Fujioka RS, Solo-Gabriele HM, Meschke JS, Amaral-Zettler LA, del Castillo E, Polz MF, Collier TK, Strom MS, Sinigalliano CD, Moeller PDR, Holland AF (2008) The coastal environment and human health: microbial indicators, pathogens, sentinels and reservoirs. Environ Health 7(2):S3

Stauber JL, Adams MS, Batley GE, Golding L, Hargreaves I, Peeters L, Reichelt-Brushett A, Simpson S (2022) A generic environmental risk assessment framework for deep-sea tailings placement using causal networks. Sci Total Environ 845:157311

Strader ME, Davies SW, Matz MV (2015) Differential responses of coral larvae to the colour of ambient light guide them to suitable settlement microhabitat. R Soc Open Sci 2:150358

Sully S, Burkepile DE, Donovan MK, Hodgson G, van Woesik R (2019) A global analysis of coral bleaching over the past two decades. Nat Commun 10:1264

Sulu R, Kumar L, Hay C, Pickering T (2004) Kappaphycus seaweed in the Pacific: review of introductions and field testing proposed quarantine protocols. Sceretariat of the Pacific Community, Noumea, p 84. Available at: ▶ https://pacificdata.org/data/hu/dataset/oai-www-spc-int-40680f4a-db48-4c3c-933b-4a81516da05a. Accessed 17 Dec 2021

Tamir R, Lerner A, Haspel C, Dubinsky Z, Iluz D (2017) The spectral and spatial distribution of light pollution in the waters of the northern Gulf of Aqaba (Eilat). Sci Rep 7:42329

Thomsen F, Erbe C, Hawkins A, Lepper P, Popper AN, Scholik-Schlomer A, Sisneros J (2020) Introduction to the special issue on the effects of sound on aquatic life. J Acoust Soc Am 148:934

Turicchia E, Hoeksema BW, Ponti M (2018) The coral-killing sponge *Chalinula nematifera* as a common substrate generalist in Komodo National Park, Indonesia. Mar Biol Res 14(8):827–833

Tuxbury SM, Salmon M (2005) Competitive interactions between artificial lighting and natural cues during seafinding by hatchling marine turtles. Biol Cons 121(2):311–316

Underwood CN, Davies TW, Queiros AM (2017) Artificial light at night alters trophic interactions of intertidal invertebrates. J Animal Ecol 86:781–789

US EPA (United States Environmental Protection Authority (2008) USEPA-Reregistration Eligibility Decision for Triclosan List B. Case No. 2340, p 98. Available at ▶ https://www3.epa.gov/pesticides/chem_search/reg_actions/reregistration/red_PC-054901_18-Sep-08.pdf. Accessed 17 Dec 2021

Vakili SV, Olcer AL, Ballini F (2020) The development of a policy framework to mitigate underwater noise pollution from commercial vessels. Mar Policy 118:104004

Vare L, Baker M, Howe J, Levin L, Neira C, Ramirez-Llodra E, Reichelt-Brushett A, Rowden A, Shimmield T, Simpson S, Soto E (2018) Scientific considerations for the assessment and management of mine tailing disposal in the deep sea. Front Mar Sci 5:17

Vermeij M, Bak R (2002) How are coral populations structured by light? Marine light regimes and the distribution of Madracis. Mar Ecol Progr Ser, 105–116

Vieira M, Beauchaud M, Amorim MCP, Fonseca PJ (2021) Boat noise affects meagre (*Argyrosomus regius*) hearing and vocal behaviour. Mar Pollut Bull 172:112824

Vollmer SV, Kline DI (2008) Natural disease resistance in threatened staghorn corals. PLoS ONE 3:e3718

Walther GR, Roques A, Hulme PE, Sykes MT, Pyšek P, Kűhn I, Zobel M, Bacher S, Botta-Dukát Z, Bugmann H, Czúcz B, Dauber J, Hickler T, Jarošík V, Kenis M, Minchin D, Moora M, Nentwig W, Ott J, Panov VE, Reineking B, Robinet C, Semenchenko V, Solarz W, Thuiller W, Vilá M, Vohland K, Settele J (2009) Alien species in a warmer world: risks and opportunities. Trends Ecol Evol 24(12):686–693

Wang H, Xie D, Bowler PA, Zeng Z, Xiong W, Liu C (2021) Non-indigenous species in marine and coastal habitats of the South China Sea. Sci Total Environ 759:143465

Weilgart LS (2007) The impacts of anthropogenic ocean noise on cetaceans and implications for management: review. Can J Zool 85:1091–1116

Williams R, Clark CW, Ponirakis D, Ashe E (2014) Acoustic quality of critical habitats for three threatened whale populations. Anim Conserv 17:174–185

Willette DA, Chalifour J, Debrot AD, Engel MS, Miller J, Oxenford HA, Short FT, Steiner SCC, Védie F (2014) Continued expansion of the trans-Atlantic invasive marine angiosperm *Halophila stipulacea* in the Eastern Caribbean. Aquat Bot 112:98–102

Witherington BE, Bjorndal KA (1991) Influences of artificial lighting on the seaward orientation of hatchling loggerhead turtles *Caretta caretta*. Biol Cons 55:139–149

Yurk H, Trites AW (2000) Experimental attempts to reduce predation by Harbor seals on out-migrating juvenile salmonids. Trans Am Fisheries Soc 129:1360–1366

Zapata MJ, Sullivan SMP, Gray S (2019) Artificial lighting at night in estuaries—implications from individuals to ecosystems. Estuaries Coasts 42(2):309–330

Marine Contaminants of Emerging Concern

Munro Mortimer and Graeme Batley

Contents

13.1   Introduction – 286
13.1.2   What is Meant by "Concern"? – 287
13.1.1   What is Meant by "Emerging"? – 287

13.2   Contaminants of Emerging Concern in the Marine Environment – 288

13.3   The Relationship Between CECs and Endocrine Disrupting Chemicals – 290

13.4   Pharmaceuticals and Personal Care Products (PPCPs) as CECs – 291

13.5   Nanomaterials – 292

13.6   PFAS (Per- and Polyfluoroalkyl Substances) – 294
13.6.1   Naming Conventions Used for PFAS – 297
13.6.2   PFAS and Precursors – 300

13.7   Summary – 300

13.8   Study Questions and Activities – 300

         References – 301

© The Author(s) 2023
A. Reichelt-Brushett (ed.), *Marine Pollution—Monitoring, Management and Mitigation*,
Springer Textbooks in Earth Sciences, Geography and Environment,
https://doi.org/10.1007/978-3-031-10127-4_13

Acronyms and Abbreviations

AFFF	Aqueous film-forming firefighting foams
ATSDR	Agency for Toxic Substances and Disease Registry (USA)
BPA	Bisphenol A
C60	Fullerenes
CAS	Chemical abstracts service
CDC	Center for Disease Control and Prevention (USA)
CEC	Contaminant of emerging concern
CFC	Chlorofluorocarbon
CNT	Carbon nanotube
DDT	Dichlorodiphenyltrichloroethane, a shortened version of a former name used for 1,1'-(2,2,2-trichloroethane-1,1-diyl)bis(4-chlorobenzene)
EDC	Endocrine disrupting chemical
EEA	European Environment Agency
FOSA	Perfluorooctane sulfonamide
Gr	Graphene, an allotrope of carbon
HCF	Hydrofluorocarbon
IMO	International Maritime Organization (a United Nations Intergovernmental Body)
IUPAC	International Union of Pure and Applied Chemistry
NOAA	National Oceanic and Atmospheric Administration (USA)
NORMAN	An international network of reference laboratories, research centres and related organizations for monitoring of emerging environmental substances
nZVI	Nanoscale zero-valent particulate iron
OECD	Organisation for Economic Co-operation and Development
PBB	Polybrominated biphenyl
PCB	Polychlorinated biphenyl
PFAA	Perfluoroalkyl acid
PFAS	Per- and polyfluoroalkyl substances
PFC	Perfluorinated chemical or perfluorocarbon. These are related but distinctly different groups of substances (see ▶ Box 13.6)
PFBA	Perfluorobutanoic acid or perfluorobutanoate (see ▶ Box 13.9)
PFCA	Perfluoroalkyl carboxylic acid
PFHxA	Perfluorohexanoic acid
PFOA	Perfluorooctanoic acid
PFOS (or PFSA)	Perfluorooctane sulfonic acid
PFOSA	Perfluorooctane sulfonamide
PFSA	Perfluoroalkane sulfonic acid
POP	Persistent organic pollutant
POSF	Perfluorooctane sulfonyl fluoride
PPCP	Pharmaceutical and personal care product
REACH	Registration, Evaluation, Authorisation and Restriction of Chemicals (a European Union Regulatory Organisation).
TBT	Tributyltin
UNESCO	United Nations Educational, Scientific and Cultural Organization
US EPA	United States Environmental Protection Agency

13.1 Introduction

Identifying and listing substances or materials as **contaminants of emerging concern (CECs)** is not a simple task, and for the marine environment specifically is a challenge for environmental regulators, managers and researchers worldwide (▶ Box 13.1) (Tornero and Hanke 2017). Some of these agencies have widely different definitions of what a CEC actually is (Halden 2015).

The meaning of the term **contaminant** is relatively well understood and is discussed in ▶ Chapter 1 of this book. Although the text used by various authors and agencies to define contamination varies, it usually includes or implies the involvement of human-related activities and results in the production of an unnatural concentration of material in a specific environment leading to an associated adverse consequence or impairment to the natural condition for one or more attributes of that environment. However, the terms **emerging** and **concern** are more subjective, and are subject to time scales and prevailing circumstances.

13.1.1 What is Meant by "Emerging"?

A meta-analysis of 143,000 publications about 12 prominent CECs ranging from the pesticide DDT to nanoparticles and microplastics (Halden 2015) showed a common time course of emergence and subsidence of concern spanning about 29 years. That study noted that a number of factors can trigger and accelerate the **emergence** of new CECs, for example, new methods of detection and lowered detection limits, paradigm shifts in scientific understanding, breakthroughs in the design and manufacture of materials and changes in marketing and consumer behaviour leading to increased chemical consumption. Each of these factors can bring long-ignored environmental contaminants into the public eye and drive an increasing **level of concern**. This increase in the level of concern about a substance or material often triggers further research and publishing activity and the development of new regulations.

Box 13.1: Definitions of Contaminants of Emerging Concern

Despite the large number of papers published in recent years on the topic of **contaminants of emerging concern**, no commonly agreed definition exists (Nilsen et al. 2018). However, the following definition used by United Nations Educational, Scientific and Cultural Organization (UNESCO) and its refinement by the NORMAN network (▶ Box 13.2) capture the essence of the definitions in common usage:

The definition used by UNESCO (UNESCO 2019).

"Emerging pollutants can be understood in a broad sense as any synthetic or naturally-occurring chemical or any microorganism that is not commonly monitored or regulated in the environment with potentially known or suspected adverse ecological and human health *effects. These contaminants include mainly chemicals found in pharmaceuticals, personal care products, pesticides, industrial and household products, metals,* surfactants, *industrial additives and solvents. Many of them are used and released continuously into the environment even in very low quantities and some may cause chronic toxicity, endocrine disruption in humans and aquatic wildlife and the development of bacterial pathogen* resistance."

The NORMAN network (▶ Box 13.2) defines the word **contaminant** as: "*Any physical, chemical, biological, or radiological substance or matter that has an adverse effect on air, water, or soil*" **and** includes the following additional criteria to define a "*contaminant of emerging concern*":

− currently not included in routine environmental monitoring programmes and which may be a candidate for future legislation due to its adverse effects and/or persistency and
− a substance for which fate, behaviour and (eco)toxicological effects are not well understood.

Important Note

Contaminants of emerging concern are **not limited to newly developed chemicals**. Many are substances that have entered and been present in the environment for years, even decades. However, their presence has only recently raised concerns.

13.1.2 What is Meant by "Concern"?

The term **concern** in the CEC context is subject to interpretation and may mean different things such as interest, importance or cause of anxiety; all of these interpretations involve factors that are difficult to measure objectively. For a typical CEC, its associated **level of concern** progresses time-wise in a common pattern shown in ◘ Figure 13.1. The level of concern associated with a particular contaminant tends to increase as potential threats and knowledge gaps are realized, and to decline as knowledge increases and risk management strategies relating to the contaminant, such as behavioural changes, exposure controls, voluntary phase-outs and as regulatory actions take effect. For some CECs, such a pattern of time-wise waxing and waning in level of concern can be repeated if novel adverse effects are observed with the contaminant.

The long human experience with the element lead (Pb) provides an example of its **emergence** as a contaminant of concern. Several millennia ago, metallic lead was readily extracted from ores by early civilizations and the metal found a multitude of uses due to the ease with which it could be cast and shaped due to its softness and low melting point relative to other metals available at the time. For example, in ancient Roman times (approx. 500 BCE through 500 CE) there was widespread use of pipes made of metallic lead for potable water supply, in wine making, and also to line copper (Cu) cooking pots, together with the use of lead-containing compounds for a range of culinary, medicinal and decorative purposes, all of which be-

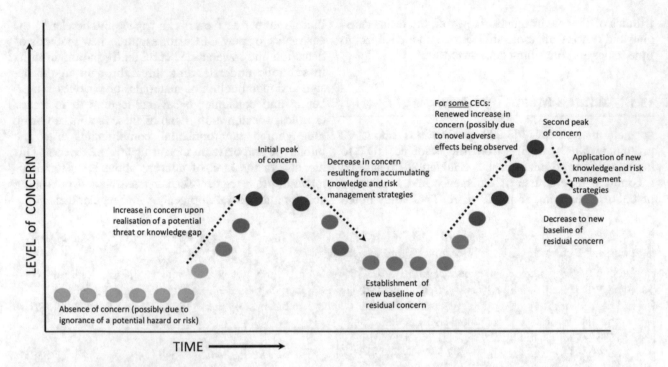

◘ Figure 13.1 The common time-wise progression that develops for a contaminant of emerging concern (CEC). Adapted from Halden (2015) by M. Mortimer

came associated with lead poisoning (Cilliers and Retief 2019), thus making lead an **emerging contaminant of concern** in those long-gone times.

However in the present time, with our better understanding of the health risks associated with lead and its uses, and the availability of non-toxic materials for potable water distribution networks, lead is no longer a CEC so far as potable water, water used for irrigation, and stock watering, or for cooking pots, is concerned. However, lead in airborne dusts is now a CEC in localities potentially impacted by lead ore processing, with recent media reporting of lead levels in children in Mount Isa in Queensland, Australia (Forbes and Taylor 2015). In addition, lead contamination of the marine environment near lead smelters, for example, at Port Pirie in South Australia is an ongoing issue relating to both toxic impacts on the marine ecosystem and adverse impacts on human health related to contamination of seafood (Lafratta et al. 2019; EPASA 2005). Likewise, in the past, many of the Persistent Organic Pollutants (POPs) (see ► Chapter 8) were once CECs (e.g. DDT and other organochlorine pesticides in the 1960s and 1970s).

Although most, if not all, of the POPs are still **contaminants of concern**, for many of them the term **emerging** is no longer applicable. There are many more historical examples of contaminants that have **emerged**, but subsequently have been managed through the acquisition of better knowledge to the stage that the reasons for **concern** are addressed and reduced or elimi-

nated and the title **contaminant of emerging concern** is no longer applicable.

13.2 Contaminants of Emerging Concern in the Marine Environment

The list of potential candidate substances to be CECs in the marine environment is very large. In excess of 100 million chemical substances are currently registered in the Chemical Abstracts Service (CAS) and about 4000 new ones are registered every day. The number of registered and pre-registered substances in REACH (the European Union legislation for the Registration, Evaluation, Authorisation and Restriction of Chemicals) lists 30,000–50,000 industrial chemicals present in daily-use products, all of which are potentially ultimately released into the environment (Dulio et al. 2018). However, not all of these chemicals are **of concern** once released to the environment, and many are unlikely to become CECs. Numerous international environmental agencies and regulators have compiled individual lists of chemicals and substances they regard as being **of concern** but there is no common list accepted by all the relevant organizations.

The European Commission Joint Research Centre has compiled a "*comprehensive list of chemical substances considered relevant*" under European Union legislation and by international organizations (Tornero and Hanke 2017). Although not all of the listed contaminants are of concern for the marine environment, this list is invaluable in presenting in one table the total of approximately 2700 **of concern** substances (or groups of substances) identified under relevant global conventions (e.g. the Stockholm Convention on POPs), European legislation (e.g. REACH), government agencies (e.g. the United States Envrionmental Protection Agency (US EPA) Priority Pollutants legislation) and international research groups (e.g. the NORMAN Network), together with the status of each contaminant on its source list.

Box 13.2: The NORMAN Network

The NORMAN Network is an international "*network of reference laboratories, research centres and related organisations for the monitoring of emerging environmental substances*". It was established by the European Union in 2005 and seeks to promote and to benefit from the synergies between research teams from different countries in the field of emerging substances.

The stated purpose of the NORMAN Network is to:

- enhance the exchange of information on emerging environmental substances;
- encourage the validation and harmonization of common measurement;
- encourage the validation and harmonization of common measurement methods and monitoring tools so that the requirements of risk assessors and risk managers can be better met; and
- ensure that knowledge of emerging pollutants is maintained and developed by stimulating coordinated, interdisciplinary projects on problem-oriented research and knowledge transfer to address identified needs.

The NORMAN Network has developed a methodology for prioritization of emerging contaminants, based on citations in the scientific literature. Examples of contaminants covered include surfactants, flame retardants, pharmaceuticals and personal care products, fuel additives and their degradation products, biocides, polar pesticides and their degradation products and various proven or suspected endocrine disrupting compounds.

The NORMAN Network systematically collects monitoring data and information on effects and hazardous properties for these substances and, on the basis of this information, allocates them to pre-defined categories (substances for which there is not yet sufficient toxicity information, substances with evidence of hazard but not yet satisfactory analytical performance, etc.). Currently the NORMAN Network website (▶ https://www.norman-network.net) lists almost 1000 substances as **emerging contaminants of concern** and provides substance fact sheets and related databases. The list is regularly updated, with particular emphasis on metabolites and transformation products that appear as relevant emerging substances but are not yet part of regular monitoring programmes.

A review of the operations of the NORMAN Network since its establishment, its organization and working groups structure is provided in Dulio et al. (2018).

Several large-scale monitoring programs in the marine environment have focused on detecting and monitoring **emerging contaminants of concern**. The most well known (and possibly the largest, longest lasting and best resourced) of these programmes is the Mussel Watch Program conducted by the National Oceanic and Atmospheric Administration (NOAA) in North America since 1986 (NOAA 2008). (See also ▶ Chapter 2, ▶ Box 2.1).

In the marine environment, a good example of an **emerging contaminant** of worldwide concern at the time (some 50 years before the present) is provided by the emergence of concern over the use of tributyltin (TBT) as an active ingredient in anti-fouling coatings applied to the hulls of ships. The published scientific material on the TBT issue is very extensive, but an overview published by the European Environment Agency (EEA) is succinct and comprehensive (Santillo et al. 2002) (see also ▶ Chapter 7). In summary, the use of antifoulant protection on the submerged portion of ship hulls is essential to minimize the growth of marine life (fouling) that causes hull damage to timber vessels and reduces speed and increases fuel consumption in all affected vessels regardless of the material from which they are constructed. Initially, wooden ships were protected with metallic copper sheathing. In later times, copper-containing paints were used on vessels of all types, and in the late 1960s, organotin compounds (in particular, TBT) were found to be a very effective ingredient in anti-fouling paints and these compounds rapidly became the active ingredient of choice in hull paints and use was widespread by the early 1970s.

However, the widespread use of TBT-based antifoulant paints by commercial shipping, including fishing fleets, and by leisure craft became associated with a marked decline in many commercially important marine mollusc fisheries (for example, mussels and oysters), characterized by declining populations of many

◼ Figure 13.2 Stripping TBT-based antifoulant paint from a ship hull during drydocking for maintenance and repaint in the Port of Brisbane, Australia during the 1990s. After work completion the drydock was re-flooded, and paint debris accumulated on the dock floor was flushed into the river. *Photo*: M. Mortimer

resource species especially where there was a high density of boat traffic. Research demonstrated the toxic consequence of the exposure of marine molluscs to low (ng/litre) concentrations of water-borne TBT was primarily imposex (the development of male sexual structures in females—leading to reproductive failure), but also shell deformities, failure of larval settlement and bioaccumulation of TBT. (See also ► Box 7.2).

Subsequently, TBT was found to be environmentally persistent, particularly in sediments (a half-life of 4 years) but much less so in waters (half-life of 6 days), and increasing concentrations were found in the tissues of a wide range of marine life including fish and marine mammals. The sources of TBT to the marine environment were not only its release from vessel coatings, but also from poorly or non-regulated disposal of TBT-containing paint residues stripped from vessels when hulls were repaired and when regularly scheduled repainting was carried out (◼ Figure 13.2).

The progressive introduction from 1982 by countries and international organizations (see timeline in Santillo et al. 2002) of restrictions on the use of TBT-based antifoulants, culminating in their effective phase-out by the International Maritime Organization (IMO), a specialized agency of the United Nations responsible for regulating shipping in its adoption of the International Convention on the Control of Harmful Antifouling Systems on Ships (IMO 2001). This convention imposed a global prohibition of the application of organotin compounds which act as biocides in anti-fouling systems on ships by 1 January 2003 and a complete prohibition of the presence of organotin compounds which act as biocides in anti-fouling systems on ships by 1 January 2008 and has succeeded in successfully managing the TBT contamination problem.

13.3 The Relationship Between CECs and Endocrine Disrupting Chemicals

Since the late 1980s, there has been growing evidence of the feminization of male fish in waters receiving sewage treatment plant discharges (e.g. Jobling et al. 1996, 1998) and this triggered concern in the general community and the attention of regulatory agencies and researchers concerning the presence of estrogenic chemicals in outfalls and receiving waters.

Common usage of the term **endocrine disruption** in the context of chemical pollution originated in 1991 as a consensus statement at a conference workshop series publication in Wisconsin, USA (Colborn and Clement 1992). The convenor of that conference, Theo Colborn along with others, subsequently published the book *Our Stolen Future* (Colborn et al. 1997), a landmark publication in raising public attention to the issues relating to endocrine disruption in wildlife and potentially humans.

The growing attention, given to **endocrine disrupting chemicals (EDCs)** phenomenon (► Box 13.3), raised the **concern** levels about contaminants in aquatic environments, and in particular chemicals that are EDCs, and as a consequence numerous chemicals and substances became CECs. However, it is important to note that the EDC phenomenon is an expression of toxic effect, and although many CECs are associated with the EDC phenomenon, many are regarded as CECs for other reasons.

Box 13.3: Endocrine Disrupting Chemicals (EDCs)

The endocrine system comprises glands that produce chemical substances (hormones) that regulate the activity of cells or organs. Thus, the endocrine system regulates the body's growth, metabolism, and sexual development and function.

EDCs interfere with the endocrine system in several ways:

- mimicking or antagonizing the action of endogenous hormones;
- interfering with the synthesis, metabolism, transport and excretion of natural hormones and
- altering the hormone receptor levels.

Pollution of marine waters by EDCs may pose adverse health effects, reproductive abnormalities and impaired development in marine life. Evidence of endocrine disruption has been reported in bivalves, crustaceans, fish, reptiles, birds and mammals (Godfray et al. 2019).

Major sources of EDC pollutants to the marine environment include sewage treatment plant discharges and runoff from intensive animal husbandry. EDCs from these sources include endogenous hormones such as estrogens, progesterone and testosterone produced in mammals, as well as synthetic hormones and industrial chemicals. Synthetic hormones are used as oral contraceptives, in hormone replacement treatment and as animal feed additives. Many industrial chemicals including phenols, halogenated substances including organochlorine pesticides and PCBs, and phthalates have EDC properties. Those of most concern have long half-lives in the marine environment. EDC effects of most concern are those at the population, community and ecosystem level, but there is limited knowledge of these affects as yet (Windsor et al. 2017), and assessing a causal link between EDCs and population-level effects in the marine environment is not an easy task because of the uncertainty generated by the (still) largely undescribed endocrinology of most marine invertebrates (Katsiadaki 2019).

A comprehensive overview of current knowledge in the field of EDCs is provided in Godfray et al. (2019).

Examples of estrogenic chemicals include dichlorodiphenyltrichloroethane (DDT), dioxins, polychlorinated biphenyls (PCBs), bisphenol A (BPA), nonylphenol, polybrominated biphenyls (PBB), phthalate esters, perfluoroalkyl and polyfluoroalkyl substances (PFAS), polybrominated diphenyl ethers, endosulfan, atrazine and triclosan (NIEHS 2022).

13.4 Pharmaceuticals and Personal Care Products (PPCPs) as CECs

The group of chemicals and substances collectively known as PPCPs includes both **pharmaceuticals and personal care products** used for personal health/well-being or for cosmetic purposes (see ▶ Chapter 12). The common usage of the term PPCPs also includes non-medicinal/non-cosmetic household products or their ingredients such as disinfectants (e.g. triclosan) and antiseptics, soaps, detergents and other cleaning products, synthetic musks and fragrances cosmetics, lotions, preservatives and sunscreen agents (e.g. oxybenzone). A recent overview of the global extent of discharges of PPCPs was provided in Dey et al. (2019).

Pharmaceuticals are defined as prescription, over-the-counter and veterinary therapeutic drugs used to prevent or treat human and animal diseases, while personal care products are used mainly to improve the quality of daily life (Boxall et al. 2012).

Pharmaceuticals can be classified by their therapeutic uses. The common uses being: anti-diabetics (e.g. alpha-glucosidase inhibitor), ß-blockers (e.g. atenolol, metoprolol), antibiotics (e.g. trimethoprim), lipid regulators (e.g. gemfibrozil), anti-epileptic (e.g. acetazolamide), tranquilizers (e.g. diazepam), anti-microbials (e.g. penicillins), anti-ulcer and anti-histamine drugs (e.g. cimetidine, famotidine), anti-anxiety or hypnotic agents (e.g. diazepam), anti-inflammatories and analgesics (e.g. ibuprofen, paracetamol, diclofenac), anti-depressants (e.g. benzodiazine-pines), anti-cancer drugs (e.g. cyclophosphamide, ifosfamide), anti-pyretics and stimulants (e.g. dexamphetamine, methylphenidate, modafinil), and estrogens and hormonal compounds (e.g. estriol, estradiol, estrone).

Currently more than 5000 manufactured pharmaceutical medicines are consumed by humans and/or domesticated animals, with an estimated total annual worldwide consumption in the range of 90,000–180,000 tonnes with the largest national consumptions being Russia, China, South Africa, India and Brazil (Van Boeckel et al. 2015; Tijani et al. 2016). A comprehensive overview of the current understanding of the extent and potential impact of contamination of the marine environment by pharmaceuticals is provided in the recent review by Ojemaye and Petrik (2019).

A large portion of medications that are ingested orally or by infusion are excreted through urine and/or faeces due to their incomplete absorption (metabolism) in humans and animals, these ultimately end up in wastewater treatment plants. Subsequently, municipal sewage treatment plants are major points of release of pharmaceuticals into the marine environment because wastewater treatment plants are not designed to decompose the vast majority of pharmaceutical compounds, which are by intent stable and robust, polar and non-volatile in nature. The most frequent and widespread pharmaceuticals in sewage and the discharge from marine outfalls are antibiotics and non-steroidal anti-inflammatory drugs (Ojemaye and Petrik 2019). Other pathways for pharmaceuticals to be delivered into the marine system are via landfill sites, septic tanks, urban wastewater, showering and bathing, industrial effluent and agricultural runoff.

Measured concentrations of pharmaceuticals from worldwide coastal environment locations in seawater, sediments and organisms (Ojemaye and Petrik, 2019) range from 0.21 to 5000 ng/L (seawater), 0.0402 ng/g dry weight to 208 ng/g wet weight (biota) and 0.2 µg/kg dry weight to 466 µg/kg wet weight (sediments). However, despite evidence of their increasing presence, little attention has been directed towards understanding the release of pharmaceuticals into coastal-marine environments and their potential negative impact on marine ecosystems. This qualifies many pharmaceuticals as CECs in the marine environment.

Since the active ingredients in pharmaceuticals are chosen on the basis that their physicochemical and biological properties can produce specific biological effects in humans and animals, they have a high potential to trigger negative impacts on non-target organisms. In addition, anti-infection agents could create an ecological hazard by advancing the spread of resistant genes in the environment (Costanzo et al. 2005).

Other concerns are that the metabolites of many pharmaceuticals are potentially active and unsafe in the environment. For example, paracetamol and amitriptyline are mostly metabolized into highly reactive compounds (Graham et al. 2013). Also, of concern is that pharmaceuticals are discharged into the marine environment from sewage treatment plants as complex mixtures thus exposing marine life to potential synergetic environmental effects. For example, a synergistic antioxidant response in fish was demonstrated in a laboratory study involving co-exposure to a mixture of fluoxetine (an antidepressant medication) and roxithromycin (an antibiotic), and also with a mixture of fluoxetine and propranolol (a β-blocker used to treat a range of cardiac disease symptoms) (Ding et al. 2016).

Similarly, some ingredients of **non-medicinal/ non-cosmetic** household products (e.g. triclosan—a widely used bactericide in healthcare products such as skin care ointments and lotions, mouthwashes and toothpastes, shower gels and shampoos) are not efficiently broken down in typical municipal sewage treatment plants, so that the end-of-treatment discharges from these facilities are major point sources of release into the marine environment (Cui et al. 2019). Triclosan is persistent and bioaccumulative in the aquatic environment and triggers a number of toxic responses (Maulvault et al. 2019).

The array of PPCPs in sewage discharges is extensive, but the potential for adverse effects is largely unknown for most of the active ingredients present (Ojemaye and Petrik 2019). The NORMAN Network (▶ Box 13.2) currently lists almost 300 PPCPs substances as CECs.

13.5 Nanomaterials

The manufacture and use of nanoparticles and **nanostructured materials** (also known as **nanomaterials**) is an expanding field of modern technology. As a consequence, the perceived risks associated with potentially toxic properties of these novel materials have resulted in their attracting attention as a new class of CECs.

By their nature, **nanoparticles** are units of particulate materials with a maximum dimension sized in nanometres (10^{-9} m). Although there is no single internationally accepted definition for **nanomaterials** (Jeevanandam et al. 2018), they are commonly defined as materials in which a single unit is sized in the range of 1–100 nm in at least one dimension. The term **aerosols** is often applied to nanoparticles when they are airborne, for example, in wind-borne dust or otherwise suspended in the atmosphere. The US EPA routinely uses the term **ultrafine particles** when discussing natural nanomaterials and aerosols. A summary of types and classifications of nanomaterials, and common technical descriptors is at ▶ Box 13.4.

Interestingly, the use and manufacture of nanomaterials are not an entirely modern phenomenon. The Ancient Egyptians used nanoparticulate lead sulfide as a hair dye some 4000 years ago (Walter et al., 2006) and more recently (400–100 BC) red enamels used by Ancient Celtic cultures were based on nanoparticulate copper oxides (Brun et al. 1991) and stained glass in medieval churches incorporated gold and silver nanoparticles (Schaming and Remita 2015).

The origin and source of nanoparticles and nanomaterials is diverse (▶ Box 13.5). Naturally occurring nanoparticles (colloids) and nanomaterials are widespread in both the living and inanimate world. In addition, nanoparticles and nanomaterials may be produced as an incidental by-product of an industrial process, or they may be manufactured explicitly by an engineered process to exploit specific features that stem from their small size.

The application of nanoparticulate and nanostructured materials has increased over the past decade be-

> **Box 13.4: Types and Classifications of Nanomaterials**
>
> **Carbon-Based Nanomaterials**: These comprise carbon and include morphologies such as hollow tubes, ellipsoids or spheres. Examples include fullerenes (C60), carbon nanotubes (CNTs), carbon nanofibers, carbon black, graphene (Gr) and carbon union.
>
> **Inorganic-Based Nanomaterials**: These comprise metal and metal oxides such as titanium dioxide (TiO_2), zinc oxide (ZnO) and zero-valent iron (nZVI).
>
> **Organic-Based Nanomaterials**: These include nanomaterials made mostly from organic matter, excluding carbon-based or inorganic-based nanomaterials.
>
> **Composite-Based Nanomaterials with One Phase of Nanoscale Dimension**: These include combinations of nanoparticles with other nanoparticles or nanoparticles combined with larger particles or with bulk-type materials (e.g. hybrid nanofibers), or more complicated structures, such as a metal–organic frameworks. Composites may be any combinations of carbon-based, metal-based or organic-based nanomaterials with any form of metal, ceramic or polymer bulk materials. Nano-objects are often categorized as to how many of their external dimensions are at the nanoscale. For example, a nano-object with:
>
> - all three external dimensions in the nanoscale is a **nanoparticle**;
> - two external dimensions in the nanoscale is a **nanofibre** and optionally with the terms **nanorods** and **nanotubes** being used if they are solid or hollow, respectively; and
> - one external dimension in the nanoscale is a **nanoplate** (if the other two dimensions are similar) or **nanoribbon** (if the other two dimensions are significantly different).
>
> Nanostructures may be categorized by the phases of their components. For example, a nanostructure comprising:
>
> - at least one physically or chemically distinct region at a nanoscale, or collection of regions with at least one at a nanoscale is a **nanocomposite**;
> - a liquid or solid matrix with at least one at a nanoscale, filled with a gaseous phase is a **nanofoam**;
> - a solid material containing pores or cavities with dimensions on the nanoscale is a **nanoporous material**; and
> - a significant fraction of crystal grains at the nanoscale is a **nanocrystalline material**.
>
> A comprehensive review of the different types of nanomaterials is provided in Jeevanandam et al. (2018).

cause they provide enhanced or unique physicochemical properties (e.g. melting point, wettability, electrical or thermal conductivity, catalytic activity, light absorbance or scattering) that are different from those of their bulk counterparts. Manufactured nanomaterials can significantly improve the characteristics of bulk materials, in terms of strength, conductivity, durability and lightness, and they can provide useful properties (e.g. self-healing, self-cleaning, anti-freezing and antibacterial) and can function as reinforcing materials for construction. By 2014, some 1814 nanotechnology-based consumer products were commercially available in over 20 countries (Vance et al. 2015). Examples of the incorporation of nanoparticles in consumer products include titanium oxide nanoparticles as a white pigment in paints, cosmetic creams and sunscreens, and silver nanoparticles used in numerous personal care products such as air sanitizers, wet wipes, shampoos and toothpastes, as well as in clothing and laundry fabric softeners (PEN 2019). Nanoscale zero-valent particulate iron (nZVI) is a widely used remediant for treating toxic wastes due to its large specific surface area and high reactivity (Stefaniuk et al. 2016).

Unfortunately, the highly sought physicochemical properties of nanomaterials that have led to their increasing applications can also have an associated environmental downside. For example, nanoparticulate zinc oxide (ZnO) used in sunscreens is toxic to marine algae largely because of its dissolution as Zn^{2+} (Franklin et al. 2007), and nZVI use in contaminant remediation presents a range of potentially harmful environmental consequences that are not well understood (Stefaniuk et al. 2016). The enhanced toxic potential of nanosized materials may arise from their capacity to penetrate and disturb the cells and cellular systems of living tissues.

The challenge for regulators is to determine whether nanomaterials should be regulated in the same way as micron-sized particles. Among metal nanomaterials, cerium dioxide (used as a diesel fuel additive) and nanosilver are more toxic than their micron-sized forms, whereas because of their solubility there is no difference in toxicity for zinc oxide nano- and micron-sized particles in freshwaters (Batley et al. 2013). The enhanced surface area of nanosized materials can result in different cellular uptake rates, oxidative mechanisms and processes including translocation relative to that of exposure to the same material when it is not nanosized (Oberdörster et al. 2005). In the environment, aggregation is a common feature of nanomaterials, and often coatings are used (e.g. citrate or polyvinylpyrrolidone (PVP)) to minimize this. Aggregation is greatest in marine waters due to their high ionic strength, leading to sizes > 100 nm and in many cases resulting in sedimentation (Klaine et al. 2008). The presence of organic particles such as those formed from extracellular polymeric substances can briefly stabilize nanomaterials (<48-h) (Gondikas et al. 2020). Seawater enhances the dissolution of silver from coated Ag nanomaterials, largely through chloride complexation, which reduces silver toxicity (Angel et al. 2013).

Some nanoparticles and nanomaterials are released directly into the environment from the use of consumer products (e.g. silicon nanoparticles in car tyres are released by abrasion in normal vehicle use), or indirectly (e.g. nanoparticles in pharmaceuticals and cosmetics can end up in sewage, and then be discharged to the marine environment).

Box 13.5: Origin and Sources of Nanomaterials

There are three main origins of nanomaterials—incidental, engineered and naturally produced.

Incidental Nanomaterials: These are produced as a by-product of industrial processes such as nanoparticles present in vehicle engine exhaust, welding fumes and other combustion processes.

Naturally Produced Nanomaterials: Dust from soil ablation by winds, volcanic eruptions and forest fires are events of natural origin that produce large quantities of nanoparticulate matter that significantly affect worldwide air quality.

Nanoparticles and nanostructures are present in living organisms ranging from microorganisms (e.g. bacteria, algae and viruses) to complex organisms, such as plants and animals. Plants accumulate nutrients extracted from the soil as biominerals in nanoform.

The natural transport of mineral aerosol particles to the oceans has an important role in supporting marine biological productivity. Iron, along with phosphorus and silica, is a limiting nutrient for most marine phytoplankton. Iron is needed for a multitude of enzymes and electron transfer proteins including those essential for photosynthesis (Bristow et al., 2017). The major source of iron input to oceanic waters far from land is deposition of wind-transported continental dust (Buseck and Posfai, 1999). Since phytoplankton form the basis of the marine food web and is responsible for approximately half of global carbon dioxide fixation, this natural transport of iron in mineral aerosol particles is an essential contributor to the removal of excess CO_2 and carbon sequestration by marine phytoplankton (Basu and Mackey, 2018).

Engineered Nanomaterials: A diverse range of nanomaterials is synthesized by both **bottom-up** (meaning the constructive build-up of material from atom to clusters to nanoparticles) and **bottom-down** (meaning the reduction of a bulk material to nanometric-scale particles) processes. These products have a multitude of applications including medical (e.g. targeted drug delivery in pharmacology), cosmetics and sunscreens, electronics, catalysis (e.g. automotive applications), food (production, processing and packaging), construction (e.g. new materials), renewable energy and environmental remediation.

A comprehensive review of the different types of engineered nanomaterials and their applications, together with methods of synthesis, is provided in Ealias and Saravanakumar (2017).

The fate and impacts of nanomaterials in the environment have been comprehensively reviewed (Klaine et al. 2008; Lead et al. 2018). It is generally agreed that the current environmental concentrations are orders of magnitude below those known to have toxic effects on aquatic biota (Batley et al. 2013).

13.6 PFAS (Per- and Polyfluoroalkyl Substances)

The term PFAS (per- and polyfluoroalkyl substances) applies to the set of more than 4700 synthetic substances manufactured and used in a variety of industries since the 1940s (OECD 2019), and some have been classified as POPs (▶ Chapter 8). All PFAS constitute an array of highly persistent environmental CECs that has triggered a global response by research and regulatory organizations over the past two decades.

PFAS comprise a set of compounds each of which has a molecular structure comprising an aliphatic moiety (i.e. a group of covalently bonded carbon atoms in a straight or branched chain, and in some cases including non-aromatic rings) that is highly fluorinated and linked to a functional group moiety. This PFAS molecular structure can be conceptualized as an alkyl tail of carbon atoms with fluorine atoms attached to a a functional group head (◘ Figure 13.3). The degree of fluorination of the aliphatic moiety in a PFAS structure can be partial or total. In polyfluoroalkyl substances, fluorine atoms replace only some of the hydrogen atoms in the aliphatic chain, whereas in perfluoroalkyl substances, fluorine atoms replace all of the hydrogen atoms in the aliphatic chain. The general formula for

Figure 13.3 Typical perfluorinated PFAS molecules showing the basic structure comprising a perfluorinated alkyl tail attached to a functional group head. Structures here are the linear isomers. A mixture of linear and branched isomers may be present in an environmental sample. Adapted from Mueller and Yingling (2017) by M. Mortimer

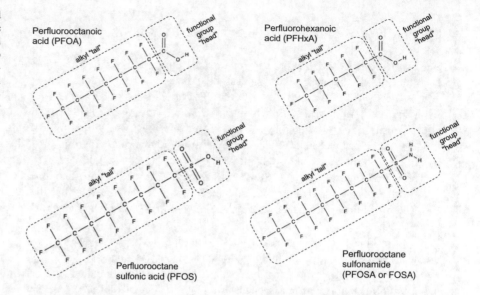

a perfluorinated PFAS is C_nF_{2n+1}-R where n is 3 or greater and -R is a functional group such as carboxylic acid (COOH), sulfonic acid (SO_3H) or sulfonamide (SO_2NH_2) (□ Figure 13.3).

Note that the term **PFAS** sometimes appears in print in the context of more than one fluorinated chemical, but the addition of the **s** is redundant since the acronym **PFAS** includes the plural (ATSDR 2017). Also, it is important to note that PFAS (per- and polyfluoroalkyl substances) are sometimes called perfluorinated chemicals and the acronym PFC is then used. However, this use of the **PFC** acronym can be confusing since **PFC** is also commonly used for a related, but distinctly different group of substances: the *perfluorocarbons* (▶ Box 13.6).

The range of structurally related compounds comprising more than 4700 member group of PFAS substances is illustrated in the **PFAS family tree** in □ Figure 13.4.

Figure 13.4 The PFAS **family tree** with examples. Adapted from Wang et al. (2017) by M. Mortimer. PFCAs = Perfluoroalkyl carboxylic acids; PFSAs = Perfluoroalkane sulfonic acids; PFPAs = Perfluoroalkyl phosphonic acids; PFPiAs = Perfluoroalkyl phosphinic acids; PFECAs and PFESAs = Perfluoroether carboxylic and sulfonic acids; PASF = Perfluoroalkane sulfonyl fluoride

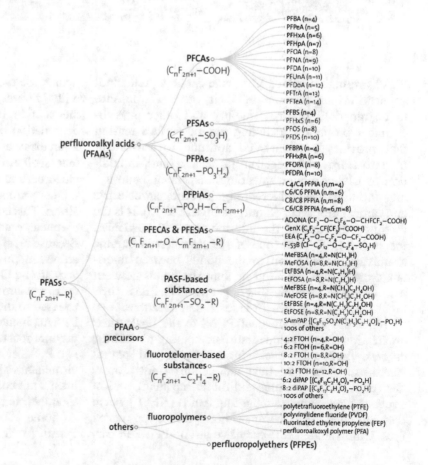

Box 13.6: What are Perfluorocarbons (PFCs) and How do they Relate to Per- and Polyfluoroalkyl Substances (PFAS)?

The acronym **PFC** is widely used for two related but distinctly different groups of substances—(1) the **perfluorocarbons** and (2) the **perfluorinated chemicals**. Examples of organizations that have used the acronym **PFC** for both groups are the USEPA, the U.S. Center for Disease Control and Prevention (CDC), and the UN Organisation for Economic Co-operation and Development (OECD). However, the USEPA now states on its website that it is trying to consistently use the acronym **PFAS** rather than **PFC** when referring to perfluorooctane sulfonic acid (PFOS), PFOA and other substances in the per- and polyfluoroalkyl group (▶ https://www.epa.gov/pfas/what-are-pfcs-and-how-do-they-relate-and-polyfluoroalkyl-substances-pfass). Likewise, the United States Agency for Toxic Substances and Disease Registry (ATSDR) website fact sheet makes a similar comment (ATSDR 2017).

Another reason for avoiding use of the PFC acronym for PFAS is that the acronym PFC has been used in official Kyoto Protocol documents since its adoption in 1997, specifically to designate greenhouse gas perfluorocarbons (United Nations 1998).

Perfluorocarbons and PFAS are closely related, in that:
− the molecules of both contain fluorine and carbon atoms; and
− both persist in the environment for long periods but are not found naturally except for the perfluorocarbon, carbon tetrafluoride that occurs in association with fluorite minerals (Mühle et al. 2010).

However, perfluorocarbons and PFAS are quite different, in that:
− perfluorocarbon molecules contain only carbon and fluorine atoms, but PFAS molecules can include many other atoms including oxygen, hydrogen, sulfur and nitrogen; and
− perfluorocarbons are used in and emitted from different applications and industries than PFAS.

Other groups of fluorinated hydrocarbon compounds that are sometimes confused with the PFAS and PFC groups are the chlorofluorocarbon (CFC) group and the hydrofluorocarbon (HFC) group. The members of the CFC group are hydrocarbon chain structures with hydrogen atoms replaced with both chlorine and fluorine atoms. An example is dichlorotetrafluoroethane ($C_2Cl_2F_4$). The HFC group is hydrocarbons with only some hydrogens replaced by fluorine. An example is tetrafluoroethane ($C_2H_2F_4$). Both CFCs and HFCs are synthetic compounds (trade name FreonsTM) used as aerosol and refrigerant gases, but being phased out since CFCs are ozone-depleting substances, and HFCs are very potent greenhouse gases.

13

PFAS are highly valued in a wide range of industrial applications on account of their extreme resistance to degradation, thermal stability and other physico-chemical properties including unique surface tension that provides a remarkable aptitude to self-assemble into sturdy thin repellent protective films, in addition to having unique spreading, dispersing, emulsifying, anti-adhesive and levelling, dielectric, piezoelectric and optical properties (Krafft and Riess, 2015). It is the unique properties of the fluorine atom, in particular, the strong C–F bond (one of the strongest in organic chemistry), in addition to the bonds between fluorinated carbons being stronger than the bonds between hydrogenated carbons, that together give PFAS their highly valued attributes for industrial applications.

Since the development of PFAS in the 1940s, their applications have included surface coating of textiles, carpets, cardboard packaging products and papers where use is made of their unique surfactant properties (both water-repelling and oil/fat-repelling), and in aqueous film-forming firefighting foams (AFFF) where they are effective in extinguishing hydrocarbon-fuelled fires. Unfortunately, the extensive use of PFAS, in particular, the use of AFFF, together with their high solubility in water, low/moderate sorption to soils and sediments and a high resistance to biological and chemical degradation has led to the global emergence of PFAS as an array of highly environmental persistent contaminants of emerging concern (CECs). Their perfluorinated carbon chains form a helical structure, in which the carbon skeleton is completely covered by fluorine atoms. This cover shields the PFAS molecule from most chemical attacks and results in highly stable molecules. Because most PFAS compounds are anthropogenic creations, and also due to the presence of the multiple and very strong C–F bonds, there is a lack of naturally occurring microbes capable of breaking them down. Once released to the environment many PFAS can degrade to PFAAs, including perfluorooctanoic acid (PFOA) and perfluorooctane sulfonic acid (PFOS), which are the two PFAS compounds most commonly found in the environment in high concentrations. Once waters and sediments are contaminated by PFAS, they present a considerable challenge to successful remediate.

Some PFAS, specifically PFOS, PFOS salts and POSF (perfluorooctane sulfonyl fluoride), are listed

POP substances in the Stockholm Convention (see ▶ Chapter 8). However, PFAS do not bioaccumulate in the same way as most other Stockholm Convention POPs such as the halogenated hydrocarbon pesticides and flame retardants that are lipophilic in nature and are preferentially accumulated in lipid-rich tissues (see ▶ Chapter 7). Notably, PFAS are not lipophilic since their alkyl tails make them both hydrophobic (water-repelling) and oleophobic/lipophobic (oil/fat-repelling)

in character. Many PFAS, including PFOA and PFOS, do have a high potential to bioconcentrate, bioaccumulate and biomagnify, and now have an ubiquitous and growing presence throughout the food chain. However, they bind to proteins and consequently bioaccumulate in blood and blood-rich tissues rather than lipid-rich tissues. The relationship between the carbon chain length of PFAS and environmental behaviour and fate is discussed in ▶ Box 13.7.

Box 13.7: The Role of PFAS Chain Length in Relation to Environmental Behaviour and Level of Concern

Sometimes carbon chain length is used to group PFAS which may behave similarly in the environment, particularly the perfluoroalkyl carboxylic acids and sulfonic acids (PFCA and PFSA), and the terms **long-chain PFAS** and **short-chain PFAS** used in relation to their potential environmental significance. For example, the Organisation for Economic Co-operation and Development (OECD 2013) uses these definitions:

Long chain refers to:

- perfluoralkyl carboxylic acids with eight or more carbons (seven or more carbons are perfluorinated);
- perfluoroalkane sulfonates with six or more carbons (six or more carbons are perfluorinated); and
- substances with the potential to degrade to PFCA or PFSA (i.e. precursors).

Short chain refers to:

- perfluoralkyl carboxylic acids with seven or fewer carbons (six or less carbons are perfluorinated); and
- perfluoroalkane sulfonates with five or fewer carbons (five or less carbons are perfluorinated).

However, caution should be applied in making generalizations about PFAS behaviour based only on chain length. Although in general terms, the potential toxicity of PFAS increases with the length of the carbon–carbon chain, as does the potential for bioconcentration, bioaccumulation and persistence in the environment, other factors besides chain length are involved, including the functional groups in the PFAS structure and the interactions involving PFAS-protein binding to form complexes (Ng and Hungerbuehler 2015). Studies also show that bioaccumulation factors reach a maximum at a carbon chain length of 11 (Ng and Hungerbühler 2014).

PFAS have been globally detected in lakes, rivers, oceans and even in precipitation water at ng/L concentrations. PFAS are significantly transported in aquatic ecosystems, including transport to remote polar regions in aerosols. Not only is PFAS contamination of marine waters occurring on a global scale, but evidence is mounting of accumulation in wildlife even at locations remote from any direct source, for example, Antarctica (Llorca et al. 2012), particularly PFAS with alkyl chain lengths of less than 10 carbon atoms, but the ecotoxicological impacts of this in the short or long term are unclear. For example, Wei et al. (2007) reported PFOS and PFOA concentrations of 21.1 and 7 pg/L, respectively, in oceanic waters hundreds of kilometres south of Tasmania, Australia. These concentrations are comparable to the range reported for the mid to southern Pacific Ocean of up to 8 and 20 pg/L in Ahrens (2011). Toxicity data obtained to date suggest that a guideline value for PFOS in freshwaters for 99% species protection is near 30 ng/L and orders of magnitude higher in marine waters (G. Batley, personal communication).

PFAS have been found at ng/kg concentrations in deep sea sediments, and at hundreds of ng/g in fish. In the absence of effective rapid breakdown of many PFAS in the environment due to their chemical structure, the global fate of most discharged PFAS is dispersal and burial in the deep ocean sediments (Ahrens 2011). However, the future impacts on wildlife from exposure to the persistently dissolved fraction of PFAS are unknown. Simpson et al. (2021) demonstrated that for PFOS, a screening value of 60 µg/kg (for 1% organic carbon) would be protective of organisms in estuarine and marine sediments.

13.6.1 Naming Conventions Used for PFAS

In common with other compounds, all PFAS have chemical names consistent with the International Union of Pure and Applied Chemistry (IUPAC) system (Favre and Powell 2013). However, many of these IUPAC names are long and somewhat impractical to use in reporting, and a simplified nomenclature and abbreviation system for PFAS has been developed in the scientific literature (Lehmler 2005). This simplified naming system for PFAS in reports and journal papers is explained further in Boxes 13.8 through 13.10 and in ☐ Tables 13.1 and 13.2.

☐ Table 13.1 Examples of simplified naming of perfluoroalkyl carboxylates and acids

X	Y	Name and (acronym)	Formula
But-(4 Carbon Chain)	Carboxylate	Perfluorobutanoate (PFBA)	$CF_3(CF_2)_2CO_2^-$
	Carboxylic acid	Perfluorobutanoic acid (PFBA)	$CF_3(CF_2)_2CO_2H$
	Sulfonate	Perfluorobutane sulfonate (PFBS)	$CF_3(CF_2)_3SO_3^-$
	Sulfonic acid	Perfluorobutane sulfonic acid (PFBS)	$CF_3(CF_2)_3SO_3H$
Pent-(5 Carbon Chain)	Carboxylate	Perfluoropentanoate (PFPeA)	$CF_3(CF_2)_3CO_2^-$
	Carboxylic acid	Perfluoropentanoic acid (PFPeA)	$CF_3(CF_2)_3CO_2H$
	Sulfonate	Perfluoropentane sulfonate (PFPeS)	$CF_3(CF_2)_4SO_3^-$
	Sulfonic acid	Perfluoropentane sulfonic acid (PFPeS)	$CF_3(CF_2)_4SO_3H$
Hex-(6 Carbon Chain)	Carboxylate	Perfluorohexanoate (PFHxA)	$CF_3(CF_2)_4CO_2^-$
	Carboxylic acid	Perfluorohexanoic acid (PFHxA)	$CF_3(CF_2)_4CO_2H$
	Sulfonate	Perfluorohexane sulfonate (PFHxS)	$CF_3(CF_2)_5SO_3^-$
	Sulfonic acid	Perfluorohexane sulfonic acid (PFHxS)	$CF_3(CF_2)_5SO_3H$
Hept-(7 Carbon Chain)	Carboxylate	Perfluoroheptanoate (PFHpA)	$CF_3(CF_2)_5CO_2^-$
	Carboxylic acid	Perfluorohexanoic acid (PFHpA)	$CF_3(CF_2)_5CO_2H$
	Sulfonate	Perfluorohexane sulfonate (PFHpS)	$CF_3(CF_2)_6SO_3^-$
	Sulfonic acid	Perfluorohexane sulfonic acid (PFHpS)	$CF_3(CF_2)_6SO_3H$
Oct-(8 Carbon Chain)	Carboxylate	Perfluorooctanoate (PFOA)	$CF_3(CF_2)_6CO_2^-$
	Carboxylic acid	Perfluorooctanoic acid (PFOA)	$CF_3(CF_2)_6CO_2H$
	Sulfonate	Perfluorooctane sulfonate (PFOS)	$CF_3(CF_2)_7SO_3^-$
	Sulfonic acid	Perfluorooctane sulfonic acid (PFOS)	$CF_3(CF_2)_7SO_3H$

– continues stepwise with additions to carbon chain… non (9), dec (10), undec (11), dodec (12), tridec (13) etc

X	Y	Name and (acronym)	Formula
Tetradec-(14 Carbon Chain)	Carboxylate	Perfluorotetradecanoate (PFTeDA)	$CF_3(CF_2)_{12}CO_2^-$
	Carboxylic acid	Perfluorotetradecanoic acid (PFTeDA)	$CF_3(CF_2)_{12}CO_2H$
	Sulfonate	Perfluorotetradecane sulfonate (PFTeDS)	$CF_3(CF_2)_{13}SO_3^-$
	Sulfonic acid	Perfluorotetradecane sulfonic acid (PFTeDS)	$CF_3(CF_2)_{13}SO_3H$

– continues stepwise with additions to carbon chain… pentadec- (15), hexadec- (16), heptadec- (17) etc

X = the name of the alkyl chain tail based on number of linked carbon atoms and Y = the functional group head

☐ Table 13.2 Examples of simplified naming for fluorotelomer-based polyfluoroalkyl substances

X	Y	Functional group	Name and (acronym)	Formula
4	2	Hydroxyl	4:2 fluorotelomer alcohol (4:2 FTOH)	$CF_3(CF_2)_3(CH_2)_2OH$
		Carboxyl	4:2 fluorotelomer carboxylic acid (4:2 FTCA)	$CF_3(CF_2)_3CH_2CO_2H$
		Sulfonyl	4:2 fluorotelomer sulfonic acid (4:2 FTSA)	$CF_3(CF_2)_3(CH_2)_2SO_3H$
6	2	Hydroxyl	6:2 fluorotelomer alcohol (6:2 FTOH)	$CF_3(CF_2)_5(CH_2)_2OH$
		Carboxyl	6:2 fluorotelomer carboxylic acid (6:2 FTCA)	$CF_3(CF_2)_5(CH_2)_2CO_2H$
		Sulfonyl	6:2 fluorotelomer sulfonic acid (6:2 FTSA)	$CF_3(CF_2)_5(CH_2)_2SO_3H$
8	2	Hydroxyl	8:2 fluorotelomer alcohol (8:2 FTOH)	$CF_3(CF_2)_7(CH_2)_2OH$
		Carboxyl	8:2 fluorotelomer carboxylic acid (8:2 FTCA)	$CF_3(CF_2)_7(CH_2)_2CO_2H$
		Sulfonyl	8:2 fluorotelomer sulfonic acid (8:2 FTSA)	$CF_3(CF_2)_7(CH_2)_2SO_3H$

X = fully fluorinated carbon atoms and Y = not fully fluorinated carbon atoms

Box 13.8: Simplified Naming System for Per- and Polyfluoroalkyl Substances (PFAS)

Many of the IUPAC names for PFAS are long and somewhat impractical to use in reporting, and a simplified nomenclature and abbreviation system for PFAS has been developed in the scientific literature (Lehmler, 2005). This practice of adopting simplified **literature names** for PFAS makes the writing (and reading) of reports and journal papers much easier. An example of this simplified naming is the use of the name **perfluorooctane sulfonate** (abbreviated as PFOS) rather than the IUPAC name for that compound, which is:

▬ 1,1,2,2,3,3,4,4,5,5,6,6,7,7,8,8,8-Heptadecafluoro-1-octanesulfonate.

Most of the IUPAC name for this compound, including the string of numbers, describes the location of the fluorine atoms along the perfluorinated alkyl tail of carbon atoms in its molecular structure.

This long name can be simplified for reporting purposes by replacing the string of numbers with the term **perfluorooctane** (in this example **octane** becomes part of the name because the alkyl tail is 8 carbons in length), followed by the name of the functional group head which in this example is **sulfonate**.

Similarly, a simplified name for a PFAS with a perfluorinated alkyl tail 10 carbons in length would use **perfluorodecane** followed by the name of the functional group (e.g. **perfluorodecane sulfonate** [abbreviated as PFDS]) if the functional group is sulfonate.

See ▶ Box 13.9 for more examples of the use of the simplified naming system for **perfluoroalkyl substances,** and ▶ Box 13.10 in the case of **polyfluoroalkyl substances**.

Box 13.9: Applying the Simplified Naming System to Perfluoroalkyl Substances

The use of the simplified naming system for **per**fluoroalkyl substances is illustrated in ◼ Table 13.1 using the perfluoroalkyl carboxylates and acids as examples.

In this system, the structure name is written in the form PFXY where:

▬ PF = perfluoro-
▬ X = the name of the alkyl carbon chain structure appropriate to the number of linked carbon atoms (for example, **but** for 4 carbons, **pent** for 5 carbons, etc.)
▬ Y = the name of the attached functional group **head**

For example, a perfluorinated (all hydrogens replaced by fluorine atoms) 4-carbon (butan-) chain, bonded to a carboxylic acid functional group is named **perfluorobutanoic acid** which can be abbreviated to the acronym **PFBA**.

However, beware that the usage of acronyms for PFAS is not standardized and many are ambiguous (for example, PFBA is used for both **perfluorobutanoic acid** and its anion **perfluorobutanoate**). Accordingly, authors and readers of reports concerning PFAS need to ensure that to avoid ambiguity, the compounds referred to using acronyms are clearly identified in the text (for example, by the CAS number and/or IUPAC name).

Note that there are both anionic (negative charged) and acid forms associated with functional groups such as carboxylate and carboxylic acid. However, except under conditions of extremely low pH, it is the anionic form that is found in the environment since the acid or salt form dissociates in solution (Buck et al. 2011).

Box 13.10: Applying the Simplified Naming System to Polyfluoroalkyl Substances

The **fluorotelomers** are a series of polyfluoroalkyl substances synthesized on an industrial scale and used in a wide range of commercial products. Two major industrial processes are used for the commercial manufacture of PFAS. These are the telomerization process and electrochemical fluorination (ECF) process. Both are described in Buck et al. (2011).

A widely used simplified naming system for fluorotelomer-based polyfluoroalkyl substances is illustrated in ◼ Table 13.2. In this system, the fluorotelomer polyfluoroalkyl structure name is written in the form of a ratio X:Y where:

▬ X = the number of fully fluorinated carbon atoms and
▬ Y = the number of carbon atoms not fully fluorinated

Thus, a fluorotelomer alcohol (FTOH) with 6 fully fluorinated carbons and 2 not fully fluorinated carbons is given the name **6:2 fluorotelomer alcohol** (abbreviated 6:2 FTOH), and similarly a fluorotelomer sulfonic acid (FTSA) with 8 fully fluorinated carbons and 2 not fully fluorinated carbons is given the name **8:2 fluorotelomer sulfonic acid** (abbreviated 8:2 FTSA or FtS 8:2).

13.6.2 PFAS and Precursors

The concept of a **precursor** is important in the PFAS contamination context. There is a wide range of poly-fluorinated PFAS comprising fluorinated structures capable of natural transformation to other more persistent fluorinated structures. Typically, such degradations follow a stepwise process ending with a perfluorinated PFAS, often PFOA or PFOS (thus sometimes termed **terminal PFAS**).

In an environment subject to PFAS contamination (for example, sites associated with PFAS manufacture, the use of AFFF, waste disposal by land-fill or wastewater treatment), there are many **precursors** of PFOA and PFOS present. This conversion of precursors enables an increase in the relative quantities of PFOA and PFOS present after wastes are released, providing an explanation for the dominance of PFOA and PFOS in the global inventory of residual PFAS in the natural environment including the seas and oceans.

Polyfluorinated PFAS structures such as the fluorotelomers are typical **precursors** since the non-fluorinated portion of the carbon chain is open to biotic degradation and modification by abiotic processes such as oxidation (▣ Figure 13.5).

▣ **Figure 13.5** Example structures of polyfluorinated precursors. The sections of the molecule with a not fully fluorinated carbon are exposed to modification by both abiotic and biotic processes. Adapted from Mueller and Yingling (2017) by M. Mortimer

13.7 Summary

There is a large body of research papers and reports concerning the topic of **Contaminants of Emerging Concern (CECs)** but the term itself is not definitive since both **emerging** and **concern** may be subjective, and the list of materials identified as CECs changes over time and in response to community perceptions of risks to health and the environment. The NORMAN Network is a key organization in identifying such materials and coordinating meaningful related research.

In the marine environment, since the late 1900s the priority focus has moved from concern over unintended impacts from the widespread use of organic tin-based antifoulants used on the hulls of sea-going vessels to impacts relating to a wide range of material types including EDCs, PPCPs, nanomaterials, PFAS compounds as well as environmental contamination by polymer and plastic debris.

Each of these current CECs covers a large number of chemical identities. Overall this is a wide-ranging and dynamic area of risk assessment, priority setting and ongoing scientific research.

13.8 Study Questions and Activities

1. In the context of the marine environment draw up a short list of up to five contaminants of emerging concern that are highlighted in recent media pub-

lications (noting that the media *may not* use the term **contaminant of emerging concern** as a descriptor) and compare this short list with contaminants which are popular topics in the programmes of recent conference presentations and journal publications. What do you suggest are reasons for similarities and differences between these two sets of CECs?

2. Identify two CECs in the marine environment that have been receiving frequent attention for a period longer than two or three years. Why are they still considered **emerging** (for example, has the baseline of residual concern changed)?

3. In this chapter, lead in the marine environment near Port Pirie, South Australia is used as an example. What other locations in Australia and other countries with territorial waters in the Pacific Ocean also have an emerging problem associated with lead mining and processing?

4. Which metallic contaminants are CECs in European marine waters?

5. What potential contaminants of marine waters are likely to become CECs as a consequence of the shift from fossil-fuel-based energy sources to renewables? What are some geographic locations where these may first emerge as CECs—explain why?

References

Ahrens L (2011) Polyfluoroalkyl compounds in the aquatic environment: a review of their occurrence and fate. J Environ Monit 13:20–31

Angel BM, Batley GE, Jarolimek C, Rogers NJ (2013) The impact of nano size on the fate and toxicity of nanoparticulate silver in aquatic systems. Chemosphere 93:359–365

ATSDR (Agency for Toxic Substances and Disease Registry) (2017) The family tree of per- and polyfluoralkyl substances (PFAS) for environmental health professionals. Available at: ▶ https://www.atsdr.cdc.gov/pfas/docs/PFAS_FamilyTree_EnvHealthPro-508.pdf. Accessed 29 Mar 2019

Basu S, Mackey K (2018) Phytoplankton as key mediators of the biological carbon pump: their responses to a changing climate. Sustainability 10(3):869

Batley GE, Kirby JK, McLaughlin MR (2013) Fate and risks of nanomaterials in the aquatic environment. Acc Chem Res 46:854–862

Boxall A, Rudd M, Brooks B, Caldwell D, Choi K, Hickmann S, Innes E, Ostapyk K, Staveley J, Verslycke T, Ankley G, Beazley K, Belanger S, Berninger J, Carriquiriborde P, Coors A, DeLeo P, Dyer S, Ericson J, Gagné F, Giesy J, Gouin T, Hallstrom L, Karlsson M, Joakim Larsson D, Lazorchak J, Mastrocco F, McLaughlin A, McMaster M, Meyerhoff R, Moore R, Parrott J, Snape J, Murray-Smith R, Servos M, Sibley P, Oliver Straub J, Szabo N, Topp E, Tetreault G, Trudeau V, Van Der Kraak G (2012) Pharmaceuticals and personal care products in the environment: what are the big questions? Environ Health Perspect 120(9):1221–1229

Bristow L, Mohr W, Ahmerkamp S, Kuypers M (2017) Nutrients that limit growth in the ocean. Curr Biol 27(11):R474–R478

Brun N, Mazerolles L, Pernot M (1991) Microstructure of opaque red glass containing copper. J Mater Sci Lett 10(23):1418–1420

Buck R, Franklin J, Berger U, Conder J, Cousins I, de Voogt P, Jensen A, Kannan K, Mabury S, van Leeuwen S (2011) Perfluoroalkyl and polyfluoroalkyl substances in the environment: terminology, classification, and origins. Integr Environ Assess Manage 7(4):513–541

Buseck P, Posfai M (1999) Airborne minerals and related aerosol particles: effects on climate and the environment. Proc Natl Acad Sci 96(7):3372–3379

Cilliers L, Retief F (2019) Lead poisoning and the downfall of Rome: reality or myth? In: Wexter P (ed) Toxicology in antiquity. Elsevier, Amsterdam, pp 135–148

Colborn T, Clement C (1992) Chemically-induced alterations in sexual and functional development: the wildlife/human connection. Princeton Scientific, Princeton, p 403

Colborn T, Dumanoski D, Myers J (1997) Our stolen future. Abacus, London, p 306

Costanzo S, Murby J, Bates J (2005) Ecosystem response to antibiotics entering the aquatic environment. Mar Pollut Bull 51(1–4):218–223

Cui Y, Wang Y, Pan C, Li R, Xue R, Guo J, Zhang R (2019) Spatiotemporal distributions, source apportionment and potential risks of 15 pharmaceuticals and personal care products (PPCPs) in Qinzhou Bay, South China. Mar Pollut Bull 141:104–111

Dey S, Bano F, Malik A (2019) Pharmaceuticals and personal care product (PPCP) contamination—a global discharge inventory. In: Prasad M, Vithanage M, Kapley A (eds) Pharmaceuticals and personal care products: waste management and treatment technology: emerging contaminants and micro pollutants. Butterworth-Heinemann, Oxford, pp 1–26

Ding J, Lu G, Li Y (2016) Interactive effects of selected pharmaceutical mixtures on bioaccumulation and biochemical status in crucian carp (*Carassius auratus*). Chemosphere 148:21–31

Dulio V, van Bavel B, Brorström-Lundén E, Harmsen J, Hollender J, Schlabach M, Slobodnik J, Thomas K, Koschorreck J (2018) Emerging pollutants in the EU: 10 years of NORMAN in support of environmental policies and regulations. Environ Sci Eur 30(1):1–13

Ealias A, Saravanakumar M (2017) A review on the classification, characterisation, synthesis of nanoparticles and their application. IOP Conf Ser Mater Sci Eng 263:032019

EPASA (Environment Protection Authority, South Australia) (2005) Heavy metal contamination in the Northern Spencer Gulf—a community summary. Available at: ▶ https://www.epa.sa.gov.au/files/477354_heavy_metal.pdf. Accessed 1 Oct 2021

Favre H, Powell W (2013) Nomenclature of organic chemistry. Royal Society of Chemistry, Cambridge, p 1568

Forbes M, Taylor M (2015) A review of environmental lead exposure and management in Mount Isa, Queensland. Rev Environ Health 30(3):183–189

Franklin NM, Rogers NT, Apte SC, Batley GE, Casey PE (2007) Comparative toxicity of nanoparticulate ZnO, bulk ZnO and ZnCl$_2$ to a freshwater microalga (*Pseudokirchnerella subcapitata*): the importance of particle solubility. Environ Sci Technol 41:8484–8490

Godfray H, Stephens A, Jepson P, Jobling S, Johnson A, Matthiessen P, Sumpter J, Tyler C, McLean A (2019) A restatement of the natural science evidence base on the effects of endocrine disrupting chemicals on wildlife. Proc Roy Soc B Biol Sci 286(1897):20182416

Gondikas A, Gallego-Urrea J, Halbach M, Derrien N, Hassellöv M (2020) Nanomaterial fate in seawater: a rapid sink or intermittent stabilization? Front Environ Sci 8:151

Graham G, Davies M, Day R, Mohamudally A, Scott K (2013) The modern pharmacology of paracetamol: therapeutic actions, mechanism of action, metabolism, toxicity and recent pharmacological findings. Inflammopharmacology 21(3):201–232

Jeevanandam J, Barhoum A, Chan Y, Dufresne A, Danquah M (2018) Review on nanoparticles and nanostructured materials: history, sources, toxicity and regulations. Beilstein J Nanotechnol 9:1050–1074

Jobling S, Sumpter J, Sheahan D, Osborne J, Matthiessen P (1996) Inhibition of testicular growth in rainbow trout (*Oncorhynchus mykiss*) exposed to estrogenic alkylphenolic chemicals. Environ Toxicol Chem 15(2):194–202

Jobling S, Nolan M, Tyler C, Brighty G, Sumpter J (1998) Widespread sexual disruption in wild fish. Environ Sci Technol 32(17):2498–2506

Klaine SJ, Alvarez PJJ, Batley GE, Fernandes TF, Handy RD, Lyon D, Mahendra S, McLaughlin MJ, Lead JR (2008) Nanomaterials in the environment: behaviour, fate, bioavailability and effects. Environ Toxicol Chem 27:1825–1851

Halden R (2015) Epistemology of contaminants of emerging concern and literature meta-analysis. J Hazard Mater 282:2–9

IMO (International Maritime Organization) (2001) International convention on the control of harmful anti-fouling systems on ships. IMO Assembly Resolution A.895 (21). Available at: ► http://www.imo.org/en/KnowledgeCentre/IndexofIMOResolutions/Assembly/Documents/A.895(21).pdf. Accessed 12 Mar 2019

Katsiadaki I (2019) Are marine invertebrates really at risk from endocrine-disrupting chemicals? Current Opin Environ Sci Health 11:37–42

Krafft M, Riess J (2015) Selected physicochemical aspects of poly- and perfluoroalkylated substances relevant to performance, environment and sustainability—part one. Chemosphere 129:4–19

Lafratta A, Serrano O, Masqué P, Mateo M, Fernandes M, Gaylard S, Lavery P (2019) Seagrass soil archives reveal centennial-scale metal smelter contamination while acting as natural filters. Sci Total Environ 649:1381–1392

Lead JR, Batley GE, Alvarez PJJ, Croteau MN, Handy RD, McLaughlin MJ, Judy JD, Schirmer K (2018) Nanomaterials in the environment: behaviour, fate, bioavailability and effects. An updated review. Environ Toxicol Chem 37:2029–2063

Lehmler H (2005) Synthesis of environmentally relevant fluorinated surfactants—a review. Chemosphere 58(11):1471–1496

Llorca M, Farre M, Tavano M, Alonso B, Koremblit G, Barceó D (2012) Fate of a broad spectrum of perfluorinated compounds in soils and biota from Tierra del Fuego and Antarctica. Environ Pollut 163:158–166

Maulvault A, Camacho C, Barbosa V, Alves R, Anacleto P, Cunha S, Fernandes J, Pousão-Ferreira P, Paula J, Rosa R, Diniz M, Marques A (2019) Bioaccumulation and ecotoxicological responses of juvenile white seabream (*Diplodus sargus*) exposed to triclosan, warming and acidification. Environ Pollut 245:427–442

Mueller R, Yingling V (2017) Naming conventions and physical and chemical properties of per- and Polyfluoroalkyl substances (PFAS). Interstate Technology Regulatory Council (ITRC), p 15. Available at: ► https://pfas-1.itrcweb.org/fact_sheets_page/PFAS_Fact_Sheet_Naming_Conventions_April2020.pdf. Accessed 13 Febr 2022

Mühle J, Ganesan A, Miller B, Salameh P, Harth C, Greally B, Rigby M, Porter L, Steele L, Trudinger C, Krummel P, O'Doherty S, Fraser P, Simmonds P, Prinn R, Weiss R (2010) Perfluorocarbons in the global atmosphere: tetrafluoromethane, hexafluoroethane, and octafluoropropane. Atmos Chem Phys 10(11):5145–5164

Ng C, Hungerbühler K (2014) Bioaccumulation of perfluorinated alkyl acids: observations and models. Environ Sci Technol 48(9):4637–4648

Ng C, Hungerbuehler K (2015) Exploring the use of molecular docking to identify bioaccumulative perfluorinated alkyl acids (PFAAs). Environ Sci Technol 49(20):12306–12314

NIEHS (2022) Endocrine disruptors. National Institute of Environmental Health Sciences. Available at: ► https://www.niehs.nih.gov/health/topics/agents/endocrine/index.cfm. Accessed 9 Mar 2022

Nilsen E, Smalling K, Ahrens L, Gros M, Miglioranza K, Picó Y, Schoenfuss H (2018) Critical review: grand challenges in assessing the adverse effects of contaminants of emerging concern on aquatic food webs. Environ Toxicol Chem 38(1):46–60

NOAA (National Oceanic and Atmospheric Administration) (2008) Mussel watch program. An assessment of two decades of contaminant monitoring in the Nation's Coastal zone. NOAA Technical Memorandum NOS NCCOS 74. Available at: ► https://repository.library.noaa.gov/gsearch?related_series=NOAA%20technical%20memorandum%20NOS%20NCCOS%20%3B%2074. Accessed 29 Mar 2019

Oberdörster G, Oberdörster E, Oberdörster J (2005) Nanotoxicology: an emerging discipline evolving from studies of ultrafine particles. Environ Health Perspect 113(7):823–839

OECD (Organisation for Economic Co-operation and Development) (2013) Synthesis paper on per- and polyfluorinated chemicals (PFCs). OECD/UNEP Global PFC Group. Available at: ► https://www.oecd.org/env/ehs/risk-management/PFC_FINAL-Web.pdf. Accessed 29 Mar 2019

OECD (Organisation for Economic Co-operation and Development) (2019) Portal on per and poly fluorinated chemicals (Online Database). Available at: ► http://www.oecd.org/chemicalsafety/portal-perfluorinated-chemicals/. Accessed 13 Febr 2022

Ojemaye C, Petrik L (2019) Pharmaceuticals in the marine environment: a review. Environ Rev 27(2):151–165

PEN (Project on Emerging Nanotechnologies) (2019) The project on emerging nanotechnologies. Available at: ► http://www.nanotech-project.tech/search/?q=silver&x=0&y=0. Accessed 13 Febr 2022

Santillo D, Johnston P, Langston W (2002) Tributyltin (TBT) antifoulants: a tale of ships, snails and imposex. In: Harremoes P, Gee D, MacGarvin M, Stirling A, Keys J, Wynne B, Vaz S (eds) Late lessons from early warnings: the precautionary principle 1896–2000. Environment Issue Report, no. 22. European Environment Agency, Copenhagen, p 135–148. Available at: ► https://www.eea.europa.eu/publications-#c7=en&c11=5&c14=&c12=&b_start=0&c13=Late+lessons+from+early+warnings%3A+the+precautionary+principle+1896-2000. Accessed 13 Febr 2022

Schaming D, Remita H (2015) Nanotechnology: from the ancient time to nowadays. Found Chem 17(3):187–205

Simpson SL, Yawen L, Spadaro DA, Kookana RS, Batley GE (2021) Chronic effects and thresholds for estuarine and marine benthic organism exposure to perfluorooctane sulfonic acid (PFOS)-contaminated sediments: influence of organic carbon and exposure routes. Sci Total Environ 776:146008

Stefaniuk M, Oleszczuk P, Ok Y (2016) Review on nano zerovalent iron (nZVI): from synthesis to environmental applications. Chem Eng J 287:618–632

Tijani J, Fatoba O, Babajide O, Petrik L (2016) Pharmaceuticals, endocrine disruptors, personal care products, nanomaterials and perfluorinated pollutants: a review. Environ Chem Lett 14(1):27–49

Tornero V, Hanke G (2017) Potential chemical contaminants in the marine environment: an overview of main contaminant lists. Office of the European Union Publications, Luxembourg. Available at: ► http://publications.jrc.ec.europa.eu/repository/bitstream/JRC108964/potential_chemical_contaminants_in_the_marine.pdf. Accessed 19 Aug 2019

UNESCO (United Nations Educational, Scientific and Cultural Organization) (2019) Emerging pollutants in water and wastewater. Available at: ► https://en.unesco.org/emergingpollutants. Accessed 12 Mar 2019

13

United Nations (1998) Kyoto protocol to the United Nations framework convention on climate change. Available at: ▶ https://unfccc.int/resource/docs/convkp/kpeng.pdf. Accessed 29 Mar 2019

Van Boeckel T, Brower C, Gilbert M, Grenfell B, Levin S, Robinson T, Teillant A, Laxminarayan R (2015) Global trends in antimicrobial use in food animals. Proc Natl Acad Sci 112(18):5649–5654

Vance M, Kuiken T, Vejerano E, McGinnis S, Hochella M, Rejeski D, Hull M (2015) Nanotechnology in the real world: redeveloping the nanomaterial consumer products inventory. Beilstein J Nanotechnol 6:1769–1780

Walter P, Welcomme E, Hallégot P, Zaluzec N, Deeb C, Castaing J, Veyssière P, Bréniaux R, Lévêque J, Tsoucaris G (2006) Early use of PbS nanotechnology for an ancient hair dyeing formula. Nano Lett 6(10):2215–2219

Wang Z, DeWitt J, Higgins C, Cousins I (2017) A never-ending story of per- and polyfluoroalkyl substances (PFASs)? Environ Sci Technol 51(5):2508–2518

Wei S, Chen L, Taniyasu S, So M, Murphy M, Yamashita N, Yeung L, Lam P (2007) Distribution of perfluorinated compounds in surface seawaters between Asia and Antarctica. Mar Pollut Bull 54(11):1813–1818

Windsor F, Ormerod S, Tyler C (2017) Endocrine disruption in aquatic systems: up-scaling research to address ecological consequences. Biol Rev 93(1):626–641

Multiple Stressors

Allyson L. O'Brien, Katherine Dafforn, Anthony Chariton, Laura Airoldi, Ralf B. Schäfer and Mariana Mayer-Pinto

Contents

14.1   Introduction – 306

14.2   The Study of Multiple Stressors – 306
14.2.1   Definitions – 307

14.3   Stressor Interactions in the Marine Environment – 308
14.3.1   Nutrients and Trace Metals – 308
14.3.2   Trace Metals and Pesticides – 309
14.3.3   Contamination and Climate Change – 309
14.3.4   Three or More Stressor Interactions – 310

14.4   Management of Multiple Stressors – 312

14.5   Summary – 312

14.6   Study Questions and Activites – 313

References – 313

© The Author(s) 2023
A. Reichelt-Brushett (ed.), *Marine Pollution—Monitoring, Management and Mitigation*,
Springer Textbooks in Earth Sciences, Geography and Environment,
https://doi.org/10.1007/978-3-031-10127-4_14

14.1 Introduction

This book has mostly considered marine contamination and the biological effects of contaminants acting as single stressors. However, marine environments are rarely exposed to a single stressor, but rather experience a complex mix of many stressors. These stressors may be contaminants, such as the ones discussed in previous chapters (nutrients, chemicals, plastics as well as carbon dioxide), or they may be other stressors, such as invasive species, built infrastructure, aquaculture or fisheries, or climatic changes which themselves can contribute to contaminant stress, for example, nutrient loading is a well-known impact of aquaculture activities. All these stressors are ubiquitous in marine environments worldwide and have the potential to interact and have very different impacts compared to if they occurred singularly.

Wastewater treatment plants and stormwater drains that discharge into coastal marine waters create **multiple stressor** conditions since they are sources of both nutrients and trace metals and metalloids. These contaminants have different modes of action, often leading to different types of ecological impacts (▶ Chapter 3). When acting separately, moderate levels of nutrients may cause an increase in a particular biological or ecological response (e.g. population growth or primary productivity) compared to control areas (e.g. Svensson et al. 2007), while trace metals may cause a relative decrease in these responses (Johnston and Roberts 2009; Mayer-Pinto et al. 2010). However, when these stressors occur simultaneously, they are likely to interact and can potentially result in no overall net effect on biological or ecological responses in the receiving environment (O'Brien et al. 2019).

Another multiple stressor situation occurs when built infrastructure (e.g. marinas, groynes or piers) is in proximity to toxic chemicals in the water column. Built infrastructure on its own can cause a significant negative effect on the diversity (i.e. number of species and abundance) of marine organisms (e.g. microbes, plankton, epibiota, and infauna) via loss or fragmentation of habitats (Bulleri and Chapman 2010; Dafforn et al. 2015; Bishop et al. 2017). This may be further exacerbated by toxic chemicals (e.g. metals; industrial wastes and agricultural runoff; McGee et al. 1995), with the overall effect of these stressors being potentially greater than if the two stressors occurred separately (a **synergistic** effect).

Understanding how and when stressors interact is critical for predicting their effects and therefore establishing **safe** and realistic guidelines to protect the environment. The consequences of **stressor interactions** are complex and pose great challenges for researchers, practitioners and decision-makers. This chapter highlights the complexity by providing some **real-world examples** and argues the need to consider the potential **interactive effects** of multiple stressors in conservation and management policies rather than focusing solely on single-stressor effects.

The first section of the chapter provides an overview of multiple stressors research, including definitions of types of multiple stressor interactions. The second section provides examples of the common types of multiple stressor interactions in the marine environment. In the final section, we briefly discuss current approaches and future directions of research on multiple stressors in the context of marine management, conservation and habitat restoration.

14.2 The Study of Multiple Stressors

Humans have exploited marine resources since at least the Palaeolithic with records of habitat modification occurring in Europe from the ninth and tenth centuries A.D (Knottnerus 2005). We know that the drivers of change are a complex synergy of anthropogenic and natural stressors, including pollution, land reclamation, coastal development, overfishing, nutrient and sediment enrichment, and that the inherent natural variability of marine ecosystems is driven by ecological processes (Airoldi and Beck 2007; Claudet and Fraschetti 2010). In addition, there is evidence that future climate scenarios will impose further pressures on the persistence and stability of these habitats (Hawkins et al. 2008; Philippart et al. 2011).

The number of stressors impacting the world's ecosystems is unprecedented. However, it is not simply their number that is of concern, but their **historical accumulation** that is driving change (Jackson et al. 2001). The **cumulative impacts** of multiple stressors can exacerbate nonlinear responses of marine and coastal systems and limit their capacity to recover (e.g. Airoldi et al. 2015). The pervasiveness of multiple stressor impacts worldwide has led to the emergence of multiple stressor research as an independent field of study (Baird et al. 2016; Van Den Brink et al. 2019; Orr et al. 2020). This research encompasses general theory and management frameworks applied across different ecosystems, including marine ecosystems (Crain et al. 2008), freshwater rivers and streams (Hale et al. 2017; Sievers et al. 2018), floodplains (Monk et al. 2019), and agricultural ditches (Bracewell et al. 2019).

The marine environment is continually exposed to multiple stressors, yet most of the research and current literature still focuses on understanding the effects of individual stressors (O'Brien et al. 2019). A global review of literature based on urban marine and estuarine environments found 93% out of the total 579 studies considered stressors, such as nutrients, chemical contaminants, non-indigenous species and built infrastruc-

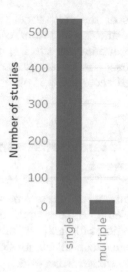

■ **Figure 14.1** The number of studies between 1990 and 2017 that assessed the effects of nutrients, chemical contaminants, non-indigenous species and/or built infrastructure in urban estuarine environments in isolation (single) or combined (multiple). Adapted from O'Brien et al. (2019) by A. O'Brien

ture, in isolation (■ Figure 14.1). Only 38 studies identified by the literature review investigated the effects of these stressors in combination, highlighting the relative gap in our understanding between multiple compared to single stressors.

14.2.1 Definitions

A useful starting point to understanding the concept of multiple stressors is to describe and define the potential interactions between two or more stressors. When more

than one stressor occurs in an environment, they may or may not interact. If the effects of a certain stressor occur **independently** of any other stressor in the environment, then we consider that the stressors are **not interacting** (no interaction; ■ Figure 14.2). This is defined as an additive (or cumulative) multiple stressor model where the overall total effects on a response variable are the sum of the single-stressor effects.

Conversely, the stressors may affect the biological system(s) in ways that are **dependent** on each other or are **interacting**. Therefore, the overall or combined effects of two or more stressors are different from the sum of the single-stressor effects. Interacting stressors are considered as either synergistic or antagonistic, which are defined as (■ Figure 14.2):

- **Synergistic Interactions**—the combined effects are greater than expected based on individual effects.
- **Antagonistic Interactions**—the combined effects are less than expected based on individual effects.

Null models can be used in multiple stressor investigations to predict the combined effect of two stressors. Two stressors are initially assumed to have an additive effect, where the combined effects of the stressors are simply summed (■ Figure 14.2). If there is deviation from the additive null model, for example, through synergistic or antagonistic effects, then more complex models can be used to predict the combined effects (Schafer and Piggott 2018).

There are issues that need to be considered with this sort of statistical approach. Deviations from additivity are mainly determined based on statistical significance, which is known to be sensitive to sample size and transformations. An increase in sample size often leads to

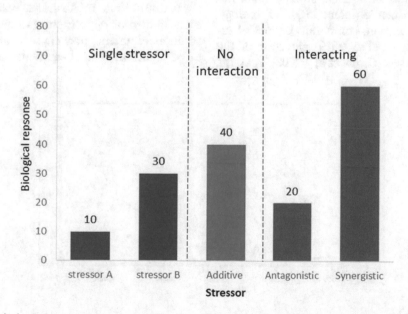

■ **Figure 14.2** Hypothetical responses to two stressors, when they occur separately (stressor A, stressor B), when they occur together with no interaction (additive) or when they occur together and interact (antagonistic or synergistic). Adapted from Côté et al. (2016) by A. O'Brien

a decrease in the p-value for the same effect size and will eventually lead to a drop below the typically chosen significance threshold of 0.05. Thus, with a higher number of samples, the same effect size can become significant and bias comparisons between studies.

Data transformations can also affect the selected null model. Griffen et al. (2016) found that 32% of 143 marine multiple stressor studies unknowingly employed a multiplicative null model because of data transformations. The selection of an additive or multiplicative null model can be influenced by data transformations and will therefore bias the assessment of the prevalence of synergism and antagonism.

Several recent studies have argued for more mechanistically informed null models for multiple stressor research (Griffen et al. 2016; Kroeker et al. 2017; De Laender 2018; Schafer and Piggott 2018). Schäfer and Piggott (2018) provide an overview of null models and guidance on their selection. They introduce two additional null models that have largely been ignored in ecological multiple stressor research and suggest the following categories that should guide the null model selection: stressor mode of action, correlation of sensitivities of organisms to the stressors, effect type (e.g. mortality, growth), effect size of individual stressors and the shape of the stressor–effect relationship.

The number of times multiple stressor interactions are mentioned in ecological and environmental science literature is increasing (◘ Figure 14.3; and see Côté et al. (2016)). In marine systems, both synergistic and antagonistic interactions between two stressors are common and have thought to have occurred more frequently than additive effects (Crain et al. 2008). However, we now know these interaction types are complicated with the addition of a third stressor (Johnson et al. 2018), and between different levels of biological organisation (population, community), types of response variables (O'Brien et al. 2019), direction of the single-stressor responses (Crain et al., 2008), and the

initial conditions or the null model (Schafer and Piggott, 2018). Nevertheless, attempts to define the different types of interactions are still important, so the severity of the impact can be assessed, and management actions prioritised accordingly (Folt et al. 1999; Cabral et al. 2019).

14.3   Stressor Interactions in the Marine Environment

The different sources of contamination discussed in this book—nutrients, trace metal, metalloids, pesticides, POPS, plastics, radioactivity, oils, CO_2, temperature and noise, etc.—can all be considered in the context of a **multiple stressor framework**.

14.3.1   Nutrients and Trace Metals

Eutrophication of marine environments is a pressing global problem and is caused by the delivery and accumulation of excess **nutrients** and/or organic matter, typically to coastal marine and estuarine habitats (▸ Chapter 4). These habitats often have high concentrations of **metals** in sediments that form the seabed. The metals can come from historical sources, such as old industrial sites, or contemporary sources delivered through stormwater drains that carry runoff from nearby urban environments into the sea. The biological effects of nutrients plus metals depend on the environmental condition of the receiving waters. Both nutrients and some metals are essential at small concentrations, but at higher concentrations are likely to exceed a threshold, causing a toxic effect. In Australia, many estuaries are nutrient limited or oligotrophic, so the threshold at which additional nutrients may cause a toxic effect is expected to be higher than when nutrients are added to already

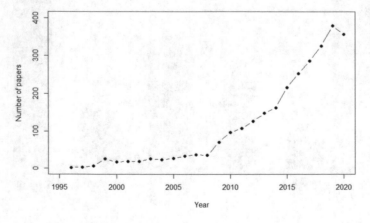

◘ **Figure 14.3**   The number of articles in the ecological and environmental sciences literature that mention 'multiple stressors' is increasing. *Image* A. O'Brien. Prior to 1996 there were not records. *Data Source*: Clarivate Analytics Web of Science Core collection database

eutrophic systems (e.g. Andersen et al. 2015). In contrast, the toxic effects of metals are less dependent on background environmental conditions and, therefore, these contaminants can reach thresholds at lower concentrations than nutrients (e.g. Samhouri et al. 2010).

Given the differing modes of action of nutrients and metals, we would predict antagonistic effects between these two stressors, with the enriching action of nutrients mitigating the toxic effect of metals up to a threshold when both become toxic. This effect has been found at different levels of biological organisation. For example, Lawes et al. (2017) found that microbial community evenness (a measure of diversity) and macrofaunal abundances decreased when exposed to metal contaminated sediments, but increased when exposed to both experimental nutrient treatments (low and high enrichment levels). Potential mechanisms underlying these measured responses include higher biological metabolic rates when exposed to nutrients that sustain detoxification processes and counteract toxic responses to metals (Sokolova and Lannig 2008). However, exceptions to this may include metal-tolerant species that can persist in highly contaminated sites regardless of nutrient availability (Mayer-Pinto et al. 2010).

14.3.2 Trace Metals and Pesticides

The combination of **metals** and **pesticides** is a common mixture of contaminants in the marine environment, especially in habitats close to catchments used for agriculture. Metal pollution can originate from point sources (e.g. stormwater drains or industrial discharges, as described above) or they may originate from diffuse sources that are comparatively difficult to identity (e.g. urban and agricultural runoff, boat harbours or historical activities) (▶ Chapter 5). Pesticides typically occur through diffuse sources, originating from agricultural activities on land before entering freshwater streams and rivers that eventually flow into estuaries and marine waters (see ▶ Chapter 7). Modern day pesticides only persist in the environment for a few days or weeks. Although they degrade quickly, there are often high concentrations of these pesticides in the environment because they are applied in large quantities over broad areas and often repeated times in a single season. The application process is ineffective, and it has been suggested that only 1% of pesticide sprayed reaches target organisms while the remaining 99% enters the surrounding soil, sediment and waterways (Stauber et al. 2016). This may partly explain the high variability in the distribution and abundance of pesticides in receiving environments, which can be characteristically high in coastal bays but only at certain times of the year or in particular **hot spots** (O'Brien et al. 2016).

The ephemeral nature of pesticides in the marine environment coupled with metals from point and diffuse sources poses challenging questions in the context of multiple stressors. For example, what is the effect of high concentrations of metals in the seabed from past industrial activities combined with the fact that seabed is now exposed to infrequent, high concentrations of pesticides from nearby river, and when does it occur? How do we know what concentrations are going to cause a biological effect?

Organisms in the seabed may be tolerant to metal exposure until a **pulse** of pesticides is delivered into the system. However, the observed biological effect may depend on the type of pesticide, and which organism or groups of organisms are likely to be affected. For example, herbicides are expected to have different biological effects on photosynthetic groups (e.g. phytoplankton, diatoms and macroalgae) compared to effects from insecticides. The timing of the contamination event, the mode of delivery into the system, the local environmental conditions, chemical characteristics and concentrations as well as potential interactions with other pre-existing contaminants are all factors that need to be considered when predicting multiple stressor effects.

14.3.3 Contamination and Climate Change

Climate change is affecting salinity regimes in marine and estuarine environments through increases in the duration, frequency and strength of rainfall and storm events (Stauber et al. 2016). Changes in global atmospheric carbon dioxide are linked to increasing ocean temperatures and decreasing pH which is termed ocean acidification (▶ Chapter 11). Marine contaminants interact with **climate-related variables**—salinity, temperature and pH, and together they can have very different effects on bioavailability, bioaccumulation and toxicity of the contaminant than if the stressors occurred separately (Cabral et al. 2019). As discussed by Alava et al. (2017) and Cabral et al. (2019), the interactive effects may be driven by changes in climate variables that cause an increased risk of exposure or susceptibility to toxic effects (climate-dominated effects; ◻ Figure 14.4) or they may be driven by contamination, where prior exposure causes increased susceptibility to climate variables (contaminant-dominated effects; ◻ Figure 14.4). Identifying dominance patterns as well as the type of interactive effects (additive, synergistic or antagonistic) is crucial in resolving such complex interactions (Cabral et al. 2019).

Nutrients and chemical toxicants are among the most studied stressors in marine environments, but there is still limited testing of these in combination with temperature and pH (Schiedek et al. 2007; Crain

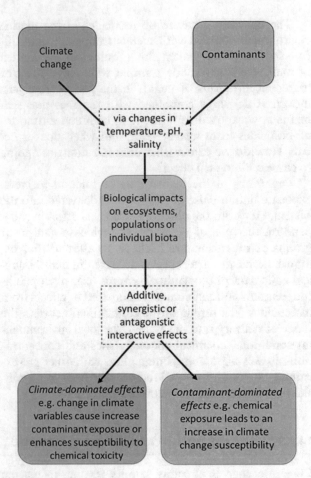

Figure 14.4 Potential interactive effects of contaminants and climate change. Adapted from Alava et al. (2017) and Cabral et al. (2019) by A. O'Brien

et al. 2008). Climate change alters the chemistry of the ocean, which can increase the bioaccumulation of chemical toxicants making organisms more susceptible to exposure (Alava et al. 2017). Temperature increases metabolic rates, food consumption increases and the risk of exposure to chemicals associated with food sources also increases (Cabral et al. 2019). In a global review of literature based on urban marine and estuarine environments, we found only 37 of a total 579 studies considered the combined effects of a climate-related variables (salinity, temperature, pH) and any other anthropogenic stressor (nutrients, chemical contaminants, non-indigenous species or built infrastructure; unpublished data from O'Brien et al. 2019). This knowledge gap is significant and needs to be addressed urgently as our climate is changing rapidly.

The few studies that have tested interactive effects of contamination and climate change highlight the need to understand these interactions at multiple biological levels (Cabral et al. 2019). For example, at the individual level, the survival of the mysid crustacean

Praunus flexuosus is reduced by nickel, chromium and zinc, but survival is further reduced at increased temperatures and salinities (McLusky and Hagerman 1987). The combined effects of heavy metals and temperature or salinity is thought to interfere with the organism's ability to osmoregulate thereby making them more susceptible to the toxic effects of the contaminants (McLusky and Hagerman 1987). Similarly, the effects of copper on brittle star (*Amphipholis squamata*) behaviour are dependent on temperature (Black et al. 2015). Interestingly, a decrease in the toxicity of copper was detected with increasing temperature from 15 °C to 25 °C (Black et al. 2015), suggesting predicted increases in sea temperatures could mitigate toxic effects of copper for these organisms. Increasing nutrients in combination with decreasing pH have resulted in antagonistic effects on corals with slower growth rates when stressors act independently compared to when they are combined (Langdon and Atkinson 2005; Holcomb et al. 2010; Chauvin et al. 2011).

At the community level, a study in Northern Ireland investigated interactive effects between nutrients, temperature (as a measure of ocean warming) and the presence of a non-indigenous seaweed (*Sargassum muticum*) (Vye et al. 2015). They found a strong antagonistic interaction between nutrient enrichment and the invasive species, with observed decreases in algal biomass when enriched with nutrients but only in the absence of *S. muticum*. However, this antagonistic interaction was no longer evident when combined with increased temperatures (Vye et al. 2015). If antagonistic interactions no longer exist under future climate scenarios it not only makes the effects difficult to predict, but could also expose these habitats to risk of algal blooms facilitated by high nutrient levels or further spread of non-indigenous species.

14.3.4 Three or More Stressor Interactions

To date, multiple stressor models have only focused on interactions between two stressors (Schafer and Piggott 2018). However, in marine environments, and particularly in urban and agricultural coastal marine habitats, **many stressors are acting simultaneously** (▶ Box 14.1). Stressors in the marine environment have been accumulating worldwide for decades and the first anthropogenic stressors were probably overfishing and untreated sewage, followed by pollution associated with industrialisation in the nineteenth Century (e.g. Jackson et al. 2001). These historical stressors coupled with modern stressors such as climate change and non-indigenous species make it difficult to disentangle the impacts of individual stressors or specific combinations of stressors.

Box 14.1: Contaminants, Boat Harbours and Non-indigenous Species

Contaminants in sheltered systems such as marinas and boat harbours are usually found at higher concentrations than in estuaries or open waters (e.g. Matthiessen et al. 1999). This is partly attributed to the construction of breakwaters and marinas as boat havens that reduce water-flow, **trapping** the contaminants that leach from antifouling paints on ship-hulls (Rivero et al. 2013; Schiff et al. 2007). In addition, boat traffic inside sheltered systems can cause the re-suspension of contaminants that have been entrapped in the sediment in these areas (e.g. Knott et al. 2009). Furthermore, boat harbours often have reduced levels of light in the water-column due to increased turbidity and artificial structures such as pier-pilings and pontoons that cause shade (e.g. Glasby 1999). Therefore, boat harbours, by reducing water circulation and light and increasing numbers of vessels, can interact synergistically with contamination, affecting local marine communities.

Boat harbours can also contribute to the establishment and spread of non-indigenous species (see work from Piola and Johnston 2008; Johnston et al. 2017). In many countries, the majority of non-indigenous marine species are associated with vessel hull fouling as a vector of introduction, and hotspots of invasion have therefore been linked to the high level of vessel activity in boat harbours and marinas (Ware et al. 2014). These systems provide significant resources for invading species in the form of artificial substrate for recruitment (Dafforn et al. 2009, 2012) and also limit the survival of less tolerant native species because of increased contaminant exposure (Rivero et al. 2013). As a result, sheltered boat harbours present a complex multi-stressor environment where contaminants and infrastructure interact to increase opportunities for non-indigenous species and negatively affect native species. The interactive effects have yet to be assessed against stressor models, however we might expect contaminants to interact additively with infrastructure to increase the prevalence of non-indigenous species up to a threshold of toxicity or resource limitation.

A common approach to studying interactions between three or more stressors is **surveys or correlation studies** that relate biological change between impact and reference sites or along a gradient of impact. This approach identifies patterns, but not specific cause-effect relationships between stressor combinations and biological responses. For example, Stuart-Smith et al. (2015) studied **community structure** of fishes and mobile invertebrates on shallow reefs that were affected by metal contamination, surrounding human population density, proximity to sewage outfalls, the city port and the distribution of invasive species. They found reefs that were the most affected by these stressors had reduced mobile invertebrate abundances and reduced fish biomass. This effect was most prominent on reefs that were invaded by **non-indigenous species** (Stuart-Smith et al. 2015). This study provides a picture of the overall impact but does not identify the specific **stressor combinations** that were causing the effect. It is suggested that community-level field experiments, which manipulate different combinations of stressors, are required to understand these underlying mechanisms (e.g. Mayer-Pinto et al. 2010; Chariton et al. 2011; O'Brien and Keough 2013; Birrer et al. 2018; Johnson et al. 2018) and inform the fundamental interactions between multiple stressors (Stuart-Smith et al. 2015).

The multiple stressor framework includes chemical and physical stressors, which have been the focus of this chapter so far, but also biological stressors, such as non-indigenous species, and built infrastructure, such as aquaculture, boat harbours, marinas, piers, jetties and groynes (see definitions in O'Brien et al. 2019). Many of these stressors are likely to act at **different temporal and spatial scales**. As a hypothetical example, a contaminant from a stormwater outfall that affects the marine environment in the immediate surrounding area may interact with a non-indigenous seaweed species that occurs across the entire coastline. The seaweed species is prolific in cooler months but dies-off in summer months when it becomes more susceptible to rising global ocean temperatures. The stressors in this example (contamination from stormwater outfall, non-indigenous species and climate change) are occurring at multiple spatial (local, regional and global) and temporal (season) scales. These issues of scale and stressor type (chemical, physical, biological or infrastructure) become more pertinent as the number of stressors increases and needs to be specifically considered in any study or monitoring program (Downes et al. 2002; Mayer-Pinto et al. 2010).

14.4 Management of Multiple Stressors

Coastal systems worldwide are threatened by multiple anthropogenic activities, including urban development, organic and inorganic pollution, over-exploitation of resources, dredging and dumping, and invasive species (Lotze et al. 2006; Airoldi and Beck 2007). When coupled with climatic instabilities, localised cumulative human perturbations create new regimes of disturbances that greatly affect the stability, resilience and productivity of ecosystems (Cimon and Cusson 2018; Piola and Johnston 2008; Sauve et al. 2016). Nonetheless, marine management has often focused on one impact at a time (Beaumont et al. 2007) instead of taking a more **holistic approach**. The shortfalls of this approach are becoming evident, with only about 7% of the world's oceans designated as protected (UNEP-WCMC and IUCN 2020). With current management processes so strongly focused on working in an impact-by-impact framework, there are entrenched scientific, cultural and institutional challenges to shifting those processes toward ecosystem-based management and marine spatial planning, which address multiple human uses of the ocean, their cumulative impacts and interactive effects (Halpin et al. 2006).

To attain sustainability, it is necessary to understand how natural systems are affected by multiple stressors and can respond to management interventions that aim to achieve multiple goals (Dafforn et al. 2015). These concepts are especially relevant when managing **ecosystem resilience**. If the ability of systems to withstand (i.e. resistance) and/or recover (i.e. resilience) from disturbances is progressively **eroded** by cumulative impacts, the system becomes vulnerable to regime shifts. These shifts are critical transitions that are characterised by different sets of structures, processes and values (Scheffer et al. 2009), which can lead to ecosystem collapse. Indeed, the European Union Marine Strategy Framework Directive (2008/56/EC) calls for the urgent establishment of coherent and coordinated programmes of measures to contain the collective pressure of human activities within sustainable levels. A philosophy of **regenerative intervention** (▶ Chapter 15) is required before sustainability can be realised.

The first step towards the management of multiple stressors is to identify **thresholds** and **trade-offs**. A threshold can be set as the level of human-induced pressure (e.g. pollution) at which small changes produce substantial improvements in protecting an ecosystem's structural (e.g. diversity) and functional (e.g. resilience) attributes (Samhouri et al. 2010). This approach is based on the detection of nonlinearities in relationships between ecosystem attributes and pressures. These relationships, however, are known in only a few cases, and they often focus on one direction only (increasing human pressures), without exploring the reverse pathways following management actions geared towards recovery.

The interactions between multiple stressors can exacerbate nonlinear responses of ecosystems to human impacts. This will limit their **recovery capacity** and reduce the likelihood that a system can retrace the same trajectory during restoration as during degradation. For example, fisheries exploitation and increased nutrient loadings jointly affect food webs and production in estuaries via reductions in fish and shellfish biomass, increased algae production and habitat degradation (Breitburg et al. 2009). As a result, there could be specific levels of fish caught per unit effort and nutrient loadings that lead to threshold responses making them resistant to restoration through fisheries and nutrient management (Breitburg et al. 2009; Scheffer et al. 2001). Maximising management outcomes relies on getting a relevant mechanistic understanding of the effects of multiple stressors at scales ranging from individuals to populations and whole ecosystems, indicators of changes, tools, and models that can be used for early identification of thresholds.

Brown et al. (2013) have examined the effectiveness of management when faced with different types of interactions between local and global (climatic) stressors in seagrass and fish communities. They showed that for additive effects, reducing the magnitude of local stressors should lead to a corresponding increase in the response of interest allowing for straightforward expectations of the response to management and conservation actions. In contrast, mitigation of stressors involved in synergistic or antagonistic interactions with global stressors will lead to greater than or less than (respectively) predicted results based on additive models. Antagonistic stressors create management challenges, as all or most stressors would need to be eliminated to see substantial recovery, except in cases where the antagonism is driven by a dominant stressor, such that mitigation of that stressor alone would substantially improve the state of species or communities. In contrast, synergisms may respond quite favourably to removal of a single stressor as long as the system has not passed a threshold into an alternative state (Hobbs et al. 2006).

14.5 Summary

There are very few ecosystems across the world that can be considered impacted by a single stressor (Van Den Brink et al. 2019). In the marine environment, the occurrence of multiple anthropogenic stressors is the new normal (Halpern et al. 2007). Pollution is one of the most important stressors affecting the marine environment, but it should no longer be considered in isolation as a single stressor (Cabral et al. 2019). The impacts of pollution and how species respond to individual pollutants and/or mixtures of pollutants depends on over-exploitation of fisheries, built infrastructure, non-indigenous species and climate change as well as the accumulation of these stressors over time (Jackson et al. 2001). Multiple stressor models of additive, synergistic and antagonistic interactions provide a good starting point for predicting impacts, but in many cases, they are not going to be able to predict the dynamic nature of interactions in the marine environment (Côté et al. 2016).

As the global environment changes, we are likely to see novel combinations of species that will change ecological processes and the way habitats function (Hobbs et al. 2006). Multiple stressors have the potential to influence species survival, community abundances and competition between species that will have implications for population growth and ecosystem functioning (Cadotte and Tucker 2017). To some extent, marine management will need to accept the impacted nature of the environment as status quo and provide solutions that focus on managing current condition, which support existing ecosystem services, rather than attempting restoration to pre-stressed conditions (e.g. Bishop et al. 2017).

Clearly, greater effort is now required to understand how multiple stressors are affecting the marine environment (O'Brien et al. 2019). This needs to be based on scientifically appropriate monitoring and study designs with replication, site selection and clear hypotheses, otherwise this knowledge gap will prevail (Mayer-Pinto et al. 2010). Terminology and concepts underlying the theory of multiple stressors, including the use of null models based on mechanistic assumptions (e.g. Schafer and Piggott 2018), are established and have been documented in the scientific literature for several decades (e.g. Folt et al. 1999; Crain et al. 2008; Van Den Brink et al. 2019). Progress is now required in understanding uncertainty and the complexity of these multiply stressed environments. How do we need to adapt our management strategies and what do policy makers need to do to ensure our marine environments remain resilient and functionally adaptable (Verges et al. 2019).

14.6 Study Questions and Activites

1. Describe an additive multiple stressor model and provide an example. This may be a hypothetical example or relate to any of the sources of contamination you have read about in this book (e.g. nutrients, trace metal, metalloids, pesticides, plastics, oils or noise).

2. Describe a synergistic interaction and an antagonistic interaction in your own words. Explain what type of interaction would be of more concern.
3. Draw a graph showing possible interactions between nutrients and trace metals. Show a response variable on the y-axis and stressors on the x-axis.
4. Explain why understanding the interaction between climate change and contamination is challenging. Include an example to illustrate your answer.

References

Airoldi L, Beck MW (2007) Loss, status and trends for coastal marine habitats of Europe. In: Gibson R, Atkinson R, Gordon J (eds) Oceanography and marine biology, vol 45. CRC, Boca Raton, p 560

Airoldi L, Turon X, Perkol-Finkel S, Rius M (2015) Corridors for aliens but not for natives: effects of marine urban sprawl at a regional scale. Divers Distrib 21:755–768

Alava J, Cheung W, Ross P, Ur S (2017) Climate change–contaminant interactions in marine food webs: toward a conceptual framework. Glob Change Biol 23:3984–4001

Andersen JH, Halpern BS, Korpinen S, Murray C, Reker J (2015) Baltic Sea biodiversity status vs. cumulative human pressures. Estuar Coast Shelf Sci 161:88–92

Baird DJ, Van Den Brink PJ, Chariton AA, Dafforn KA, Johnston EL (2016) New diagnostics for multiply stressed marine and freshwater ecosystems: integrating models, ecoinformatics and big data. Mar Freshw Res 67:391–392

Beaumont NJ, Austen MC, Atkins JP, Burdon D, Degraer S, Dentinho TP, Derous S, Holm P, Horton T, Van Ierland E, Marboe AH, Starkey DJ, Townsend M, Zarzycki T (2007) Identification, definition and quantification of goods and services provided by marine biodiversity: implications for the ecosystem approach. Mar Pollut Bull 54:253–265

Birrer SC, Dafforn KA, Simpson SL, Kelaher BP, Potts J, Scanes P, Johnston EL (2018) Interactive effects of multiple stressors revealed by sequencing total (DNA) and active (RNA) components of experimental sediment microbial communities. Sci Total Environ 637:1383–1394

Bishop MJ, Mayer-Pinto M, Airoldi L, Firth LB, Morris RL, Loke LHL, Hawkins SJ, Naylor LA, Coleman RA, Chee SY, Dafforn KA (2017) Effects of ocean sprawl on ecological connectivity: impacts and solutions. J Exp Mar Biol Ecol 492:7–30

Black JG, Reichelt-Brushett AJ, Clark MW (2015) The effect of copper and temperature on juveniles of the eurybathic brittle star *Amphipholis squamata*—exploring responses related to motility and the water vascular system. Chemosphere 124:32–39

Bracewell S, Verdonschot RCM, Schäfer RB, Bush A, Lapen DR, Van Den Brink PJ (2019) Qualifying the effects of single and multiple stressors on the food web structure of Dutch drainage ditches using a literature review and conceptual models. Sci Total Environ 684:727–740

Breitburg DL, Craig JK, Fulford RS, Rose KA, Boynton WR, Brady DC, Ciotti BJ, Diaz RJ, Friedland KD, Hagy JD, Hart DR, Hines AH, Houde ED, Kolesar SE, Nixon SW, Rice JA, Secor DH, Targett TE (2009) Nutrient enrichment and fisheries exploitation: interactive effects on estuarine living resources and their management. Hydrobiologia 629:31–47

Brown CJ, Saunders MI, Possingham HP, Richardson AJ (2013) Managing for interactions between local and global stressors of ecosystems. PLoS ONE 8(6):e65765

Bulleri F, Chapman MG (2010) The introduction of coastal infrastructure as a driver of change in marine environments. J Appl Ecol 47:26–35

Cabral H, Fonseca V, Sousa T, Costa Leal M (2019) Synergistic effects of climate change and marine pollution: an overlooked interaction in coastal and estuarine areas. Int J Environ Res Public Health 16:2737

Cadotte MW, Tucker CM (2017) Should environmental filtering be abandoned? Trends Ecol Evol 32:429–437

Chariton AA, Maher WA, Roach AC (2011) Recolonisation of translocated metal-contaminated sediments by estuarine macrobenthic assemblages. Ecotoxicology 20:706–718

Chauvin A, Denis V, Cuet P (2011) Is the response of coral calcification to seawater acidification related to nutrient loading? Coral Reefs 30:911–923

Cimon S, Cusson M (2018) Impact of multiple disturbances and stress on the temporal trajectories and resilience of benthic intertidal communities. Ecosphere Ecosphere 9(10):e02467

Claudet J, Fraschetti S (2010) Human-driven impacts on marine habitats: a regional meta-analysis in the Mediterranean Sea. Biol Cons 143:2195–2206

Côté IM, Darling ES, Brown CJ (2016) Interactions among ecosystem stressors and their importance in conservation. Proc Roy Soc B Biol Sci 283

Crain CM, Kroeker K, Halpern BS (2008) Interactive and cumulative effects of multiple human stressors in marine systems. Ecol Lett 11:1304–1315

Dafforn KA, Johnston EL, Glasby TM (2009) Shallow moving structures promote marine invader dominance. Biofouling 25:277–287

Dafforn KA, Glasby TM, Johnston EL (2012) Comparing the invasibility of experimental "Reefs" with field observations of natural Reefs and artificial structures. PLoS ONE 7(5):e38124

Dafforn KA, Glasby TM, Airoldi L, Rivero NK, Mayer-Pinto M, Johnston EL (2015) Marine urbanization: an ecological framework for designing multifunctional artificial structures. Front Ecol Environ 13:82–90

De Laender F (2018) Community- and ecosystem- level effects of multiple environmental change drivers: beyond null model testing. Glob Change Biol 24:5021–5030

Downes BJ, Barmuta LA, Fairweather PG, Faith DP, Keough MJ, Lake PS, Mapstone BD, Quinn GP (2002) Monitoring ecological impacts: concepts and practice in flowing waters. Cambridge University Press, Cambridge, p 434

Folt CL, Chen CY, Moore MV, Burnaford J (1999) Synergism and antagonism among multiple stressors. Limnol Oceanogr 44:864–877

Glasby TM (1999) Effects of shading on subtidal epibiotic assemblages. J Exp Mar Biol Ecol 234:275–290

Griffen BD, Belgrad BA, Cannizzo ZJ, Knotts ER, Hancock ER (2016) Rethinking our approach to multiple stressor studies in marine environments. Mar Ecol Prog Ser 543:273–281

Hale R, Piggott JJ, Swearer SE (2017) Describing and understanding behavioral responses to multiple stressors and multiple stimuli. Ecol Evol 7:38–47

Halpern BS, Selkoe KA, Micheli F, Kappel CV (2007) Evaluating and ranking the vulnerability of global marine ecosystems to anthropogenic threats. Conserv Biol 21:1301–1315

Halpin PN, Read AJ, Best BD, Hyrenbach KD, Fujioka E, Coyne MS, Crowder LB, Freeman SA, Spoerri C (2006) OBIS-SEAMAP: developing a biogeographic research data commons for the ecological studies of marine mammals, seabirds, and sea turtles. Mar Ecol Prog Ser 316:239–246

Hawkins S, Moore P, Burrows M, Poloczanska E, Mieszkowska N, Herbert R, Jenkins Sr, Thompson R, Genner M, Southward A (2008) Complex interactions in a rapidly changing world: responses of rocky shore communities to recent climate change. Clim Res 37:123–133

Hobbs RJ, Arico S, Aronson J, Baron JS, Bridgewater P, Cramer VA, Epstein PR, Ewel JJ, Klink CA, Lugo AE, Norton D, Ojima D, Richardson DM, Sanderson EW, Valladares F, Vila M, Zamora R, Zobel M (2006) Novel ecosystems: theoretical and management aspects of the new ecological world order. Glob Ecol Biogeogr 15:1–7

Holcomb M, Mccorkle DC, Cohen AL (2010) Long-term effects of nutrient and CO_2 enrichment on the temperate coral *Astrangia poculata* (Ellis and Solander, 1786). J Exp Mar Biol Ecol 386:27–33

Jackson JBC, Kirby MX, Berger WH, Bjorndal KA, Botsford LW, Bourque BJ, Bradbury RH, Cooke R, Erlandson J, Estes JA, Hughes TP, Kidwell S, Lange CB, Lenihan HS, Pandolfi JM, Peterson CH, Steneck RS, Tegner MJ, Warner RR (2001) Historical overfishing and the recent collapse of coastal ecosystems. Science 293:629–638

Johnson GC, Pezner AK, Sura SA, Fong P (2018) Nutrients and herbivory, but not sediments, have opposite and independent effects on the tropical macroalga, Padina boryana. J Exp Mar Biol Ecol 507:17–22

Johnston EL, Dafforn KA, Clark GF, Rius M, Floerl O (2017) How anthropogenic activities affect the establishment and spread of non-indigenous species post-arrival. In: Hawkins S, Evans A, Dale C, Firth L, Hughes D, Smith I (eds) Oceanography and marine biology: an annual review, vol 55, pp 2–33

Johnston EL, Roberts DA (2009) Contaminants reduce the richness and evenness of marine communities: a review and meta-analysis. Environ Pollut 157:1745–1752

Knott NA, Aulbury JP, Brown TH, Johnston EL (2009) Contemporary ecological threats from historical pollution sources: impacts of large-scale resuspension of contaminated sediments on sessile invertebrate recruitment. J Appl Ecol 46:770–781

Knottnerus OS (2005) History of human settlement, cultural change and interference with the marine environment. Helgol Mar Res 59:2–8

Kroeker KJ, Kordas RL, Harley CDG (2017) Embracing interactions in ocean acidification research: confronting multiple stressor scenarios and context dependence. Biol Lett 13(3):20160802

Langdon C, Atkinson MJ (2005) Effect of elevated pCO(2) on photosynthesis and calcification of corals and interactions with seasonal change in temperature/irradiance and nutrient enrichment. J Geophys Res Oceans 110(9):1–16

Lawes JC, Dafforn KA, Clark GF, Brown MV, Johnston EL (2017) Multiple stressors in sediments impact adjacent hard substrate habitats and across biological domains. Sci Total Environ 592:295–305

Lotze HK, Lenihan HS, Bourque BJ, Bradbury RH, Cooke RG, Kay MC, Kidwell SM, Kirby MX, Peterson CH, Jackson JBC (2006) Depletion, degradation, and recovery potential of estuaries and coastal seas. Science 312:1806–1809

Matthiessen P, Reed J, Johnson M (1999) Sources and potential effects of copper and zinc concentrations in the estuarine waters of Essex and Suffolk, United Kingdom. Mar Pollut Bull 38:908–920

Mayer-Pinto M, Underwood AJ, Tolhurst T, Coleman RA (2010) Effects of metals on aquatic assemblages: what do we really know? J Exp Mar Biol Ecol 391:1–9

McGee BL, Schlekat CE, Boward DM, Wade TL (1995) Sediment contamination and biological effects in a Chesapeake Bay marina. Ecotoxicology 4:39–59

Mclusky DS, Hagerman L (1987) The toxicity of chromium, nickel and zinc: effects of salinity and temperature, and the osmoregulatory consequences in the mysid *Praunus flexuosus*. Aquat Toxicol 10:225–238

Monk WA, Compson ZG, Choung CB, Korbel KL, Rideout NK, Baird DJ (2019) Urbanisation of floodplain ecosystems: weight-of-evidence and network meta-analysis elucidate multiple stressor pathways. Sci Total Environ 684:741–752

14

O'Brien AL, Keough MJ (2013) Detecting benthic community responses to pollution in estuaries: a field mesocosm approach. Environ Pollut 175:45–55

O'Brien D, Lewis S, Davis A, Gallen C, Smith R, Turner R, Warne MStJ, Turner S, Caswell S, Mueller JF, Brodie J (2016) Spatial and temporal variability in pesticide exposure downstream of a heavily irrigated cropping area: application of different monitoring techniques. J Agric Food Chem 64:3975–3989

O'Brien AL, Dafforn KA, Chariton AA, Johnston EL, Mayer-Pinto M (2019) After decades of stressor research in urban estuarine ecosystems the focus is still on single stressors: a systematic literature review and meta-analysis. Sci Total Environ 684:753–764

Orr JA, Vinebrooke RD, Jackson MC, Kroeker KJ, Kordas RL, Mantyka-Pringle C, Van Den Brink PJ, De Laender F, Stoks R, Holmstrup M, Matthaei CD, Monk WA, Penk MR, Leuzinger S, Schafer RB, Piggott JJ (2020) Towards a unified study of multiple stressors: divisions and common goals across research disciplines. Proc Roy Soc B-Biol Sci 287(1926):20200421

Philippart CJM, Anadón R, Danovaro R, Dippner JW, Drinkwater KF, Hawkins SJ, Oguz T, O'sullivan G, Reid PC (2011) Impacts of climate change on European marine ecosystems: observations, expectations and indicators. J Exp Mar Biol Ecol 400:52–69

Piola RF, Johnston EL (2008) Pollution reduces native diversity and increases invader dominance in marine hard-substrate communities. Divers Distrib 14:329–342

Rivero NK, Dafforn KA, Coleman MA, Johnston EL (2013) Environmental and ecological changes associated with a marina. Biofouling 29:803–815

Samhouri JF, Levin PS, Ainsworth CH (2010) Identifying thresholds for ecosystem-based management. PLoS ONE 5(1):e8907

Sauve AMC, Fontaine C, Thebault E (2016) Stability of a diamond-shaped module with multiple interaction types. Thyroid Res 9:27–37

Schafer RB, Piggott JJ (2018) Advancing understanding and prediction in multiple stressor research through a mechanistic basis for null models. Glob Change Biol 24:1817–1826

Scheffer M, Carpenter S, Foley JA, Folke C, Walker B (2001) Catastrophic shifts in ecosystems. Nature 413:591–596

Scheffer M, Bascompte J, Brock WA, Brovkin V, Carpenter SR, Dakos V, Held H, Van Nes EH, Rietkerk M, Sugihara G (2009) Early-warning signals for critical transitions. Nature 461:53–59

Schiedek D, Sundelin B, Readman JW, Macdonald RW (2007) Interactions between climate change and contaminants. Mar Pollut Bull 54:1845–1856

Schiff K, Brown J, Diehl D, Greenstein D (2007) Extent and magnitude of copper contamination in marinas of the San Diego region, California, USA. Mar Pollut Bull 54:322–328

Sievers M, Hale R, Parris KM, Swearer SE (2018) Impacts of human-induced environmental change in wetlands on aquatic animals. Biol Rev 93:529–554

Sokolova IM, Lannig G (2008) Interactive effects of metal pollution and temperature on metabolism in aquatic ectotherms: implications of global climate change. Climate Res 37:181–201

Stauber JL, Chariton A, Apte S (2016) Global change. In: Blasco J, Chapman P, Campana O, Hampel M (eds) Marine ecotoxicology: current knowledge and future issues, pp 273–313

Stuart-Smith RD, Edgar GJ, Stuart-Smith JF, Barrett NS, Fowles AE, Hill NA, Cooper AT, Myers AP, Oh ES, Pocklington JB, Thomson RJ (2015) Loss of native rocky reef biodiversity in Australian metropolitan embayments. Mar Pollut Bull 95:324–332

Svensson CJ, Pavia H, Toth GB (2007) Do plant density, nutrient availability, and herbivore grazing interact to affect phlorotannin plasticity in the brown seaweed *Ascophyllum nodosum*. Mar Biol 151:2177–2181

UNEP-WCMC and IUCN (United Nations Environment Program-World Conservation Monitoring Centre and International Union for Conservation of Nature) (2020). Protected planet. Available at: ▶ https://www.protectedplanet.net/en/news-and-stories/the-lag-effect-in-the-world-database-on-protected-areas. Accessed 19 Feb 2022

Van Den Brink PJ, Bracewell SA, Bush A, Chariton A, Choung CB, Compson ZG, Dafforn KA, Korbel K, Lapen DR, Mayer-Pinto M, Monk WA, O'Brien AL, Rideout NK, Schäfer RB, Sumon KA, Verdonschot RCM, Baird DJ (2019) Towards a general framework for the assessment of interactive effects of multiple stressors on aquatic ecosystems: results from the Making Aquatic Ecosystems Great Again (MAEGA) workshop. Sci Total Environ 684:722–726

Verges A, Mccosker E, Mayer-Pinto M, Coleman MA, Wernberg T, Ainsworth T, Steinberg PD (2019) Tropicalisation of temperate reefs: Implications for ecosystem functions and management actions. Funct Ecol 33:1000–1013

Vye SR, Emmerson MC, Arenas F, Dick JTA, O'connor NE (2015) Stressor intensity determines antagonistic interactions between species invasion and multiple stressor effects on ecosystem functioning. Oikos 124:1005–1012

Ware C, Berge J, Sundet JH, Kirkpatrick JB, Coutts ADM, Jelmert A, Olsen SM, Floerl O, Wisz MS, Alsos IG (2014) Climate change, non-indigenous species and shipping: assessing the risk of species introduction to a high-Arctic archipelago. Divers Distrib 20:10–19

Pollution Mitigation and Ecological Restoration

Amanda Reichelt-Brushett

Contents

15.1   Introduction – 318

15.2   What is Restoration? – 318

15.3   Key Principles of Practices in Ecological Restoration – 319

15.4   Cost and Success of Restoration – 320

15.5   Marine Pollution Mitigation and Reduction – 320
15.5.1 Mitigating Coastal Catchment Discharges – 322

15.6   Marine Habitat Restoration – 325
15.6.1 Oyster Reefs – 325
15.6.2 Coral Reefs – 326
15.6.2 Seagrasses – 329
15.6.3 Mangroves – 329
15.6.4 Saltmarsh – 330
15.6.5 Engineering, Technology and Marine Ecosystem Restoration – 332

15.7   Marine Species as Bioremediators – 332

15.8   Summary – 332

15.9   Study Questions and Activities – 333

        References – 333

© The Author(s) 2023
A. Reichelt-Brushett (ed.), *Marine Pollution—Monitoring, Management and Mitigation*,
Springer Textbooks in Earth Sciences, Geography and Environment,
https://doi.org/10.1007/978-3-031-10127-4_15

Acronyms and Abbreviations

NOAA   National Oceanic and Atmospheric Administration
SER   Society for Ecological Restoration
UNEP   United Nations Environment Program
USA   United States of America
USD   United Stated Dollar

15.1 Introduction

▶ Chapter 1 presented to you the problem of marine pollution and through the book we explored the wide range of polluting substances with many chapters highlighting specific management approaches. ▶ Chapter 1 also highlighted that we are all potentially part of the solution to marine pollution. While pollution prevention must be considered a primary goal, research and practice that focuses on successful habitat improvement is a rapidly expanding area (e.g. Edwards et al. 2013). This chapter provides a general understanding of the **restoration of marine ecosystems** and includes the important role that pollution reduction (or mitigation) plays in order to gain positive outcomes.

Restoration ecology is a relatively new discipline area, particularly for marine ecosystems and has gained increased attention since the 1990s (e.g. Geist and Hawkins 2016; Basconi et al. 2020). The establishment of societies and organisations has helped to develop key principles and standards and ensure scientific rigour. For example, The Society for Ecological Restoration (SER) was established in 1988 to:

》 *"bring together academics, researchers, practitioners, artists, economists, advocates, legislators, regulators, and others who support restoration to define and deliver excellence in the field of ecological restoration"*

(SER 2021) (▶ https://www.ser.org/). It is an international society with branches in many countries and there are other similar societies, networks and organisation established in many countries of the world that help enable local on ground activities (e.g. Australian Coastal Restoration Network: ▶ https://www.acrn.org.au/). Of further interest, in 2021 the United Nations Environment Program (UNEP) launched the United Nations Decade on Ecosystem Restoration 2021–2030, which globally encompasses environmental restoration of all degraded ecosystems including in coastal and marine environments (▶ https://www.decadeonrestoration.org/).

Public and private partnerships and collaborations are important elements in successful restoration programs. The partnerships may be between large organisations such as The Nature Conservancy and the National Oceanic and Atmospheric Administration (NOAA) and help to connect communities, expertise and funding. Grant opportunities also help to grow partnership and develop skills and expertise. There are many existing programs for coastal and marine ecosystem restoration in numerous countries and they most commonly focus on improvements to oyster reefs, clam beds, seagrasses, saltmarshes, mangroves, macroalgae forests and coral reefs (e.g. Bayraktarov et al. 2016; Basconi et al. 2020). Interestingly, there is now even interest in deep-sea ecosystem restoration (e.g. degraded canyons impacted by illegal dumping, litter and waste in the Mediterranean Sea (O'Conner et al. 2020). Importantly evaluating success and ecological outcomes needs to consider the desired goal and monitoring that shows a trajectory to reaching that desired goal. This evaluation helps to refine techniques, understand ecosystem services and economic benefits of the restoration (e.g. Abelson et al. 2015; Adame et al. 2019).

There is much more depth that can be explored in expert texts and a wide range of journal articles that are dedicated to marine habitat restoration. This chapter and the reference list of this chapter provides a helpful start.

15.2 What is Restoration?

There are many words used to describe habitat improvement, **restoration** relates to the active re-creation of favourable conditions and is similar to **rehabilitation** and **remediation** (Geist and Hawkins 2016). Rehabilitation and remediation have been suggested to represent less comprehensive restorations actions but there are many detailed definitions and arguments (Geist and Hawkins 2016). Like Geist and Hawkins (2016), this chapter will use the term **restoration** in a broad sense and readers are invited to explore the semantics on concepts and terminology in the wider literature for themselves.

Importantly when discussing degraded habitats, the stressors that have caused the impacts are not always pollution, they may be related to overfishing, destructive fishing such as dynamite fishing in coral reef areas, or changed physical conditions from coastal development to mention a few. In order for an ecosystem to recover, the stressors need to be alleviated and, in some cases, removing these pressures might be all that is required. These limited measures that support **passive recovery** are sometimes considered separate to **active recovery** (e.g. Elliot et al. 2007). Restoration in general should not be considered a one-off event but as an ongoing process over a time scale of years which is likely to need adaptive management (e.g. Edwards and Gomez 2007).

Sometimes ecosystems are so degraded that actions cannot re-create favourable conditions for restoration. In these instances, investment might be made in creating a **replacement** (or novel) ecosystem which is some form of acceptable new ecosystem that restores some ecological integrity, ecosystem services, amenity and recreational opportunities.

15.3 Key Principles of Practices in Ecological Restoration

To understand the processes required to improve ecosystem health the **specific stressors** acting on the site need to be identified along with the history of the site and the degree of perturbation—these can be the keys to effective decision-making to ensure success of restoration efforts (Laegdsgaard 2006). It is important to have some understanding of how restoration efforts are going to affect the ecosystems in need of improvement (i.e. Will the effort work? How does the ecosystem recover? Is it capable of recovery? How will it function? What is a measure of successful recovery?). Some of these questions can be answered with a thorough understanding of the mechanisms behind recovery and the ecology of the ecosystem. For example, knowledge of successional patterns, plant and animal physiology, environmental conditions for recruitment of keystone species, establishment and growth, diversity, amongst other ecological functions is required. These features should be incorporated into monitoring studies to assess improvement in the ecosystem condition.

As noted earlier, **restoration begins with mitigating the stressors**, after this, the chemical, physical and structural properties (e.g. hydrodynamics) need to be considered. Once these conditions are suitable, biological attributes generally follow. Some natural biological recovery may occur if the restoration site has connectivity with other similar habitats but active restoration is assisted by transplantation of keystone or foundation species.

As the science and practice of marine ecosystem restoration has developed, it has become evident that successful restoration and the ability to measure success requires many factors which are summarised in ◘ Table 15.1.

◘ **Table 15.1** Principles for success ecological restoration (*Detail sourced from*: Gann et al. 2019; Basconi et al. 2020)

Principles of ecological restoration	Detail
Engagement of stakeholders	Restoration is carried out to satisfy not only conservation values but also socioeconomic values, including cultural ones (e.g. of indigenous people).
Draws on many types of knowledge	Bring multidisciplinary scientists, practitioners, local community, indigenous knowledge together for projects inception, implementation and monitoring. Include socioeconomic concepts.
Practice is informed by native reference ecosystems while considering environmental change	Key attributes of a reference ecosystem: •physical condition (suitability and similarity with restoration site) •species composition •community structure (food webs) •ecosystem function (processes) •external exchange (interaction with surrounding environment •absence of stressors or threats
Supports ecosystem recovery processes	Ensure restorative practices enhance the natural recovery process. Pre-planning assessment to reinstate the missing biotic or abiotic elements. Consider climate change implications. Consider ecosystem services.
Assessed against clear goals and objectives using measurable indicators	Each project should define a set of goals that can be measured and used to assess the short-term and long-term success of the project.
Seeks the highest level of recovery attainable	It is important to bear in mind that the desired outcome may take a long time to achieve (e.g. years to decades). Managers should adopt a policy of continuous improvement informed by sound monitoring (e.g. five-star system of ecological recovery wheel described in McDonald et al. 2016).
Gains cumulative value when applied at large scales	Small projects can be beneficial but many ecological processes function at landscape, watershed, and regional scales. Degradation occurring at larger scales can overwhelm smaller restoration efforts. In some cases, investing in gradual improvements at larger scales (e.g. catchment runoff) may achieve greater results than more intense work at smaller scales or over shorter periods of time.
Part of a continuum of restorative activities	Progress evaluation. Formal field experiments can also be incorporated into restoration practice, generating new findings to both inform adaptive management and provide valuable insights for the natural sciences.

15.4 Cost and Success of Restoration

Average reported costs for one hectare of marine coastal habitat restoration were between US$80,000 and US$1,600,000, varying widely between ecosystem types and noting that projects may be up to 30 times cheaper in developing economies compared to developed economies (Bayraktarov et al. 2016). The categories of developing and developed economies have most recently been defined by United Nations (UN 2021).

The reviews of costs and feasibilities of marine restoration by Bayraktarov et al. (2016) and Basconi et al. (2020) are summarized in ◘ Table 15.2. Techniques are evolving and attributes of success noted in ◘ Table 15.2 may change over time. Most marine restoration projects reported in the literature have been conducted in countries with developed economies, in particular Australia, Europe and USA, although there are likely many unreported projects in countries with developing economies (Bayraktarov et al. 2016). They are mostly funded by government and private companies (as compensatory habitat) (Basconi et al. 2020). Partnerships with the government and other private, community and/or non-government entities and the development of markets for ecosystem services may provide incentives for financial investments into marine restoration projects (Murtough et al. 2002; Basconi et al. 2020).

Suitable site selection is essential for the success of restoration projects, and low survivorship of transplantations of seagrass, coral reef and mangroves has been attributed to poor site selection (Bayraktarov et al. 2016; Sheaves et al. 2021), lack of habitat-based research and limited reliable success metrics (Basconi et al. 2020). There is very limited long-term data on the success of restoration projects and long-term monitoring (e.g. 15–20 years) yet this has been commonly recommended (e.g. Hawkins et al. 2002; Bayraktarov et al. 2016; Basconi et al. 2020; Pollack et al. 2021). Although there is a cost associated with long-term monitoring, it provides valuable data to support adaptive management and improve techniques.

15.5 Marine Pollution Mitigation and Reduction

Marine ecosystems become degraded by a wide variety of threats. Degrading factors can be physical, biological or chemical (◘ Table 15.3) and may occur simultaneously or sequentially at any one site. If these degrading factors are not mitigated the likely success of restoration projects is compromised (e.g. Sheaves et al. 2021). Mitigating measures need to target the source of the degradation. Mitigation steps in restoration projects are initiated for many reasons including marine pollution accidents (e.g. oil spills), unexpected pollution (e.g. tributyltin) and more broadly because of diffuse source inputs (e.g. catchment runoff) and coastal development (Hawkins et al. 2002). The different sources need to be managed differently and in general it is less complicated to manage point source discharges and one-off events than complex diffuse sources with numerous polluting substances (see also ► Chapter 1). This section introduces you to some tools and approaches that are used to mitigate pollution (◘ Table 15.3). Where appropriate, some of these tools and approaches may be incorporated into restoration programs in coastal catchments and marine ecosystems.

◘ **Table 15.2** A summary of the relative costs and success of marine restoration projects in the published literature (*Data sources*: Bayraktarov et al. 2016 and Basconi et al. 2020)

Ecosystem	Relative cost of restoration	Attributes of success[a]	Relative scale of sites
Coral reefs	High	Transplanting, coral gardening and coral farming projects	Small scale
Seagrasses	High	Transplanting seedlings, sprogs, shoots and rhizomes	Small scale
Mangroves	Low	Facilitation of natural recovery through planting of seeds, seedlings or propagules	Largest scale
Macroalgae forests	Unknown	Transplantation of adults, sporophyte, seedlings, germlings or juveniles	Increasing
Saltmarshes	Medium	Construction and planting, seeds, seedlings or sods	Small-medium scale
Oyster reefs	Medium	Establishment of no-harvest zones and transplanting hatchery raised juveniles	Unknown

[a] Success based on survival was more dependent on ecosystems, site selection and techniques rather than money spent

◻ **Table 15.3** Marine pollution mitigation strategies

Treat or stressor	Mitigation strategies (current and recommended)	Further reading
Chemical		
Polychlorinated biphenyls (PCB)s	Stockholm convention Capacity building for inventory and destruction facilities	Chapters 8 and 16 Stuart-Smith and Jebson (2017)
Tributyltin (TBT)	International bans Development of suitable and low toxic alternatives	Chapters 6 and 13
Metals	Bioremediation Biosorption	▶ Chapter 5 Michalak (2020)
Brine (desalination waste)	Brine mining (recovery) Reduce liquid waste discharge Dilution	▶ Chapter 12 Panagopoulos and Haralambous (2020)
Illegal ship waste oil dumping	Reduction -onboard pyrolysis technology Improved disposal facilities in ports Improved policy and regulations	Mazzoccoli et al. (2020)
Nutrients	Catchment management Wastewater treatment Bioremediation Multitrophic aquaculture Water quality off-sets	▶ Chapter 4 Lang et al. (2020) Michalak (2020)
Pesticides	Pesticide use regulation Catchment management Enhanced microbial degradation Ecological risk assessment	Chapters 7 and 8
Oil spills	Double hull tankers Rapid implementation oils spill response programs	Chapters 6 and 16
Physical		
Plastic	Ecolabeling for informed consumer decisions Reduction, reuse, recycling Bans and imposed fees Policy and Conventions (e.g. OSPAR Convention 1998) Clean up strategies Behavioural change strategies Biotechnology (bioplastics) Extended producer responsibility Credit system Waste to energy Life cycle assessment of products and packaging	▶ Chapter 9 Ogunola et al. (2018) Lee (2021) Li et al. (2021)
Turbidity	Silt curtains Catchment riparian vegetation reinstatement Catchment management	▶ Chapter 12
Development of urban and port infrastructure	Rescue and relocation of species Development strategies	Liñán-Rico et al. (2019)
Noise	Rerouting of vessels and noise generating activities in area during high animal density and biologically important areas Noise reduction programs [e.g. SILENV (Ships oriented innovative soLutions to rEduce noise and vibrations 2009–2012)] Acoustic deterrent devices Reducing ship speed Vessel quieting technologies Voluntary agreements Passage planning Optimising ship handling and maintenance	▶ Chapter 12 Chou et al. (2021) Vakili et al. (2021)
Biological		
Introduced species	International agreements (e.g. Convention of biological diversity) Quarantine regulations Containment and eradication Precautionary approach (avoid the economic cost of invasion)	▶ Chapter 12 Occhipinti-Ambrogi (2021)

(continued)

◨ **Table 15.3** (continued)

Treat or stressor	Mitigation strategies (current and recommended)	Further reading
Harmful algae blooms	Nanoparticle treatment technology See nutrient mitigation strategies	► Chapter 12 Gonzalez-Jartin et al. (2020)
Disease	Quarantine regulations	► Chapter 12 Sampaio et al. (2015)

15.5.1 Mitigating Coastal Catchment Discharges

Catchment runoff is a major source of pollution to coastal environments and includes a combination of point and non-point sources which may be a result of both current and legacy (historic) activities. Not all pollutants generated in catchments reach the marine environment (e.g. Waterhouse et al. 2012), in general, and logically, lower transport rates to the ocean occur for pollutants generated further upstream in catchments (e.g. Star et al. 2018). The type and amount of pollutants that reach the ocean from catchments depends on the land use, rainfall intensity and duration, geomorphology, integrity of the riparian zone, chemical behavior of specific pollutants, and other physiochemical properties of the environment (see ► Section 7.5.1, Chapter 7).

Mitigating Inputs from Agriculture
Agricultural activities in coastal catchments create diffuse sources of eroded soils, nutrients and pesticides that are delivered to the marine environment (Chapters 4 and 6). Management actions to mitigate inputs from agriculture have had scalability issues and sometimes limited results (e.g. Cook et al. 2013; Creighton et al. 2021; Waltham et al. 2021). However, it is important to note that mitigating activities may take several years to show measurable differences in inputs at the catchment scale (e.g. Star et al. 2018) and groundwater transport of pollutants to the ocean needs to be considered in the pathways of inputs (e.g. Carroll et al. 2021). ► Box 15.1 shows an example of a long-term water quality improvement plan for the Great Barrier Reef, Australia, to mitigate the effects of land-based human activities including agriculture.

Diffuse nutrient runoff from agriculture can be managed directly through best practice farm management including a reduction in fertiliser use and by using tools such as cover crops (e.g. Vilas et al. 2022). However, the elimination of fertilisers is a highly unlikely proposition. Therefore, treating drainage water before it enters river systems and the ocean is an important mitigation strategy. There are several approaches used to reduce the nutrient loading in drainage water including constructed wetlands, water retention ponds, denitrifying bioreactors, riparian buffer zones and/or a combination of these. Some approaches capture the benefits that ecosystem services offer for nutrient uptake and storage (e.g. Carstensen et al. 2020; Hsu et al. 2021). Constructed wetlands and riparian buffer zones also provide biodiversity values and are forms of ecosystem restoration in their own right.

In situations where the sources are difficult to manage (e.g. low lying, low-productivity land as a source of dissolved inorganic nitrogen) land-use conversion may be appropriate (Waltham et al. 2021). Land-use conversion may include support to farmers for developing alternative crops and grazing, aquaculture opportunities or forestry, or may require buy-back to reinstate natural vegetation (Waltham et al. 2021).

The selection of the approach or combination of approaches used requires stakeholder involvement, cost benefit assessment, and consideration of the local geographical and climatic conditions including the integration of future changes such as climate and land use (e.g. Carstensen et al. 2020).

Box 15.1: Reef 2050 Water Quality Improvement Plan—A Mitigation Strategy

Associate Professor Michael St. J. Warne, Ecotoxicologist.
University of Queensland, Australia; Queensland Department of Environment and Science, Australia; Centre for Agroecology, Water and Resilience, Coventry University, United Kingdom.
The Great Barrier Reef (GBR) is the world's largest reef running for over 2500 km along the east coast of Queensland, Australia. It is under threat from a range of stressors including: climate change; coral bleaching; crown of thorn starfish outbreaks; commercial and recreational fishing; mining; urban development; commercial and recreational shipping; agriculture and the quality of water entering the GBR lagoon. In terms of water quality, suspended solids from soil erosion, nutrients (nitrogen and phosphorus) and pesticides have been identified as the key pollutants. These pollutants all originate from land-based human activities and are predominantly transported to the lagoon via surface and ground-

Pollution Mitigation and Ecological Restoration

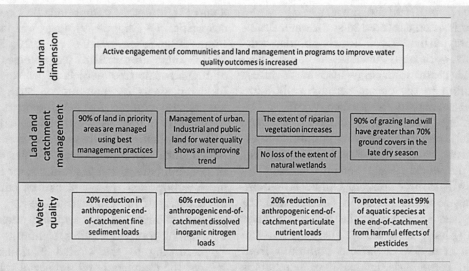

▣ **Figure 15.1**   ▶ Box 15.1: The targets in the reef 2050 water quality improvement plan. Adapted from AGQG, 2018 by M St. J. Warne

water. To address this, the Australian and Queensland governments developed a series of plans to improve the quality of water entering the lagoon and thus improve the health and resilience of the reef—these can be considered mitigation strategies. The current plan is the Reef 2050 Water Quality Improvement Plan 2017–2022 (▶ https://www.reefplan.qld. gov.au/__data/assets/pdf_file/0017/46115/reef-2050-water-quality-improvement-plan-2017-22.pdf) and a new plan will be released in 2023. The underpinning assumption of these plans is that by reducing the ecological stress from poor water quality the overall stress will decrease and the reef ecosystems will have a greater ability to deal with other stressors including climate change.

Each water quality improvement plan has had a series of targets that aim to improve water quality and land management practices. The targets have been modified in the plans to reflect improved scientific knowledge of what is required to increase the health and resilience of the reef. The current targets are presented in ▣ Figure 15.1 and the aim is that they should be met by 2025.

The Paddock to Reef (P2R) Integrated Monitoring, Modelling and Reporting Program was developed to implement the plans. The P2R uses an adaptive management approach (also termed a Monitoring, Evaluation, Reporting and Improving (MERI) framework) to drive progress towards meeting the targets. Monitoring is done on land, in waterways and in the GBR lagoon. As the magnitude of the pollutant loads (mass) are highly correlated to climate, Source Catchments Models are used to remove the climatic signal and to estimate annual progress towards meeting the water quality targets. Progress to meeting the targets and the current ecological condition of the GBR is summarised and presented in the semi-annual Reef Water Quality Report Card (▶ https://reportcard.reefplan.qld.gov.au/home?report=condition%year=611f443aba3074128316cb07).

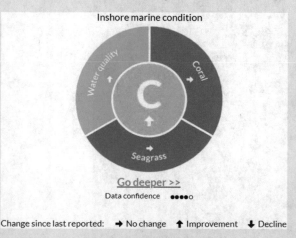

▣ **Figure 15.2**   ▶ Box 15.1: The reported inshore conditions of the GBR ecosystems in 2020. Condition ranges from A—very good to E—very poor. D is poor condition. *Source*: Queensland Government CC BY 4.0

Research conducted since the previous Water Quality Improvement Plan is synthesised approximately every five years in the Reef Scientific Consensus Statement which is then combined with the results of the Reef Report Card (e.g. ▣ Figure 15.2) and other information to determine the targets in the next Water Quality Improvement Plan.

Until recently the governments have been encouraging adoption of Best Management Practices and have co-invested (50:50) with farmers to purchase improved equipment (e.g. hooded spraying rigs) or infrastructure (e.g. water retention ponds or artificial wetlands). But recently the Queensland Government has introduced mandatory measures to drive further improvement of land management practices—the Reef Protection Regulations which address the issue of fine suspended solids and dissolved inorganic nitrogen.

Mitigating Inputs from Urban Stormwater

Rainwater water is often captured in stormwater drainage infrastructure, particularly in heavily populated urban environments with hard surfaces and limited permeability. Urban stormwater may also be a diffuse source of pollution to the ocean, through infiltration and groundwater movement and surface runoff.

Various solutions have been developed to mitigate stormwater from urban areas carrying pollutants to estuaries and marine waters. These tools have different names around the world; Water Sensitive Urban Design (WSUD) in Australia and the United Kingdom, Low Impact Development (LID) in Canada and the USA, Nature-Based Solutions in the EU and Sponge Cities in China (Zhang et al. 2020). These systems are usually effective in removing various pollutants from stormwater, and some jurisdictions have regulations that require their installation as part of infrastructure development (e.g. New South Wales, Australia; State Environmental Planning Policy (Building Sustainability Index: BASIX 2004). Furthermore, stormwater management has additional benefits such as flood mitigation, microclimate improvement, improvement in the amenity values in urban landscapes and harvested stormwater can be a valuable water resource (Zhang et al. 2020 and references therein).

Mitigating Inputs from Municipal and Industrial Wastewater

Sewage and industrial wastewater **discharges** are complex mixtures including organic compounds (Chapters 8 and 12), inorganic compounds (▶ Chapter 5) and microplastics (▶ Chapter 9) (e.g. Mintenig et al. 2016; Prata 2018; Schernewski et al. 2020; Sridharan et al. 2021). There are excellent technologies through large- and small-scale treatment facilities to reduce the flow of chemicals to the environment. Such facilities are often legally required for developments and activities, particularly in countries with developed economies. As with all infrastructure these facilities need to be maintained since leaking and broken pipes can be a source of contaminants through groundwater inputs. Furthermore, suboptimal treatment can be caused by exceeding capacity of built infrastructure (e.g. when an urban population increases more rapidly than infrastructure updates) or poorly operating facilities.

According to the United Nations (2017), about 70% of the municipal and industrial wastewater generated by high-income countries is treated. In upper middle-income countries and lower middle-income countries that ratio drops to 38% and 28%, respectively. In low-income countries, only 8% is treated in anyway. Globally, 80% of wastewater is discharged untreated (UN 2017). Where there is limited use of treatment facilities, it is often related to a lack of financial resources (◙ Figure 15.3). The United Nations Sustainable Development Goals highlight the importance of clean water and sanitation (Goal 6) and life below the water (Goal 14) and may potentially be drawn upon to invoke action to upgrade and deliver municipal services in developing economics and reduce wastewater discharges to the marine environment.

◙ **Figure 15.3**   Waterways carry waste through cities to the ocean. Open drains, like the one pictured, are often used to dispose of unwanted wastes and no treatment occurs before the waterways reach the ocean. *Photo* A. Reichelt-Brushett

15.6 Marine Habitat Restoration

Keystone and foundation species are essential for particular types of ecosystem structure. These species may be plants (e.g. mangroves) or animals (e.g. scleractinain corals) and we often name ecosystems after their keystone species. In essence, without these species present the ecosystems do not function. Indeed, marine ecosystem restoration attracts large amounts of funding. In the USA many coastal and marine habitat projects are funded by NOAA with an annual budget of around US$10 million (2019) that is distributed through a competitive grant submission process. In this section of the chapter, some types of marine habitat restoration are discussed. Restoration projects can be developed with basic tools and good knowledge of the ecosystem requirements but at times engineering and technology can support and enhance restoration outcomes.

15.6.1 Oyster Reefs

Oyster reefs and beds may be intertidal or subtidal biogenic structures formed by oysters living at high densities and building a habitat with significant surface complexity (Baggett et al. 2014 and references therein). Historically, most oyster restoration efforts focused on the recovery of oyster fisheries and mitigating losses from natural and anthropogenic effects. More recently there has been recognition of the **valuable ecosystem services** provided by oyster beds such as water biofiltration, benthic habitat for biodiversity (e.g. for epibenthic invertebrates), nutrient sequestration, shoreline stabilisation and enhanced secondary production (Baggett et al. 2014). Many of these values are now included in the goals of restoration projects (Baggett et al. 2014 and references therein). According to Bayraktarov et al. (2016), harvest sanctuaries and transplanting juvenile oysters from hatcheries achieve positive results. An example of a large-scale oyster reef restoration project is the Billion Oyster Project in New York Harbour

(▶ https://www.billionoysterproject.org/) and smaller scale work includes Lau Fau Shan and Tolo Harbour in Hong Kong (▶ https://www.tnc.org.hk/en-hk/what-we-do/hong-kong-projects/oyster-restoration/). However, oyster bed restoration projects still have limited monitoring, even in well-known projects like in Chesapeake Bay, USA, monitoring from 1990 to 2007 was limited and project goals were not well defined (e.g. Kennedy et al. 2011). This omission has reduced adaptive management and development of standard methodologies.

The *Oyster Habitat Restoration-Monitoring and Assessment Handbook* by Baggett et al. (2014) was produced to address the shortfall of previous programs and to support programs to demonstrate successful outcomes. The handbook provides standard techniques (named Universal Metrics) that can be used for comparisons among sites and to help develop performance criteria. This focus on monitoring and assessment enables an understanding of the basic project performance and how the performance meets ecosystem services-based restoration goals (Baggett et al. 2014).

More recently, enhanced approaches are being considered to include, **focused site selection**, potential use of **artificial substrates**, and **oyster species and selection of genotypes** for seeding to support oyster survival and delivery of ecosystem services (Howie and Bishop 2021; Pollack et al. 2021). The consideration of the most suitable growth form is important because it influences ecosystem service delivery (Howie and Bishop 2021); however, trade-offs might be required depending on the goals (e.g. high elevation reefs are most effective at attenuating waves) (Hogan and Reidenbach 2022). Furthermore, oyster species and genotypes should be selected according to their environmental suitability, resilience to environmental change, and the size and shape of reefs they form (which influences ecosystem services) (Howie and Bishop 2021) (▶ Box 15.2). Choosing stock from aquaculture or wild populations also needs to be a key consideration and will sometimes depend on availability.

Box 15.2: Assess Before you Invest: The Need for Careful Site Selection in Shellfish Reef Restoration

Professor Kirsten Benkendorff, Marine Biologist.
National Marine Science Centre, Southern Cross University, Australia.
It is estimated that over 85% of oyster reef ecosystems have been lost globally (Beck et al. 2011; Ford et al. 2016), due to a range of human activities including unsustainable harvest, destructive trawling and bottom dredging, increased sedimentation from clearing of riparian vegetation, decreased water quality and disease. Oyster reefs were once extensive in many estuaries, but are now reduced to remnant reef areas or in some cases are considered functionally extinct (Beck et al. 2011; Ford et al. 2016; Gillies et al. 2018). However, oysters are being increasingly recognised as ecosystem engineers that play an integral role in benthic-pelagic coupling, water clarification, carbon sequestration, habitat provision for invertebrates, fish and algae, and the protection of shorelines (Coen et al. 2007; Grabowski et al. 2012). This has triggered significant efforts to restore degraded oyster reef habitats at key locations, in at least seven countries (Fitzsimons et al. 2020).

□ Figure 15.4   ▶ Box 15.2 Leaf oyster reefs provide good habitat to other invertebrates and fish (left) and can improve water quality as part of an active catchment management plan. When in decline due to significant runoff from intensive agriculture, with pesticides, high sediment and nutrient loads smothering by algae growth but can occur (right). *Photos*: K. Benkendorff

Restoring oyster reefs on the scale required to recover ecosystem services requires significant infrastructure and financial investment. The return on investment for oyster restoration has been shown to vary widely but tends to increase with the scale of the project (Bersoza Hernández et al. 2018). Consequently, the first stage in oyster reefs restoration programs must be to undertake a thorough assessment of the proposed location and develop a feasibility plan (Fitzsimons et al. 2020). It is essential that the causes of the original decline are well understood and effectively mitigated. Persistent problems with water quality, pollutants and sedimentation will cause chronic stress, reducing the resilience of oysters and increasing the likelihood of disease and mortality. Unfortunately, habitat suitability indexes for oyster restoration (Theuerkauf and Lipcius 2016) don't consider water quality beyond the basic physicochemical parameters or the surrounding land use practices that influence the likelihood of ongoing exposure to aquatic pollution. A catchment wide assessment is required to determine the likelihood of chronic exposure to contaminants, such as pesticides that are known to impact oyster health (e.g. Ewere et al. 2020).

For biosecurity reasons, the use of local species is also essential for oyster reef restoration. Oysters sourced from near-by populations are also more likely to have adapted to the local conditions. We have been investigating the potential for including the large reef-forming leaf oyster *Isognomon ephippium* (□ Figure 15.4) in oyster reef restoration programs (Benthotage et al. 2020). These leaf oysters occur in slow moving estuarine creeks and bays often covered in silt. We have recorded populations in areas with high agricultural nutrient runoff and fluctuating pH reaching as low as 5 from acid sulphate soil runoff. However, these are long lived oysters and some populations appear to be in decline. Further research is required to understand the tolerance range of these and other oysters and match these to environmental conditions at locations proposed for oyster reef restoration. In some cases, a whole of catchment approach will be required to manage terrestrial runoff to ensure the future viability of oyster reefs and their inherent ecological value.

15

15.6.2 Coral Reefs

Coral reef degradation results from many different stressors, some of which are caused by polluting substances such as nutrients (▶ Chapter 4), metals (▶ Chapter 5), pesticides (▶ Chapter 7), sedimentation (▶ Chapter 12) and atmospheric gases (▶ Chapter 11). Other stressors such as coastal development, over harvesting, destructive fishing, invasive species, outbreaks of predatory organisms such as Crown of Thorns Starfish (▶ Chapter 4), prolonged elevated water temperatures leading to coral bleaching and impacts from recreational activity need to be included in mitigation strategies as there may be a multitude of stressors to address at any one site (Pandolfi et al. 2003). As with all restoration projects the removal of the stressors is a key mitigation step required at the very first stage of restoration. As discussed,

catchment management and sewage treatment can help remove polluting impacts such as sedimentation and chemical loads. Mitigating effects of anthropogenic temperature change and ocean acidification are more challenging undertakings and may require specific interventions such as assisted evolution (van Oppen et al. 2017). Considerations of **socio-economic contexts** are required to optimise recovery (Gouezo et al. 2021). Restoration of coral reefs has not yet resulted in fully functional reefs but some success has occurred on the scale of up to a few hectares (Edwards and Gomez 2007). The field of coral reef restoration has advanced rapidly over the past 10–15 years and continues to evolve.

Coral transplantation has been used in coral reef restoration efforts for many years (e.g. Ferse et al. 2021). In this method fragments of coral are taken from donor reefs and secured at the restoration sites. This

◘ Figure 15.5   Small coral transplants are taken from donor reefs and attached mid water to enable grow out before transplanting to restoration sites. *Photo*: "Coral nursery, Coral Restoration Foundation" by kareneglover CC BY-NC 2.0

strategy creates impacts at donor reefs. To help mitigate these impacts sometimes these donor colonies are taken as small fragments and then used in coral gardening or coral farming which provides more space to grow up colonies in mid water (◘ Figure 15.5) or in benthic gardens before use at the restoration site (e.g. Feliciano et al. 2018). Other programs have **grown corals from** spawning **in laboratory conditions and out-planted the juveniles** (Guest et al. 2014; Bayraktarov et al. 2016). More recently collection of gametes from wild coral spawning events has been successfully trialled, with larvae reared in the laboratory or in floating larval pools on reefs (Harrison et al. 2021). The approximately 5-day old larvae (that are ready to settle) are then distributed by various methods directly onto target reef areas. This process is known as mass larval settlement (dela Cruz and Harrison 2017; Harrison et al. 2021) (▶ Box 15.3). By **collecting slicks of broadcast** spawning **corals** many millions of potential recruits, that in natural conditions would not survive, are utilised. This approach takes the pressure off donor reefs that occurs with transplantation and coral gardening.

Coral genotypes that can survive extreme conditions including temperature and pH anomalies may be used as sources for selective breeding to support assisted evolution and focus recruitment strategies (van Oppen et al. 2017; Basconi et al. 2020; Rinkevich 2021). These techniques are evolving rapidly.

Box 15.3: Scaling up Coral Restoration for Reef Recovery

Professor Peter Harrison and Dr. Dexter dela Cruz, Coral Reef Ecologists.
Marine Ecology Research Centre, Southern Cross University, Australia.

Accelerating loss of foundation reef corals in most reef regions around the world is impairing the natural resilience of coral communities and resulting in reef degradation (Burke et al. 2011). Consequently, increasing attention is being focused on active coral restoration interventions on degraded but recoverable reef areas where the previous impacts and immediate threats are being managed (Harrison et al. 2021). Reef corals have two primary modes of reproduction in their life cycles: **asexual** budding of genetically identical polyps to create complex colonies or solitary individuals, and in some cases growth forms that enable breakage and fragmentation of colonies to produce new corals; and **sexual** reproduction involving broadcast spawning or gametes and planktonic larval development, or internal brooding of larvae that are released at an advanced stage of development (Harrison and Wallace 1990). These two modes of coral reproduction have enabled the development of two different approaches to coral restoration using asexual fragmentation and cloning, or sexual production of millions of coral larvae for settlement on degraded reefs.

◘ Figure 15.6   ▶ Box 15.3: Asexual fragmentation and coral gardening enhanced coral recovery at smaller scales in the Philippines. *Photo*: D. dela Cruz

Coral fragmentation and production of genetically identical colonies with subsequent direct transplantation on the reef has been the most common asexual method for restoration (◘ Figure 15.6). The methods have been refined to include an intermediate nursery phase to produce larger quantity of nubbins and reduce the high rates of mortality of coral fragments during the early phase of outplanting onto reefs (Rinkevich 1995; Shaish et al. 2008; Edwards 2010). Advantages of fragmentation, coral gardening and outplanting approaches include relatively simple training and engagement of diverse stakeholder groups, varied approaches for different reef environments, rapid increases in coral colonies and cover on degraded reefs, and potential for healthy fragments to grow quickly if environmental conditions on the reef are still suitable (Young et al. 2012; dela Cruz et al. 2014; Omori 2019; Howlett et al. 2021). Disadvantages of asexual propagation include damage to healthy parent donor colonies, increased diseases from damaged tissues, low genetic diversity among coral colonies from few parental genotypes leading to low resilience to different stressors such as temperature stress and mass bleaching events, and high costs associated with manual collection and outplanting on reefs plus increased costs from establishing and maintaining coral nurseries (Edwards 2010; Bostrom-Einarsson et al. 2020). Consequently, coral gardening approaches are considered to be relatively expensive and more suitable for smaller-scale restoration projects such as increasing coral cover on damaged high value reef patches important for tourism (Bostrom-Einarsson et al. 2020; Howlett et al. 2021).

In contrast, sexual propagation promotes increased genetic diversity of restored coral populations and communities. The production of genetically diverse larvae from cross-fertilisation of eggs and sperm from many different colonies, increases the potential for rapid evolution of heat-tolerance and other traits that may enhance survival and resilience in rapidly changing reef environments (Baums 2008; Harrison et al. 2016, 2021; Randall et al. 2020). However, most corals are broadcast-spawners characterised by high production of gametes but low survival and settlement of planktonic larvae coupled with high post-settlement mortality during early life stages, which can create a bottleneck in reproductive success (Harrison 2011, 2021; Randall et al. 2020). Studies have used sexual larval propagation methods and two main approaches have been trialled. First, larvae can be cultured in tanks and settled onto tiles and other devices and reared in laboratory hatchery systems or in in-situ nurseries prior to outplanting on reefs (Guest et al. 2014; Chamberland et al. 2017). Alternatively, larvae can be directly settled ('seeded') onto reef areas with or without the use of larval mesh enclosures (Heyward et al. 2002; Edwards et al. 2015; dela Cruz and Harrison 2017; 2020; Harrison et al. 2021). Larval settlement onto tiles and devices and laboratory nursery rearing has some advantages. It reduces post-settlement Mortality, but significantly increases production costs per coral (Guest et al. 2014), and may select for genotypes that are maladapted to degraded reef environments. In contrast, mass larval production and direct larval settlement on degraded reefs is more cost-efficient and can produce breeding populations within two to three years (dela Cruz and Harrison 2017; Harrison et al. 2021) (◘ Figure 15.7). However, post-settlement survival can be low during the first few weeks and months after settlement due to strong selective pressures operating in degraded reef environments (dela Cruz and Harrison 2017, 2020).

Reef restoration activities and methods are now rapidly expanding in many regions and include innovative approaches to increase scales of larval production and reproductive success across many stages of the coral life cycle. Re-

15

☐ **Figure 15.7** ▶ Box 15.3 Mass supply of branching *Acropora tenuis* coral larvae significantly increased coral cover and restored breeding populations within a few years on badly degraded reef systems in the Philippines. *Photo*: P. Harrison

cent developments include direct capture of large spawn slicks from surviving healthy corals using floating spawn catchers and mass culture of many millions of larvae in floating pools moored on reefs (Harrison et al. 2021), hybridisation to enhance environmental tolerance and climate resilience (van Oppen et al. 2017; Chan et al. 2018), cryopreservation of gametes and artificial breeding for assisted gene flow (Daly et al. 2022), and selective breeding and provision and uptake of heat-tolerant Symbiodiniaceae microalgal symbionts and the use of probiotics (van Oppen et al. 2017; Quigley et al. 2020). These sexual propagation approaches and combination of culture techniques have great potential for massively increasing the scale and success of coral restoration to enable the recovery of degraded coral communities and reef systems around the world, but reef restoration will only be successful in the longer-term if effective action is taken to reduce global greenhouse gas emissions and global warming.

15.6.2 Seagrasses

Seagrasses are submerged vascular plants known to support marine biodiversity with an historic total global cover of 171,000 km² (Green and Short 2003). Human population expansion has been considered the most serious cause of seagrass habitat loss particularly increasing contaminant inputs to the coastal oceans (Short and Wyllie-Echeverria 1996; Zenone et al. 2021). Efforts at restoration have occurred in Australia, Florida, India, Indonesia, Italy, Sweden and New Zealand (Nadiarti et al. 2021 and citations there in) and probably in other areas too that have more limited reporting. Early restoration projects occurred in Florida in the 1980s and the resource value of seagrasses was well recognised before then (Fonseca et al. 1996 and references therein). Unfortunately, global seagrass loss has been dramatic and was estimated at about 7% a year in 2009 (Waycott et al. 2009).

Well restored seagrass sites have shown longevity for many decades in both tropical and subtropical areas (Thorhaug et al. 2020). Data that can show such long-term success is a testament to well-planned restoration programs and continued funding for monitoring and on-going restoration work to counter effects from

extreme weather events. However, data documenting restored ecosystem services have not been collected consistently and frequently enough to provide marine resource managers with hard data as to the ecosystem services returned, except in the Atlantic USA where fisheries food webs and carbon sequestration assessment were included in monitoring (Thorhaug et al. 2020).

The highest survival of seagrass in restoration projects used a range of techniques including transplantation of seedlings, sprigs, shoots and rhizomes (Bayraktarov et al. 2016) with methodologies and success somewhat dependant on location and species used (e.g. *Zostera marina* the most commonly transplanted species in temperate regions) (see also Thorhaug et al. 2020). However, reduced genetic diversity has been identified in planted seagrass beds compared to natural ones (Williams and Davis 1996) and this could lead to longer term vulnerabilities.

15.6.3 Mangroves

Global mangrove forest cover is an estimated 84,000 km² spread across 105 countries (Hamilton and Casey 2016). **Deforestation** is one of the main causes of mangrove loss, however, they exist in depositional environments acting as traps for fine particles, organic matter and associated chemical and physical pollutants (see Chapters 5 and 6). For this reason, restoration projects must consider the site contamination and risk of pollutants to diversity and structure. The main reasons for restoring mangrove ecosystems include conservation and landscaping, economic security, food security and coastal protection (Field 1998).

Mangrove restoration can be conducted relatively cheaply and easily and is **arguably the most established marine ecosystem restoration activity**. It is relatively easy to engage community groups in planting programs and this gains similar community engagement to tree planting programs on land (◘ Figure 15.8). Most mangrove restoration projects that achieve high survival rates include facilitation of natural recovery by planting of seeds, seedlings and propagules, investment in the planting of saplings and small trees, hydrological restoration and weed management (Bayraktarov et al. 2016).

Since 1965 Singapore has lost > 90% of its mangrove forest and attempts to restore these have had limited success (Ellison et al. 2020). However, some sites of Mangrove rehabilitation in Singapore have provided new knowledge on how to enhance ecological diversity and ecosystem services in an urbanised coastal setting. For example, the Pulau Tekong hybrid engineering project demonstrated how mangrove vegetation can be incorporated into engineered coastal defence structures (Friess 2017) and highlighted the value of multiple species plantings and matching species traits to prevailing environmental conditions (e.g. Field 1998).

Mangrove forests also sequester carbon (blue carbon) (see ► Chapter 11). However, estimates of above ground and underground carbon storage are variable between studies and depend upon different scenarios (e.g. Moritsch et al. 2021). More research is required to understand long-term carbon storage potential.

15.6.4 Saltmarsh

Saltmarsh are found in 99 countries throughout the world (particularly mid and high latitudes and) in the **upper tidal limits of lower estuaries** (Mcowen et al. 2017). The saltmarsh environment is harsh, as the community is exposed to extreme salinity, desiccation, and tidal flooding. For this reason, saltmarsh plants are known as halophytes with specialised adaptations to grow in salty conditions. **Micro-elevation and the tidal inundation** regime strongly influence the gradation between saltmarsh (on the landward side) and mangroves (to the water side) (Adam 2000; Green et al. 2009a). Saltmarsh require fewer tidal inundations per year compared to mangroves. The species composition is mostly contributed to by plants, but fauna groups consist of terrestrial species (e.g. birds, and bats) and aquatic species (e.g. fish, molluscs and crustaceans), with some being specialized salt marsh dwellers (Laegdsgaard 2006). The most conspicuous invertebrate fauna in saltmarshes are crustaceans and molluscs and in a comprehensive study of 65 saltmarshes around Tasmania, Australia, Richardson et al. (1997) found over 50 species.

◘ **Figure 15.8** Community collaborations can be small scale. This site is near Pattimura University (Ambon, Maluku, Indonesia) will be monitored over time by students. This collaboration was between staff and students of Southern Cross University and the University of Pattimura (led by Y. Male), **a** the site prior to any activity, **b** litter removal, **c** planting mangrove seedlings and **d** celebration of working together for positive environmental outcomes. *Photos*: A. Reichelt-Brushett

Pollution Mitigation and Ecological Restoration

Saltmarsh habitats have been degraded in the past due to their lack of perceived value and usefulness, being disregarded and used as illegal dump sites, off-road motorbiking and four-wheel driving as well as being at risk from the encroachment of urban, industrial, and agricultural development and localised runoff (e.g. Bucher and Saenger 1991; Green et al. 2009a) (▶ Box 15.4). Furthermore, they are vulnerable to floating pollutants such as oil and plastics that are transported and deposited through tidal inundations. Today saltmarshes are valued ecological communities providing fish feeding habitat during flood tides, carbon sequestration, coastal protection and other ecological services (Mcowen et al. 2017). In some countries, they are protected habitats.

Actions such as fencing to remove cattle and recreational vehicles from saltmarsh areas, diversion of stormwater and weed removal are the most common first steps in rehabilitation for saltmarsh. Large-scale saltmarsh restoration projects have been undertaken in North America since the late 1980s (e.g. Sinicrope et al. 1990; Fell et al. 1991; Frenkel and Morlan 1991). In Australia, saltmarsh restoration occurred at the Sydney Olympic Park among other sites in the late 1990s and related research improved knowledge of germination and establishment of saltmarsh species (Burchett et al. 1998; Laegdsgaard 2006).

Box 15.4: Case Study: Fingal Wetland Rehabilitation Project, New South Wales, Australia
Dr. Joanne Green, Restoration Ecologist.

The aim of the Fingal Wetland Rehabilitation Project was to reverse ongoing degradation of a saltmarsh area due to sand mining, exotic weeds, rubbish dumping (including old cars and trail bikes (◻ Figure 15.9)) and four-wheel drive recreational activity. The project encompassed an agreement between Tweed Shire Council and the Tweed Byron Local Aboriginal Land Council, plus an initiative developed by Wetland Care Australia with assistance from NSW Fisheries and The Fish Unlimited Project (funded by Federal Government through the Sustainable Regions Program). The area was characterised by fragmented patches of remnant saltmarsh dominated by three plant species, Saltcouch (*Sporobolus virginicus*), Sea Blite (*Suaeda australis*) and Samphire (*Sarcocornia quinqueflora*).

After the removal of cars and other rubbish, the natural topography was restored by connecting the patches of remnant saltmarsh with suitable fill and allowing natural regeneration to occur. Surface sediments were stripped back so the topsoil could be used to inoculate the new surface thus providing a source of silt, nutrients and the micro-fauna assemblages that were already occupying this niche. Saltcouch was also planted at low tide using 1 m quadrats made of PVC conduit. The conduit quadrats allowed accurate spacing and layout across the site for maximum use of donor material and future counting of success.

An associated research program (Green et al. 2009a, b; Green et al. 2010) measured changes in the soil carbon, algae first colonisers, plant coverage and invertebrate colonisation for several years after restoration work. Variables measured included soil moisture, pH, electrical conductivity, Total Organic Carbon and Total Nitrogen. Other measurements included soil algal abundances (Chlorophyll *a*), diatom abundance, and flora and fauna colonisation. Chlo-

◻ **Figure 15.9** ▶ Box 15.4 Cars removed from the Fingal Wetland Rehabilitation site prior the restoration works. *Photo*: T. Alletson

rophyll *a* results showed that the restored saltmarsh sites were progressing towards, but were not equivalent to, the reference site two years after restoration despite the fast growth rates of algae and its role as a primary coloniser. The analyses of variables showed that solar radiation, rainfall and tidal inundation were influential to micro algal growth. Measurements of the flora and fauna at restoration sites showed that the sites were moving towards a saltmarsh ecosystem but climatic conditions can affect short-term measures. For this reason, seasonal and longer term sampling is recommended.

The project success to date is the result of strong collaboration between all the stakeholders with a focus on a common goal: the removal of threatening processes and the restoration of the saltmarsh vegetation. The ongoing commitment by the project partners culminated in a successful grant from the NSW Government Environmental Trust to undertake additional works in the area.

15.6.5 Engineering, Technology and Marine Ecosystem Restoration

Artificial habitats are sometimes developed using science and engineering technologies to support restoration. An artificial reef is "*a submerged structure placed on the seafloor deliberately to mimic some characteristics of a natural reef*" (OSPAR 1999). Seaman (2007) highlighted the use of artificial structures in restoration projects in four case studies: kelp beds (California, USA), coral reefs (Florida, USA), oyster beds (Chesapeake Bay, USA), fisheries populations (Hong Kong, China). Engineering and technology are being used in multidisciplinary approaches to ecological restoration and collaborations help to support innovation (NRC 1994), some examples include

- ecological engineering and augmented evolution for coral resilience to climate change (e.g. van Oppen et al. 2017; Rinkevich 2021);
- cathodically protected steel mats to replace plastic for reseeding oyster reefs (Hunsucker et al. 2021);
- sustainable cementitious composite substrate for oyster reef restoration using recycled oyster shells and low cement content (Uddin et al. 2021);
- development of a lattice structure made out of a biodegradable potato starch to support seagrass restoration (MacDonnell et al. 2022); and
- biodegradation of micro- and nano-plastics in liquid and solid waste (Zhou et al. 2022).

Successful engineering and technology solutions will likely result when biotic needs are strongly connecting with engineering and technology solutions in a feasible and cost effects manner.

15.7 Marine Species as Bioremediators

Another angle of environment improvement and contaminant removal from the environment includes bioremediation activities. The process is similar to land-based phytoremediation and other bioremediation research except using marine species. Clearly, there are

ecosystems service provisions that help to mitigate pollution, such as water quality improvement from oyster beds, but there is also targeted research on particular species. Brown marine algae (*Sargassum natans* and *Fucus vesiculosus* and *Turbinaria ornata*) and green algae (*Cladophora fascicularis*, *Enteromorpha prolifera* and *Ulva reticulata*) show promising bio-sorbant properties for some metals (Brinza et al. 2007; Mudhoo et al. 2012 and references there in; Areco et al. 2021). Marine diatoms can play a role in the degradation, speciation and detoxification of chemical wastes and hazardous metals using mechanisms both external to the cell and internally (Marella et al. 2020). Marine bacteria show promise in helping to develop biotechnology for ocean clean-up of metal contaminants (Fulke et al. 2020) and plastics (Jenkins et al. 2019; Wei and Wierckx 2021). These developments provide an exciting field of discovery that focuses on environmental remediation.

15.8 Summary

There are numerous important ecological habitats in marine environments and many have been impacted by human activities, including pollution. Marine ecosystem restoration has been gaining increasing attention since the 1990s and those ecosystems that have had committed restoration works include **coral reefs, seagrasses, mangroves, macroalgae forests, saltmarshes** and **oyster reefs**. Each of these requires specific conditions for habitats to thrive and discussion and examples are provided.

Mitigating pollution and other stressors is an important first step in ecological restoration and may take several years to achieve measurable improvements, particularly for diffuse source inputs such as agricultural activities. It is important to follow the major principles of successful ecological restoration explained in ◘ Table 15.1. ► Section 15.5 describes important pollution mitigation practices and highlights the importance of mitigating land-based sources of stressors including nutrients, metals, pesticides, and turbidity. Other human activities such as shipping and infrastructure development also create stressors such as oil spills and noise as well as acting as vectors for invasive species.

Engineering and technology solutions play a developing role in marine pollution mitigation and ecosystems restoration activities.

15.9 Study Questions and Activities

1. Describe ecological restoration in your own words.
2. Create a table that highlights ecosystem features and considerations for successful coral reef, seagrass, salt marsh, mangroves and oyster reef restoration. If you think you have done a great job, send it to the editor and we may discuss including it in the next edition of this book.
3. Select one of the types of pollutants shown in ◖ Table 15.3 and expand on the mitigation strategies through literature searches of your own.
4. Consider the United Nations Sustainability Goals and discuss how they may be used to invoke action to upgrade and delivery municipal services in developing economies and reduce wastewater discharges to the marine environment.

References

Abelson A, Halpern BS, Reed DC, Orth RJ, Kendrick GA, Beck MW, Belmaker J, Krause G, Edgar GJ, Airoldi L, Brokovich E, France R, Shashar N, De Blaeij A, Stambler N, Salameh P, Shechter M, Nelson PA (2015) Upgrading marine ecosystem restoration using ecological-social concepts. Bioscience 66(2):156–163

Adam P (2000) Saltmarshes in a time of change. Environ Conserv 29:39–61

Adame MF, Roberts ME, Hamilton DP, Ndehedehe CE, Reis V, Lu J, Griffiths M, Curwen G, Ronan M (2019) Tropical coastal wetlands ameliorate nitrogen export during floods. Front Mar Sci 6:00671

AGQG (Australian Government and Queensland Governments) (2018) Reef 2050 water quality improvement plan 2017–2022. Reef water quality protection plan Secretariat, Queensland, Australia. P 40. Available at: ▶ https://www.reefplan.qld.gov.au/. Accessed 7 Feb 2022

Areco MM, Salomone VN, dos Santos Afonso M (2021) *Ulva lactuca*: a bioindicator for anthropogenic contamination and its environmental remediation capacity. Mar Environ Res 171:105468

Baggett LP, Powers SP, Brumbaugh R, Coen LD, DeAngelis B, Green J, Hancock B, Morlock S (2014) Oyster habitat restoration monitoring and assessment handbook. The Nature Conservancy, Arlington, p 96. Available at ▶ http://www.oyster-restoration.org/oyster-habitat-restoration-monitoring-and-assessment-handbook/. Accessed 12 Oct 2021

Baraktarov E, Saunders MI, Abdullah S, Mills M, Beher J, Possingham HP, Mumby PJ, Lovelock CE (2016) The cost and feasibility of marine coastal restoration. Ecol Appl 26(4):1055–1074

Basconi L, Cadier C, Guerrero-Limón G (2020) Challenges in marine restoration ecology: how techniques, assessment, metrics and ecosystem valuation can lead to improved restoration success. In: Jungblut S, Liebuch V, Bode-Dalby M (eds) YOUMARES

9 -the oceans: our research, our future -proceedings of the 2018 conference for the YOUng MArine RESearcher in Oldenburg, Germany. Springer, Cham, pp 83–100

Baums IB (2008) A restoration genetics guide for coral reef conservation. Mol Ecol 17(12):2796–2811

Beck MW, Brumbaugh RD, Airoldi L, Carranza A, Coen LD, Crawford C, Defeo O, Edgar GE, Hancock B, Kay MC, Lenihan HS, Luckenbach MW, Totopova CL, Zhang G, Guo X (2011) Oyster reefs at risk and recommendations for conservation, restoration, and management. Bioscience 61(2):107–116

Benthotage C, Cole VJ, Schulz KG, Benkendorff K (2020) A review of the biology of the genus *Isognomon* (Bivalvia; Pteriidae) with a discussion on shellfish reef restoration potential of *Isognomon ephippium*. Molluscan Res 40(4):286–307

Bersoza Hernández A, Brumbaugh RD, Frederick P, Grizzle R, Luckenbach MW, Peterson CH, Angelini C (2018) Restoring the eastern oyster: how much progress has been made in 53 years? Front Ecol Environ 16(8):463–471

Brinza L, Dring M, Gavrilescu M (2007) Marine micro and macro algal species as biosorbents for heavy metals. Environ Eng Manag J 6(3):237–251

Boström-Einarsson L, Babcock RC, Bayraktarov E, Ceccarelli D, Cook N, Ferse SC, Hancock B, Harrison P, Hein M, Shaver E, Smith A, Suggett D, Stweart-Simclair PJ, Vardi T, McLeod IM (2020) Coral restoration -a systematic review of current methods, successes, failures and future directions. PLoS ONE 15(1):e0226631

Burke L, Reytar K, Spalding M, Perry A (2011) Reefs at risk revisited. World Resource Institute, Washington DC, p 114. Available at: ▶ https://www.wri.org/research/reefs-risk-revisited. Accessed 11 Feb 2022

Burchett MD, Allen C, Pulkownik A, MacFarlane G (1998) Rehabilitation of saline wetland, Olympic 2000 site, Sydney (Australia). II: saltmarsh transplantation trials and application. Mar Pollut Bull 37(8–12):526–534

Bucher D, Saenger P (1991) An inventory of Australian estuaries and enclosed marine waters: an overview of results. Aust Geogr Stud 29:370–381

Carroll JM, Kelly JL, Treible LM, Bliss T (2021) Submarine groundwater discharge as a potential driver of eastern oyster, *Crassostrea virginica*, populations in Georgia: effects of groundwater on oysters. Mar Environ Res 170:105440

Carstensen MV, Hashemi F, Hoffman CC, Zak D, Audet J, Kronvang B (2020) Efficiency of mitigation measures targeting nutrient losses from agricultural drainage systems: a review. Ambio 49(11):1820–1837

Chamberland VF, Petersen D, Guest JR, Petersen U, Brittsan M, Vermeij MJ (2017) New seeding approach reduces costs and time to outplant sexually propagated corals for reef restoration. Sci Rep 7:1–12

Chan WY, Peplow LM, Menendez P, Hoffmann AA, Van Oppen MJ (2018) Interspecific hybridization may provide novel opportunities for coral reef restoration. Front Mar Sci 5:00160

Chou E, Southall BL, Robards M, Rosenbaum HC (2021) International policy, recommendations, actions and mitigation efforts of anthropogenic underwater noise. Ocean Coast Manag 202:105426

Coen LD, Brumbaugh RD, Bushek D, Grizzle R, Luckenbach MW, Posey MH, Powers SP, Tolley SG (2007) Ecosystem services related to oyster restoration. Mar Ecol Prog Ser 341:303–307

Cook FJ, Knight JH, Silburn DM, Kookana RS, Thorburn PJ (2013) Upscaling from paddocks to catchments of pesticide mass and concentration in runoff. Agr Ecosyst Environ 180:136–147

Creighton C, Waterhouse J, Brodie J (2021) Criteria for effective regional scale catchment to reef management: a case study of Australia's great barrier reef. Mar Pollut Bull 173:112882

Daly J, Hobbs RJ, Zuchowicz N, O'Brien JK, Bouwmeester J, Bay L, Quigley K, Hagedorn M (2022) Cryopreservation can assist gene flow on the Great Barrier Reef. Coral Reefs 41:455–462

dela Cruz DWD, Villanueva RD, Baria MVB (2014) Community-based, low-tech method of restoring a lost thicket of *Acropora* corals. ICES J Mar Sci 71(7):1866–1875

dela Cruz DW, Harrison PL (2017) Enhanced larval supply and recruitment can replenish reef corals on degraded reefs. Sci Rep 7:13985

dela Cruz DW, Harrison PL (2020) Enhancing coral recruitment through assisted mass settlement of cultured coral larvae. PLoS One 15(11):e0242847

Edwards AJ, Gomez ED (2007) Reef restoration concepts and guidelines: making sensible management choices in the face of uncertainty. Coral Reef targeted research and capacity building for management programme. St. Luica, Australia, p 38. Available at: ▶ http://www.reefbase.org/pacific/pub_A0000003675.aspx. Accessed 11 Feb 2022

Edwards AJ (2010) Reef rehabilitation manual. The coral reef targeted research & capacity building for management program. St Lucia, Australia, p 166. Available at: ▶ https://pipap.sprep.org/content/reef-rehabilitation-manual. Accessed 11 Feb 2022

Edwards PET, Sutton-Grier AE, Coyle GE (2013) Investing in nature: restoring coastal habitat blue infrastructure and green job creation. Mar Policy 38:65–71

Edwards AJ, Guest JR, Heyward AJ, Villanueva RD, Baria MV, Bollozos IS, Golbuu Y (2015) Direct seeding of mass-cultured coral larvae is not an effective option for reef rehabilitation. Mar Ecol Prog Ser 525:105–116

Ellison AM, Felson AJ, Friess DA (2020) Mangrove rehabilitation and restoration as experimental adaptive management. Front Mar Sci 7:327

Elliot M, Burdon D, Hemmingway KL, Apitz SE (2007) Estuarine, coastal and marine ecosystem restoration: confusing management and science -a review of concepts. Estuar Coast Shelf Sci 74:349–366

Ewere EE, Reichelt-Brushett A, Benkendorff K (2020) The neonicotinoid insecticide imidacloprid, but not salinity, impacts the immune system of Sydney rock oyster, *Saccostrea glomerata*. Sci Total Environ 742:140538

Feliciano GNR, Mostrales TPI, Acosta AKM, Luzon K, Bangsal JCA, Licuanan WY (2018) Is gardening corals of opportunity the appropriate response to reverse Philippine reef decline? Restor Ecol 26(6):1091–1097

Fell PE, Murphy KA, Peck MA, Recchia ML (1991) Reestablishment of *Melampus-Bidentatus* (Say) and other macroinvertebrates on a restored impounded tidal marsh—comparison of populations above and below the impoundment dike. J Exp Mar Biol Ecol 152(1):33–48

Ferse SCA, Hein MY, Rölfer L (2021) A survey of current trends and suggested future directions in coral transplantation for reef restoration. PLoS One 16:e0249966

Field CD (1998) Rehabilitation of mangrove ecosystems: an overview. Mar Pollut Bull 37(8–12):383–392

Fitzsimons JA, Branigan S, Gillies CL, Brumbaugh RD, Cheng J, DeAnelis BM, Geselbracht L, Hancock B, Jeffs A, McDonald IM, Pofoda B, Theuerkauf SJ, Thomas M, Westby S, Zu Ermgassen PSE (2020) Restoring shellfish reefs: global guidelines for practitioners and scientists. Conserv Sci Pract 2:e198

Fonseca MS, Meyer DL, Hall MO (1996) Development of planted seagrass beds in Tampa Bay, Florida, USA. II. Faunal components. Mar Ecol Progr Ser 132(1–3):141–156

Ford JR, Hamer P (2016) The forgotten shellfish reefs of coastal Victoria: documenting the loss of a marine ecosystem over 200 years since European settlement. Proc R Soc Victoria 128:87–105

Frenkel RE, Morlan JC (1991) Can we restore our saltmarshes? Lessons from the Salmon River, Oregon. Northwest Environ J 7:119–135

Friess DA (2017) Mangrove rehabilitation along urban coastlines: a Singapore case study. Reg Stud Mar Sci 16:279–289

Fulke AB, Kotian A, Giripunje MD (2020) Marine microbial response to heavy metals: mechanism, implications and future prospect. Bull Environ Contam Toxicol 105:182–197

Gann GD, McDonald T, Walder B, Aronson J, Nelson CR, Jonson J, Hallett JG, Eisenberg C, Guariguata MR, Liu J, Hua F, Echeverría C, Gonzales E, Shaw N, Decleer K, Dixon KW (2019) International principles and standards for the practice of ecological restoration, 2nd edn. Ecol Restor 27(1):S1–S46

Geist J, Hawkins S (2016) Habitat recovery and restoration in aquatic ecosystems: current progress and future challenges. Aquat Conserv Mar Freshwat Ecosyst 26:942–962

Gillies CL, McLead IM, Alleway HK, Cook P, Crawford C, Creighton C, Diggles B, Ford J, Hamer P, Heller-Wanger G, Lebrault E, Le Port A, Russell K, Sheaves M, Warnock B (2018) Australian shellfish ecosystems: past distribution, current status and future direction. PLoS One 13:e0190914

González-Jartín JM, de Castro Alves L, Alfonso A, Piñeiro Y, Yáñez Vilar S, Rodríguez I, González Gomez M, Vargas Osorio Z, Sainz M, Vieytes MR, Rivas J, Botana LM (2020) Magnetic nanostructures for marine and freshwater toxins removal. Chemosphere 256:127019

Gouezo M, Fabricius K, Harrison P, Golbuu Y, Doropoulos C (2021) Optimizing coral reef recovery with context-specific management actions at prioritized reefs. J Environ Manage 295:113209

Green EP, Short FT (2003) World atlas of seagrasses, Berkley, University of California Press, pp 5–26. Available at ▶ https://archive.org/details/worldatlasofseag03gree/page/n5/mode/2up. Accessed 18 Jan 2022

Green J, Reichelt-Brushett A, Jacobs SWL (2009a) Re-establishing a saltmarsh vegetation structure in a changing climate. Ecol Manag Restor 10(1):20–30

Green J, Reichelt-Brushett A, Jacobs SWL (2009b) Investigating gastropod habitat associations in saltmarsh. Wetlands (Australia) 25(1):25–37

Green J, Reichelt-Brushett A, Brushett D, Squires P, Brooks L, Jacobs SWL (2010) An investigation of soil algal abundance using chlorophyll a in a subtropical saltmarsh after surface restoration. Wetlands 30:87–98

Guest JR, Baria MV, Gomez ED, Heyward AJ, Edwards AJ (2014) Closing the circle: is it feasible to rehabilitate reefs with sexually propagated corals? Coral Reefs 33:45–55

Grabowski JH, Brumbaugh RD, Conrad RF, Keeler AG, Opaluch JJ, Peterson CH, Piehler MF, Powers SP, Smyth AR (2012) Economic valuation of ecosystem services provided by oyster reefs. Bioscience 62(10):900–909

Hamilton SE, Casey D (2016) Creation of a high spatio-temporal resolution global database of continuous mangrove forest cover for the 21st century (CGMFC-21). Glob Ecol Biogeogr 25:729–738

Harrison PL, Wallace CC (1990) Reproduction, dispersal and recruitment of scleractinian corals. In: Dubinsky Z (ed) Ecosystems of the world, 25. Coral Reefs. Elsevier Science Publishing Company, Inc., Amsterdam, pp 133–207

Harrison PL, Villanueva R, dela Cruz DW (2016) Coral reef restoration using mass coral larval reseeding. Final report to Australian Centre for International Agricultural Research, Project SRA FIS/2011/031, June 2016, P 60

Harrison PL, dela Cruz DW, Cameron KA, Cabaitan PC (2021) Increased coral larval supply enhances recruitment for coral and fish habitat restoration. Front Mar Sci 8:750210

15

Harrison PL (2011) Sexual reproduciton of scleractinian corals. In: Dubinsky Z, Stambler N (eds) Coral reefs: an ecosystem in transition. Springer, New York, pp 59–86

Harrison PL (2021) More sex on the reef: can coral spawning help save reefs? Feature Article. Ocean Geogr Soc 56:25–33

Hawkins SJ, Gibbs PE, Pope ND, Burt GR, Chesman BS, Bray S, Proud SV, Spence SK, Southward AJ, Langston WJ (2002) Recovery of polluted ecosystems: the case for long-term studies. Mar Environ Res 54:215–222

Heyward AJ, Smith LD, Rees M, Field SN (2002) Enhancement of coral recruitment by in situ mass culture of coral larvae. Mar Ecol Prog Ser 230:113–118

Hogan S, Reidenbach MA (2022) Quantifying tradeoffs in ecosystem services under various oyster reef restoration designs. Estuar Coasts 45:677–690

Howie AH, Bishop MJ (2021) Contemporary oyster reef restoration: responding to a changing world. Front Ecol Evol 9:689915

Howlett L, Camp EF, Edmondson J, Henderson N, Suggett DJ (2021) Coral growth, survivorship and return-on-effort within nurseries at high-value sites on the Great Barrier Reef. PLoS One 16(1):e0244961

Hsu C-Y, Yan G-E, Pan K-C, Lee K-C (2021) Constructed wetlands as a landscape management practice for nutrient removal from agricultural runoff -a local practice case on the east coast of Taiwan. Water (Switzerland) 13(21):2973

Hunsucker K, Melnikov A, Gilligan M, Gardner H, Erdogan C, Weaver R, Swain G (2021) Cathodically protected steel as an alternative to plastic for oyster restoration mats. Ecol Eng 164:106210

Jenkins S, Quer AM, Fonseca C, Varrone C (2019) Microbial degradation of plastics: new plastic degraders, mixed cultures and engineering strategies. In: Kumar J, Batool R (eds) Soil microenvironment for bioremediation and polymer reduction. Scrivener Publishing, Beverly, pp 213–238

Kennedy VS, Breitburg DL, Christman MC, Luckenbach MW, Paynter K, Kramer J, Sellner KG, Dew-Baxter J, Keller C, Mann R (2011) Lessons learned from efforts to restore oyster populations in Virginia and Maryland, 1990 to 2007. J Shellfish Res 30:1–13

Lang Z, Zhou M, Zhang Q, Yin X, Li Y (2020) Comprehensive treatment of marine aquaculture wastewater by a cost-effective flow-through electro-oxidation process. Sci Total Environ 722:137812

Laegdsgaard P (2006) Ecology, disturbance and restoration of coastal saltmarsh in Australia: a review. Wetland Ecol Manag 14:379–399

Lee MKK (2021) Plastic pollution mitigation net plastic circularity through a standardized credit system in Asia. Coast Ocean Manag 210:105733

Li L, Zuo J, Duan X, Wang S, Hu K, Chang R (2021) Impacts and mitigation measures of plastic waste: a critical review. Environ Impact Assess Rev 90:106642

Liñán-Rico V, Cruz-Ramírez A, Michel-Morfín JE, Liñán-Cabello MA (2019) Rescue and relocation of benthic organisms during an urban-port development project: Port of Manzanillo, case study. Aquat Conserv Mar Freshwat Ecosyst 30(6):1137–1148

MacDonald T, Gann GD, Jonson J, Dixon KW (2016) International standards for the practice of ecological restoration—including principles and key concepts. Society for ecological restoration. Available at: ▶ https://cdn.ymaws.com/sites/www.ser.org/resource/resmgr/docs/SER_International_Standards.pdf. Accessed 15 Jan 2022

MacDonnell C, Tiling K, Encomio V, van der Heide T, Teunis M, Wouter L, Didderen K, Bouma TJ, Inglett PW (2022) Evaluating a novel biodegradable lattice structure for subtropical seagrass restoration. Aquat Bot 176:103463

Marella TK, Saxena A, Tiwari A (2020) Marine bioremediators Diatom mediated heavy metal remediation: a review. Biores Technol 350:123068

Mazzoccoli M, Altosole M, Vigna V, Bosio B, Arato E (2020) Marine pollution mitigation by waste oil recycling onboard ships: technical feasibility and need for new policy and regulations. Front Mar Sci 7:566363

Mcowen CJ, Weatherdon LV, Van Bochove J-W, Sullivan E, Blyth S, Zockler C, Stanwell-Smith D, Kingston N, Martin CS, Spalding M, Fletcher S (2017) A global map of saltmarshes. Biodivers Data J 21(5):e11764

Michalak I (2020) Chapter four -the application of seaweeds in environmental biotechnology. Adv Bot Res 95:85–111

Mintenig SM, Int-Veen I, Löder MGJ, Primpke S, Gerdts G (2016) Identification of microplastic in effluents of waste water treatment plants using focal plane array-based micro-Fourier-transform infrared imaging. Water Res 108:365–372

Moritsch MM, Young M, Carnell P, Macreadie PI, Lovelock C, Nicholson E, Raimondi PT, Wedding LM, Ierodiaconou D (2021) Estimating blue carbon sequestration under coastal management scenarios. Sci Total Environ 777:145962

Mudhoo A, Garg VK, Wang S (2012) Removal of heavy metals by biosorption. Environ Chem Lett 10:109–117

Murtough G, Aretino B, Matysek A (2002) Creating markets for ecosystem services. AusInfo, Canberra, Australia. Available at: ▶ https://www.pc.gov.au/research/supporting/ecosystem-services. Accessed 15 Jan 2022

Nadiarti N, La Nafie YA, Priosambodo D, Umar MT, Rahim SW, Inaku DF, Musfirah NH, Paberu DA, Moore AM (2021) Restored seagrass beds support macroalgae and sea urchin communities. In: 4th international symposium on marine science and fisheries, vol 860, p 012014

NRC (National Research Council) (1994) Restoring and protecting marine habitat: the role of engineering and technology. The National Academics Press, Washington, p 212. Available at: ▶ http://nap.edu/2213. Accessed 19 Nov 2021

Occhipinti-Ambrogi A (2021) Biopollution by invasive marine non-indigenous species: a review of potential adverse ecological effects in a changing climate. Int J Environ Res Public Health 18:4268

O'Conner E, Hynes S, Chen W (2020) Estimating the non-market benefit value of deep-sea ecosystem restoration: evidence from a contingent valuation study of the Dohrn Canyon in the Bay of Naples. J Environ Manage 275:1111180

Ogunola OS, Onada OA, Falaye AE (2018) Mitigation measures to avert the impacts of plastics and microplastics in the marine environment (a review). Environ Sci Pollut Res 25:9293–9310

Omori M (2019) Coral restoration research and technical developments: what we have learned so far. Mar Biol Res 15(7):377–409

OSPAR (1999) OSPAR guidelines on artificial reefs in relation to living marine resources. OSPAR Commission, London. OSPAR 99/15/1-E, Annex 6. Available at ▶ https://www.miteco.gob.es/es/costas/temas/proteccion-medio-marino/OSPAR_Artificial%20Reefs%20Guidelines_tcm30-157010.pdf. Accessed 18 Jan 2022

Panagopoulos A, Haralambous K-J (2020) Environmental impacts of desalination and brine treatment -challenges and mitigation measures. Mar Pollut Bull 161:111773

Pandolfi JM, Bradbury RH, Sala E, Hughes TP, Bjorndal KA, Cooke RG, McArdle D, McClenachan L, Newman MJH, Paredes G, Warner RR, Jackson JBC (2003) Global trajectories of the long-term decline coral reef ecosystems. Science 301:955–958

Prata JC (2018) Microplastics in wastewater: state of the knowledge on sources, fate and solutions. Mar Pollut Bull 129:262–265

Pollack JB, Palmer TA, Williams AE (2021) Medium-term monitoring reveals effects of El Niño Southern oscillation climate variability on local salinity and faunal dynamics on a restored oyster reef. PLoS ONE 16:e0255931

Quigley KM, Randall CJ, Van Oppen MJH, Bay LK (2020) Assessing the role of historical temperature regime and algal symbionts on the heat tolerance of coral juveniles. Biol Open 9:bio047316

Randall CJ, Negri AP, Quigley KM, Foster T, Ricardo GF, Webster NS, Bay LK, Harrison PL, Babcock RC, Heyward AJ (2020) Sexual production of corals for reef restoration in th Anthropocene. Mar Ecol Prog Ser 635:203–232

Richardson AMM, Swain R, Wong V (1997) The crustacean and molluscan fauna of Tasmanian saltmarshes. Proc R Soc Tasmania 131(21):30

Rinkevich B (2021) Augmenting coral adaptation to climate change via coral gardening (the nursery phase). J Environ Manage 291:112727

Sampaio FDF, Freire CA, Sampaio TVM, Vitule JRS, Fávaro LF (2015) The precautionary principle and its approach to risk analysis and quarantine related to the trade of marine ornamental fishes in Brazil. Mar Policy 51:163–168

Schernewski G, Radtke H, Hauk R, Baresel C, Olshammar M, Osinski R, Oberbeckmann S (2020) Transport and behavior of microplastics emissions from urban sources in the Baltic Sea. Front Mar Sci 8:579361

Seaman W (2007) Artificial habitats and the restoration of degraded marine ecosystems and fisheries. Hydrobiologia 580(1):143–155

Shaish L, Levy G, Gomez E, Rinkevich B (2008) Fixed and suspended coral nurseries in the Philippines: establishing the first step in the "gardening concept" of reef restoration. J Exp Mar Biol Ecol 358:86–97

Sheaves M, Waltham NJ, Benham C, Bradley M, Mattone C, Deidrich A, Sheaves J, Sheaves A, Hernandez S, Dale P, Banhalmi-Zakar Z, Newlands M (2021) Restoration of marine ecosystems: understanding possible futures for optimal outcomes. Sci Total Environ 796:148845

Short FT, Wyllie-Echeverria S (1996) Natural and human-induced disturbance of seagrasses. Environ Conserv 23(1):17–27

Sinicrope TL, Hine PG, Warren RS, Niering WA (1990) Restoration of an impounded salt marsh in New England. Estuaries 13(1):25–30

SER (Society for Ecological Restoration) (2021) Our impact. Available at: ▶ https://www.ser.org/page/Impact. Accessed 19 Nov 2021

Sridharan S, Kumar M, Bolan NS, Singh L, Kumar S, Kumar R, You S (2021) Are microplastics destabilizing the global network of terrestrial and aquatic ecosystem services? Environ Res 198:111243

Star M, Rolfe J, McCosker K, Smith R, Ellis R, Waters D, Waterhouse J (2018) Targeting for pollutant reductions in the Great Barrier Reef river catchments. Environ Sci Policy 89:365–377

Stuart-Smith SJ, Jepson PD (2017) Persistent threats need persistent counteraction: responding to PCB pollution in marine mammals. Mar Policy 84:69–75

Theuerkauf SJ, Lipcius RN (2016) Quantitative validation of a habitat suitability index for oyster restoration. Front Mar Sci 3:00064

Thorhaug A, Belaire C, Verduin JJ, Schwarz A, Kiswara W, Prathep A, Gallagher JB, Huang XP, Berlyn G, Yap T-K, Dorward S (2020) Longevity and sustainability of tropical and subtropical restored seagrass beds among Atlantic, Pacific, and Indian Oceans. Mar Pollut Bull 160:111544

Uddin MJ, Smith KJ, Hargis CW (2021) Development of pervious oyster shell habitat (POSH) concrete for reef restoration and living shorelines. Constr Build Mater 295:123685

UN (United Nations) (2017) 2017 UN world water development report, wastewater: the untapped resource. Available at: ▶ http://www.unesco.org/new/en/natural-sciences/environment/water/wwap/wwdr/2017-wastewater-the-untapped-resource/. Accessed 15 Jan 2022

UN (United Nations) (2021) World economic situation and prospects as of mid-2021. Available at ▶ https://www.un.org/development/desa/dpad/publication/world-economic-situation-and-prospects-as-of-mid-2021/. Accessed 21 Nov 2021

Vakili S, Ölçer AI, Ballini F (2021) The development of a transdisciplinary policy framework for shipping companies to mitigate underwater noise pollution from commercial vessels. Mar Pollut Bull 171:112687

van Oppen MJH, Gates RD, Blackall LL, Cantin N, Chakravarti LJ, Chan WY, Cormick C, Crean A, Damjanovic K, Epstein H, Harrison PL, Jones TA, Miller M, Pears RJ, Peplow LM, Raftos DA, Schaffelke B, Stewart K, Torda G, Wachenfeld D, Weeks AR, Putnam HM (2017) Shifting paradigms in restoration of the world's coral reefs. Glob Change Biol 23(9):3437–3448

Vilas MP, Shaw M, Rohde K, Power B, Donaldson S, Foley J, Silburn M (2022) Ten years of monitoring dissolved inorganic nitrogen in runoff from sugarcane informs development of a modelling algorithm to prioritise organic and inorganic nutrient management. Sci Total Environ 803:150019

Waltham NJ, Wegscheidl C, Volders A, Smart JCR, Hasan S, Lédée E, Waterhouse J (2021) Land use conversion to improve water quality in high DIN risk, low-lying sugarcane areas of the Great Barrier Reef catchments. Mar Pollut Bull 167:112373

Waterhouse J, Brodie J, Lewis S, Mitchell A (2012) Quantifying the sources of pollutants in the Great Barrier Reef catchments and the relative risk to reef ecosystems. Mar Pollut Bull 65(4–9):394–406

Waycott M, Duarte CM, Carruthers TJB, Orth RJ, Dennison WC, Olyarnik S, Calladine A, Fourqurean JW, Heck KL Jr, Hughes AR, Kendrick GA, Kenworthy J, Short FT, Williams SL (2009) Accelerating loss of seagrasses across the globe threatens coastal ecosystems. Proc Natl Acad Sci USA 106(30):12377–12381

Wei R, Wierckx N (2021) Editorial: microbial degradation of plastics. Front Microbiol 12:635621

Williams SL, Davis CA (1996) Population genetic analyses of transplanted eelgrass (Zostera marina) beds reveal reduced genetic diversity in southern California. Restor Ecol 4(2):163–180

Woodhead AJ, Hicks CC, Norström AV, Williams GJ, Graham NA (2019) Coral reef ecosystem services in the Anthropocene. Funct Ecol 33:1023–1034

Young CN, Schopmeyer SA, Lirman D (2012) A review of reef restoration and coral propagation using the threatened genus Acropora in the Caribbean and Western Atlantic. Bull Mar Sci 88:1075–1098

Zenone A, Pipitone C, D'anna G, La Porta B, Bacci T, Bertasi F, Bulleri C, Cacciuni A, Calvo S, Conconi S, Flavia Gravina M, Mancusi C, Piazzi A, Targusi M, Tomasello A, Badalamenti F (2021) Stakeholders' attitudes about the transplantations of the mediterranean seagrass Posidonia oceanica as a habitat restoration measure after anthropogenic impacts: a Q methodology approach. Sustainability (Switzerland) 13(21):12216

Zhang K, Bach PM, Mathios J, Dotto CBS, Deletic A (2020) Quantifying the benefits of stormwater harvesting for pollution mitigation. Water Res 171:115395

Zhou Y, Kumar M, Sarsaiya S, Sirohi R, Awasthi SK, Sindhu R, Binod P, Pandey A, Bolan NS, Zhang Z, Singh L, Kumar S (2022) Challenges and opportunities in bioremediation of micro-nano plastics: a review. Sci Total Environ 802:149823

Zockler C, Stanwell-Smith D, Kingston N, Martin CS, Spalding M, Fletcher S (2017) A global map of saltmarshes. Biodivers Data J 5:e11764

15

Regulation, Legislation and Policy—An International Perspective

Edward Kleverlaan and Amanda Reichelt-Brushett

Contents

16.1 **Introduction** – 341

16.2 **The Global Setting** – 342
16.2.1 Global Regulatory Structure of Marine Pollution – 342

16.3 **Shipping** – 345
16.3.1 The International Maritime Organization (IMO) – 345
16.3.2 Hierarchy of Legalization and Responsibilities – 345
16.3.3 Benefits of IMO Responsibility to Prevent Marine Pollution – 348
16.3.4 Limitations of the IMO and the London Convention and London Protocol – 349

16.4 **Other Global Instruments that Relate to Marine Pollution** – 352
16.4.1 The Paris Agreement – 352
16.4.2 Other Conventions – 352
16.4.3 The International Seabed Authority – 354
16.4.4 International Atomic Energy Agency – 354
16.4.5 Convention of Biological Diversity – 354
16.4.6 Global Legislation on Plastic Waste? – 354
16.4.7 The Precautionary Principle – 355

16.5 **Summary** – 355

16.6 **Study Questions and Activities** – 356

References – 356

© The Author(s) 2023
A. Reichelt-Brushett (ed.), *Marine Pollution—Monitoring, Management and Mitigation*,
Springer Textbooks in Earth Sciences, Geography and Environment,
https://doi.org/10.1007/978-3-031-10127-4_16

Abbreviations

ABIDJAN	Convention for Cooperation in the Protection, Management and Development of the Marine and Coastal Environment of the Atlantic Coast of the West and Central Africa Region
AHEG	Ad Hoc Open-Ended Expert Group on Marine Litter and Microplastics
AFS	Anti-Fouling Systems Convention, 2001
AOA	Assessment of Assessments
BAMAKO	African regional treaty equivalent to the Basel Convention
BARCELONA	Convention for the Protection of the Mediterranean Sea
BWM	Ballast Water Management Convention, 2004
BASEL	Control of Transboundary Movement of Hazardous Wastes and Their Disposal, 1989
BUCHAREST	Convention on the Protection of the Black Sea Against Pollution
CARTAGENA	Convention for the Protection and Development of the Marine Environment of the Wider Caribbean Region
CBD	Convention on Biological Diversity
CCAMLR	Convention for the Conservation of Antarctic Marine Living Resources
CITES	Convention on the International Trade in Endangered Species of Wild Fauna and Flora
CMS	Bonn Convention on Migratory Species
COLREGS	Convention on the International Regulations for Preventing Collisions at Sea, 1972
CPPS	Permanent Commission for the South Pacific
DDT	Dichlorodiphenyltrichloroethane
EAST ASIA SEAS	Coordinating Body on the Seas of East Asia (COBSEA) (▶ https://www.unenvironment.org/cobsea/)
EEZ	Exclusive Economic Zone (defined in UNCLOS)
FAO	Food and Agriculture Organization of the United Nations
GESAMP	Joint Group of Experts on Scientific Aspects of Marine Environmental Protection
GPA	Global Programme of Action for the Protection of the Sea from Land-based Activities (of UNEP)
HELCOM	Helsinki Commission
HELSINKI	Convention on the Protection of the Marine Environment of the Baltic Sea Area
IAEA	The International Atomic Energy Agency
IOC	Intergovernmental Oceanographic Commission of UNESCO
IMO	International Maritime Organization
ITOPF	International Tanker Owners Pollution Federation
ISA	International Seabed Authority of the UN
London Convention	International Convention on the Prevention of Marine Pollution by Dumping of Wastes and Other Matter, 1972
London Protocol	Protocol to the Convention on the Prevention of Marine Pollution by Dumping of Wastes and Other Matter, 1996
LC/LP	London Convention and Protocol
LME	Large Marine Ecosystem
MARPOL	International Convention for the Prevention of Pollution from Ships
MEPC	Marine Environment Protection Committee (of IMO)
NAIROBI	Nairobi Convention for the Protection, Management and Development of the Marine and Coastal Environment of the Western Indian Ocean (▶ https://www.unenvironment.org/nairobiconvention/)
NDC	Nationally determined contributions
NOUMEA	Noumea Convention for the Protection, Management and Development of the Marine and Coastal Environment of the Pacific
OSPAR	Convention for the Protection of the Marine Environment of the North-East Atlantic
OPRC/HNS	The International Convention on Oil Preparedness, Response and Cooperation (1990)/ Hazardous Noxious Substances Protocol (1996)
PERSGA	Regional Organization for the Protection of the Red Sea and Gulf of Aden
PCBs	Polychlorinated Biphenyls

POPs	Persistent Organic Pollutants
ROPME	Regional Organization for the Protection of the Marine Environment (Persian Gulf)
RSP	UNEP Regional Seas Programme
SOLAS	International Convention for the Safety of Life at Sea
SPREP	Secretariat of the Pacific Regional Environment Programme
STOCKHOLM	Convention on Persistent Organic Pollutants
STWC	International Convention on Standards of Training, Certification and Watchkeeping for Seafarers 1978
UN	United Nations
UNCED	United Nations Conference on Environment and Development
UNCLOS	United Nations Convention of the Law of the Sea
UNCTAD	United Nations Conference on Trade and Development
UNEA	United Nations Environmental Assembly
UNEP	United Nations Environment Programme
UNESCO	United Nations Education, Scientific and Cultural Organization
UNFCCC	United Nations Framework Convention on Climate Change
WOA I	World Ocean Assessment I
WOA II	World Ocean Assessment II

16.1 Introduction

In accordance with Part XII of the **United Nations Law of the Sea Convention (UNCLOS)** and various other related international agreements, parties are obliged to **prevent**, **reduce** and **control** pollution of the marine environment. The **responsibility** to implement these agreements or other non-regulatory codes or standards rests primarily on **national governments**. They in turn give effect to the agreements by developing domestic legislation in various forms and enforcing them within their fields of jurisdiction. The jurisdictional extent and scope of activities to which national governments can regulate is also defined by UNCLOS.

Implementing national legislation can be a very effective method to control a range of human activities in the marine environment; however, this is not necessarily the only method, as education and voluntary actions are essential components to achieve the desired outcomes and objectives of international agreements. From the discussion below, it can be noted that many stressors in the marine environment resulting from human activities actually come from land-based activities and therefore fall outside the UNCLOS framework. Nonetheless, legislation of activities such as shipping or waste disposal can have dramatic positive effects on marine pollution prevention and control. For example, in the 1960s, several major oil spill catastrophes focused on the minds of many actors in the marine environment and highlighted the need for stronger international rules to reduce and prevent such major devastations to marine and coastal environments. This led the **International Maritime Organization (IMO)**,

located in London, United Kingdom, to develop and subsequently adopt (by consensus) a number of international agreements, such as, the International Convention for the Prevention of Pollution from Ships, 1973, as modified by the Protocols of 1978 and 1997 relating thereto (MARPOL) that would apply to all ships. MARPOL was designed to cover all operational ship-generated pollution and ensured that ships are adequately equipped, certified and inspected by Contracting Governments. In subsequent years, modifications to MARPOL resulted in the phase out of single-hulled tankers and changes to other critical design and operational activities to achieve the dramatic reduction in oil spills from the early 1970s onwards. MARPOL in concert with other IMO agreements has made major inroads in tackling marine pollution arising from shipping. Currently, there are 160 Contracting Governments that enforce MARPOL through their legislative frameworks.

Similarly, with the adoption of the Convention on the Prevention of Marine Pollution by Dumping of Wastes and Other Matter, 1972 (**London Convention**), and later the Protocol to the Convention on the Prevention of Marine Pollution by Dumping of Wastes and Other Matter, 1996 (**London Protocol**), the dumping at sea of industrial and radioactive wastes is now prohibited, and only a few waste categories (such as dredge spoil) may be considered for dumping at sea following a stringent impact assessment and licensing process. This reversed a centuries-old practice that used the world's oceans as a dumping ground for wastes generated by people, with little thought given to the consequences of such actions. It was not until the 1960s that communities began to have an increased awareness of the impact of such reckless action on the marine environment, on seafood and on other living marine resources. As of 2023, there are 87 Contracting Parties to the London Convention who agree to enforce the regulations through their own legislative frameworks.

16.2   The Global Setting

Major pollutants of global concern have been identified in a number of studies and reports in recent years. The most authoritative global study that examined the state of knowledge of the world's ocean and the ways in which humans benefit from and impact it through, inter alia, direct and in-direct sources of marine pollution was completed in 2015 under the United Nations in the First Global Integrated Marine Assessment, 2016 **World Ocean Assessment I** (WOA I) (UN 2017). A good overview of hazardous pollutants may be found in Chapter 20—Coastal, Riverine and Atmospheric Inputs from Land of the WOA I. More recently, the **World Ocean Assessment II** (WOA II) (UN 2021) provides additional information and trends on important aspects of the ocean and more on its relationships with humans.

Another key source of information was the 2009 report by the Joint Group of Experts on the Scientific Aspects of Marine Environmental Protection (GESAMP 2009), which was prepared as a contribution to the Assessment of Assessments (AoA) start-up phase for the WOA I. This has since been updated by GESAMP in 2015 (GESAMP 2015) as part of the **Transboundary Water Assessment Programme** (TWAP) funded by the Global Environment Facility (GEF) and conducted through the Intergovernmental Oceanographic Commission (IOC) of the United Nations Educational, Scientific and Cultural Organization (UNESCO). The final TWAP deliverables released in July 2016, include a set of technical assessment reports for LMEs and Open Ocean as well as a summary for decision-makers, and a data portal where indicators can be visualized and data downloaded (► http://onesharedocean.org). A succinct overview of key pollutants and their status as a hazard is given in ◻ Table 16.1. The table also provides an indication of trends in environmental levels or loads of the contaminants and GESAMP's perspective regarding their relative, overall environmental significance.

Major source countries were identified through a study carried out by Lebreton et al. (2016) by synthesizing reports by national governments and expert opinion, as shown in ◻ Figure 16.1. The amounts shown were derived from a global model of plastic inputs from rivers into oceans based on waste management, population density and hydrological information. The model itself was calibrated against measurements available in the literature.

16.2.1   Global Regulatory Structure of Marine Pollution

In response to the wide range of threats to the marine environment, a global regulatory framework has been developed at different times to address pollution from a range of sources by global and regional arrangements complemented with local regulations.

The overall primary legal instrument to protect the marine environment is the United Nations Law of the Sea Convention (UNCLOS) and in particular Part XII, thereof, which was adopted in 1982. Further information on UNCLOS can be found at: ► http://www.un.org/depts/los/.

Following the increasing catastrophic oil spillages from ships, the large-scale incineration and disposal of industrial waste, as well as the disposal of radioactive wastes at sea between the 1950s and 1970s, calls were made to reduce and eliminate such accidents and damaging activities. The two global instruments mentioned above, MARPOL and the London Convention, together with a range of environmental policies were spawned following the United Nations Conference on the Human Environment, held in Stockholm, Sweden in 1972 (more information on the "Stockholm Conference" can be found at: ► http://www.un.org/ga/search/view_doc.asp?symbol=A/CONF.48/14/REV.1).

The Stockholm Conference was the UN's first major conference on international environmental issues and marked a turning point in the development of environmental politics. The United Nations Environment Programme (UNEP) was also established after the Conference and the UNEP Regional Seas Programme (RSP) was initiated in 1974 to:

» *"address the accelerating degradation of the world's oceans and coastal areas through a shared seas approach—namely, by engaging neighbouring countries in comprehensive and specific actions to protect their common marine environment. Today, more than 146 countries participate in 18 Regional Seas Conventions and Action Plans for the sustainable management and use of the marine and coastal environment. In most cases, the Action Plan is underpinned by a strong legal framework in the form of a regional Convention and associated Protocols on specific problems"*.

Other individual Conventions and Action Plans reflect a similar approach, yet each is tailored by its own gov-

Table 16.1 Major marine pollutants and trends. Adapted from: GESAMP 2015

Topic	Natural Occurrence	Human Impact	Demonstrable effects (from human inputs)	Trend Load	High status as hazard?
Oil	Y	Y++	Y	→	Y
Debris	N	Y++	Y	↗	Y
Radioactivity	Y	Y+	N	→	N
Carbon					
CO_2/ocean acidification	Y	Y+++	Y	↗	Y
POPs/PBTs	N	Y+++	Y	↗	Y
DDE	N	Y+++	Y	↓	N
Nutrients/metals					
Nitrogen	Y	Y+++	Y	↗	Y
Phosphorus	Y	Y+	N	↗	N
Iron (soluble)	Y	Y++	N	↗	Y
Lead	Y	Y++	N	↘	N
Copper	Y	Y++	Y	↗	Y
Other trace metals	Y	Y++	N	↗	
Mercury	Y	Y+++	Y	↗	Y
Noise	Y	Y+++	Y	↗	Y

Y=Yes N=No   + Low   ++Moderate   +++High

Confidence levels		
HIGH	MEDIUM	LOW

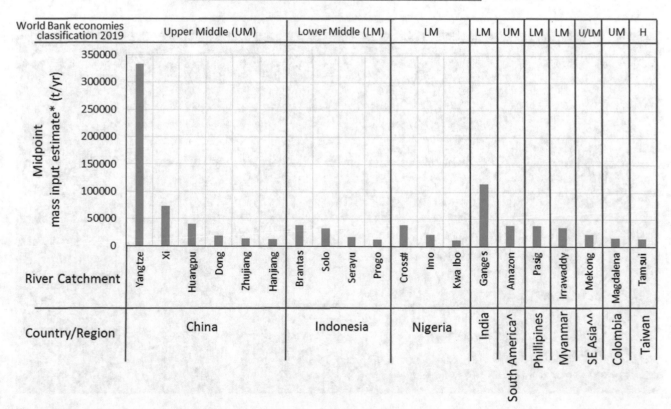

Figure 16.1 Top 20 river plastic emissions to the world's oceans showing that the most polluting rivers are from countries classified by the World Bank as upper- or lower-middle income economies. This suggests that while wealth and standard of living is reasonable investment on waste management and related infrastructure is limited *Data Sources*: Lebreton et al. (2016); # = also includes Cameroon; ^includes Brazil, Peru, Columbia, Ecuador; ^^ includes Thailand, Cambodia, Laos, China, Myanmar, Vietnam. *Image*: E. Kleverlaan

ernments and institutions to suit their particular environmental challenges. UN Environment coordinates the eleven UNEP Regional Seas Programme (11 RSPs), based at the Nairobi headquarters. A full list of the regional seas programme and related regional conventions with direct links to each website can be found at: ▶ https://www.unenvironment.org/explore-topics/oceans-seas/what-we-do/working-regional-seas/why-does-working-regional-seas-matter.

Shortly after the Stockholm Conference, several important marine species-related agreements were developed and adopted, including the Convention on the International Trade in Endangered Species of Wild Fauna and Flora (CITES 1973) and the Bonn Convention on Migratory Species (CMS 1979). The former monitors, regulates or bans trade in at-risk species with over 30,000 species protected and the CMS enables countries to make binding agreements to protect 120 migratory species. More details can be found at: ▶ https://www.cites.org/eng/disc/what.php and ▶ https://www.cms.int respectively.

Impacts to the marine environment from atmospheric inputs were recognized in the United Nations Framework Convention on Climate Change (UNFCCC 1992) (also see ▶ https://unfccc.int), which was adopted at the United Nations Conference on Environment and Development (UNCED), Earth Summit, held in 1992 at Rio de Janeiro in Brazil. Since then, through the work being taken under the auspices of the

Sustainable Development Goals (see ▶ https://sustainabledevelopment.un.org) and the Paris Agreement under the UNFCCC strong action on ocean acidification is being developed.

The activities being undertaken as a result of the adoption of the Convention on Biological Diversity (CBD) also adopted in 1992 at the Rio Conference (see: ▶ https://www.cbd.int/history/) have led to crucial work to protect marine areas as well as marine species. The CBD is the first international treaty to address all threats to biodiversity and ecosystem.

From the perspective of land-based sources of marine pollution, the Washington Declaration in 1995 led to the launching of the **Global Programme of Action (GPA)** to protect the marine environment from land-based sources of pollution. It was adopted by 108 governments and the European Union, and is the only global initiative to address terrestrial, freshwater, coastal and marine ecosystems. While not a legally binding treaty, it has launched crucial national programmes across the globe and via the regional seas conventions to address land-based sources of pollution. ◘ Figure 16.2 provides an illustrative overview of the relationship between various global and regional treaties or programmes that aim to protect the marine environment. Ultimately, all inputs, that are atmospheric, land- or sea-based pollutants, find their way to the oceans and seas.

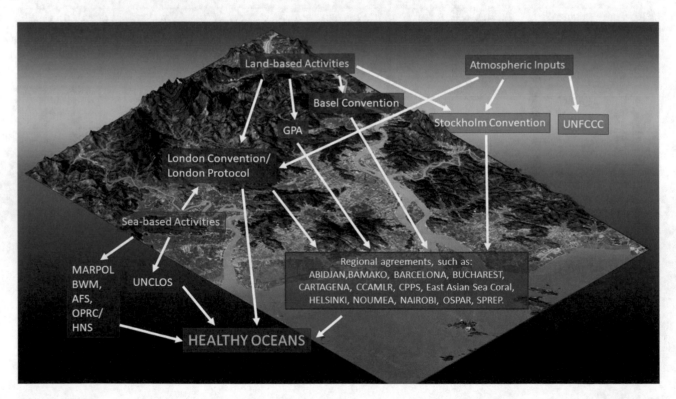

◘ **Figure 16.2** Context and overview of the various global and regional treaties of programmes aimed to protect the marine environment. *Image*: E. Kleverlaan, A. Reichelt-Brushett and K. Petersen

16.3 Shipping

In the following paragraphs, we will look more closely at the various components of the international legal framework to protect the marine environment.

16.3.1 The International Maritime Organization (IMO)

The mandate of the IMO, as a United Nations specialized agency, is to

» *"promote safe, secure, environmentally sound, efficient and sustainable shipping. This is accomplished by adopting the highest practicable standards of maritime safety and security and prevention and control of pollution from ships, as well as through consideration of the related legal matters and effective implementation of IMO's instruments with a view to their universal and uniform application".*

While IMO's original mandate was principally concerned with maritime safety, soon after it began in 1959, it also assumed responsibility for pollution issues and has since adopted a wide range of measures to prevent and control pollution caused by ships and mitigate the effects of any damage that may occur as a result of maritime operations. These include:

- The International Convention for the Prevention of Pollution from Ships, 1973, as modified by the Protocols of 1978 and 1997 relating thereto (MARPOL) that covers accidental and operational oil pollution as well as pollution by chemicals, goods in packaged form, sewage, garbage and air pollution (▶ https://www.imo.org/en/OurWork/Environment/Pages/Pollution-Prevention.aspx);
- The International Convention Relating to Intervention on the High Seas in Cases of Oil Pollution Casualties, 1969 affirms the right of a coastal State to take measures on the high seas to prevent, mitigate or eliminate danger to its coastline from a maritime casualty. The 1973 Protocol extends the Convention to cover noxious substances other than oil (▶ https://www.imo.org/en/About/Conventions/Pages/International-Convention-Relating-to-Intervention-on-the-High-Seas-in-Cases-of-Oil-Pollution-Casualties.aspx);
- The International Convention on Oil Pollution Preparedness, Response and Co-operation (OPRC), 1990 provides a global framework for international co-operation in combating major incidents or threats of marine pollution. The Protocol on Preparedness, Response and Co-operation to Pollution Incidents by Hazardous and Noxious Substances, 2000, covers marine pollution by hazardous and noxious substances (▶ https://www.imo.org/en/OurWork/Environment/Pages/Pollution-Response.aspx);
- The Convention on the Prevention of Marine Pollution by Dumping of Wastes and Other Matter, 1972, and the 1996 Protocol thereto, prohibit the dumping of hazardous materials and other wastes or matter in the sea (▶ https://www.imo.org/en/OurWork/Environment/Pages/London-Convention-Protocol.aspx);
- The International Convention on the Control of Harmful Anti-fouling Systems on Ships, 2001 (AFS) prohibits the use of harmful organotins in anti-fouling paints used on ships and established a mechanism to prevent the potential future use of other harmful substances in anti-fouling systems (▶ https://www.imo.org/en/OurWork/Environment/Pages/Anti-fouling.aspx);
- The International Convention for the Control and Management of Ships' Ballast Water and Sediments, 2004 (BWM) aims to prevent, minimize and ultimately eliminate the transfer of harmful aquatic organisms and in ships' ballast water and sediments (▶ https://www.imo.org/en/OurWork/Environment/Pages/BallastWaterManagement.aspx);
- The Hong Kong International Convention for the Safe and Environmentally Sound Recycling of Ships, 2009, when it enters into force, it will provide regulations on the design, construction, operation and preparation of ships so as to facilitate safe and environmentally sound recycling. This approach is to be taken without compromising ships' safety and operational efficiency

» *"the operation of ship recycling facilities in a safe and environmentally sound manner; and the establishment of an appropriate enforcement mechanism for ship recycling, incorporating certification and reporting requirements".*

(▶ https://www.imo.org/en/OurWork/Environment/Pages/Ship-Recycling.aspx); and
- In addition, a range of mandatory and voluntary Guidelines and Codes have been developed and adopted to provide international standards for the safe transport, storage and handling of harmful substances and for operating in sensitive areas (e.g. Polar Regions) (e.g. ▶ https://www.imo.org/en/MediaCentre/HotTopics/Pages/Polar-default.aspx and ▶ https://www.imo.org/en/MediaCentre/PressBriefings/Pages/02-Polar-Code.aspx).

16.3.2 Hierarchy of Legalization and Responsibilities

From an IMO perspective, being a global industry regulator, new or amendments to existing regulations originate from discussions based on a number of factors. These factors may include responding to an emergency (e.g. shipping incident or disaster), acting on an innovation to improve existing procedures or standards, alleviating a chronic issue such as ongoing pollution (invasive species transmission) or taking in to account a technological or knowledge advancement important to the industry or environment.

IMO will, following an internal process, involving all members of IMO and all NGO observers, adopt global treaties and guidelines at the intergovernmental level. All rules and standards are agreed by consensus; however, Member Governments and those that have officially signed and have agreed to be bound by the new regulations, are responsible for implementing and enforcing the adopted regulatory framework. Comprehensive Flag, Port and Coastal State enforcement mechanisms will be part of the agreements and IMO will oversee implementation through inter alia, a mandatory audit scheme of all Member Governments. IMO does not have a policing or enforcing mandate.

The precise manner by which, and the timing of, the entry into force of IMO international agreements comes into effect, is unique for each agreement. The protocols a State needs to follow are also laid out within each agreement.

A thorough overview of IMO's treaty-making process and the legal steps a State must follow to ratify or accede to a new treaty is set out under: ► https://www.imo.org/en/About/Conventions/Pages/Default.aspx.

◩ Figure 16.3 illustrates the steps a State can follow to be become legally bound to a treaty, such as the London Convention or the London Protocol. In both cases, it is important that States give effect to the obligations under the treaty, through domestic law. Each State usually follows a different approach in giving effect, varying from using the text of the agreement as the basis of its domestic law, or it may develop, following an analysis of benefits and costs (such as in Australia), domestic law as a new Act of Parliament, or by modifying existing legislation or adding new sub-ordinate legislation or regulations of an existing Act. For further details about the process in Australia see: ► https://www.legislation.gov.au/Home and ► https://www.ag.gov.au/Internationalrelations/InternationalLaw/Pages/default.aspx.

The following sections will focus on marine pollution prevention agreements such as MARPOL, the AFS Convention and the London Convention and London Protocol. While the London Convention and London Protocol are not directly related to pollution arising from shipping operations and address land-based wastes that are dumped at sea, from vessels, aircraft, platforms or other manmade structures, they are both administered by the IMO.

International Convention for the Prevention of Pollution from Ships (MARPOL)

Adoption: 1973 (Convention), 1978 (1978 Protocol), 1997 (Protocol—Annex VI); Entry into force: 2 October 1983 (Annexes I and II)

The MARPOL Convention is the main international convention covering prevention of pollution of the marine environment by ships from operational or accidental causes and currently has six separate annexes, which set out regulations dealing with pollution from ships by oil; by noxious liquid substances carried in bulk; harmful substances carried by sea in packaged form; sewage, garbage; and the prevention of air pollution from ships. MARPOL has been updated by amendments through the years and has laid the foun-

16

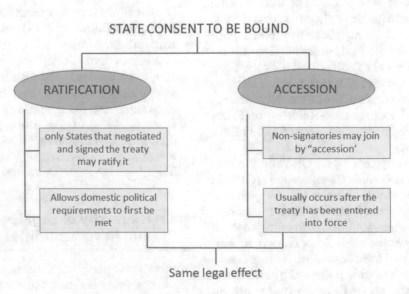

◩ **Figure 16.3**   Steps a State can follow to become legally bound to a treaty. *Image*: E. Kleverlaan and A. Reichelt-Brushett

dation for substantial and continued reductions in pollution from ships despite a considerable increase in world seaborne trade (◨ Table 16.2).

International Convention on the Control of Harmful Anti-fouling Systems on Ships, 2001 (AFS)
Adoption: 2001; Entry into force: 17/09/2008. Parties: 91 (95.93% World tonnage)

The AFS Convention is a good example of an international agreement that was agreed following scientific evidence that a particular chemical used by the shipping industry was highly detrimental to the marine environment and resulted in the banning of the use of organotins as an anti-fouling agent on ships.

The harmful environmental effects of organotin compounds were recognized by IMO in 1989. The IMO's Marine Environment Protection Committee (MEPC) adopted a resolution recommending Governments adopt measures to eliminate the use of antifouling paints *"containing TBT on non-aluminium hulled vessels of less than 25 m in length and eliminate the use of anti-fouling paints with a leaching rate of more than four µg of TBT per day"*. Later in November 1999, IMO adopted an Assembly resolution that called on the MEPC to develop an instrument, legally binding throughout the world, to address the harmful effects of anti-fouling systems used on ships (see also Chapters 7 and 13).

IMO subsequently adopted a global prohibition on the application of organotin compounds which act as biocides in antifouling systems on ships in 2001 and a complete prohibition of the use of organotins as an anti-fouling agent on ships came into force in 2008.

Under the terms of the Convention *"Parties are required to prohibit and/or restrict the use of harmful anti-fouling systems on ships flying their flag, as well as ships not entitled to fly their flag but which operate under their authority and all ships that enter their ports, shipyards or offshore terminals. Anti-fouling systems to be prohibited or controlled are listed in an annex to the Convention, which is updated as and when necessary"*.

London Convention and London Protocol
Convention on the Prevention of Marine Pollution by Dumping of Wastes and Other Matter, 1972 and 1996 Protocol
Adoption: 1975; Entry into Force 30/08/1975; Parties: 87 (57.71% World Tonnage)
Protocol Adopted 1996; Entry into Force 24/03/2006; Parties: 53 (40.47% World Tonnage)

The London Convention is one of the first global conventions to protect the marine environment from human activities and has been in force since 1975. Its objective is to promote the effective control of all sources of marine pollution and to take all practicable steps to prevent pollution of the sea by dumping of wastes and other matter.

This international agreement was in response to long-term historical dumping into the ocean, dredging and incineration at sea. Its creation gained traction from the 1972 United Nations Stockholm Conference on the Human Environment, and its provisions influenced the negotiation of the ocean dumping provisions of the United Nations Convention on the Law of the Sea, particularly Articles 210 and 216. Currently, 87 states are Parties to this Convention.

The purpose of the London Convention

» *"is to control all sources of marine pollution and prevent pollution of the sea through regulation of dumping into the sea of waste materials. A so-called [black- and grey-list] approach is applied for wastes, which can be considered for disposal at sea according to the hazard they present to the environment. For the blacklist items dumping is prohibited. Dumping of the grey-listed materials requires a special permit from a designated national authority under strict control and provided certain conditions are met. All other materials or substances can be dumped after a general permit has been issued"*.

The Convention recognizes a change in approach

» *"In recognizing the need for a more precautionary and preventative approach, the [Contracting] Parties undertook a comprehensive review of the Convention*

◨ **Table 16.2** MARPOL Annexes and their uptake by the world's shipping fleet

Annexes I & II	Annex III	Annex IV	Annex V	Annex VI
Oil and noxious liquid substances	Harmful substances carried at sea in packaged form	Sewage from ships	Garbage from ships	Air pollution from ships
160 parties	150 parties	146 parties	155 parties	101 parties
98.86% of world tonnage	98.33% of world tonnage	96.32% of world tonnage	98.49% of world tonnage	96.75% of world tonnage

Further information: ► https://www.imo.org/en/OurWork/Environment/Pages/Default.aspx
For current status of ratifications see: ► https://wwwcdn.imo.org/localresources/en/About/Conventions/StatusOfConventions/StatusOfTreaties.pdf

leading to the London Protocol, a new, free-standing treaty which entered into force in 2006 and is intended to replace the Convention. The [Contracting Parties] to the London Protocol have responded to new activities such as carbon capture and storage and marine geoengineering through amendments to the London Protocol adopted in 2006, 2009 and 2013".

There are currently 53 Parties to the London Protocol.

The purpose of the London Protocol is similar to that of the London Convention, but the London Protocol is more restrictive: application of a **precautionary approach** is included as a general obligation; a **reverse list** approach is adopted, which implies that all dumping is prohibited unless explicitly permitted; incineration of wastes at sea is prohibited; and export of wastes for the purpose of dumping or incineration at sea is prohibited. Extended compliance procedures and technical assistance provisions have been included, while a so-called transitional period allows new Contracting Parties to phase in compliance with the London Protocol over a period of five years, provided certain conditions are met.

Further information can be found at: ▶ https://www.imo.org/en/OurWork/Environment/Pages/London-Convention-Protocol.aspx

16.3.3 Benefits of IMO Responsibility to Prevent Marine Pollution

The measures that IMO have developed have been shown to be successful in reducing vessel-sourced pollution and illustrate the commitment of the Organization and the shipping industry towards protecting the environment. This is best demonstrated by reviewing the number and scale of oil spillages over time along with the increased intensity of sea trade of these commodities.

According to the United Nations Conference on Trade and Development's (UNCTAD) Maritime Review 2017, world seaborne trade has more than tripled in the forty-year period from 1969 to 2016 (◼ Table 16.3). Yet estimates of the quantity of oil spilt during the same pe-

◼ **Table 16.3** Growth in international sea-borne trade across the main types of goods (1970–2016) for selected years. *Data source*: UNCTAD 2017

Year	Tanker trader[a]	Main bulk[b]	Other dry cargo[c]	Total (all cargo)
1970	1440	448	717	2605
1980	1871	608	1225	3704
1990	1755	988	1265	4008
2000	2163	1186	2635	5984
2005	2422	1579	3108	7109
2006	2698	1676	3328	7702
2007	2747	1811	3478	8036
2008	2742	1911	3578	8231
2009	2641	1998	3218	7857
2010	2752	2232	3423	8408
2011	2785	2346	3626	8775
2012	2840	2564	3791	9195
2013	2828	2734	3951	9513
2014	2825	2964	4054	9842
2015	2932	2930	4161	10,023
2016	3058	3009	4228	10,295

Source: Compiled by the UNCTAD secretariat, based on data supplied by reporting countries and as published on government and port industry websites, and by specialist sources. Data for 2006 onwards have been revised and updated to reflect improved reporting, including more recent figures and better information regarding the breakdown by cargo type. Figures for 2016 are estimates, based on preliminary data or on the last year for which data were available

Notes:

[a] Tanker trade includes crude oil, refined petroleum products, gas and chemicals

[b] Main bulk includes iron ore, grain, coal, bauxite/alumina and phosphate. With regard to data as of 2006, main bulk includes iron ore, grain and coal only. Data relating to bauxite/alumina and phosphate are included under dry cargo other than main bulk.

[c] Includes minor bulk commodities, containerized trade and general cargo

riod show a steady reduction. Data from the Independent Tanker Owners Pollution Federation (ITOPF) reveal that, despite the rare major accident which can cause a spike in the annual statistics, the overall trend shows a continuing improvement, both in the number of oil spills and the quantity of oil spilt each year (ITOPF 2022).

The average number of oil spills over 700 tonnes has shrunk from over 25 in the 1970s to just 3.7 in the 2000s (◘ Figure 16.4). It is interesting to note, in this context, that the biggest single decade-to-decade reduction was from the 1970s to the 1980s, coinciding with the adoption and entry into force of the MARPOL Convention, which is credited with having had a substantial positive impact in decreasing the amount of oil that enters the sea from maritime transportation activities—both as a result of accidents or from the normal operation of ships.

Similarly, if we review the effect of IMO measures to reduce greenhouse gases, namely carbon dioxide (CO_2) from ships we see a downward trend of emissions (◘ Table 16.4). The 2014 IMO Greenhouse Study found that shipping, in total, accounted for approximately 3.1% of annual global CO_2 emissions for the period 2007–2012. For international shipping, the CO_2 estimate dropped from 2.8% in 2007 to 2.2% in 2012.

It is worthy to note that IMO has also adopted instruments such as the safety of ships at sea (International Convention for the Safety of Life at Sea, SOLAS), collision avoidance provisions (Convention on the International Regulations for Preventing Collisions at Sea, 1972, COLREGS), and standards and training of watchkeeping (International Convention on Standards of Training, Certification and Watchkeeping for Seafarers 1978, STWC) that provide a basis for area-based management of ships' operations engaged in international voyages with a view to control and prevent pollution of the marine environment arising from accidents/collisions or human-error and therefore directly protect marine biological diversity. These instruments also act indirectly as environmental protective measures by reducing the likelihood of incidents or casualties.

A second strong argument to have a single global regulator responsible for international shipping as this industry, similar to the airline industry, is a unique truly global industry with vessels from almost every country on the planet voyaging across all water bodies and oceans. A ship may well be owned by a company in Greece yet registered in Panama (Flag State) and operate with a multinational crew/staff, transporting goods owned by a multitude of persons/companies from different countries travelling from various countries and traversing through a range territorial waters and arriving at a number of ports on a single voyage. The industry is therefore very complex and requires a single global regulator to ensure a universal and consistent approach.

When we review the large number of ratifications and subsequent implementation via domestic legislation of the AFS Convention, we can conclude that the IMO ban of organotin anti-fouling systems on ships is another successful outcome for the marine environment. The implementation is guaranteed through the stringent port state control regime that IMO has promoted throughout the globe coupled with the **non-favourable treatment** provision in the Convention of ships when they enter ports.

The effectiveness of the London Convention and London Protocol agreements is further discussed in the next section.

16.3.4 Limitations of the IMO and the London Convention and London Protocol

There are a number of general limitations that IMO has in relation to bringing adopted agreements or resolutions into force. This is often a political and or economic matter for many States and is beyond IMO's jurisdiction. This leads to long delays in the uptake of the new regulations leading to a continuing risk to the environment. Many safety-related amendments often come into force via a tacit acceptance process, which means that the new arrangement comes into force at a particular time unless before that date, objections to the amendment are received from a specified number of Parties.

Equally important is the lack of control over implementation levels by those that have ratified or accepted the agreed regulatory framework.

IMO is largely dependent on its Member Governments to implement and enforce the agreements; however, there are mechanisms, such as Port State Control inspections, that allow national authorities to verify that ships calling at their ports are in compliance (this is applied equally to all ships regardless of flag and regardless of whether the flag state has actually ratified the Convention) this is the no favourable treatment clause found in most IMO Conventions.

For the London Convention and London Protocol, the objective and purpose of the agreements limit their scope to wastes or other matter disposed at sea from a vessel, aircraft, platform or other manmade structures. They do not address pipelines from land (such as those used for sewage outfalls or submarine disposal of mine tailings from mining operations on land) (see ▶ Chapter 5).

Having noted this, recent discussions suggest that the scope of the London Protocol has been modified to

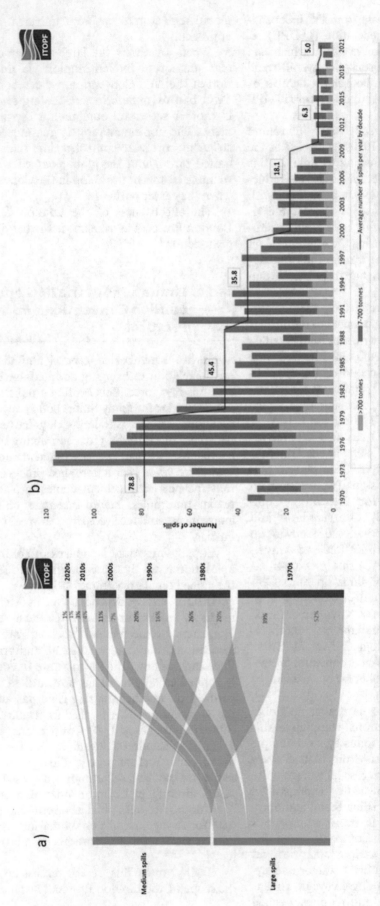

Figure 16.4 **a** Number of large oil spills (over 700 tonnes) from 1970 to 2021 and **b** the decline in the number of oil spills (over 7 tonnes) against the increase in crude oil and gas loaded. Adapted from ITOPF 2022 by A. Reichelt-Brushett

■ Table 16.4   Table depicting the trend of CO_2 emissions from ships. *Data Source*: IMO 2014

Year	Global CO_2	Total shipping	% of global	International shipping	% of global
2007	31,409	1100	3.5	885	2.8
2008	32,204	1135	3.5	921	2.9
2009	32,047	978	3.1	855	2.7
2010	33,612	915	2.7	771	2.3
2011	34,723	1022	2.9	850	2.4
2012	35,640	938	2.6	796	**2.2**
Average	33,273	1015	3.1	846	2.6

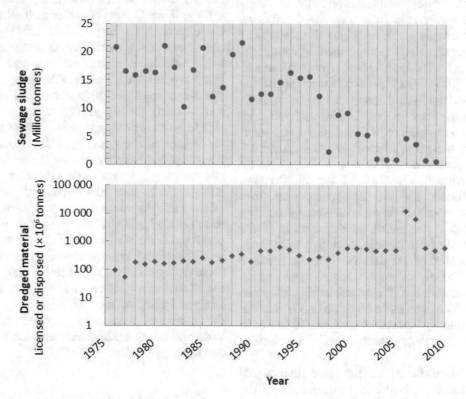

■ **Figure 16.5**   Trends in volumes of sewage sludge and dredged spoil submitted by contracting parties to the London Convention and the London protocol over the period 1975 to 2010. *Data Source*: Pers. comm. pending IMO 2021. *Image*: A. Reichelt-Brushett

address a particular limitation and has resulted in the inclusion of CO_2 waste streams for sequestration in the sea-bed pumped from land to a platform for injection into the sea-bed. The Contracting Parties to the London Protocol have also agreed that certain new technologies with the potential to cause harm to the marine environment, such as marine geoengineering (e.g. ocean fertilization) was within the scope of the agreement. In other words, while there may be limitations, some can be overcome through agreement by the Parties.

There are currently ongoing discussions under the London Convention and London Protocol about managing the submarine disposal of mine tailings through the development of guidance or best practice manuals to minimize, reduce or eliminate impacts on the marine environment (See GESAMP 2016 and: ► https://www. imo.org/en/OurWork/Environment/Pages/newandemer-gingissues-default.aspx).

A case study of trends in dumping volumes of sewage sludge and dredged spoil under the London Convention and London Protocol demonstrates the effectiveness of the agreements (■ Figure 16.5). Volumes of sewage sludge disposal to oceans have been steadily decreasing since 1975. For dredged material, the data show that there is no upward trend even though an increase in port development and maintenance has occurred in the same period.

Furthermore, the Parties to the Protocol, having noted that the practice of dumping sewage sludge had declined considerably and that alternatives exist for the use of the waste, agreed that there was sufficient evidence and justification for amending the Protocol to remove sewage sludge from the list of permissible wastes. This issue is currently being considered.

16.4 Other Global Instruments that Relate to Marine Pollution

As highlighted in ▶ Chapter 1, the major threats to the health, productivity and biodiversity of marine environments result from human activities on land—in coastal areas and further inland. Around 80% of pollution in the oceans originates from land-based activities. Many of these pollutants can be found all through the ocean from the shallowest waters to the deepest depths (e.g. Angiolillo et al. 2021) and most remote polar seas (e.g. Isla et al. 2018). They predominantly affect the productive coastal areas and many can be transportable globally via the atmosphere. The intense pressures put on the coastal systems require serious commitment and preventative action at all levels of governance: local, national, regional and global.

As noted previously, The Global Program of Action for the Protection of the Marine Environment from Land-Based Activities (GPA) is a non-legally binding instrument, aimed at preventing the degradation of the marine environment from land-based activities by facilitating the realization of the duty of States to preserve and protect the marine environment. It proposes action primarily at the national and regional levels with some coordination tasks at the global level. The GPA is designed to be a source of practical guidance to States in taking actions within their respective policies, priorities and resources.

Additional to land-based sources, there are dissipating pollutants that are transported to the ocean via the atmosphere. The text below provides some examples that link to topics covered in this textbook and highlight relevant agreement, conventions and protocols.

16.4.1 The Paris Agreement

The Paris Agreement is a legally binding international treaty on climate change. It was adopted by 196 Parties at the Conference of the Parties (COP 21) in Paris, on 12 December 2015 and was entered into force on 4 November 2016. Its goal is to limit global warming to well below 2.0 °C, preferably to 1.5 °C, compared to pre-industrial levels with countries aiming to reach global peaking of greenhouse gas emissions as soon as possible to achieve a climate neutral world by mid-century. The Paris Agreement is a landmark in the multilateral climate change process because, for the first time, a binding agreement where nations made pledges of nationally determined contributions (NDCs) to show how they will combat climate change and adapt to its effects. These are to be updated every 5 years.

There is a series of marine mitigation and adaptions in NDCs which have been defined into categories by Gallo et al. (2017) (◘ Table 16.5), as well, 39 countries have NDCs for doing additional research related to the marine environment. Interestingly, of those Parties that do not include the oceans in their NDCs, 14 are coastal, some with very large Exclusive Economic Zones (EEZs) such as Australia, Brazil, the European Union, Micronesia, New Zealand, Norway, the Russian Federation, and the United States of America (Gallo et al. 2017).

According to Bopp et al. (2013), the four marine stressors for marine ecosystems are ocean warming, acidification, deoxygenation and changes in primary productivity. The NDCs address these to some extent but not as multiple stressors. Furthermore, Harrould-Kolieb (2019) argues that ocean acidification should be a core obligation along with the central focus on temperature targets in Paris Agreement. However, because it has been framed by the scientific community as a concurrent threat to climate change rather than an effect of it, it now falls outside of the direct purpose of the Paris Agreement (Harrould-Kolieb 2019). Gallo et al. (2017) highlight that the few Parties that address ocean acidification in NDCs mostly are from small island developing countries, and there is even less interest in deoxygenation, suggesting a lack of knowledge at the international policy level about causes of deoxygenation and acidification. ▶ Chapter 11 explains in detail the important connections between greenhouse emissions and ocean acidification, and Chapters 1 and 4 address deoxygenation.

16.4.2 Other Conventions

The 1989 **Basel Convention** on the Control of Transboundary Movements of Hazardous Wastes and Their Disposal, the 1998 Rotterdam Convention on the Prior Informed Consent Procedure for certain Hazardous Chemicals and Pesticides in International Trade, the 2001 Stockholm Convention on Persistent Organic Pollutants, and the 2013 **Minamata Convention on Mercury** all aim to protect human health and the environment from hazardous chemicals and wastes. They do not directly relate to marine pollution, however, they are applicable. The four conventions have joined together to highlight the impacts of pollution on biodiversity in a recent report titled *"Interlinkages between the chemicals and waste multilateral environmental agreements and biodiversity: Key Insights"*. It seeks to enable the four conventions to contribute to discussions on and imple-

◻ **Table 16.5** Marine mitigations and adaption categories and the number of Parties of the Paris Agreement with NDCs. *Data Source*: Gallo et al. 2017

Categories	Number of countries with NDCs
Mitigation	
Mangrove restoration and conservation	19
Fisheries management	15
Maritime transport	15
Ocean renewable energy	14
Wetland restoration and conservation	8
Marine ecosystem management	7
Ocean carbon storage	6
Seagrass resorption and conservation	4
Coral reef conservation	2
Offshore energy production	1
Marine impacts and adaption	
Coastline impacts	95
Ocean warming	77
Fisheries impacts	72
Marine ecosystem impacts	62
Mangrove management	35
Marine tourism impacts	32
Marine biodiversity protection	28
Ecosystem-based management	24
Coral reef impacts	21
Creation of marine protected areas (MPAs)	17
Watershed management	16
Ocean acidification	14
Marine fauna distribution change	13
Seawater desalination	11
Coral bleaching	9
Marine pollution management	4
Reef ecosystem resilience	3
Blue economy	3
Harmful algal blooms	2
Ocean deoxygenation	1

mentation of the post-2020 biodiversity framework and the future work of biodiversity-related instruments. It highlighted that pollution was a major driver for biodiversity loss and several conclusions were directly relevant to the marine environment including:

— the increasing anthropogenic mercury emissions have severe consequences for human health and the environment, particularly biodiversity;

— Persistent Organic Pollutants (POPs) that are persist in the air, water, and soil, (e.g. PCBs and DDT) are continuing to be found in biota and PCBs are associated with declines in killer whale populations;

— global food security is at risk from threats to pollinators and the deterioration of soil ecosystems, with agricultural runoff including pesticides being a major source of water pollution and contamination of groundwater aquifers;

— plastics negatively affect marine species through entanglement, ingestion, contamination, and transport, and have potential to also threaten terrestrial ecosystems, including soils; and

— climate change amplifies the effects of chemicals and is expected to contribute to the re-volatilization of both mercury and POPs (e.g. melting permafrost and ice are expected to release significant quantities of both into the environment).

16.4.3   The International Seabed Authority

The **International Seabed Authority (ISA)** is made up of 167 Member States, and the European Union. It is mandated under the UN Convention on the Law of the Sea to organize, regulate and control all mineral-related activities in the international seabed area for the benefit of mankind as a whole. To date, no deep-sea mineral extraction has occurred globally; however, there has been much exploration and there is great interest in extracting these resources.

ISA has the duty to ensure the effective protection of the marine environment from harmful effects that may arise from deep-seabed-related activities. It is currently developing regulations for seabed mining but one of the greatest challenges is conducting risk assessment of ecosystems that are poorly understood and still being discovered. As part of the regulations, the legal owner of the resource needs to be clarified. Those located within the EEZ of countries are within national jurisdictions that extend up to 200 nautical miles from the coast and beyond the EEZ there are international waters (or the Area) that fall under international regulatory arrangements that are the responsibility of the United Nations International Seabed Authority (ISA) (e.g. Reichelt-Brushett et al. 2022). The South Pacific nation of Nauru is currently working towards active seabed mining operations. In June 2021 Nauru notified the ISA on its intention to invoke Section 1(15) of the 1994 Implementing Agreement (1994 Agreement Relating to the Implementation of Part XI of the United Nations Convention on the Law of the Sea (New York, 28 July 1994, in force 28 July 1996) 1836 UNTS 3.)) and intends to apply for the approval of a plan of work for the exploitation of seabed minerals in the Area. This effectively triggered the "two-year rule" during which the ISA has two years to finalize regulations governing the deep-sea mining industry Singh, 2022).

16.4.4   International Atomic Energy Agency

The **International Atomic Energy Agency (IAEA)** involves many treaties, which play an important role in establishing legally binding international rules in the areas that they cover and in relation to atomic energy. They do not directly relate to the marine environment but because of their existence the marine environment has gained protection from intentional radioactive pollution.

16.4.5   Convention of Biological Diversity

The **Convention on Biological Diversity (CBD)** entered into force on 29 December 1993 and is signed by 150 government leaders at the 1992 Rio Earth Summit. It has three main objectives

» *"the conservation of biological diversity, the sustainable use of the components of biological diversity, the fair and equitable sharing of the benefits arising out of the utilization of genetic resources".*

The Convention helps to address the spread of invasive species which are considered to be a main driver of biodiversity loss (see ▶ Chapter 12).

16.4.6   Global Legislation on Plastic Waste?

▶ Chapter 9 provides an in-depth discussion about the problem of plastics in our oceans. It highlights the magnitude of the global problem and ineffectual solutions given that plastic pollution is a growing concern, albeit now getting serious research attention. Some inter-country agreements have been made on a regional scale that addresses plastics along with other land-based pollutants (e.g. The Convention for the Protection of the Marine Environment of the North-East Atlantic (the **OSPAR Convention 1998**), but to date no global agreements exist.

Given the global nature of the problem, it suggests that a global legal instrument is necessary. Indeed, some attempts have been made to consider this, for example in December 2017 at the United Nations Environmental Assembly (UNEA) of the United Nations Environment Programme (UNEP) member states supported actions to eliminate the discharge of plastic litter and microplastics to the oceans. The actions included preventing plastic waste, increasing reuse and recycling and avoiding the unnecessary use of plastic, and highlighted the role of the extended producer responsibilities. To develop an ongoing and co-ordinated international action the Ad Hoc Open-Ended Expert Group on Marine Litter and Microplastics (AHEG) was established in 2017 to examine options for combatting marine plastic litter and microplastics from all sources, including through globally legally binding mechanisms. In 2019, the UNEA adopted the resolution (UNEP/EA.4/Res.6) (see also UNEP/EA.3/Res.7), in which

» *"noting with concern that the high and rapidly increasing levels of marine litter, including plastic litter and microplastics, represent a serious environmental problem at a global scale".*

In March 2022, the UNEA passed a resolution (UNEP/EA5/L23/REV.1) to end plastic pollution and forge an international legally binding agreement by 2024. Heads of State, Ministers of environment and other representatives from 175 nations endorsed this landmark agreement that addresses the full lifecycle of plastic from source to sea.

According to the UNEP, plastic production has risen exponentially in the last decades and now amounts to some 400 million tonnes per year—a figure set to double by 2040.

For further reading:

► https://www.unep.org/news-and-stories/story/what-you-need-know-about-plastic-pollution-resolution.

In 2018, IMO's Marine Environment Protection Committee (MEPC) adopted the IMO Action Plan to address marine plastic litter from ships, which aims to enhance existing regulations and introduce new supporting measures to reduce marine plastic litter from ships. The Action Plan provides IMO with a mechanism to identify specific outcomes, and actions to achieve these outcomes, in a way that is meaningful and measurable. It builds on existing policy and regulatory frameworks, identifies opportunities to enhance these frameworks and introduces new supporting measures to address the issue of marine plastic litter from ships.

Since 2020, IMO and the Food and Agriculture Organization of the United Nations (FAO) have been co-implementing a global project, called GloLitter, which aims to prevent and reduce marine plastic litter from the shipping and fisheries sectors.

For further reading: ► https://www.imo.org/en/MediaCentre/HotTopics/Pages/marinelitter-default.aspx, and ► https://www.imo.org/en/OurWork/PartnershipsProjects/Pages/GloLitter-Partnerships-Project-.aspx.

Another step in global legislative frameworks to manage marine plastics occurred in 2019, at the 14th meeting of the Conference of the Parties (COP) to the Basel Convention. In this meeting, the COP adopted two important decisions to address plastic waste: Decision BC-14/12 by which the COP amended Annexes II, VIII and IX to the Convention in relation to plastic waste, and decision BC-14/13 on further actions to address plastic waste. These actions provide recognition of the importance of enhanced cooperation in tackling plastic waste. Furthermore, the COP requested the Secretariat through decision BC-14/21 among others to continue to work closely with other international organizations on activities related marine plastic litter and microplastics.

Various POPs may also be contained in plastic waste (e.g. brominated flame retardants and short-chain chlorinated paraffins). Research continues to understand if the leaching of POPs from plastic particles may have significant adverse effect on the health of both terrestrial and marine wildlife (► Chapter 9).

Plastic debris can also adsorb POPs such as PCBs, DDT and dioxins which, if ingested, exhibit a wide range of adverse chronic effects in marine organisms. The Stockholm Convention controls various POPs and through decision BC-14/13, the COP welcomed the work of the Stockholm Convention to eliminate or control the production or use of POPs in plastic products that may reduce the presence of such pollutants in plastics waste. This further contributes to reducing the environmental risks associated with marine plastic litter and microplastics at the global level.

For further reading: ► http://www.basel.int/Implementation/Plasticwaste/Cooperationwithothers/tabid/8335/Default.aspx

16.4.7 The Precautionary Principle

On a final note, the **precautionary principle** is a guiding principle to encourage decision-makers to consider the likely harmful effects of their proposed activities on the environment. It has emerged as a principle of law, requiring that polluters use appropriate burden of proof to demonstrate that their activities are not causing damage to the environment (Cameron and Abouchar 1991). It has increased global consciousness of the political importance of protecting the environment and is evident as an underlying concept of the London Convention and London Protocol and more generally by the United Nations Environment Program (UNEP) Governing Council.

16.5 Summary

Implementing national **legislation can be a very effective method to control a range of human activities in the marine environment**; however, this is not necessarily the only method, as education and voluntary actions are essential components to achieve the desired outcomes and objectives of international agreements. The first global agreement that tackled marine pollution from shipping was established by the IMO and is known as International Convention for the Prevention of Pollution from Ships, 1973, as modified by the Protocols of 1978 and 1997 relating thereto (MARPOL) that would apply to all ships. The Convention requires the 160 Parties or signatories to develop legislation to implement the Convention within their jurisdictions. For this reason, the IMO is largely dependent on its Member Governments to implement and enforce the agreements.

Similarly, with the adoption of the Convention on the Prevention of Marine Pollution by Dumping of Wastes and Other Matter, 1972, and later the 1996 London Protocol, the dumping at sea of industrial and radioactive wastes is now prohibited, and only a few waste categories (such as dredge spoil) may be considered for dumping at sea following a stringent impact assessment and licensing process.

Over the years, a wide range of measures to prevent and control pollution to the marine environment have been established. Some main ones include:

- the Global Program of Action for the Protection of the Marine Environment from Land-Based Activities (GPA), a United Nations Environment was adopted by over 108 governments on the 3 November 1995;
- the UNEP Regional Seas Programme (RSP) and the related regional seas conventions and action plans currently being implemented by more than 146 countries addressing the degradation of the oceans and seas at a regional level;
- the Paris Agreement is a legally binding international treaty on climate change; and
- the International Seabed Authority (ISA) is made up of 167 Member States, and the European Union. It is mandated under the UN Convention on the Law of the Sea to organize, regulate and control all mineral-related activities in the international seabed area for the benefit of mankind as a whole.

In March 2022 Heads of State, Ministers of environment and other representatives from UN Member States committed to developing a legally binding agreement by 2024 to

» *End Plastic Pollution and forge an international legally binding agreement by 2024. The resolution addresses the full lifecycle of plastic, including its production, design and disposal.*

This resolution was endorsed at the UN Environment Assembly (UNEA-5).

16.6 Study Questions and Activities

1. Look up the national (and possibly state) Acts or legislative instruments that are used in your country to implement MARPOL and the London Convention and London Protocol (if your country is a signatory or has acceded to the treaties).
2. Plastic wastes from land-based and sea-based sources (maritime and fishing sectors) in the marine environment are becoming an increasingly alarming problem across the globe, affecting the marine habitats, marine organisms and the livelihoods of people—can you indicate what your country is doing to address this problem and can you give some new ideas on how this can be improved, also in terms of using international laws and programmes, not governing activities but also manufacturing, use and disposal?
3. List the non-governmental organizations or civil society groups that are making a difference on marine environment protection in your country and what specific international agreements they are supporting.

References

Angiolillo M, Gérigny O, Valente T, Fabri M-C, Tambute E, Rouanet E, Claro F, Tunesi L, Vissio A, Daniel B, Galgani F (2021) Distribution of seafloor litter and its interactions with benthic organisms in deep waters of the Ligurian Sea (Northwestern Mediterranean). Sci Total Environ 788:147745

Bopp L, Resplandy L, Orr J, Doney S, Dunne J, Gehlen M, Halloran P, Heinze C, Ilyina T, Séférian R, Tjiputra J, Vichi M (2013) Multiple stressors of ocean ecosystems in the 21st century: projections with CMIP5 models. Biogeosciences 10:6225–6245

Cameron J, Abouchar J (1991) The precautionary principle: a fundamental principle of law and policy for the protection of the global environment. Int Comp Law Rev 1, 14(1):1–27

CITES (Convention on International Trade in Endangered Species of Wild Fauna and Flora) (1973) Convention on international trade in endangered species of wild fauna and flora. Available at: ▶ https://cites.org/eng/disc/text.php. Accessed 14 Oct 2021

CMS (Convention on Migratory Species of Wild Animals) (1979) Convention on migratory species of wild animals. Available at: ▶ https://www.cms.int/en/convention-text. Accessed 14 Oct 2021

Gallo N, Victor D, Levin L (2017) Ocean commitments under the Paris Agreement. Nat Clim Chang 7:833–840

GESAMP (Joint Group of Experts on the Scientific Aspects of Marine Environmental Protection IMO/FAO/UNESCO-IOC/UNIDO/WMO/IAEA/UN/UNEP) (2009) Pollution in the open ocean: a review of assessments and related studies. Rep. Stud. GESAMP No. 79, P 64. Available at: ▶ http://www.gesamp.org/publications. Accessed 10 Aug 2021

GESAMP (Joint Group of Experts on the Scientific Aspects of Marine Environmental Protection IMO/FAO/UNESCO-IOC/UNIDO/WMO/IAEA/UN/UNEP) (2015) In: Boelens R, Kershaw P (eds) Pollution in the open oceans 2009–2013—a report by a GESAMP task team. Rep. Stud. GESAMP No. 91, P 87. Available at ▶ http://www.gesamp.org/publications. Accessed 10 Aug 2021

GESAMP (Joint Group of Experts on the Scientific Aspects of Marine Environmental Protection IMO/FAO/UNESCO-IOC/UNIDO/WMO/IAEA/UN/UNEP) (2016) Proceedings of the GESAMP international workshop on the impacts of mine tailings in the marine environment. Rep. Stud. GESAMP No. 94, P 84. Available at ▶ http://www.gesamp.org/publications. Accessed 10 Aug 2021

Harrould-Kolieb E (2019) (Re)Framing ocean acidification in the context of the United Nations Framework Convention on Climate Change (UNFCCC) and Paris Agreement. Clim Policy 19(10):1255–1238

IMO (International Maritime Organisation) (2014) Third IMO GHG study 2014. P 326. Available at ▶ https://www.imo.org/en/OurWork/Environment/Pages/Greenhouse-Gas-Studies-2014.aspx. Accessed 10 Aug 2021

IMO (International Maritime Organisation) (2021) Study on the current practices of managing or dumping of sewage sludge at sea. LC 43/10, and reviewed as document LC/SG 43/8/1

ITOPF (International Tanker Owners Pollution Federation) (2022) Oil tanker spill statistics 2021. Available at ▶ https://www.itopf.org/knowledge-resources/data-statistics/statistics/. Accessed 15 Nov 2022

Isla E, Pérez-Albaladejo E, Porte C (2018) Toxic anthropogenic signature in Antarctic continental shelf and deep sea sediments. Sci Rep 8(1):9154

Lebreton LCM, van der Zwet J, Damsteeg J-W, Slat B, Andradt A, Reisser J (2016) River plastic emissions to the world's oceans. Nat Commun 15611

Reichelt-Brushett A, Hewitt J, Kaiser S, Kim R, Wood R (2022) Deep seabed mining and communities: a transdisciplinary approach to ecological risk assessment in the South Pacific. Integr Environ Assess Manage 18(3):664–673

Singh PA (2022) The Invocation of the 'Two-Year Rule' at the International Seabed Authority: Legal Consequences and Implications. Int J Mar Coast Law. 375–412

UNCTAD (United Nations Conference on Trade and Development) (2017) Review of maritime transport 2017. UNCTAD/RMT/2017. P 128. Available at: ► https://unctad.org/system/files/official-document/rmt2017_en.pdf. Accessed 15 Aug 2020

UNFCCC (United Nations Framework Convention on Climate Change) (1992) United Nations framework convention of climate change. Available at ► https://unfccc.int/resource/docs/convkp/conveng.pdf. Accessed 15 Oct 2021

UN (United Nations) (2017) The first global integrated marine assessment: world ocean assessment I. Available at ► https://www.un.org/regularprocess/. Accessed 15 Oct 2021

UN (United Nations) (2021) The second world ocean assessment (World ocean assessment II). Available at ► https://www.un.org/regularprocess/. Accessed 15 Oct 2021

Supplementary Information

Appendix I – 360

Appendix II – 362

Index – 363

Appendix I

Units and Conversion Tales

Adapted from International Society of Automation. Available at: ▶ https://www.isa.org/ [Accessed 12 January 2022].

Prefix	Symbol	Multiplying Factor	
exa	E	10^{18}	1 000 000 000 000 000 000
peta	P	10^{15}	1 000 000 000 000 000
tera	T	10^{12}	1 000 000 000 000
giga	G	10^{9}	1 000 000 000
mega	M	10^{6}	1 000 000
kilo	k	10^{3}	1 000
hecto*	h	10^{2}	100
deca*	da	10	10
deci*	d	10^{-1}	0.1
centi	c	10^{-2}	0.01
milli	m	10^{-3}	0.001
micro	u	10^{-6}	0.000 001
nano	n	10^{-9}	0.000 000 001
pico	p	10^{-12}	0.000 000 000 001
femto	f	10^{-15}	0.000 000 000 000 001
atto	a	10^{-18}	0.000 000 000 000 000 001

*these prefixes are not normally used

Millimetre square mm^2	Centimetre square cm^2	Metre square m^2	Inch square in^2	Foot square ft^2	Yard square yd^2
1	0.01	0.000 001	0.001 55	0.000 011	0.000 001
100	1	0.000 1	0.155	0.001 076	0.000 12
1 000 000	10 000	1	1 550.003	10.763 91	1.195 99
645.16	6.451 6	0.000 645	1	0.006 944	0.000 772
92 903	929.030 4	0.092 903	144	1	0.111 111
836 127	8 361.274	0.836 127	1296	9	1

Centimetre cube cm^3	Meter cube m^3	Litre ltr	Inch cube in^3	Foot cube ft^3	US gallons US gal	Imperial gallons Imp. gal	US barrel (oil) US brl
1	0.000 001	0.001	0.061 024	0.000 035	0.000 264	0.000 22	0.000 006
1 000 000	1	1000	61 024	35	264	220	6.29
1 000	0.001	1	61	0.035	0.264 201	0.22	0.006 29
16.4	0.000 016	0.016 387	1	0.000 579	0.004 329	0.003 605	0.000 103
28 317	0.028 317	28.316 85	1 728	1	7.481 333	6.229 712	0.178 127
3 785	0.003 785	3.79	231	0.13	1	0.832 701	0.023 81
4 545	0.004 545	4.55	277	0.16	1.20	1	0.028 593
158 970	0.158 97	159	9 701	6	42	35	1

| Grams | Kilograms | Metric tonnes | Short ton | Long ton | Pounds | Ounces |
g	kg	tonne	shton	Lton	lb	oz
1	0.001	0.000 001	0.000 001	9.84e-07	0.002 205	0.035 273
1 000	1	0.001	0.001 102	0.000 984	2.204 586	35.273 37
1 000 000	1 000	1	1.102 293	0.984 252	2 204.586	35 273.37
907 200	907.2	0.907 2	1	0.892 913	2 000	32 000
1 016 000	1 016	1.016	1.119 929	1	2 239.859	35837.74
453.6	0.453 6	0.000 454	0.000 5	0.000 446	1	16
28	0.028 35	0.000 028	0.000 031	0.000 028	0.062 5	1

| Gram/millilitre | Kilogram/metre cube | Pound/foot cube | Pound/inch cube |
| g | kg | tonne | shton |
g/ml	kg/m^3	lb/ft^3	lb/in^3
1	1000	62.421 97	0.036 127
0.001	1	0.062 422	0.000 036
0.016 02	16.02	1	0.000 579
27.68	27 680	1 727.84	1

Degree Celsius (°C)	(°F -32) x 5/9
	(K -273.15)
Degree Fahrenheit (°F)	(°C x 9/5) + 32
	(1.8 x K) – 459.67
Kelvin (K)	(°C +273.15)
	(°F +459.67) ÷ 1.8

Appendix II

Periodic Table of the Elements

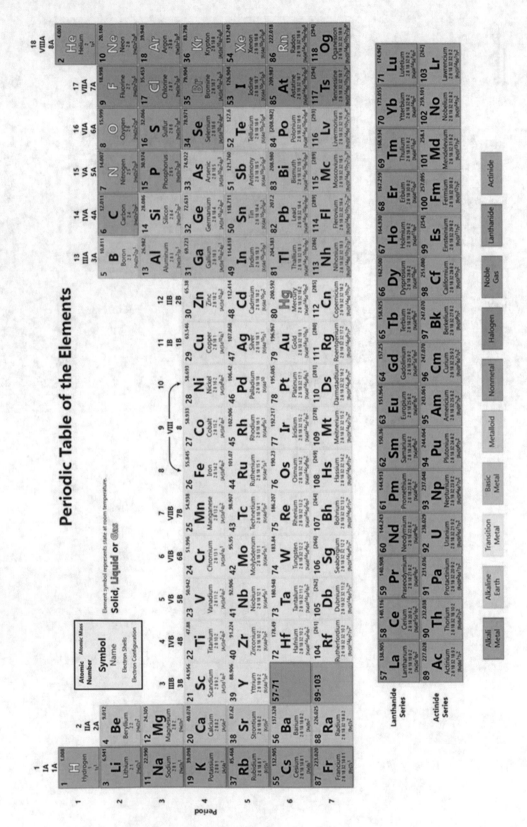

Source: Available at: ▶ sciencenotes.org/printable-periodic-table/ [Accessed 18 December 2021]

Index

A

Abiotic 12, 20, 39, 40, 42, 55, 56, 64, 67, 103, 166, 167, 215, 263, 266, 272, 300, 319
Abiotic sounds 263
Absorbed dose 233
Aceoprole 160
Acetylcholinesterase 159
Acetylcholinesterase activity (AChE) 120
Acid Sulfate Soils (ASS) 112
Acid Volatile Sulfides (AVS) 116
Active transport 117
Acute 58, 141
Acute toxicity 66
Additive effect 307
Adsorption 11, 32–34, 38, 137, 174, 219, 220, 236, 272
Aerobic bacteria 77
Aerosols 292, 297
Africa 47, 79, 88–90, 138, 148, 168, 173, 175, 276, 291
Agricultural activities 309, 322, 332
Agricultural runoff 76, 111
Agriculture 167
Air-breathing animals 200
Alaska 144, 145
Aldicarb carbofuran oxamyl methomyl 160
Aldrin 158, 167–170, 190
Algae 84
Algal biomass 84
Allethrin resmethrin permethrin cyfluthrin esfenvalerate 160
Alpha particles 231
Aluminium (Al) 9, 33, 103, 138, 217, 231
Amazon River 88
American Samoa 92
Ametryn 159, 161
Ammonium 78
Animal ethics 61
Anoxic 32, 33, 82, 86, 94, 174, 276
Anoxic zone 115
Antagonistic 242, 307–310, 312, 313
Antagonistic interactions 307
Antarctica 297
Anthropocene 2
Anthropogenic 130, 310
Anthropogenic noise 265
Antifoulants 174
Antifouling 347
Antimicrobial agents 273
Aquaculture 11, 16, 17, 80, 88, 89, 94, 95, 143, 147, 174, 175, 210, 217, 220, 266, 275, 277, 306, 311, 321, 322, 325
Aquaculture operations 80
Aquatic animals 190, 200
Arcachon Bay 175
Arctic 200
Argentina 162, 169, 170
Arsenic 161
Artificial Light At Night (ALAN) 266
Artisanal and Small-scale Mining (ASM) 109
Assessment Factor (AF) 65
Atlantic Ocean 79, 89, 170, 171, 210, 213, 230, 236, 239, 277
Atmosphere 8, 9, 29, 33, 64, 77, 80–82, 92, 94, 131, 137, 167, 180, 214, 248, 251, 252, 256, 257, 262, 269, 292, 352
Atmospheric carbon dioxide 309
Atrazine 159, 161, 168

Australia 106
Austria 162

B

Background concentration 65
Bacteria 16, 17, 41, 85, 136, 138, 144, 148, 159, 166, 174, 215, 219, 248, 251, 267, 269, 271, 272, 294
Ballast water 141, 272, 345
Baltic Sea 85
Bangladesh 4, 138, 221
Basel Convention 352
Bauxite 104
Behaviour 119
Benthic 8, 17, 33, 40, 42, 59, 64, 68, 69, 76–78, 83, 85, 86, 94, 143, 146, 169, 174, 217, 218, 239, 241, 268, 270, 277, 325, 327
Beta particles 231
Bilge water 132
Bioaccumulation 39, 120
Bioacculumative, Very Persistent chemical (vBvP) 166
BioAccumulation Factor (BAF) 165
Bioassay 114
Bioavailability 39, 112
Biochemical biomarker 120
Biocides 174
Bioconcentration 39, 120
BioConcentration Factor (BCF) 165
Biodegradation 166
Biodiversity 108
Biofouling 174
Biological effects 58, 219, 292, 306, 308
Biomagnification 39
BioMagnification Factor (BMF) 165
Biomagnify 118
Biomarkers 42
Biomass 10, 59, 76, 77, 79, 82, 85–90, 94, 144, 169, 248, 252, 255, 310–312
Bioremediation 321, 332
Bioremediators 332
Biota 12, 26–28, 38, 39, 42, 47, 63, 69, 136, 143, 167, 169, 172, 180, 192, 194, 199, 219, 230, 239, 241–243, 269, 274, 292, 294
Biota-Sediment Accumulation Factor (BSAF) 166
Biotic interactions 40
Biotic Ligand Model (BLM) 114
Biotic sound 263
Bioturbation 35, 59
Birds 12, 61, 143, 144, 146, 157, 159, 165, 170, 176, 177, 190, 191, 193, 265, 268, 291, 330
Bivalves 13, 14, 26, 43, 59, 60, 68, 85, 143, 178, 200, 201, 217, 220, 274, 291
Bonn Convention on Migratory Species 344
Brazil 88, 89, 162, 170, 276, 291, 343, 344, 352
Brittany 89
Brominated fire-retardant compounds 189
Buru Island 109

C

Cadmium (Cd) 16, 103, 161
Calcification 253
Calcifiers 254
Calcium (Ca) 8, 14, 236, 237, 251, 253, 255, 275

Canada 57, 66, 148, 175, 176, 256, 324
Canadian 191, 222
Canyons 108
Carbamates 157
Carbaryl 160
Carbon 248
Carbonate chemistry 253
Carbonate counter pump 251
Carbon dioxide 8, 248
Carbon pumps 251
Carbon reservoir 248
Caribbean 88
Carrier-mediated transport 117
Catchment 11, 171
Cation exchange capacity 114
Cetacean 107, 177, 264, 265
Chernobyl 237
Chesapeake Bay 87
Chile 106
China 47, 79, 80, 82, 87, 162, 168–170, 203, 213, 221, 236, 276, 291, 324,
    332, 343
Chlordane 160
Chlorine (Cl) 160, 194–199, 234, 296
Chlorpyrifos 160, 168
Chromium (Cr) 9, 103, 161, 237, 310
Chronic 141
Chronic exposure 26, 146, 326
Chronic toxicity 121
Clay minerals 114
Clays 26, 114
Climate change 173, 309
Climate-related variables 309
Clothianidin 160
Coal 105
Coal-fired power stations 111
Coastal development 92
Coastal ecosystems 76, 84, 92–94, 275
Coastal systems 82, 94, 235, 306, 312, 352
Cobalt (Co) 9, 103, 237
Coccolithophore 255
Combustion 81
Community structure 311
Compartment/s 15, 27, 28, 42, 166, 167, 180
Complexation 12, 59, 294
Congeners 196
Contaminants of emerging concern 286
Continental margins 108
Control 69
Convention on Biological Diversity (CBD) 344, 354
Convention on the International Trade in Endangered Species of Wild
    Fauna and Flora 344
Copepods 143, 145, 217
Copper (Cu) 120, 161, 287
Coral farming 320, 327
Corals 267
Coral triangle 108
Coring devices 38
Corrosion 268
Crabs 60, 143, 170, 178
Criteria 65
Criteria Continuous Concentrations (CCCs) 47
Criteria Maximum Concentrations (CMCs) 47
Crown of Thorns Starfish (CoTS) 84
Crude oil 130
Crust 248, 256
Crustaceans 65, 85, 117, 118, 120, 121, 143, 144, 158, 170, 177, 190, 191,
    193–195, 200, 220, 291, 310, 330

Cumulative impacts 306
Curie 232
Cyanazine 161

D

Dead zones 76
Debris 4, 6, 18, 29, 35, 36, 93, 94, 140, 208–214, 216–220, 222, 270, 275,
    290, 300, 355
Deep seabed mining 109
Deep-sea communities 108
Deep-Sea Tailings Placement (DSTP) 106
Deep water horizon 2
Deepwater horizon (oil spill) 136, 143, 145
Definitive test 57
Degradates 166
Degradation 47, 166
Degradation products 166
Denmark 79, 143, 162, 176
Depuration 117
Desalination 112
Desorption 33
Detergents 76, 209
Detoxification 117
Diatoms 255
Diazion 160
1,1-Dichloro-2,2-bis(p-chlorophenyl)ethane (DDD) 157
1,1-Dichloro-2,2-bis(p-chlorophenyl)ethylene (DDE) 157
Dichloro-diphenyl-trichloroethane (DDT) 157, 188
Dictyosphaeria cavernosa 91
Dieldrin 160
Diffuse sources 2, 94, 309, 320, 322, 332
Digestion 119
Dilute-acid extraction 115
Dioxin-like PCBs 198
Dioxins 198
Dirty dozen, The 186
Disease 85
Dispersants 17, 136, 139–141, 143, 145, 147
Dissolution 8, 46, 138, 139, 230, 251, 254, 293, 294
Dissolved metals 105, 107, 112, 114, 116
Dissolved Organic Carbon (DOC) 31, 169
Dissolved Organic Matter (DOM) 174, 180, 269
Dissolved Organic Nitrogen (DON) 78
Dissolved Organic Phosphorus (DOP) 78
Dissolved oxygen 115
Diuron 159
Diversity 6, 42, 55, 64, 143, 144, 216, 268, 269, 272, 275, 277, 306, 309,
    312, 319, 321, 328–330, 349, 354
Dolphins 107, 137, 144, 170, 176, 241, 263–265
Doses 56, 57, 59, 64, 70, 163, 233, 241, 242, 274
Dredge 341
Dredging 111
Drill cuttings 110
Drilled cuttings 134
Dumping 111, 347

E

Early life stages 12, 59, 64, 266, 328
Earth's crust 104
EC10 120
EC50 57, 59, 65, 177–179, 197
Echolocation 263

Ecological Risk Assessment (ERA) 62, 64, 321
Ecological services 87
Ecosystem resilience 312
Ecosystem services 11, 208, 220, 313, 318–320, 322, 325, 326, 329, 330
Ecotoxicology 54
Effective dose 233
Effects-Directed Analysis (EDA) 63
Egypt 221
Electrochemical proton gradient 254
Electronics industry 103
Embryonic development 119
Emergency response 148
Emerging pollutants 272
Emulsification 138, 139
Endocrine Disrupting Chemicals (EDCs) 290
Endocrine disruption 290
Endosulfan 168
Entanglement 216
Environmental Risk Limits (ERLs) 47
Estrogenic 275, 290, 291
Estrogens 291
Estuaries 9, 11, 14, 34, 38, 39, 46, 67, 68, 80, 91, 94, 169, 200, 234, 308, 309, 311, 312, 324, 325, 330
Ethylenediaminetetraacetic acid 31
Euphotic zone 108
European Union 288
Eutrophic 77
Eutrophication 76, 77
Evaporation 7, 8, 130, 138, 139, 172, 173
E-waste 103
Excretion 119
Extreme weather 91
Exxon Valdez (oil spill) 2, 144

F

Factorial design 68
Faecal 38
Farming 162
Feeding 119
Fenoprop 161
Fertiliser 76
Field studies 67
Fine sediments 11, 18, 109, 270
Fipronil 160
Fishes 5, 13, 20, 26, 27, 42, 58, 59, 61, 64, 65, 67, 68, 77, 78, 85, 86, 92, 136, 143, 144, 146, 157, 159, 165, 169, 170, 176, 177, 180, 190, 191, 193–195, 198, 200, 202, 217–220, 238, 241, 242, 255, 263, 266–268, 270, 274, 277, 290–292, 297, 311, 312, 325, 326, 330
Fisheries 11, 80, 84, 86, 89, 136, 137, 143, 145–147, 218, 239, 266, 276, 289, 306, 312, 313, 325, 331, 332, 353, 355
Fisheries food webs 329
Fishing 12, 17, 18, 20, 26, 42, 82, 89, 146, 209, 210, 214, 217, 218, 220, 265, 289, 318, 322, 326, 356
Fishing gear 210
Fission 230
Flame retardants 16, 189, 193–196, 209, 215, 219, 289, 297, 355
Flocculation 11, 39
Flood plumes 85
Florida 91
Flow-through 58
Fluorine (F) 237, 294, 296, 299
Food webs 68, 76, 79, 94, 192, 253, 294, 312, 319
Fossil fuels 17, 80, 81, 83, 138, 148, 209, 234, 249, 252, 301
Foundation species 54, 59, 60, 319, 325
France 108

Fulvic 38
Fulvic acids 10, 112, 269
Fungi 85, 158, 159, 166, 191, 248, 271, 272
Fungicides 16, 111, 157–159, 162, 176, 178, 194
Fusion 8

G

Gamma particles 18
Gamma rays 231, 232
Gas 8, 15, 17, 32, 37, 45, 89, 130, 131, 134–138, 141, 148, 150, 250–252, 256, 257, 265, 296, 329, 348, 350, 352
Genetic 41, 241, 242, 248, 277, 328, 329, 354
Geochemical models 114
Germany 143, 162, 196
Ghost fishing 220
Ghost nets 217
Gills 13, 39, 40, 59, 114, 200, 201, 220, 270
Global carbon cycle 252
Global distillation 172
Global Programme of Action (GPA) 344
Glutathione 120
Glutathione S-Transferase (GST) 120
Golden tides 88
Gradient studies 68
Grasshopper effect 172
Great Barrier Reef 171
Greece 134, 157, 349
Green algae 91
Greenland 203
Groundwater 32, 79
Group of Experts on the ScientificAspects of Marine Environmental Protection (GESAMP) 12, 131, 141, 212, 342, 343, 351
Growth 119
Guidelines 65
Guideline values 121

H

Half-life 167, 233
Halogen 199, 273
Halogenated 16, 291, 297
Harmful Algal Blooms (HABs) 76
Harmful concentrations 66
Hawaii 42, 91, 92, 213, 230, 275–277
Healthcare products 292
Heat 7–9, 18, 39, 103, 138, 148, 166, 191, 214, 240, 268, 328, 329
Heavy metal 103
Helsinki Commission 86
Heptachlors 158, 160, 167–170, 191, 198
Hexazinone 159, 168
Holistic approach 312
Homeostasis 117
Homologous series 196
Homologue 196
Hong Kong 169, 332, 345
Hormones 161, 176, 268, 291
Household products 292
Human health 12, 15, 25, 47, 48, 180, 188, 192, 195, 208, 216, 219, 220, 239, 273, 277, 287, 288
Humic 10, 38, 112, 269
Hydrocarbons 17, 39, 61, 130–132, 134, 137–139, 141, 143, 144, 148, 157, 162, 189, 209, 296, 297
Hydrogen (H) 7, 8, 82, 114, 130, 240, 296, 299
Hydrogen atoms 7, 294, 296

Hydrogen bonding 7
Hydrogen ions 8, 250
Hydrogen peroxide 240
Hydrophilic 163
Hydrophobic 163
Hypoxia 76, 78, 83, 86, 91

I

Imidacloprid 160
Immunotoxic 193
Immunotoxicity 198
Imposex 175
Independent Tanker Owners Pollution Federation (ITOPF) 349
India 4, 138, 162, 170, 236, 272, 276, 277, 291, 329
Indian Ocean 213, 251, 253
Indonesia 4, 64, 83, 106, 137, 138, 170, 212, 220, 221, 270, 276, 329, 330
Industrial discharges 80, 94, 110, 131, 309
Industrial Revolution 251
Industrial wastes 91
Ingestion 216
In situ surveys 67
Interactive effects 306
International Atomic Energy Agency (IAEA) 354
International Convention for the Prevention of Pollution from Ships (MARPOL) 19, 132, 133, 143, 210, 221, 341, 342, 345, 346, 349, 355, 356
International Maritime Organization (IMO) 341
International Oil Pollution Compensation Funds 130
International Seabed Authority (ISA) 110, 354
Invasive marine species 277
Invasive species 6, 218, 278, 306, 310–312, 326, 333, 346
Ion 114, 117, 236, 255
Ionic 8, 34, 38, 294
Ionising radiation 232
Ireland 267, 310
Irish Sea 170
Isotopes 231
Italy 162, 169, 329

J

Japan 2, 4, 20, 83, 134, 168–170, 175, 196, 237–239

K

Kāne'ohe Bay 91
Kerosene 139
Keystone species 144, 177, 257, 325
Kinetics 64
Koa 165
Koc 165
Kow 163
Kuwait 133

L

Labile metal 114
Land-based 341
Landfills 16, 112, 192, 196, 292
Larval development 59, 121, 327
Larvae 40, 59, 61, 84
LC50 57, 58, 197

Lead 287
Lethal concentrations 58, 59
Life stages 12, 56–58, 64, 116, 120, 121, 242, 266, 328
Ligands 113
Limpets 274
Lindane 160
Line of Evidence (LOE) 55
Lipids 10, 15, 39, 47, 64, 161, 163, 166, 178, 200, 201, 269, 291
Lipophilic 163, 200
Lipophobic 163
London Convention 341
London Protocol 341
Long range transport 192, 193, 195
Louisiana 11

M

Macquarie Island 144
Macroplastic 209
Magnesium (Mg) 114
Malaysia 64, 138, 221
Mammals 61, 64, 143, 144, 150, 157, 170, 175, 176, 190, 191, 201, 204, 217, 263–265, 290, 291
Management 312
Manganese (Mn) 103
Mangrove 13, 33, 80, 83, 138, 143, 210, 217, 276, 318, 320, 325, 329, 330, 332, 353
Mariculture 15, 16, 141, 159, 276
Marine Litter 354
Marine mammals 200
Marine sediments 115
MCAP 157
Megaplastic 209
Mercury 120, 161
Mesocosms 55, 56, 167, 255, 256
Mesoplastic 209
Mesotrophic 77
Metal ion assimilation 116
Metalloids 103
Metallothionein proteins 117
Metals 103, 308
Metal of environmental concern 122
Metal speciation 112
Methane 131, 197
Methylmercury 16, 20, 39, 41
Mexico 2, 3, 88, 90, 136, 138, 144, 169, 170, 230, 276, 277
Microalgae 83, 235
Microbeads 211
Microbial 9, 16, 17, 41, 70, 82, 83, 85, 166, 218–220, 309, 321
Microfibres 211
Microplastics 209, 354
Minamata 19, 20, 48
Minamata Bay 2
Minamata Convention on Mercury 352
Minamata disease 20
Mineral processing 110
Mississippi 136, 143
Mitigation 15, 62, 88, 94, 130, 221, 276, 318, 320, 322, 323, 326, 332, 333, 352, 353
Mixing zone 11, 38
Modes of Action (MoA) 158, 159
Montara (oil spill) 137
Mortality 57, 83, 84, 86, 145, 146, 176, 178, 272, 275, 308, 326, 328
Mud flat 235
Multiple Line of Evidence (LOE) 54
Multiple stressor 306

Multiple stressor framework 308
Mussels 42, 59, 94, 143, 146, 150, 170, 217, 242, 274, 289
Mussel watch 42, 43, 289
Mysticetes 264, 265

N

Nanomaterials 292
Nanoparticles 292
Nanoplastic 209
Nanostructured materials 292
Natural gas 138, 148, 150, 209
Naturally Occurring Radioactive Materials (NORM) 234
Naturally seeped oil 131
Neonicotinoids 158
Netherlands 47, 143, 170, 222
Nets 6, 18, 42, 89, 217, 218, 220
Neurotoxicity 59
Neutrons 231, 237
New Caledonia 106
New York 325
New Zealand 47, 66, 175, 177, 188, 329, 352
Nickel (Ni) 70, 103, 310
Nicotinic acetylcholine receptor 159
Nigeria 221
Niskin 32
Nitrate 78
Nitrite 78
Nitrogen 76
Nitrogen cycle 76
Nitrogen-fixing microbes 76
Nitrogen fluxes 93
Noise 262
Non-indigenous species 311
Non-Native Species (NNS) 275
Non-polar 8
No Observed Effect Concentration (NOEC) 65
Normalise 47
North America 81, 83, 143, 289, 331
North Carolina 68
Northern Pacific Gyre Garbage Patch 2
North Korea 236, 276
Norway 108, 352
N:P ratio 76
Nuclear accident 2
Null models 307
Nutrients 76, 308
Nutrification 77

O

Objectives 65
Ocean acidification 253, 254
Octanol-air partition coefficient 165
Octanol-water partition coefficient 163
Octopus 210, 264
Odontocetes 263–265
Oil 130
Oil production platforms 134
Oil spill response 2, 148
Oligotrophic 77
Organic carbon pump 251
Organic carbon-water partition coefficient 165
Organic ligands 16

Organic matter 82, 85, 161, 163, 180, 248, 251, 269, 270, 272, 293, 308, 330
Organic particulate matter 269
Organochlorines 159
Organometallics 10, 16, 39, 30
Organophosphates 159
Organotin, TBT 161, 347
Organotin Compounds (OGTCs) 174
Orthophosphate 78
Osmoregulation 119
Our Stolen Future 290
Over fishing 20, 306, 318
Oxic zone 115
Oxygen-containing ligands 113
Oxygen deficit 82
Oyster reefs 325
Oysters 42, 47, 59, 158, 289, 325, 326
Ozone 296

P

Pacific Ocean 251
Panama 83, 277, 349
Papua New Guinea (PNG) 64, 106, 269
Paris Agreement 252, 344, 352
Particle size 114
Particulate Inorganic Nitrogen (PIN) 78
Particulate Inorganic Phosphorus (PIP) 78
Particulate Organic Carbon (POC) 251
Particulate Organic Matter (POM) 83, 256, 257, 269
Particulate Organic Nitrogen (PON) 78
Particulate Organic Phosphorus (POP) 78, 186
Partition coefficient 163
Passive diffusion 117
Pathogens 271
Per- and polyfluoroalkyl substances (PFAS) 294
Per-fluorinated compound 189
Perfluorooctanoic acid 31
Persistent 16, 18, 39, 64, 130, 135, 139, 161, 166, 167, 172, 173, 177, 186–195, 200, 203, 208, 219, 273, 275, 290, 292, 296, 300, 326
Persistent, Bioaccumulative, Toxic chemical (PBT) 166
Persistent Organic Pollutants (POPs) 64, 157, 158, 160, 161, 172, 186–189, 195–198, 200–204, 219, 288, 289, 294, 355
Persistent Organic Pollutants Review Committee (POPRC) 189
Personal Care Products (PCPs) 272, 291
Pesticides 156
Petroleum 131
Petroleum products 130
Pharmaceuticals 291
Phenylpyrazoles 159, 160
Philippines 64, 143, 170, 221, 277, 328
Phosphates 78, 82, 235, 236, 348
Phosphorus 76
Photodegradation 273
Photosynthesis 248
Photosystem II 159
Picoeukaryote 255
Plankton 10, 42, 76, 79, 143, 306
Plastic 208, 342
Plastic debris 216
Plastic ingestion 216
Plasticisers 195, 209, 215, 219
Plastic polymers 209, 212, 215, 216
Plastisphere 218
Plutonium (Pu) 230, 236, 240
Point sources 3, 11, 64, 91, 94, 292, 309, 320

Polarity 7, 163
Polar regions 297
Pollution 121
Polybrominated hydrocarbons 189
Polycarbonate 30
Polychaetes 60, 68, 85, 144
Polychlorinated Biphenyl (PCB) 18, 191, 192, 194, 196–198, 200, 291, 321, 355
Polycyclic Aromatic Hydrocarbons (PAHs) 30, 39, 130, 132
Polyphosphates 78
Pore water 32, 115
Power stations 147
Precautionary approach 348
Precautionary principle 355
Prestige (oil spoil) 133
Primary production 78, 248
Prince William Sound 144, 145
Propazine simazine 161
Protected areas 140, 353
Protective concentrations 54, 66
Protons 231, 250, 254
Pulse exposure 58, 177
Pyrethrins 157, 159
Pyriprole 160

Q

Quality Assurance and Quality Control (QA/QC) 25

R

Rachel Carson 157
Radioactive decay 231
Radioactive waste 2, 230, 233, 239–241, 243, 341, 342, 355
Radioactivity 231
Radioecology 235
Radioisotopes 231
Radium (Ra) 231, 232, 234, 236
Radon (Rn) 234
Range finder 57
Rare earth elements 103
Rceiving waters 62, 290, 308
Reactive Oxygen Species (ROS) 162
Recovery capacity 312
Recreation 11, 17, 47, 87, 220, 272
Recreational 12, 84, 146, 210, 220, 265, 272, 275, 319, 322, 326, 331
Recycling 208
Redfield-Brzezinski ratio 79
Redfield ratios 78
Redox 9, 29, 30, 34, 37, 44, 236
Red tides 84, 94
Reference sites 46, 68
Regenerative intervention 312
Regulation concerning the Registration, Evaluation, Authorisation and Restriction of Chemicals (REACH) 167, 288, 289
Rehabilitation 318
Remediation 318
Reproduction 12, 13, 40, 58–60, 63, 143, 144, 217, 218, 263, 265, 268, 272, 327
Reproduction and development 119
Resistance 57, 166, 268, 272, 287, 296, 312
Respiration and metabolism 119
Restoration 318
Restoration ecology 318
Restoration goals 91

Risk assessments 3, 28, 34, 42, 56, 61, 63, 64, 71, 147, 148, 198, 300, 354
Riverine 80, 83, 86, 94, 342
River systems 11, 322
Rocky Shores 13, 132, 143, 144
Roentgen 233

S

Sacrificial anodes 111
Salinity 7
Saltmarsh 330
Saltwater wedge 11, 38
Sampling methods 33, 48
Sampling program 27
Sargassum 88
Seabirds 267
Sea Empress (oil spill) 146
Sea floor 108
Seagrasses 13, 35, 68, 76, 77, 80, 83–85, 87, 91, 137, 143, 170, 177, 210, 217, 254, 276, 312, 318, 320, 329, 332, 333, 353
Seawater (properties) 6
Sedentary 12, 26, 40, 42, 43
Sediment cores 33
Selective extraction schemes 115
Selenium (Se) 103
Semi-static 58
Sessile 12, 14, 26, 40, 68, 212, 217, 218, 239, 270, 274
Sewage 81, 292
Sewage treatment 110
Sewage Treatment Plant (STP) 76
Shipping 110, 130
Shipwrecks 111
Siderophores 117
Sievert 233
Silent Spring 157, 187
Silica 78
Simazine 159
Single stressors 306
Solid waste 2, 332
Solubility 7, 8, 10, 15, 16, 26, 33, 39, 43, 57, 103, 162–166, 173, 174, 180, 251, 255, 257, 272, 293, 296
Soluble 8, 26, 27, 41, 79, 139, 160, 161, 163, 255
Sound 262
South Korea 170, 276
Spain 133, 169, 267
Spawning 13, 14, 60, 86, 266, 327
Species protection 121
Species Sensitivity Distribution (SSD) 55, 121
Spiked studies 69
Sri Lanka 221
Standards 65
Static systems 58
Stockholm Convention 157, 158, 186–189, 195, 196, 203, 204, 289, 297, 321, 355
Stormwater 91, 92, 110, 131, 148, 169, 272, 306, 308, 309, 311, 324, 331
Stressor combinations 311
Stressor interactions 306
Sublethal 119, 177
Submarine Tailings Disposal (STD) 108, 271
Submerged Aquatic Vegetation (SAV) 87
Sulfides 36, 82, 84, 292
Sunscreens 273
Surfactants 17, 30, 36, 177, 287, 289, 296
Surficial sediments 33
Suspended particles 113
Suspended sediments 38, 270

Sustainability 312
Sustained management actions 87
Synergistic 306
Synergistic interactions 307
Synthetic chemicals 54
Synthetic polymers 208, 216
Synthetic substances 12, 46, 294

T

Tailings dams 106
Tampa Bay 91
Tar balls 130
Tasmania 144, 276, 297, 330
Tebuthiuron 159
Technologically Enhanced Naturally Occurring Radioactive Materials
    (TENORM) 234
Terrestrial runoff 76
Thailand 106
Thermal pollution 268
Thermal tolerances 268
Thiacloprid 160
Tin 161
Titanium oxide 293
Torrey Canyon (oil spill) 2
Total acid digestion 115
Total Organic Carbon (TOC) 31, 34, 36, 48, 331
Tourism 11, 80, 84, 87, 89, 146, 210, 220, 265, 328, 353
Toxic Equivalent (TEQ) 198
Toxic Equivalent Factor (TEF) 198
Toxicity Identification Evaluations (TIEs) 62
Trace metals 29, 48, 103, 306, 308, 309, 313
Translocation studies 69
Triazines 159
Tributyltin (TBT) 174, 289, 347
Trophic levels 15, 39–42, 203, 216, 219, 255
Tuna canneries 92
Turbidity 271
Turbidity maximum 39
Turkey 108
Turtles 266
2,4,5-T 157
2,4-D 157

U

Ultrafine particles 292
Ulva prolifera 87
Underwater noise 265

United Kingdom 56, 80, 87, 90, 92, 133, 222, 236, 324, 341
United Nations Conference on Trade and Development's (UNCTAD) 348
United Nations Environment Programme 342
United Nations Framework Convention on Climate Change 344
United Nations Law of the Sea Convention (UNCLOS) 341
United States of America (USA) 42, 57, 65, 80, 143, 157, 162, 167, 170,
    175, 176, 187, 191, 196, 203, 222, 230, 236, 237, 239, 276, 277, 290,
    320, 324, 325, 329, 332, 352
Upwelling 82, 108
Uranium (U) 234
Urban runoff 16, 131, 148
Urban settlements 11
Urea 80

V

Valathion 160
Vanadium (V) 103
Van Veen grab 33
Vapour pressure 166
Vietnam 157, 169, 170, 221, 267, 343
Volatile Organic Compounds 137
Volatility 166

W

Wadden Sea 143
Wastewaters 17, 20, 79, 82, 91, 92, 131, 169, 211, 212, 275, 292, 300, 306,
    321, 324, 333
Water clarity 269
Water Quality Criteria (WQC) 47
Water sampling 29
Weight Of Evidence (WOE) 56
Whales 143, 170, 200, 217, 218, 220, 263–265
World Health Organisation (WHO) 198, 199, 201, 202
World Heritage 144, 171, 177
World Ocean Assessment I 342
World Ocean Assessment II 342

X

Xenobiotics 54

Z

Zinc (Zn) 9, 16, 103, 237

Printed in the United States
by Baker & Taylor Publisher Services